STATISTICAL RELIABILITY ENGINEERING

STATISTICAL RELIABILITY ENGINEERING

BORIS GNEDENKO
Moscow State University

IGOR PAVLOV
Moscow Technical University

IGOR USHAKOV
QUALCOMM, Inc.

Edited by **SUMANTRA CHAKRAVARTY**
QUALCOMM, Inc.

JOHN WILEY & SONS, INC.
New York / Chichester / Weinheim / Brisbane / Singapore / Toronto

This book is printed on acid-free paper. ♾

Library of Congress Cataloging in Publication Data:

Gnedenko, Boris Vladimirovich, 1912–
Statistical reliability engineering / by Boris Gnedenko, Igor Pavlov, Igor Ushakov ; edited by Sumantra Chakravarty.
p. cm.
"A Wiley-Interscience publication."
Includes index.
ISBN 0-471-12356-0 (cloth : alk. paper)
1. Reliability (Engineering)—Statistical methods. I. Pavlov, I. V. II. Ushakov, I. A. (Igor' Alekseevich) III. Chakravarty, Sumantra. IV. Title.
TA169.G59 1999
620′.00452—dc21 98-38904

Printed in the United States of America

10 9 8 7 6 5 4 3 2 1

CONTENTS

PREFACE

This book complements a previously published volume titled *Probabilistic Reliability Engineering,* which was published by Wiley in 1995. Academician Boris Gnedenko initiated both book projects some ten years ago. He noticed that there was a well-developed probabilistic theory of system reliability. At the same time the statistical part of reliability engineering was covered only with standard approaches that did not reflect its unique aspects.

The main difference between this book and other monographs on reliability is that this work contains new results in statistical analysis of complex systems. Russian authors have mainly developed this part of reliability engineering. We refer readers to two reviews on the subject: Belyaev, Gnedenko, and Ushakov (1983) and Rukhin and Hsieh (1987) for a more detailed reference.

The authors of this book have actively researched in the area of reliability engineering application to designing and testing telecommunication systems. Almost all approaches appearing in this book were initiated by real practical problems.

All three authors have worked in the industry and taught at Universities. This book, although not the simplest for the first reading, targets both practical industrial engineers and university students.

We would like to thank two of our friends who contributed material to the book: Dr. Eugene Gordienko from the Universidad Autonoma Metropolitana, Mexico, and Dr. Mark Kaminsky from QUALCOMM, San Diego. Eugene wrote the Appendix to Chapter 11 and Mark presented Chapter 5.

We hope that this book will be useful as a textbook for University post-graduate and doctoral students as well as for industrial applied statisticians dealing with practical reliability problems.

We finished this book without our teacher and co-author, Professor Boris Vladimirovich Gnedenko. He passed away in December of 1995 after the first volume was published. His participation in this book cannot be underestimated: It started by his initial idea and finished with his last comments on the manuscript.

The role of Professor Gnedenko in the theory of mathematical statistics is known worldwide. His influence on the development of the reliability theory is invaluable. Completion of this book is our tribute to the memory of this outstanding mathematician.

Igor Pavlov
Moscow Bauman Technical University
Moscow, Russia

Igor Ushakov
QUALCOMM, Inc.
San Diego, California

STATISTICAL RELIABILITY ENGINEERING

PART I

UNIT RELIABILITY ESTIMATION

CHAPTER 1

MAIN KNOWLEDGE OF STATISTICS

1.1 INTRODUCTION

In reliability one deals with objects of various complexity. It is usual to speak about *systems* and *units*. We begin with consideration of a unit. We will call a unit an indivisible object in the frame of current reliability analysis. So, systems consist of units. Of course, in an engineering sense, a system can be considered as a "unit" if it is taken as a whole, that is, if there is no need to consider its structure and its constituent parts.

Statistical reliability analysis of a unit predominantly consists of standard statistical inferences. Since this material is covered by many excellent books on statistics, here we will present only a brief introduction of these statistical methods. The main goal of this book is statistical analysis of system reliability, which will be given in subsequent chapters.

1.1.1 Reliability Indices

The engineering practice of reliability characterization requires indices that are probabilistic in nature. We deal with random variables (e.g., time to failure or repair time) and their characterization. Detailed analysis of these indices is given in *Probabilistic Reliability Engineering* (*PRE;* Gnedenko and Ushakov, 1995). Here we only describe main indices along with brief explanations.

Any object fails after a random period of operation. We call this period *time to failure* (TTF) and denote this nonnegative random variable (r.v.) by ξ. A complete mathematical description of TTF is given by the TTF *distribution function* (d.f.):

$$F(t) = P\{\xi \le t\}.$$

A continuous r.v. can also be characterized by the *density function*

$$f(t) = \frac{d}{dt} F(t).$$

In particular, if an analyzed random variable is time to failure, we will also call this function a *failure density.* From the definition of the density function, it follows that

$$F(t) = \int_0^t f(t)\, dt.$$

In reliability, one frequently uses the *reliability function,* or *survival function,* defined as

$$R(t) = 1 - F(t) = P\{\xi > t\}.$$

A quantile t_p of level p of a continuous distribution $F(t)$ is defined as the solution of equation $F(t_p) = p$. A quantile of level p shows that $(1 - p) \times 100\%$ of objects are expected to survive during time t_p.

One of the main reliability indices for unrepairable objects is the *mean time to failure* (MTTF) T, which is defined as

$$T = \int_0^\infty tf\,(t)\, dt.$$

As shown in *PRE,* the MTTF can be expressed in another (equivalent) form:

$$T = \int_0^\infty R(t)\, dt.$$

For repairable objects, one also introduces the *mean time between failures* (MTBF), which is similar to the MTTF. The MTBF is the mathematical expectation of *time between failures* (TBF). The difference between MTTF and MTBF is explained in *PRE* in detail. We notice that in many engineering probabilistic models repair is considered "ideal"; that is, repaired item is assumed to be identical with a new one.

The *mean repair time* τ is defined via the distribution of random repair time in a similar way. In practice the mean repair time is usually not derived on the basis of tests but rather is based on an expert evaluation.

In addition to the indices mentioned above, repairable objects are characterized by either the *availability coefficient* or the *operational availability coefficient.* The first index represents the probability that a repairable object will be in the operational state at specified moment of time. The second one is the probability that an object will be operational at some moment and will have been operating without failures during a specified interval of time. (For details, see *PRE.*) Here we will deal only with the *stationary availability coefficient,* that is, the probability that an object will be operational at "a time far into the future." This index can be determined as

$$K = \frac{T}{T + \tau}.$$

In reliability engineering one often refers to the *failure rate,* or *hazard rate,* formally defined as

$$\lambda(t) = \frac{f(t)}{R(t)}.$$

In other words, the failure rate at an instant is the conditional density of TTF if it has survived up to moment t. In addition, note that the "element of probability" that the object that has survived up to t will have failed before moment $t + \Delta$ is $\lambda(t)\Delta$.

The dependence of the failure rate on time is a helpful qualitative characterization of life distributions. The *increasing failure rate* (*IFR*) and *increasing failure rate average* (*IFRA*), introduced by Barlow and Proschan (1975), relate to a wide class of distributions important in practical applications. Such distributions characterize "aging" objects whose reliability properties worsen in time. The *decreasing failure rate* (*DFR*) and *decreasing failure rate average* (*DFRA*) relate to the "younging" objects whose reliability properties improve over time. Such a phenomenon takes place in burn-in testing and in some specific situations with hardening.

The numerical values of the reliability indices introduced above can be experimentally checked by special testing or from the analysis of field data. Although reliability indices for units and systems are similar by their sense, statistical methods for their estimation might be different. In particular, a very special statistical reliability problem is a system indices estimation on the basis of testing its units.

This chapter will present a brief review of statistical methods, which will be necessary to understand the material presented in this text.

1.1.2 Main Tasks of Mathematical Statistics

If you perform reliability tests of an object or observe its utilization, you might collect some statistical data and use them for characterization of the object's reliability. These collected data are realizations of some r.v.'s $X_1, X_2, \ldots, X_n$.

Such data might represent TTF, repair time, number of cycles before failure, number of spare units used for repair and preventive maintenance, and so on. These values can be continuous or discrete depending on their nature. They are usually used for obtaining the sample mean or sample variance. The same data can be transformed into order statistics and be useful for plotting histograms and/or empirical distributions.

Another problem arises when you intend to construct confidence intervals for parameters of the distribution. You should usually have some prior information about the r.v. and possess special mathematical methods of statistical inferences.

Observing random events, you collect the number of outcomes of different types and the total number of trials. In probability theory one usually calls events "successes" or "failures." We can introduce the indicator δ_k of the kth event: $\delta_k = 1$ if success has been observed and $\delta_k = 0$ otherwise. So, we formally consider it as a discrete r.v. taking two meanings: 0 or 1.

Let us illustrate the role of statistics with an example of a classical Bernoulli trial. Remember that this is a series of n independent, identical trials, each of which might be a success with probability p or a failure with probability $1 - p$. In this case we can prescribe 1 to indicate success and 0 failure. This model is completely defined by the value of parameter p for a trial. Examples of such a situation are often met in practice. Under some specified conditions, mass production of some items is characterized by an almost stable percentage of items with fixed quality. A group of practically homogeneous objects, tested in similar conditions, is characterized by some stable frequency of successful operation. Of course, real life differs from mathematical models: A sequence of Bernoulli trials is only an approximation for the description of these practical schemes.

If real value p is unknown but there are experimental data, we can use methods of mathematical statistics to find various probabilistic characteristics, for instance, such expected number of successes or the probability of m successes in n trials. In this case of Bernoulli trials, the experimental data are the observed number of successes m in a series of n independent trials. On the basis of these data we need to make a conclusion about the value of an unknown parameter p, which is the unknown probability of success.

In mathematical statistics the following main problems are studied:

- *Point Estimation of an Unknown Parameter.* We wish to find a function of the experimental results (m successes in n trials), which allows us to obtain a "good" estimate of the unknown parameter p. A standard estimator of the probability is the observed frequency of success: $\hat{p} = m/n$.
- *Interval Estimation of an Unknown Parameter.* In this case, we need to construct an interval $[\underline{p}, \bar{p}]$ that will cover an unknown real value of param-

eter p with the specified probability

$$P\{\underline{p} \le p \le \bar{p}\} \ge \gamma, \tag{1.1}$$

where γ is the confidence coefficient, usually chosen close to 1.

We should emphasize that the limits of the confidence interval are random because they are functions of random variables: $\underline{p} = \underline{p}(m)$, $\bar{p} = \bar{p}(m)$. Moreover, the confidence interval covers unknown parameter p but gives no information about its "real" position within this interval. Expression (1.1) says that in $\gamma \times 100\%$ of cases this confidence interval will cover an unknown parameter and in $(1 - \gamma) \times 100\%$ of cases the parameter will lie outside these limits.

- *Test of Hypothesis.* One needs to check some hypothesis, for instance, that an unknown value of parameter p satisfies inequality $p \le p_0$ or equality $p = p_0$ (or other conditions), where p_0 is given. These statements are also made with some guaranteed probabilities. The specifics of this test of a hypothesis will be considered later.

These types of statistical inferences comprise the main body of applied statistics.

1.1.3 Sample

Probability theory deals with a d.f. based either on measure theory (following Kolmogorov) or on the frequency of an event occurring with a potentially arbitrary number of observations (following von Mises). In statistics, one always has a *sample* of a finite size, say, n. Usually, we define a sample from a distribution as n *independent and identically distributed* (i.i.d.) r.v.'s

$$X_1, X_2, \ldots, X_n. \tag{1.2}$$

The problem in mathematical statistics is to make some inferences about the distribution to which extracted X_k's belong.

We extract a sample from a finite homogeneous population characterized by some probabilistic properties. In statistics this group of objects is called a *general population.* Usually, in practice (in sociology, econometrics, medicine, telecommunications), the size of a general population, N, is assumed to be large. One takes a random sample of size n from a general population and makes a conclusion about a general population as a whole.

A sample from a general population can be taken with or without replacement. If sampling is performed with replacement, then the general population remains without changes at each extraction of a new item. In this case observations of r.v.'s are assumed to be independent. If an extracted item has not been returned to the general population, the latter can change its probabilistic properties. (The general population is assumed finite in this case.) Samples without replacement are frequently used in special problems of quality control, especially in cases where tests are destructive.

For sampling without replacement, each current trial depends on the results of all previous trials. For instance, let us have a general population of 100 items among which there are exactly 3 failed items. We make a sample of size 5. Let the first pick be a failed unit. Then at the second step the probability to choose a failed item at random equals 2/99. But if at the first step we have picked up a good item, then at the second step the probability to choose a failed item at random equals 3/99. Further, if at the first three steps we have picked up 3 failed items, the probability to choose a failed item at any next step equals zero.

If the size of a general population is very large (but finite), there is practically no difference between these two types of sampling.

Furthermore, we will almost exclusively consider samples from distributions, that is, from infinite general populations.

If sample (1.2) is placed in ascending order as

$$X_{(1)} \leq X_{(2)} \leq \cdots \leq X_{(n-1)} \leq X_{(n)},$$

then $X_{(k)}$ is called *k*th-*order statistic.*

Example 1.1 Five independent measurements of TTF give the following records (in hours): $X_1 = 104$, $X_2 = 95$, $X_3 = 93$, $X_4 = 101$, and $X_5 = 107$. Then the order statistics are $X_{(1)} = 93$, $X_{(2)} = 95$, $X_{(3)} = 101$, $X_{(4)} = 104$, and $X_{(5)} = 107$. □

The function

$$\hat{F}_n(x) = \frac{r(x)}{n}$$

based on a sample (1.2) is called an *empirical distribution function* [here $r(x)$ is a number of observations X_i that are smaller than x]. An empirical d.f. can also be written with the help of the order statistic as

$$\hat{F} = \begin{cases} 0 & \text{if } x < X_1, \\ \dfrac{j}{n} & \text{if } X_{(j)} \leq x < X_{j+1}, j = 1, \ldots, n, \\ 1 & \text{if } X_{(n)} \leq x. \end{cases}$$

Example 1.2 Construct the empirical d.f. for the data given in Example 1.1. The result is depicted in Figure 1.1. □

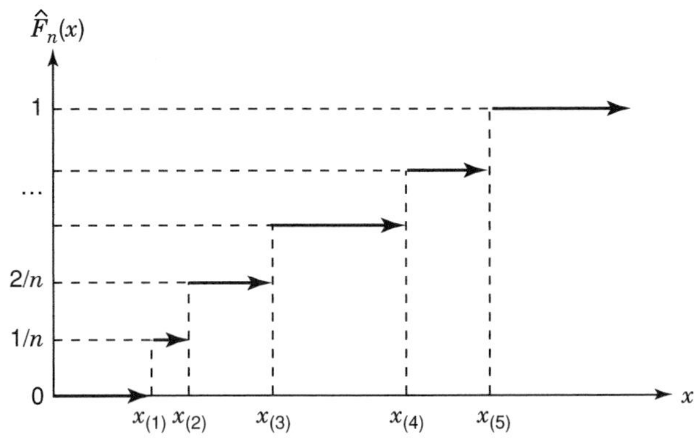

Figure 1.1 Empirical distribution function.

Let $F(x)$ be a true (or theoretical) d.f. Then in accordance with the Glivenko–Cantelli theorem, well known in probability theory, the condition

$$\sup_x |\hat{F}_n(x) - F(x)| \to 0$$

holds with probability 1 for $n \to \infty$. This means that an empirical d.f. stochastically converges to its theoretical value when the sample size infinitely increases. This fact could be explained with the following simple arguments. Each value of an empirical d.f. is a frequency of the event. An r.v. is smaller than a corresponding fixed value. The frequency converges (in a probabilistic sense) to the probability with increasing number of trials, and so does the set of such frequencies (i.e., the empirical d.f.).

The following *empirical,* or *sample, characteristics* can be constructed on the basis of sample data. The value of

$$\overline{X} = \frac{1}{n} \sum_{1 \le i \le n} X_i$$

is called the *empirical* (or *sample*) *mean.* The value of

$$s^2 = \frac{1}{n} \sum_{1 \le i \le n} (X_i - \overline{X})^2$$

is called the *empirical* (or *sample*) *variance.* If the sample mean indicates a "location" of the distribution, the sample variance characterizes the sample dispersion around its mean. Note that the mean is analogous to the center of mass in mechanics, and the variance to the moment of inertia. The value of $\hat{\sigma} = S$ is called the *empirical* (or *sample*) *standard deviation.* In some sense, the standard deviation is more convenient for sample characterization because it has the same dimension as the mean (and, consequently, the observed r.v.).

Another characteristic describing the spread of the sample is a *range of the sample,* defined as $R_n = X_{(n)} - X_{(1)}$. The deficiency of this value is in its main property: With increasing sample size, this value is increasing monotone. (In principle, for distributions with an infinite area of domain, the range increases to infinity.)

Empirical (sample) moments $\hat{\mu}_r$ and *empirical central moments* $\hat{\nu}_r$ of the rth order are calculated by the formulas

$$\hat{\mu}_r = \frac{1}{n} \sum_{1 \le i \le n} X_i^r, \qquad \hat{\nu}_r = \frac{1}{n} \sum_{1 \le i \le n} (X_i - \overline{X})^r.$$

Obviously, the sample mean and the sample variance are particular cases of $\overline{X} = \hat{\mu}_1$, $S^2 = \hat{\nu}_2$.

Example 1.3 Find the sample mean, variance, standard deviation, and range for the data presented in Example 1.1. In this particular case

$$\overline{X} = (\tfrac{1}{5})(93 + 95 + 101 + 104 + 107) = 100,$$

$$S^2 = (\tfrac{1}{5})(7^2 + 5^2 + 1^2 + 4^2 + 7^2) = 28,$$

$$\hat{\sigma} = S = \sqrt{28} \approx 5.3,$$

$$R_5 = 107 - 93 = 14. \square$$

Remark 1.1 If value X_i repeats in the sample n_i times, $i = 1, \ldots, m$, and $n = n_1 + n_2 + \cdots + n_m$, where m is the number of different X_i's, then the formulas for the sample mean and variance can be rewritten in the forms

$$\overline{X} = \frac{1}{n} \sum_{1 \le i \le m} n_i X_i, \qquad s^2 = \frac{1}{n} \sum_{1 \le i \le m} n_i (X_i - \overline{X})^2. \square$$

Example 1.4 Find the sample mean and variance of the following 100 observations:

X_i	1250	1270	1280	1290
n_i	20	25	50	5

Note that one has $m = 4$ different types of sample values. In this case

$$\overline{X} = \tfrac{1}{100}(20 \times 1250 + 25 \times 1270 + 50 \times 1280 + 5 \times 1290) = 1272;$$

$$S^2 = \tfrac{1}{100}(20 \times 22^2 + 25 \times 2^2 + 50 \times 8^2 + 5 \cdot 18^2) \approx 125;$$

$$\hat{\sigma} = S = \sqrt{125} \approx 11.2. \square$$

1.2 MAIN DISTRIBUTIONS

We consider here only the main probability distributions used in reliability statistical problems. More detailed descriptions of distributions as well as interrelations between various distributions are in *PRE.*

1.2.1 Continuous Distributions

Normal Distribution A normal distribution occupies a special place in probability theory and its applications. This distribution is often called Gaussian because the German mathematician Karl Friedreich Gauss studied its main properties and widely applied it in practice.

A number of phenomena in nature, engineering, and science can be modeled with the help of this distribution. A normally distributed random value appears where a large number of independent factors influence a considered parameter.

In probability theory there is the central limit theorem, which is known in several forms. All forms state that the sum of independent r.v.'s (usually, although not necessarily, assumed to be identically distributed) has the asymptomatically normal distribution. Lindeberg's formulation of the central limit theorem states that the distribution of the sample mean asymptotically converges to the normal distribution.

The density of the normal d.f. and its main characteristics are presented in Table 1.1.

For practical use, one applies the so-called standard normal distribution. This distribution has the mean equal to 0 and variance equal to 1; that is, its density is

$$f(x) = \frac{1}{\sqrt{2\pi}} \exp(-\tfrac{1}{2}x^2).$$

It is clear that a general normally distributed r.v., say, η, is subjected to two transformations in the form of a standard normally distributed r.v. ξ. First, this r.v. must be centered, that is, represented as $\eta - \mu$. Second, it must be normalized; that is, its scale must be changed in accordance with the standard deviation: $\xi = (\eta - \mu)/\sigma$. Thus, any normally distributed r.v. can be easily transformed to the standard form.

The cumulative standard normal distribution (sometimes called the Laplace function) has the form

$$\Phi(x) = \frac{1}{\sqrt{2\pi}} \int_{-\infty}^{x} \exp(-\tfrac{1}{2}t^2)\, dt.$$

There are detailed numerical tables for the standard normal distribution (one is given in Table 1 of the Appendix).

Table 1.1 Main Continuous Distributions

Name	Density	Mean	Variance	Domain
Normal	$f(x) = \frac{1}{\sqrt{2\pi}\sigma} e^{-(x-\mu)^2/2\sigma^2}$	μ	σ	$-\infty < x < +\infty$
Gamma	$f(x) = \frac{\lambda^a}{\Gamma(a)} x^{a-1} e^{-\lambda x}$	a/λ	a/λ^2	$x \geq 0$
Exponential	$f(x) = \lambda e^{-\lambda x}$	$1/\lambda$	$1/\lambda^2$	$x \geq 0$
Erlang	$f_k(x) = \lambda \frac{(\lambda x)^{k-1}}{(k-1)!} e^{-\lambda x}$	k/λ	k/λ^2	$x \geq 0$
Weibull–Gnedenko	$f(x) = \frac{\alpha}{\beta^\alpha} x^{\alpha-1} \exp\left[-\left(\frac{x}{\beta}\right)^\alpha\right]$	$\beta\Gamma\left(1 + \frac{1}{\alpha}\right)$	$\beta^2 \left\{\Gamma\left(1 + \frac{2}{\alpha}\right) - \left[\Gamma\left(1 + \frac{1}{\alpha}\right)\right]^2\right\}$	$x \geq 0$
Uniform	$f(x) = \frac{x}{b-a}$	$\frac{1}{2}(b-a)$	$\frac{1}{12}(b-a)^2$	$a \leq x \leq b$

This distribution is symmetrical around $x = 0$, and its domain is $(-\infty, +\infty)$. Due to the symmetry of the distribution, the quantile of the level q of the standard normal distribution satisfies the equation $U_{1-q} = -U_q$, and the cumulative d.f. satisfies the equation $\Phi(x) = 1 - \Phi(-x)$.

Sum of Normally Distributed Random Variables Let $\xi_1, \xi_2, \ldots, \xi_n$ be normally distributed r.v.'s and assume parameters μ_i and σ_i are known for each ξ_i, $i = 1, 2, \ldots, n$. Then the sum $\xi_1 + \xi_2 + \cdots + \xi_n$ also has a normal distribution with mean $\mu = \mu_1 + \mu_2 + \cdots + \mu_n$ and variance $\sigma^2 = \sigma_1^2 + \sigma_2^2 + \cdots + \sigma_n^2$.

This property of the normal distribution is true for any number of summands n.

Gamma Distribution The density of this distribution is presented in Table 1.1. This distribution can be considered as the "head of a family" of distributions that are very important in reliability theory and other applications.

An essential property of the gamma distribution is presented below.

Sum of Gamma Distributed Random Variables Let $\xi_1, \xi_2, \ldots, \xi_n$ be gamma distributed r.v.'s and assume parameters λ and a_i are known for each ξ_i, $i = 1, 2, \ldots, n$. Then the sum $\xi_1 + \xi_2 + \cdots + \xi_n$ also has the gamma distribution with parameters λ and $a = a_1 + a_2 + \cdots + a_n$.

Exponential Distribution Exponential distribution has extremely wide application in reliability theory. This is explained by two main reasons.

First, from a theoretical viewpoint, the exponential distribution allows us to obtain simple analytical results for many mathematical models. But this alone does not explain its extensive use. Second, and more important, from a practical viewpoint, this distribution is an appropriate reflection of many real physical phenomena. As practice shows, electronic equipment often has TTF distributed exponentially. In addition, this distribution has a close relation to a Poisson process (see *PRE* for details). Remember that the Poisson process can be considered a sequence of point events separated from each other by independent exponentially distributed random time intervals. The Poisson process is a convenient model for a flow of failures of complex systems consisting of a large number of highly reliable units. This fact is confirmed by the Khinchin, Renyi, and Grigelionis-Pogozhev theorems, procedures used to analyze the thinning and superposition of point stochastic processes (for details see *PRE*).

Notice that the exponential distribution formally is a particular case of the gamma distribution for $a = 1$. The density of the exponential distribution is presented in Table 1.1.

The exponential distribution possesses the *lack of memory* (or *Markov*) property: The conditional probability $P(x + y|x)$ does not depend on x; that is, $P(x + y|x) = P(y)$.

Erlang Distribution This distribution bears its name after A. K. Erlang, the Danish telephone engineer who introduced it and widely used it in telecommunication problems. An r.v. is said to have an Erlang distribution of kth order if it can be represented as a sum of K i.i.d. r.v.'s each of which has an exponential distribution. The form of the density function with its mathematical expectation and variance is presented in Table 1.1. Sometimes for special tasks one considers the generalized Erlang distribution, in which r.v.'s are not identical.

Sum of Erlang Random Variables The sum of M Erlang r.v.'s of orders $n_1, n_2, \ldots, n_M$ has the Erlang distribution of order $n = n_1 + n_2 + \cdots + n_M$. This fact follows immediately from the definition of the Erlang distribution.

Weibull–Gnedenko Distribution This distribution plays an important role in many applications because its form allows one to use it in many practical cases. Weibull introduced this distribution to analyze wearing failures. A year later Gnedenko published a paper dedicated to limit distributions of extremum statistics. The distribution introduced by Weibull was a particular class of limit distributions. So, excellent engineering intuition and first-class mathematical research generated a new distribution with a wide area of application.

The density of this distribution and its mean and variance are presented in Table 1.1.

This distribution is widely used in engineering practice because of the obvious convenience: Two parameters—one of scale and another of shape—allow one to approximate many empirical distributions. In particular, such different classes of distributions as IFR and DFR can be expressed with the help of the Weibull–Gnedenko distribution.

Notice that a particular case for $\alpha = 1$ corresponds to the exponential distribution.

Uniform Distribution An r.v. η is said to have the uniform distribution if it might take any value from some closed interval with equal probability (see Table 1.1). This distribution is essentially used in Monte Carlo simulation. To generate an r.v. ξ with given arbitrary d.f. $F(t)$, one has to solve the equation $F(\xi) = \eta$; that is, $\xi = F^{-1}(\eta)$. Such a transformation is easily performed on a computer.

1.2.2 Discrete Distributions

Binomial Distribution The binomial distribution characterizes samples of identical and independent events each of which has two possible outcomes: success or failure. Let n Bernoulli trials be performed. The probability of success in a single trial is p. Then the number of successes m in n trials has a binomial distribution (see Table 1.2). This distribution is often used in sam-

Table 1.2 Main Discrete Distributions

Name	Probability	Mean	Variance	Domain
Binomial	$P_n(m) = \binom{n}{m} p^m q^{n-m}$	np	npq	$0, 1, \ldots, n$
Hypergeometric	$P_m = \dfrac{\binom{M}{m}\binom{N-M}{n-m}}{\binom{N}{n}}$	$n\dfrac{M}{N}$	$\dfrac{M(N-M)n(N-n)}{N^2(N-1)}$	$0, 1, \ldots, \min(M, n)$
Poisson	$P_m = \dfrac{\lambda^m}{m!} e^{-\lambda}$	λ	λ	$0, 1, 2, \ldots$
Geometric	$P_m = pq^{m-1}$	$\dfrac{1}{p}$	$\dfrac{q}{p^2}$	$0, 1, 2, \ldots$
Negative binomial	$P_m = \binom{-n}{m} (-q)^m p^{n-m}$	$\dfrac{m}{p}$	$\dfrac{mq}{p^2}$	$m, m+1, \ldots$

ple quality control of mass production. Some redundant systems are also described by this distribution.

Joint Sample of Several Subsamples Consider M series of Bernoulli trials of sizes $n_1, n_2, \ldots, n_M$. Let the probability of success in each sample be the same and equal p. In this case, the probability to observe m successes in all M samples is $\binom{n}{m} p^m q^{n-m}$, where $n = n_1 + n_2 + \cdots + n_M$. This statement follows directly from the reformulation of the problem: m samples of different sizes n_i can be considered as a single sample of size n.

Geometric Distribution Consider again a sequence of Bernoulli trials. Let ν denote the random number of successes until the first failure has occurred. This random number has the geometric distribution (see Table 1.2). This distribution describes, for instance, a random number of successful cycles of operation or switching when each of them occurs independently and with the same probability.

A geometric distribution can be considered as a discrete analog of an exponential distribution.

Negative Binomial Distribution This distribution is, in some sense, a generalization of the geometric distribution. Assume that a sequence of Bernoulli trials is performing. The question of interest is, What is the probability of observing the mth failure at the nth trial? This event is equivalent to the product of the two following events: In $n - 1$ first trials one observes $m - 1$ failures in any order, and then with probability q the mth failure might occur at the very last trial. The probability of this combined event is

$$P_m = q \binom{n-1}{m-1} p^{n-m} q^{m-1}.$$

Some combinatorial transformations (see *PRE*) leads to the standard form (see Table 1.2). This distribution is often used in quality control.

Poisson Distribution Consider a sequence of independent events in time such that a time interval between two neighbor events, τ, is exponentially distributed with parameter λ. Such a sequence of events is called a Poisson process. In engineering practice the flow of complex system failures forms a Poisson process. The number of failures m occurring during a fixed interval t is a discrete r.v. with the Poisson distribution with parameter $\Lambda = \lambda t$ (see Table 1.2), where λ is the intensity of a corresponding Poisson process. The Poisson distribution is the asymptotic form of the binomial distribution if $n \to \infty$, $p \to 0$, and at the same time $np = \Lambda$.

Sum of Poisson Random Variables Let us consider a Poisson process with intensity λ. Take two nonintersecting intervals t_1 and t_2 and consider the numbers of events on each one of them. The number of events on the kth interval has the Poisson distribution with parameter λt_k. Due to the Markov property, there is no difference if these intervals are neighbors or not. The condition of nonintersecting intervals means that the number of events on these intervals are independent. Let us consider a new interval $t = t_1 + t_2$. It is clear that the number of events on this joint interval has the Poisson distribution with parameter λt.

Obviously, the same rule expands on an arbitrary number of nonintersected intervals t_k.

Multinomial Distribution This distribution is a generalization of a binomial distribution. In this case, one of $k \geq 2$ different events can be observed in each single trial. Let p_i be the probability of the event of type j, $j = 1, 2, \ldots, n$, $p_1 + p_2 + \cdots + p_n = 1$. Let ν_j be the number of observations of the jth event in a series of n trials, $\nu_1 + \nu_2 + \cdots + \nu_k = n$. Then the vector $\boldsymbol{\nu} = (\nu_1, \nu_2, \ldots, \nu_k)$ has a multinomial distribution with parameters n and ν.

1.2.3 Special Distributions

The following distributions are frequently used in solving various statistical problems.

χ^2 Distribution Let $\xi_1, \xi_2, \ldots, \xi_n$ be i.i.d. r.v.'s each of which has a standard normal distribution. Then the sum of squares of these r.v.'s has a χ^2 distribution with m *degrees of freedom.* The density of this sum is presented in Table 1.3.

Notice that this distribution is a particular case of a gamma distribution with parameters $\lambda = \frac{1}{2}$, $a = \frac{1}{2}m$.

Table 1.3 Main Distribution Used for Statistical Inferences

Name	Density
χ^2	$f(x) = \dfrac{x^{(k/2)-1}e^{-(x/2)}}{2^{k/2}\Gamma(k/2)}$
Student	$f(x) = \dfrac{1}{\sqrt{m}B(1/2, m/2)}\left(1 + \dfrac{x^2}{m}\right)^{-(m+1)/2}$
Fisher	$f(x) = \dfrac{(n/m)^{n/2}x^{(n/2)-1}}{B(n/2, m/2)(1 + nx/m)^{n+m/2}}$
Kolmogorov	$K(t) = 1 - 2 \sum_{1 \leq j < \infty} (-1)^j \exp(-2j^2t^2)$

Due to this relation between gamma and χ^2 distributions, their quantiles can be expressed via each other as

$$\Gamma_q(\lambda, m) = \frac{\chi_q^2(2m)}{2\lambda}, \qquad m = 1, 2, \ldots. \tag{1.3}$$

where $\Gamma_q(\lambda, m)$ is the quantile of level q of a gamma distribution with parameters λ and m and $\chi_q^2(2m)$ is the quantile of level q of a χ^2 distribution with m degrees of freedom.

Sum of χ^2 Distributed Random Variables Let $\xi_1, \xi_2, \ldots, \xi_n$ be χ^2 distributed independent r.v.'s each of which has degree of freedom m_i, $i = 2, \ldots, n$. Then the sum $\xi_1 + \xi_2 + \cdots + \xi_n$ also has a χ^2 distribution with degree of freedom $m = m_1 + m_2 + \cdots + m_n$.

Student Distribution This distribution is applied to find confidence limits of the sample mean of a normal distribution if the variance of the distribution is unknown.

Let r.v.'s η and ξ be independent and η have a standard normal distribution with parameters 0 and 1 and ξ a χ^2 distribution with m degrees of freedom. Then r.v. $t = (\xi/\sqrt{\eta})\sqrt{m}$ has the Student distribution with m degrees of freedom. The density of a Student distribution is shown in Table 1.3. There

$$B(y, z) = \int_0^1 u^{y-1}(1 - u)^{z-1}\, du$$

is the beta function (for details see, e.g., Rao, 1965, Chapter 3).

Remark 1.2 If a sample $X_1, X_2, \ldots, X_n$ is chosen from a normal distribution, then the r.v.'s

$$\xi = \frac{1}{\sigma}\sqrt{n}(\overline{X} - \mu), \qquad \eta = \frac{nS^2}{\sigma^2} = \frac{1}{\sigma^2}\sum_{1 \le i \le n}(X_i - \overline{X})^2$$

are independent and have, respectively, standard normal distribution and χ^2 distribution with $n - 1$ degrees of freedom (Rao, 1965). So, the r.v.

$$T = \frac{\xi}{\sqrt{\eta}}\sqrt{n-1} = \left(\frac{\overline{X} - \mu}{S}\right)\sqrt{n-1}$$

has the Student distribution with $n - 1$ degrees of freedom. This is the basis for the construction of the standard confidence interval for an unknown mathematical expectation of a normal distribution if the variance of the distribution, σ^2, is also unknown. □

Fisher Distribution This distribution is used, in particular, for constructing the confidence limits of the ratio of variances of two normal distributions. In reliability problems, the Fisher distribution is used for constructing the confidence limits of the availability coefficient.

If r.v.'s η and ξ are independent and have χ^2 distributions with n and m degrees of freedom, respectively, then the r.v. $\varphi = m\xi/n\eta$ has a Fisher distribution (see the density given in Table 1.3).

Kolmogorov Distribution This distribution characterizes the asymptotic (for $n \to \infty$) behavior of the r.v.

$$T_n = \sqrt{n} \sup_x |\hat{F}_n(x) - F(x)|,$$

where $\hat{F}_n(x)$ is an empirical d.f. based on sample size n and $F(x)$ is a true (theoretical) d.f. This d.f. is given in Table 1.3.

1.3 POINT ESTIMATION

1.3.1 Introduction

Let the r.v. ξ have d.f. $F(x, \Theta)$ and density $f(x, \Theta)$ that depend on some parameter Θ whose true value is unknown. (This parameter might be a vector.) We would like to estimate this parameter on the basis of n independent observations of r.v. ξ. In other words, there is a sample $X_1, X_2, \ldots, X_n$ of size n from d.f. $F(x, \Theta)$.

The problem of constructing the point estimator of

$$\hat{\Theta} = \hat{\Theta}(X_1, \ldots, (X_n) \tag{1.4}$$

of parameter Θ consists of finding a function of observations such that the r.v. (or random vector) $\hat{\boldsymbol{\theta}}$ in some sense is close to an unknown true value of the parameter Θ of the corresponding distribution.

A function of observations $\varphi = \varphi(X_1, \ldots, X_n)$ is called a *statistic*. For instance, a point estimator is a statistic. This particular statistic guarantees the closeness to the unknown true parameter.

1.3.2 Properties of Estimators

Unbiased Estimator An estimator $\hat{\boldsymbol{\theta}}$ is called an unbiased estimator of a parameter Θ if its mathematical expectation $E\hat{\Theta} = \Theta$ for all Θ. In other words, it means the following. Let us estimate an unknown parameter for a sample of arbitrary fixed size. Let us repeat this procedure until the number of such samples becomes sufficiently large. Then the mean of these estimates will be approximately equal to the unknown parameter.

Example 1.5 Consider n independent Bernoulli trials with probability of success p and probability of failure $q = 1 - p$. Suppose m successes have occurred.

The standard estimator of the unknown probability of success is the frequency $\hat{p} = m/n$. This estimator is unbiased because

$$E\hat{p} = E\left(\frac{m}{n}\right) = \frac{1}{n}Em = \frac{1}{n}np = p. \square$$

Example 1.6 Consider an estimation of the mean, $\mu = E\xi$, of some r.v. ξ by n independent observations $X_1, X_2, \ldots, X_n$. The sample mean $\overline{X}$ is the unbiased estimator of ν because

$$E\overline{X} = E\left\{\frac{1}{n}\sum_{1\le i\le n} x_i\right\} = \frac{1}{n}\sum_{1\le i\le n} Ex_i = \frac{1}{n}n\mu = \mu. \square$$

Example 1.7 For data of the previous example, determine whether the sample variance S^2 is an unbiased estimator of the variance $\sigma^2 = E(\xi - \nu)^2$ of r.v. ξ. For this purpose, let us write the sample variance in the form

$$S^2 = \frac{1}{n}\sum_{1\le i\le n}(X_i - \overline{X})^2 = \frac{1}{n}\sum_{1\le i\le n}(X_i^2 - 2\overline{X}\,X_i + \overline{X}^2) = \frac{1}{n}\sum_{1\le i\le n} X_i^2 - \overline{X}^2. \square$$

Now use the well-known formula that connects the mathematical expectation and variance of any arbitrary r.v. η as

$$E\eta^2 - (E\eta)^2 = \text{Var}\,\{\eta\}$$

In the considered case, we obtain

$$ES^2 = \frac{1}{n}\sum_{1\le i\le n} EX_i^2 - (E\overline{X})^2 = \frac{1}{n}\sum_{1\le i\le n}(\mu^2 + \sigma^2) - [(E\overline{X})^2 + \text{Var}\,\overline{X}].$$

The expected value of the sample mean is $E\{\overline{X}\} = \mu$ and its variance is

$$\text{Var}\,\overline{X} = \text{Var}\left\{\frac{1}{n}\sum_{1\le i\le n} X_i\right\} = \frac{1}{n^2}\sum_{1\le i\le n}\text{Var}\{X_i\} = \frac{1}{n^2}n\sigma^2 = \frac{\sigma^2}{n}.$$

From here it follows that

$$ES^2 = \frac{n-1}{n}\sigma^2.$$

So, the sample variance is a biased estimator of the true variance. Instead of S^2, therefore, we frequently use an estimator S_1^2,

$$S_1^2 = \frac{n}{n-1} S^2 = \frac{1}{n-1} \sum_{1 \le i \le n} (X_i - \overline{X})^2,$$

because the latter estimator is unbiased.

Asymptotically Unbiased Estimators An estimator $\hat{\theta}_n$ is called asymptotically unbiased if

$$\lim_{n \to \infty} E\hat{\Theta}_n = \Theta$$

for all possible values of Θ. In other words, if we produce a large number of trials, then the asymptotically unbiased estimate will approximately coincide with the unknown value of the parameter.

Example 1.8 In the Bernoulli trials considered above, any estimator for the probability of success of the form

$$\hat{p} = \frac{m + C}{n + C}$$

is asymptotically unbiased. (Here C is an arbitrary positive constant.) □

Example 1.9 In the conditions of Example 1.6 the sample variance S^2 is an asymptotically unbiased estimator of the population variance σ^2. □

Consistent Estimators Consider the dependence of an estimator $\hat{\theta}_n = \hat{\theta}_n(X_1, \ldots, X_n)$ on a sample size. Such an estimator is called consistent if it converges in probability to the true value of parameter Θ as $n \to \infty$; that is,

$$\lim_{n \to \infty} \mathrm{P}(|\hat{\Theta}_n - \Theta| > \varepsilon) = 0$$

for any $\varepsilon > 0$ and for any possible Θ.

In practical terms, this is close to being asymptotically unbiased.

Example 1.10 Let us show that the estimator $\hat{p} = m/n$ in Example 1.5 is consistent for the probability p.

For an arbitrary r.v. η with finite mathematical expectation $E\eta$ and variance $\mathrm{Var}\{\eta\}$, the well-known Chebyshev inequality

$$P(|\eta - E\eta| > \varepsilon) \le \frac{\mathrm{Var}\{\eta\}}{\varepsilon^2} \tag{1.5}$$

holds. Utilizing this inequality and taking into account that

$$E\hat{p} = p, \qquad \mathrm{Var}\{\hat{p}\} = \mathrm{Var}\left\{\frac{m}{n}\right\} = \frac{1}{n^2}\mathrm{Var}\{m\} = \frac{p(1-p)}{n},$$

one obtains that, as $n \to \infty$,

$$\Pr(|\hat{p} - p| > \varepsilon) \le \frac{\mathrm{Var}\{\hat{p}\}}{\varepsilon^2} = \frac{p(1-p)}{n\,\varepsilon^2} \to 0,$$

and the consistency of the estimator follows. □

Example 1.11 In an analogous way, we can verify that the sample mean is a consistent estimator for the mathematical expectation $\mu = E\xi$.

We again use the Chebyshev inequality (1.5) and the formula for variance $\mathrm{Var}\{\overline{X}\}$ to obtain

$$P(|\overline{X} - \mu| > \varepsilon) \le \frac{\mathrm{Var}\,\overline{X}}{\varepsilon^2} = \frac{\sigma^2}{n\,\varepsilon^2} \to 0. \ \square$$

With the help of the Chebyshev inequality, we can prove a more general result concerning the consistency of asymptotically unbiased estimators.

Theorem 1.1 Let $\hat{\theta}_n$ be an asymptotically unbiased estimator of parameter Θ such that

$$\lim_{n\to\infty} \mathrm{Var}\,\hat{\Theta}_n = 0.$$

Then $\hat{\theta}_n$ is a consistent estimator. □

It is easy to see that in the case of Bernoulli trials the consistent estimator of parameter p is not only a value $\hat{p} = m/n$ but also any estimator of the form $\hat{p} = (m + C)/(n + C)$. For the variance σ^2, both S^2 and S_1^2 are consistent.

Efficient Estimators Assume that we have two unbiased estimators $\hat{\theta}$ and $\hat{\theta}'$ for a parameter Θ. It is natural to say that the estimator $\hat{\theta}$ is more efficient than $\hat{\theta}'$ if the variance of the first is smaller than the variance of the second,

$$\mathrm{Var}\{\hat{\Theta}\} = E(\hat{\Theta} - \Theta)^2 \le E(\hat{\Theta}' - \Theta)^2 = \mathrm{Var}\{\hat{\Theta}'\}, \tag{1.6}$$

for all possible values of Θ. If in some classes of estimators there exists an estimator $\hat{\boldsymbol{\theta}}$ for which (1.6) holds for all of the members of this class, one says that this estimator is efficient in the chosen class of estimators.

The major approach for finding efficient estimators is based on the *Cramer–Rao inequality.*

Cramer–Rao Inequality Let us introduce the function $I(\Theta)$ of parameter Θ defined as

$$I(\Theta) = E\left[\frac{\partial \ln f(x, \Theta)}{\partial \Theta}\right]^2 = \int_{-\infty}^{\infty} \left[\frac{\partial \ln f(x, \Theta)}{\partial \Theta}\right]^2 f(x, \Theta)\, dx,$$

where $f(x, \Theta)$ is the density function. This function is called the *Fisher information.* Under some general conditions of regularity, for any unbiased estimator $\hat{\theta}_n$ of parameter Θ, the following Cramer–Rao inequality for the variance of the estimator holds for any values of Θ:

$$\mathrm{Var}\{\hat{\Theta}_n\} \geq \frac{1}{n \cdot I(\Theta)}.$$

The value

$$e(\Theta) = \frac{1}{n \cdot I(\Theta) \cdot \mathrm{Var}\{\hat{\Theta}_n\}} \tag{1.7}$$

is called the *efficiency* of the unbiased estimator. By the Cramer–Rao inequality, any estimator satisfies inequality $0 \leq e(\Theta) \leq 1$. An unbiased estimator $\hat{\theta}_n$ is called *efficient* if $e(\Theta) = 1$, or in other words, if its variance $\mathrm{Var}\{\hat{\theta}_n\}$ exceeds the Cramer–Rao lower limit for any Θ.

Example 1.12 For Bernoulli trials from Example 1.5, the unknown parameter is the probability of success $\Theta = p$. Since in this case $f(x, \Theta) = p$ for $x = 1$ and $f(x, \Theta) = 1 - p$ for $x = 0$, the Fisher information can be written as

$$\begin{aligned} I(p) &= E\left[\frac{\partial \ln f(x, p)}{\partial p}\right]^2 \\ &= P(x = 0) \cdot \left[\frac{\partial \ln f(0, p)}{\partial p}\right]^2 + P(x = 1) \cdot \left[\frac{\partial \ln f(1, p)}{\partial p}\right]^2 \\ &= (1 - p) \cdot \left(\frac{1}{1 - p}\right)^2 + p \cdot \left(\frac{1}{p}\right)^2 = \frac{1}{p(1 - p)}. \end{aligned}$$

The variance of the estimator $\hat{p} = m/n$ equals $(1/n)p(1 - p)$ (see Example 1.10), and thus it follows that the estimator $\hat{p}$ is efficient. □

Example 1.13 Consider a sample of size n from normal distribution with the known variance σ^2 and density

$$f(x, \mu) = \frac{1}{\sqrt{2\pi\sigma}} e^{(x-\mu)^2/2\sigma^2}.$$

In this case $\Theta = \mu$ and the Fisher information is expressed as

$$I(\mu) = E\left[\frac{\partial \ln f(x, \mu)}{\partial \mu}\right]^2 = E\left[\frac{(x-\mu)^2}{\sigma^2}\right]^2 = \frac{1}{\sigma^2}.$$

But the variance of the estimator $\hat{\mu} = (1/n)(X_1 + \cdots + X_n)$ equals σ^2/n. Hence from the Cramer–Rao inequality, we see that this estimator of parameter μ is efficient. □

Example 1.14 Consider a sample of size n from exponential distribution with density

$$f(x, \Theta) = \frac{1}{\Theta} e^{-x/\Theta}, \qquad x > 0.$$

In this case the Fisher information is

$$I(\Theta) = E\left[\frac{\partial \ln f(x, \Theta)}{\partial \Theta}\right]^2 = E\left[\frac{(x-\Theta)^2}{\Theta^4}\right]^2 = \frac{1}{\Theta^2}.$$

The variance of the estimator $\hat{\mu} = (1/n)(X_1 + \cdots + X_n)$ equals

$$\mathrm{Var}\{\hat{\Theta}\} = \frac{1}{n^2} \sum_{1 \le i \le n} \mathrm{Var}\{X_i\} = \frac{\Theta^2}{n},$$

so this estimator of the exponential parameter is efficient. □

Example 1.15 Consider a Poisson distribution

$$f(x, \Theta) = \frac{\Theta^x}{x!} e^{-\Theta}, \qquad x = 1, 2, \ldots .$$

The Fisher information in this case is

$$I(\Theta) = E\left[\frac{\partial}{\partial \Theta}(x \ln \Theta - \Theta - \ln x!0\right]^2 = E\left(\frac{x}{\Theta} - 1\right)^2 = \frac{1}{\Theta},$$

and the variance of the estimator of parameter $\hat{\theta} = (1/n^2)(x_1 + \cdots + x_n)$ equals

$$\text{Var } \hat{\Theta} = \frac{1}{n^2} \sum_{1 \le i \le n} \text{Var } X_i = \frac{1}{n^2} \cdot n \cdot \Theta = \frac{\Theta}{n}.$$

Thus the estimator $\hat{\boldsymbol{\theta}}$ is efficient. □

1.3.3 Methods of Estimation

The two most frequently used methods of parameter estimation are the *method of moments* and the *method of maximum likelihood.*

Method of Moments Let $X_1, \ldots, X_n$ be a sample from a distribution with density $f(x, \Theta)$ that depends on a single unknown parameter Θ. The moment estimate of Θ, found by the method of moments, is the solution $\hat{\boldsymbol{\theta}}$ of the equation derived by setting the theoretical first moment equal to the sample first moment $\overline{X}$. That is, we solve

$$\mu_1 = \mu_1(\Theta) = EX(\Theta) = \overline{X}.$$

Let us take the solution of the equation

$$\mu_1(\Theta) = \overline{X}. \tag{1.8}$$

In other words, we take as an estimate of the parameter such a value for which the true (or "theoretical") value of the first moment (expressed as a function of Θ) coincides with its value found from experimental data.

In an analogous way the method of moments is applied for the case of multiple unknown parameters. If $\boldsymbol{\theta} = (\Theta_1, \ldots, \Theta_k)$ is a k-dimensional parameter, then estimators $\hat{\theta}_1, \ldots, \hat{\theta}_k$ can be found from the solution of the system of k equations

$$\begin{gathered} \mu_1(\Theta_1, \ldots, \Theta_k) = \hat{\mu}_1, \\ \vdots \\ \mu_k(\Theta_1, \ldots, \Theta_k) = \hat{\mu}_k, \end{gathered}$$

where

$$\mu_r(\theta_1, \ldots, \theta_r) = \int_{-\infty}^{\infty} x^r f(x, \theta_1, \ldots, \theta_r)\, dx$$

is the theoretical moment of order r, and

$$\hat{\mu}_r = \frac{1}{n} \sum_{1 \le i \le n} x_i^r$$

is its empirical value found from the sample.

For $k = 2$, the above system of equations can be written in the form

$$\mu_1(\Theta_1, \Theta_2) = \overline{X}, \qquad \text{Var}\{\Theta_1, \Theta_2\} = S^2, \tag{1.9}$$

where $\text{Var}\{\Theta_1, \Theta_2\}$ is the true variance and S^2 the empirical variance.

Example 1.16 Consider the estimation of an unknown parameter (probability of success p) in Bernoulli trials. Let m be the number of observed successes in n trials. The first moment, or the expectation of m, equals

$$\mu_1 = \mu_1(p) = Em = \sum_{0 \le i \le n} m \binom{n}{m} p^m(1-p)^{n-m} = np.$$

Corresponding to the method of moments, the estimator $\hat{p}$ of parameter p is thus found from the equation $np = m$. It follows that $\hat{p} = m/n$. □

Example 1.17 Consider a sample from the exponential distribution with density $f(x, \lambda) = \lambda e^{-\lambda x}$, $x > 0$, where parameter λ is unknown. In this case the first moment equals

$$\mu_1 = \mu_1(\lambda) = \int_0^{\infty} x\lambda e^{-\lambda x}\, dx = \frac{1}{\lambda}.$$

Thus equation (1.8) has the form $1/\lambda = \overline{x}$, from which it follows that $\hat{\lambda} = 1/\hat{x}$. □

Example 1.18 For the Erlang distribution of order r with density

$$f(x, \lambda) = \frac{\lambda^r x^{r-1}}{(r-1)!} e^{-\lambda x}, \qquad x > 0,$$

the first moment is derived as

$$\mu_1(\lambda) = \int_0^{\infty} xf\,(x, \lambda)\, dx = \frac{r}{\lambda}.$$

So, the estimator of parameter λ by the method of moments is found from the equation $r/\lambda = \overline{x}$ and, consequently, $\hat{\lambda} = r/\overline{x}$. □

Example 1.19 Consider a sample from the normal distribution with two unknown parameters μ and σ. The system of equations (1.9) in this case has the simple form

$$\mu = \overline{X}, \qquad \sigma^2 = S^2,$$

which gives $\hat{\mu} = \bar{x}$, $\hat{\sigma} = S$. □

Example 1.20 Consider a sample from the gamma-distribution with density

$$f(x, \lambda, \alpha) = \frac{\lambda^\alpha x^{\alpha-1}}{\Gamma(\alpha)} e^{-\lambda x}, \qquad x > 0,$$

which includes two unknown parameters, λ, α. Using the well-known expression for the gamma function

$$\Gamma(\alpha) = \int_0^\infty t^{\alpha-1} e^{-t}\, dt$$

and recurrence relation $\Gamma(\alpha + 1) = \alpha\Gamma(\alpha)$, one obtains the following expressions for the first and second moments and the variance:

$$\mu_1(\lambda, \alpha) = \int_0^\infty \frac{\lambda^\alpha x^\alpha}{\Gamma(\alpha)} e^{-\lambda x}\, dx = \frac{\Gamma(\alpha + 1)}{\lambda\Gamma(\alpha)} = \frac{\alpha}{\lambda},$$

$$\mu_2(\lambda, \alpha) = \int_0^\infty \frac{\lambda^\alpha x^{\alpha+1}}{\Gamma(\alpha)} e^{-\lambda x}\, dx = \frac{\Gamma(\alpha + 2)}{\lambda^2\Gamma(\alpha)} = \frac{\alpha(\alpha + 1)}{\lambda^2},$$

$$\mathrm{Var}(\lambda, \alpha) = \mu_2(\lambda, \alpha) - \mu_1^2(\lambda, \alpha) = \frac{\alpha}{\lambda^2}.$$

In this case the system of equations for finding estimators of the parameters has the form

$$\frac{\alpha}{\lambda} = \overline{X}, \qquad \frac{\alpha}{\lambda^2} = S^2.$$

Thus and it follows that $\hat{\lambda} = \bar{x}/S^2$, $\hat{\alpha} = \bar{x}^2/S^2$. □

The method of moments gives consistent estimators of the parameters but these estimators are not always good from an efficiency viewpoint.

Method of Maximum Likelihood Consider a continuous distribution with density $f(x, \Theta)$. Let $X_1, \ldots, X_n$ be a sample of size n from this distribution. The joint density of all sample data, written as a function of Θ,

$$L(X_1, \ldots, X_n, \Theta) = f(X_1, \Theta) \cdot f(X_2, \Theta) \cdots f(X_n, \Theta), \tag{1.10}$$

is called the *likelihood function.*

The *maximum-likelihood estimator* (MLE) $\hat{\boldsymbol{\theta}} = \hat{\boldsymbol{\theta}}(X_1, \ldots, X_n)$ is found as a value of parameter Θ for which the likelihood function reaches the maximum in Θ under the condition that results of observation $X_1, \ldots, X_n$ are fixed. In other words, the MLE is found from the equation

$$L(X_1, \ldots, X_n, \hat{\Theta}) = \max_{\Theta} L(X_1, \ldots, X_n, \Theta).$$

It is typically more convenient to search for the maximum of the logarithm of the likelihood function rather than the maximum of the function itself. (The maximum of the logarithm and the function itself coincide.) Thus, under the assumption that function $f(x, \Theta)$ is differentiable with respect to Θ, one can find the MLE $\hat{\boldsymbol{\theta}}$ from the equation

$$\frac{\partial}{\partial \Theta} \ln L(X_1, \ldots, X_n, \Theta) = 0. \tag{1.11}$$

Equation (1.11) is called the *likelihood equation.*

Remark 1.3 Equation (1.11) is the necessary but not sufficient condition for obtaining the maximum. However, for many distribution families used in practice, it happens that the solution is unique and delivers the desired MLE.

In an analogous way, in the case of a vector parameter $\boldsymbol{\theta} = (\Theta_1, \ldots, \Theta_k)$, to find the MLE of the parameters $\hat{\theta}_1, \ldots, \hat{\theta}_l$, one needs to solve a system of equations (with respect to $\Theta_1, \ldots, \Theta_k$ for fixed observed values $X_1, \ldots, X_n$):

$$\begin{aligned} \frac{\partial}{\partial \Theta_1} \ln L(X_1, \ldots, X_n, \Theta_1, \ldots, \Theta_k) &= 0, \\ &\vdots \\ \frac{\partial}{\partial \Theta_k} \ln L(X_1, \ldots, X_n, \Theta_1, \ldots, \Theta_k) &= 0. \end{aligned}$$

The important property of the MLE is the following: If an efficient estimator exists, the maximum-likelihood method delivers this estimator. In general, under some conditions, the maximum-likelihood method delivers asymptotically unbiased and asymptotically efficient estimates. □

Example 1.21 Let us apply the maximum-likelihood method to find the estimator of an unknown parameter p in a set of Bernoulli trials. In this case, the likelihood function for the probability of observing m successes in a series of n trials equals

$$L(m, n) = \binom{n}{m} p^m(1 - p)^{n-m}.$$

The equation of likelihood has the form

$$\frac{\partial}{\partial p} [m \ln p + (n - m)\cdot\ln(1 - p)] = \frac{m}{p} - \frac{n - m}{1 - p} = 0,$$

so we obtain $\hat{p} = m/n$. □

Example 1.22 Let us find an estimator of parameter λ of the exponential distribution using the method of maximum likelihood. The likelihood function in this case is

$$L(X_1, \ldots, X_n, \lambda) = \lambda e^{-\lambda x_1}, \ldots, \lambda e^{-\lambda x_n} = \lambda^n \exp\left(-\lambda \sum_{1\le i\le n} x_i\right).$$

The likelihood equation has the form

$$\frac{\partial}{\partial \lambda}\left(n \ln \lambda - \lambda \sum_{1\le i\le n} x_i\right) = \frac{n}{\lambda} - \sum_{1\le i\le n} x_i = 0,$$

from where

$$\hat{\lambda} = \frac{n}{\sum_{1\le i\le n} x_i}. \ \square$$

Example 1.23 Using the method of maximum likelihood, find the point estimators of parameters μ, σ of the normal density

$$f(x, \mu, \sigma) = \frac{1}{\sqrt{2\pi}\sigma} e^{-(x-\mu)^2/2\sigma^2}.$$

In this case, the likelihood function and its logarithm are given as

$$L = \frac{1}{(\sqrt{2\pi}\sigma)^n} \exp\left[-\frac{1}{2\sigma^2} \sum_{1\le i\le n} (x_i - \mu)^2\right]$$

and

$$\ln L = -n \ln\sqrt{2\pi} - n \ln \sigma - \frac{1}{2\sigma^2} \sum_{1\le i\le n} (x_i - \mu)^2.$$

There are two unknown parameters in this problem, and the system of equations for finding this MLE has the form

$$\frac{\partial}{\partial \mu} \ln L = \frac{1}{\sigma^2} \sum_{1 \le i \le n} (X_i - \mu) = 0,$$

$$\frac{\partial}{\partial \sigma} \ln L = -\frac{n}{\sigma} + \frac{1}{\sigma^3} \sum_{1 \le i \le n} (X_i - \mu)^2 = 0.$$

It thus follows that

$$\hat{\mu} = \frac{1}{n} \sum_{1 \le i \le n} X_i, \qquad \hat{\sigma}^2 = \frac{1}{n} \sum_{1 \le i \le n} (X_i - \overline{X})^2.$$

So the MLE for the expectation μ and the variance σ^2 of the normal distribution coincide with the sample mean $\overline{X}$ and sample variance S^2. □

1.3.4 Sufficient Statistics

Let $X_1, \ldots, X_n$ be a random sample of size n from a distribution with density $f(x, \Theta)$. For the sake of simplicity, let us restrain ourselves to the case where r.v.'s are discrete. In this case, $f(x, \Theta)$ represents the probability that r.v. ξ has value x.

Let $T = T(X_1, \ldots, X_n)$ be some statistic, that is, some function of the observations. Assume that after an experiment we do not know the total sample $(X_1, \ldots, X_n)$ but know only the value of the statistic

$$T(X_1, \ldots, X_n) = t \tag{1.12}$$

Statistic T is called a *sufficient statistic* for parameter Θ [or more precisely, for the parametric family of distributions $f(x, \Theta)$] if, for any event A, the conditional probability of occurrence of this event under condition (1.12),

$$P\{A | T(x_1, \ldots, x_n) = t\},$$

does not depend on the value of parameter Θ for any possible value of statistic t. In the discrete case the conditional probability above can be calculated by the formula

$$\Pr\{A | T(x_1, \ldots, x_n) = t\} = \frac{\Sigma_{T=t, \overline{x} \in A} L(X_1, \ldots, X_n, \Theta)}{\Sigma_{T=t} L(X_1, \ldots, X_n, \Theta)},$$

where $L(X_1, \ldots, X_n, \Theta)$ is the likelihood function, the sum in the denominator is taken over all possible $\mathbf{x} = (X_1, \ldots, X_n)$ for which $T(\underline{\mathbf{x}}) = t$, and the sum in the numerator is taken over all $\mathbf{x}$ such that $T(\mathbf{x}) = t$, $\overline{X} \in A$.

The sense of the definition given above is the following: For some known value of a sufficient statistic, changing parameter Θ does not influence the probabilities of one or another events [or, more precisely, the conditional distribution of the sample under condition (1.12)]. This means that statistic T delivers the complete information about parameter Θ. The following known result gives a simple criterion for verifying whether a statistic is sufficient or not.

Theorem 1.2 (Criterion of Factorization) Statistic $T(X_1, \ldots, X_n)$ is sufficient for parameter Θ if and only if the likelihood function has the form

$$L(X_1, \ldots, X_n, \Theta) = h(X_1, \ldots, X_n) \cdot g[T(X_1, \ldots, X_n), \Theta]. \ \square$$

In other words, the likelihood function can be written as a product of two factors, one of them depending only on the results of the observations (not on parameter Θ) and another depending on parameter Θ and on observations $X_1, \ldots, X_n$ only via statistic T.

For continuous distributions with multidimensional parameter Θ and multidimensional sufficient statistic T, the principles are analogous. From the criterion of factorization, it follows that the method of maximum likelihood always leads to the expression of the parameter's estimator via the sufficient statistic. Another important property of sufficient statistics lies in the fact that if an efficient estimator of a parameter exists, then it is expressed via a sufficient statistic.

In other words, a sufficient statistic contains all necessary information about observations.

Example 1.24 Consider the exponential distribution with density

$$f(x, \Theta) = \frac{1}{\Theta} e^{-x/\Theta}, \qquad x > 0.$$

In this case, the likelihood function is

$$L(X_1, \ldots, X_n, \Theta) = f(X_1, \Theta) \cdots f(X_k, \Theta) = \frac{1}{\Theta^n} \exp\left(-\frac{1}{\Theta} \sum_{1 \le i \le n} x_i\right).$$

From the factorization criterion, it follows that statistic $T = X_1 + \cdots + X_n$ is sufficient. In this case, an efficient estimator of parameter Θ, expressed via a sufficient statistic, exists (see Example 1.6). $\square$

Example 1.25 For the normal distribution with two unknown parameters μ, σ, the likelihood function has the form

$$L(X_1, \ldots, X_n, \mu, \sigma) = \prod_{1 \le i \le n} \frac{1}{\sqrt{2\pi}\sigma} \exp\left[-\frac{(x_i - \mu)^2}{2\sigma^2}\right]$$

$$= \left(\frac{1}{\sqrt{2\pi}\sigma}\right)^n \exp\left(-\frac{n\mu^2}{2\sigma^2}\right)$$

$$\exp\left(\frac{1}{2\sigma^2} \sum_{1 \le i \le n} x_i^2 + \frac{\mu}{\sigma^2} \sum_{1 \le i \le n} x_i\right).$$

In accordance with the factorization criterion, we see that two-dimensional sufficient statistic for the two-dimensional parameters (μ, σ) is $T = (T_1, T_2)$, where

$$T_1 = \sum_{1 \le i \le n} x_i, \qquad T_2 = \sum_{1 \le i \le n} x_i^2. \ \square$$

1.4 CONFIDENCE INTERVALS

1.4.1 Introduction

Confidence intervals are used to estimate unknown population characteristics with a certain level of guarantee. The confidence interval is such that an unknown parameter occurs within this interval with some guaranteed probability (confidence probability). Let us give a verbal explanation of this probability. If we increase the numbers of homogeneous samples (i.e., samples of identical and independent objects) and construct the confidence interval for each sample, the relative frequency of cases in which the unknown parameter will be covered by these confidence intervals converges to the confidence probability.

In practice we often construct a symmetrical confidence interval. It makes an impression that the confidence interval "surrounds" the real value of the investigated parameter. It is a wrong impression. We do not know the real position of the unknown parameter within the confidence interval. Moreover, with some nonzero probability the real parameter might occur outside of the confidence interval.

Let us consider the problem in strict terms. Let $X_1, \ldots, X_n$ be a sample of size n from d.f. $F(x, \Theta)$ that depends on parameter Θ in a prior unknown way. Let us now assume an interval $\underline{\theta}, \overline{\theta}]$ such that the lower and upper limits are functions of test results

$$\underline{\theta} = \underline{\theta}(X_1, \ldots, X_n), \qquad \overline{\theta} = \overline{\theta}(X_1, \ldots, X_n)$$

and the inequality

$$P(\underline{\Theta} \leq \Theta \leq \overline{\Theta}) = \gamma \tag{1.13}$$

holds for all possible values of Θ. Both these limits are random because they depend on a set of r.v.'s. These limits will change from sample to sample. (Samples are assumed homogeneous and of the same size.) Interval $[\underline{\theta}, \overline{\theta}]$ is called the *confidence interval with confidence coefficient* γ (or, briefly, γ-*confidence interval*) for parameter Θ. In practice, one chooses the level of γ close to 1, for instance, 0.9, 0.95, or 0.99.

Based on the same statistical data, we can build a set of different confidence intervals with different confidence probabilities. Common sense hints that, for the same statistical data, the better is the level of guarantee (i.e., the higher the confidence probability), the wider is the confidence interval. On the basis of given statistical data, let us build a symmetrical confidence interval $[\underline{\theta}, \overline{\theta}]$ for confidence probability $\gamma' = 0.9$. Using the same data, we can construct confidence interval $[\underline{\theta}, \overline{\theta}]$ for confidence probability $\gamma'' = 0.99$. If we choose the higher confidence probability, we must sacrifice accuracy: The confidence interval will be wider.

Thus, confidence interval $[\underline{\theta}, \overline{\theta}]$ is an interval with random limits constructed on the basis of test results. This interval covers an unknown true value of parameter Θ with probability γ. So, in contrast to the point estimator, the confidence interval gives guaranteed information about the unknown true value of a parameter though this information is "fuzzy."

In some cases (e.g., for discrete r.v.'s) it is possible to satisfy only the inequality

$$P(\underline{\Theta} \leq \Theta \leq \overline{\Theta}) \geq \gamma \tag{1.14}$$

for all possible Θ instead of equality (1.13) because there are no lower and upper limits that deliver an exact value of γ. In this case, we say that interval $[\underline{\theta}, \overline{\theta}]$ is a confidence interval with confidence coefficient not less than γ.

Sometimes we need to find only a one-sided interval for parameter Θ, from below or from above. If the inequality

$$P(\underline{\Theta} \leq \Theta) \geq \gamma$$

holds, then $\overline{\theta}$ is called the lower γ-confidence interval of parameter Θ. Analogously, if

$$P(\overline{\Theta} \geq \Theta) \geq \gamma$$

holds, then $\overline{\ddot{}}$ is called the upper γ-confidence interval of parameter Θ. For example, if we are interested in the MTTF of some equipment, we should be sure that this value is not smaller than some specified level but are less interested in the upper limit. If we are considering the probability of failure, on the contrary, we are interested in the fact that this value is not larger than some given level.

1.4.2 Construction of Confidence Intervals

A frequently used method of constructing confidence intervals is based on use of some central statistic that is a function depending on parameter Θ and observations

$$T = T(X_1, \ldots, X_n, \Theta) \tag{1.15}$$

such that its d.f. $F(t) = P(T \leq t)$ does not depend on Θ.

The value K_q defined from the relation

$$P(T \leq K_q) = F(K_q) = q$$

is called the quantile of level q of d.f. $F(t)$ of r.v. T. Let us choose two small enough values of α and β and find values t_1, t_2 from the following conditions:

$$P(T \leq t_1) = F(t_1) = \alpha,$$

$$P(T > t_2) = 1 - F(t_2) = \beta.$$

For these purposes one needs to set $t_1 = K_\alpha$, $t_2 = K_{1-\beta}$. Then for the central statistic, equality

$$\Pr\{t_1 \leq T(X_1, \ldots, X_n, \Theta) \leq t_2\} = \gamma \tag{1.16}$$

holds for $\gamma = 1 - \alpha - \beta$. Further, the lower and upper limits of the confidence interval for parameter Θ are defined, respectively, as minimum and maximum values among all Θ satisfying inequality

$$\tau_1 \leq T(X_1, \ldots, X_n, \Theta) \leq \tau_2.$$

Therefore, from (1.16), it follows that

$$\Pr(\underline{\Theta} \leq \Theta \leq \overline{\Theta}) = \gamma;$$

that is, the interval defined in such a manner is a confidence interval for parameter Θ with confidence coefficient equal to $\gamma = 1 - \alpha - \beta$:

$$T(X_1, \ldots, X_n, \underline{\Theta}) = t_2, \qquad T(X_1, \ldots, X_n, \overline{\Theta}) = t_1. \tag{1.17}$$

If the central statistic monotonically increases in Θ, then the confidence limits are found from the following equations:

$$T(X_1, \ldots, X_n, \underline{\Theta}) = t_1, \qquad T(X_1, \ldots, X_n, \overline{\Theta}) = t_2 \tag{1.18}$$

Usually, in practice, a confidence interval is chosen to be symmetrical, that is, $\alpha = \beta = \frac{1}{2}(1 - \gamma)$.

General Method of Confidence Interval Construction An appropriate central statistic may not always be found. This obstacle leads to the use of another, more general, method sometimes called the method of confidence sets. We consider this case for a one-dimensional parameter Θ.

Let $S = S(X_1, \ldots, X_n)$ be some initial statistic. Most often for these purposes we use an unbiased point estimator, that is, $S = \hat{\theta}$. For a given Θ, a distribution function of the statistic is denoted as $F(t, \Theta) = P(S \leq t)$. For the sake of simplicity, this function will be assumed continuous and strictly increasing in t and strictly decreasing in Θ.

Let us set values $t_1 = t_1(\Theta)$, $t_2 = t_2(\Theta)$ corresponding to any possible value of parameter Θ. These values are chosen from conditions

$$F(t_1, \Theta) = \alpha, \qquad F(t_2, \Theta) = 1 - \beta, \tag{1.19}$$

so that $t_1(\Theta)$, $t_2(\Theta)$ are, at the same time, the quantiles of levels α and $1 - \beta$ of d.f. $F(t, \Theta)$ of statistic S. So if $\gamma = 1 - \alpha - \beta$, equality

$$P\{t_1(\Theta) \leq S \leq t_2(\Theta)\} = \gamma$$

holds. The set of values of statistic S belonging to the interval $[t_1(\Theta), t_2(\Theta)]$ is denoted as H_Θ and is called a γ-zone of parameter Θ (see Figure 1.2). For any possible value of parameter Θ, the probability that statistic S will belong to the γ-zone equals γ by construction.

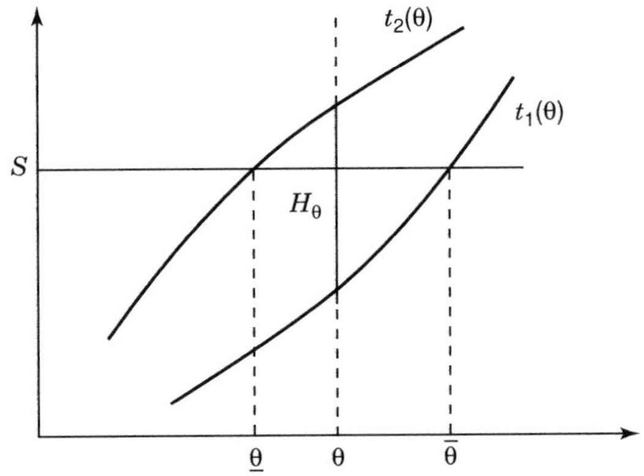

Figure 1.2 Form of γ-zone of parameter Θ.

Further, in correspondence to an observed value of statistic S, let us define an interval of all values of Θ for which this value of S belongs to the γ-zone (see Figure 1.2). The lower and upper limits of this interval, $\underline{\Theta}$ and $\overline{\Theta}$, are found from the conditions

$$t_2(\underline{\Theta}) = S, \qquad t_1(\overline{\Theta}) = S$$

or, by (1.29), from the equivalent conditions

$$F(S, \overline{\Theta}) = \alpha, \qquad F(S, \underline{\Theta}) = 1 - \beta. \tag{1.20}$$

The interval constructed above is a γ-confidence interval for parameter Θ. Indeed, for any possible value of parameter Θ (including an unknown true value), the interval $[\underline{\theta}, \overline{\Theta}]$ covers Θ if and only if an observed value of statistic S belongs to the γ-zone H_Θ for the specified value of Θ. So, by construction of the γ-zone, the equality

$$\Pr\{\underline{\Theta} \le \Theta \le \overline{\Theta}\} = \gamma$$

holds.

If d.f. $F(t, \Theta)$ is monotone increasing in parameter Θ, then the limits of the γ-zone $t_1(\Theta)$, $t_2(\Theta)$ are monotone decreasing in Θ. Repeating the arguments above, we find that the lower and upper limits of the confidence interval in this case are defined as

$$F(S, \underline{\Theta}) = \alpha, \qquad F(S, \overline{\Theta}) = 1 - \beta. \tag{1.21}$$

Confidence Interval for Discrete Random Variables The above method is used in an analogous way for the case of discrete r.v.'s. Consider, for example, a case, often met in practice, where statistic S takes on integer values $0, 1, 2, \ldots$.

Let $F(n, \Theta) = P(S \le n)$, $n = 0, 1, 2, \ldots$, be the d.f. of statistic S. Differing from a continuous case, the limits of the γ-zone here have a steplike form (see Figure 1.3).

For a specified fixed value of parameter Θ, the lower limit $t_1(\Theta)$ of the γ-zone is defined as the maximal number among n for which inequality

$$P(S \ge n) = 1 - F(n - 1, \Theta) \ge 1 - \alpha$$

holds. The lower limit $\tau_2(\Theta)$ of the γ-zone is defined as the minimal number among n for which inequality

$$P(S \le n) = F(n, \Theta) \ge 1 - \beta$$

holds. After this, the lower and upper limits of the confidence interval for

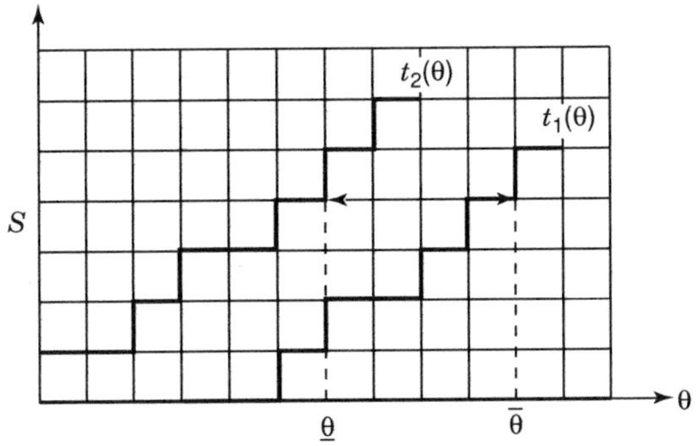

Figure 1.3 Steplike form of γ-zone of parameter Θ in discrete case.

parameter Θ with confidence coefficient not less than $\gamma = 1 - \alpha - \beta$ is defined by the minimum and maximum values among all Θ on the basis of statistic S. This satisfies inequalities

$$t_1(\Theta) \leq S \leq t_2(\Theta), \tag{1.22}$$

that is, among all Θ that belong to the γ-zone for a specified value of statistic S. Inequalities (1.32) are equivalent to

$$F(S, \Theta) \geq \alpha, \qquad F(S - 1, \Theta) \leq 1 - \beta.$$

From the above inequalities, it follows that if d.f. $F(n, \Theta)$ of statistic S is monotone decreasing in Θ, then the lower and upper limits for the γ-confidence interval for parameter Θ can be found from the solution of the system of equations

$$F(S - 1, \underline{\Theta}) = 1 - \beta, \qquad F(S, \overline{\Theta}) = \alpha. \tag{1.23}$$

Analogously, if d.f. $F(n, \Theta)$ is monotone increasing in Θ, then the γ-confidence interval for parameter Θ can be found from

$$F(S - 1, \underline{\Theta}) = \alpha, \qquad F(S, \overline{\Theta}) = 1 - \beta, \tag{1.24}$$

where, again, $\gamma = 1 - \alpha - \beta$.

Confidence Sets for Vector of Parameters In a similar way, we can construct the confidence sets for the multidimensional parameter $\mathbf{\Theta} = (\Theta_1, \ldots, \Theta_m)$. Let $\mathbf{x}$ be a vector of observations and $P_\Theta(\mathbf{x})$ be the d.f. of $\mathbf{x}$ for a given Θ. Let C_Θ of values $\mathbf{x}$ be defined for each possible Θ in such a way that inequality

$$P_\Theta(\mathbf{X} \in C_\Theta) \geq \gamma \tag{1.25}$$

holds for all of possible Θ's. Then for each observation $\mathbf{x}$ we find a set $D_\mathbf{x}$ of parameter Θ such that $\mathbf{x} \in C_\Theta$. This procedure is a straightforward extension of the procedure above for a one-dimensional parameter. By constructing set $D_\mathbf{x}$ for each fixed Θ, events $A = \{\mathbf{x} \in C_\Theta\}$, $B = \{\mathbf{x} \in D_\mathbf{x}\}$ are equivalent. From here, taking into account (1.25), it follows that

$$P_\Theta\{\Theta \in D_\mathbf{x}\} \geq \gamma \tag{1.26}$$

for all possible values of parameter Θ. A collection of all sets $D_\mathbf{x}$ for all possible values $\mathbf{x}$ satisfying (1.26) is called a system of γ-confidence sets for parameter Θ.

Further, if we need to construct a γ-confidence interval for some function $f = f(\Theta)$ of the vector of parameters Θ, then such an interval can be constructed on the basis of confidence sets D_x in the following way. Let us choose the lower and upper limits of f as values

$$\underline{f} = \min f(\Theta), \quad \overline{f} = \max f(\Theta),$$

where the minimum and maximum are taken over all Θ belonging to sets D_x. Then it follows directly from (1.26) that

$$\Pr_\Theta(\underline{f} \leq f(\Theta) \leq \overline{f}) \geq \gamma$$

for all Θ; that is, interval $[\underline{f}, \overline{f}]$ is the γ-confidence interval for $f(\Theta)$.

This type of problem will be considered below in more detail for the confidence estimation of reliability of a complex system on the basis of test results of its units.

We now consider constructing confidence intervals for parameters of the most frequently used distribution functions.

1.4.3 Confidence Estimation of Exponential Distribution

Consider an exponential distribution with density $f(x, \lambda) = \lambda e^{-\lambda x}$, $x > 0$. As a central statistic, we choose

$$T = 2\lambda \sum_{1 \leq i \leq n} \mathbf{X}_i.$$

This statistic has standard χ^2 distribution with $2n$ degrees of freedom (see Section 1.2). Equations (1.18) in this case take the form

$$2\underline{\lambda} \sum_{1\le i\le n} X_i = t_1 = \chi^2_\alpha(2n),$$

$$2\overline{\lambda} \sum_{1\le i\le n} X_i = t_2 = \chi^2_{1-\beta}(2n),$$

where $\chi^2_q(2n)$ is the quantile of level q for a χ^2 distribution with $2n$ degrees of freedom. From here, it follows that the lower and upper limits of the confidence interval with confidence coefficient $\gamma = 1 - \alpha - \beta$ for parameter λ have the forms

$$\underline{\lambda} = \frac{\chi^2_\alpha(2n)}{2\Sigma_{1\le i\le n} X_i} \quad \text{and} \quad \overline{\lambda} = \frac{\chi^2_{1-\beta}(2n)}{2\Sigma_{1\le i\le n} X_i}.$$

1.4.4 Normal Distribution, Known Variance σ^2

Consider the confidence interval for mean m of the normal distribution with known variance σ^2. Choose the central statistic of the form

$$T = \frac{\overline{X} - \mu}{\sigma/\sqrt{n}},$$

which has the standard normal distribution with mean 0 and variance 1. In this case, system (1.17) takes the form

$$\frac{\overline{X} - \overline{\mu}}{\sigma/\sqrt{n}} = u_{1-\beta}, \qquad \frac{\overline{X} - \underline{\mu}}{\sigma/\sqrt{n}} = u_\alpha,$$

where u_α is the quantile of level $1 - \alpha$ of the standard normal distribution. Taking into account that $u_{1-\beta} = -u_\beta$ for a normal distribution, we have the following lower and upper limits for the γ-confidence interval for parameter μ:

$$\underline{\mu} = \overline{X} - u_\alpha\left(\frac{\sigma}{\sqrt{n}}\right), \qquad \overline{\mu} = \overline{X} + u_\beta\left(\frac{\sigma}{\sqrt{n}}\right).$$

Normal Distribution, Unknown Variance σ^2 Now consider the confidence interval for the mean of a normal distribution with unknown variance σ^2. For this case, the central statistic is

$$T = \frac{\overline{X} - \mu}{S/\sqrt{n-1}}$$

where $\overline{X}$ and S^2 are the sample mean and variance, respectively.

This statistic has a Student d.f. with $n - 1$ degrees of freedom (see Section 1.2). The system of equations (1.17) in this case takes the form

$$\frac{\overline{X} - \underline{\mu}}{S/\sqrt{n-1}} = t(n-1, \alpha), \qquad \frac{\overline{X} - \overline{\mu}}{S/\sqrt{n-1}} = t(n-1, 1-\beta),$$

where $t(n-1, \alpha)$ is the quantile of level $1 - \alpha$ of the Student distribution with $n - 1$ degrees of freedom. Since the Student distribution is symmetrical, $t(n-1, 1-\beta) = -t(n-1, \beta)$. It follows therefore that the lower and upper limits of the confidence interval with confidence coefficient $\gamma = 1 - \alpha - \beta$ for parameter μ (when the variance is unknown) can be found by the formulas

$$\underline{\mu} = \overline{X} - t(n-1, \alpha)\frac{S}{\sqrt{n-1}},$$

$$\overline{\mu} = \overline{X} + t(n-1, \beta)\frac{S}{\sqrt{n-1}}.$$

Normal Distribution, Known Mean Consider confidence interval for the standard deviation of a normal distribution with known expectation μ. The central statistic in this case is

$$T = \frac{1}{\sigma^2}\sum_{1 \le i \le n}(x_i - \mu)^2,$$

which has χ^2 distribution with n degrees of freedom (see Section 1.2). Analogous to the previous case, we obtain the following lower and upper confidence limits with confidence coefficient $\gamma = 1 - \alpha - \beta$ for parameter σ:

$$\underline{\sigma} = \sqrt{\frac{\Sigma_{1 \le i \le n}(x_i - \mu)^2}{\chi^2_{1-\alpha}(n)}}, \qquad \overline{\sigma} = \sqrt{\frac{\Sigma_{1 \le i \le n}(x_i - \mu)^2}{\chi^2_{\beta}(n)}},$$

where $\chi^2_q(n)$ is the quantile of level q for a χ^2 distribution with n degrees of freedom.

Normal Distribution, Unknown Mean Now consider the confidence interval for the standard deviation of a normal distribution with unknown expectation μ. The central statistic in this case is

$$T = \frac{nS^2}{\sigma^2} = \frac{1}{\sigma^2}\sum_{1 \le i \le n}(X_i - \overline{X})^2,$$

which has a χ^2 distribution with $n - 1$ degrees of freedom. This leads to the following lower and upper limits of the confidence interval with confidence coefficient $\gamma = 1 - \alpha - \beta$ for parameter σ:

$$\underline{\sigma} = \frac{\sqrt{n}S}{\sqrt{\chi^2_{1-\alpha}(n-1)}}, \qquad \overline{\sigma} = \frac{\sqrt{n}S}{\sqrt{\chi^2_{\beta}(n-1)}}.$$

1.4.5 Approximation Based on Levy–Lindeberg Theorem

Consider a simple and constructive approximation for the confidence interval of the mean based on the Levy–Lindeberg theorem. Let $X_1, \ldots, X_n$ be a sample of n independent observations of some r.v. ξ with a finite and unknown mathematical expectation $\mu = E\xi$ and variance $\sigma^2 = E(\xi - \mu)^2$. We also assume that a d.f. of the observed r.v. is unknown.

Consider statistic $T = (\overline{X} - \mu)/(\sigma/\sqrt{n})$. In accordance with the Levy–Lindeberg form of the central limit theorem, this statistic has asymptotically normal distribution. This fact allows us to consider a normal approximation if n is sufficiently large. In this case, the inequalities

$$-u_\beta \leq \frac{\overline{X} - \mu}{\sigma}\sqrt{n} \leq u_\alpha \tag{1.27}$$

hold with probability close to $\gamma = 1 - \alpha - \beta$. Inequalities (1.27) are equivalent to the inequalities

$$\overline{X} - u_\alpha\left(\frac{\sigma}{\sqrt{n}} - \right) \leq \mu \leq \overline{X} + \mu_\beta\left(\frac{\sigma}{\sqrt{n}}\right).$$

These inequalities still do not give a confidence interval for parameter μ because the left and right parts contain an unknown parameter σ. Using another approximation, namely, substituting into these inequalities the estimate $\hat{\sigma} = S$ instead of σ, we obtain the approximate lower and upper limits of the confidence interval with the confidence coefficient $\gamma = 1 - \alpha - \beta$ for the mathematical expectation μ:

$$\underline{\mu} = \overline{X} - u_\alpha\frac{S}{\sqrt{n}}, \qquad \overline{\mu} = \overline{X} + u_\beta\frac{S}{\sqrt{n}}.$$

1.4.6 Clopper–Pearson Confidence Intervals

Consider construction of the confidence limits for the parameter of a binomial distribution. For evaluation of unit reliability, we often use the following procedure. A sample of size n is taken randomly from a homogeneous general population. (Sometimes this is not a sample from a population but special trial items manufactured before mass production.) This sample is tested under some specified condition. After the test is completed, one observes that m items have survived. For this case, the sequence of Bernoulli trials is considered as an appropriate mathematical model.

The distribution function of statistic m (the number of successes in a series of n independent Bernoulli trials) has the form

$$F(m, p) = \sum_{0 \le j \le m} \binom{n}{j} p^j (1 - p)^{n-j}.$$

Note that this function is decreasing in p. Applying the general formulas (1.23) obtained above, we find that the lower and upper limits of the confidence interval with the confidence coefficient $\gamma = 1 - \alpha - \beta$ for parameter p are found from the equations

$$\sum_{m \le j \le n} \binom{n}{j} \underline{p}^j (1 - \underline{p})^{n-j} = \beta,$$

$$\sum_{0 \le j \le m} \binom{n}{j} \overline{p}^j (1 - \overline{p})^{n-j} = \alpha.$$

Of course, solution of these equations is not a simple task, especially for high-order polynomials. In practice, for this purpose one uses tables of incomplete beta functions or standard computer programs.

For $m = 0$, the lower limit equals 0, and for $m = n$ the upper limit equals 1.

1.4.7 Approximation for Binomial Distribution

Although accurate calculation of confidence intervals using the Clopper–Pearson method is possible with a computer, sometimes it is useful to have a simple approximate method of confidence limit construction.

Let m be the number of observed successes in a series of n Bernoulli trials with an unknown parameter: the probability of success in a single trial, p. To construct confidence intervals for p, take

$$T = \frac{m - np}{\sqrt{np(1 - p)}}$$

as an initial statistic. In accordance with the DeMoivre–Laplace limit theorem (see *PRE*), this statistic is asymptotically normal. Thus, for large n, we can use the inequalities

$$-u_\beta \le \frac{m - np}{\sqrt{np(1 - p)}} \le u_\alpha,$$

which hold approximately with probability close to $\gamma = 1 - \alpha - \beta$. These inequalities can be rewritten in the form

$$\frac{m}{n} - u_\alpha \sqrt{\frac{p(1-p)}{n}} \leq p \leq \frac{m}{n} + u_\beta \sqrt{\frac{p(1-p)}{n}}.$$

These inequalities still do not give the confidence interval for p because the left and right sides contain unknown parameter p. Therefore, in practice, one often substitutes the estimate $\hat{p} = m/n$ instead of p. As a result, one has the following lower and upper limits of the confidence interval with the confidence coefficient $\gamma = 1 - \alpha - \beta$ for parameter p:

$$\underline{p} = \frac{m}{n} - u_\alpha \sqrt{\frac{m}{n^2}\left(1 - \frac{m}{n}\right)}, \qquad \overline{p} = \frac{m}{n} + u_\beta \sqrt{\frac{m}{n^2}\left(1 - \frac{m}{n}\right)}.$$

These confidence intervals are approximate and can be used only if n, the size of a sample, is large.

1.4.8 Confidence Interval for Parameter of Poisson Distribution

A flow of failures occurring in time for a complex system can be successfully described by the Poisson process. As shown in *Probability Reliability Engineering* (*PRE;* Gnedenko and Ushakov, 1995), the random number of failures occurring within an interval of fixed length has a Poisson distribution. The problem is to construct the confidence limits for an unknown parameter of the failure flow or for a parameter of the Poisson distribution. Consider this problem in formal terms.

Let $d = 0, 1, 2, \ldots$ be an integer r.v. with the Poisson distribution (see Section 1.2) with unknown parameter Λ. We need to construct the confidence interval for parameter Λ on the basis of the observed value of d. The distribution function of r.v. d is defined as

$$F(d, \Lambda) = \sum_{0 \leq j \leq d} \frac{\Lambda^j}{j!} e^{-\Lambda}.$$

Note that this function is decreasing in Λ. Applying formulas (1.23), as before, we obtain that the lower and upper limits of the confidence interval for Λ with confidence coefficient equal to $\gamma = 1 - \alpha - \beta$ are found from the equations

$$\exp(-\underline{\Lambda}) \sum_{0 \leq j \leq d-1} \frac{\underline{\Lambda}^j}{j!} = 1 - \beta, \qquad \exp(-\overline{\Lambda}) \sum_{0 \leq j \leq d} \frac{\overline{\Lambda}^j}{j!} = \alpha.$$

For $d = 0$ the lower limit is $\underline{\Lambda} = 0$. Notice that the left-hand sides of these equations can be expressed via a gamma distribution (see Section 1.2). Taking into account the relation between gamma and χ^2 distributions [see formula (1.3)], the confidence interval obtained above can be rewritten in the form

$$\underline{\Lambda} = \tfrac{1}{2}\chi^2_{\beta}(2d), \qquad \overline{\Lambda} = \tfrac{1}{2}\chi^2_{1-\alpha}(2d + 2),$$

where $\chi^2_q(n)$ is the quantile of level q for a χ^2 distribution with n degrees of freedom.

1.5 TEST OF HYPOTHESIS

1.5.1 Introduction

In previous sections we considered the two main statistical inferences related to test results. They are point and confidence estimations. However, sometimes the objective is not to estimate an unknown parameter but to make a decision about some claims regarding that investigated parameter. For instance, we can formulate two contradictory claims about the parameter, and the problem is to find which one is correct (with some probability, of course). Such claims might be (a) unknown parameter α relates to some specified level A as $\alpha < A$ (or $\alpha \leq A$, $\alpha > A$, $\alpha \geq A$, $\alpha = A$), or (b) unknown parameter α relates to another unknown parameter β as $\alpha < \beta$ (or $\alpha \leq \beta$, $\alpha > \beta$, $\alpha \geq \beta$, $\alpha = \beta$). Two examples of the first case are a comparison of the empirical probability of call completion obtained during some test of telephone network with a specified baseline and an analysis of an unknown and random strength of some mechanical construction under a fixed load. Analogous examples of the second case are two competing telephone companies comparing their probabilities of call completion and two constructions compared by their strength under some certain load. Each such claim is called a *statistical hypothesis.* Procedures for executing this statistical inference are called *tests of hypothesis.*

A test of a hypothesis includes two contradictory hypotheses: one, believed to be true, is called the *null hypothesis* and is denoted by H_0, and the other is called the *alternative hypothesis* and is denoted by H_a. For instance, the null hypothesis H_0: $\theta = \theta_0$ means that we suggest that the true hypothesis is that parameter θ is equal to some specified number θ_0 (which is called the *null value*). For this null hypothesis, there might be the following three possible alternative hypotheses:

(a) H_1: $\theta < \theta_0$,
(b) H_1: $\theta > \theta_0$, and
(c) H_1: $\theta \neq \theta_0$.

Hypothesis H: $\theta = \theta_0$ (i.e., θ_0 is a single value) is called a *simple* hypothesis. Hypothesis H: $\Theta \in D$ is called a *composite hypothesis* if D is some set of values of parameter θ including more than a single point. Thus, the alternative hypotheses (a), (b), and (c) considered above are composites. Of course, the null hypothesis might be a composite hypothesis and the alternative hypothesis might be, on the contrary, a simple one.

The hypotheses considered above are called *parametric,* in contrast to *nonparametric.* The latter will be considered in Section 1.5.6.

1.5.2 Two Simple Hypotheses

Let us have a sample of size n of independent observations of r.v.'s $X_1, \ldots, X_n$ with density function $f(\mathbf{x}, \boldsymbol{\theta})$ where $\mathbf{x} = (X_1, \ldots, X_n)$ and $\boldsymbol{\theta}$ can also be a vector (multidimensional parameter), $\boldsymbol{\theta} = (\theta_1, \ldots, \theta_k)$.

Let us consider the simplest case where two simple hypotheses

$$H_0: \boldsymbol{\theta} = \theta_0 \quad \text{and} \quad H_1: \boldsymbol{\theta} = \theta_1$$

are tested.

The test procedure allows us to accept or reject the null hypothesis. This procedure has two constituents: (1) a *test statistic* that is a function of the sample data on which the decision of acceptance (or rejection) is based and (2) a *rejection region* W representing the set of all test statistic values for which the null hypothesis will be rejected. The rule used for hypothesis rejection is also called a *criterion of the test of a hypothesis.*

We might make the two following types of errors:

a *type I error* (α) consisting of rejecting H_0 when it is true and
a *type II error* (β) consisting of accepting H_0 when it is false.

The errors of the first and second types are defined as follows:

$$\alpha = P\{(X_1, \ldots, X_n) \in W | H_0\},$$

$$\beta = P\{X_1, \ldots, X_n) \in \overline{W} | H_1\},$$

where $\overline{W}$ is the complementary set to W and $P\{A|H_j\}$ is the probability of event A under the condition that hypothesis H_j is true, $j = 0, 1$.

Let us introduce the *likelihood function*

$$L(X_1, \ldots, X_n, \boldsymbol{\theta}) = f(X_1, \theta) \cdots f(X_n, \boldsymbol{\theta}).$$

Errors of types I and II can be written as

$$\alpha = \int \underset{W}{\cdots} \int L(x_1, \ldots, x_n, \theta_0)\, dx_1 \cdots dx_n,$$

$$\beta = \int \underset{\bar{W}}{\cdots} \int L(x_1, \ldots, x_n, \theta_1)\, dx_1 \cdots dx_n.$$

The error of type α is also called the *level of significance* of the criterion. The value of $1 - \beta$, equal to the probability of rejecting hypothesis H_0 when it is true, is called the *power of the criterion.*

1.5.3 Neyman–Pearson Criterion

The most frequent problem of criterion construction is formulated as follows: For a fixed level of significance α, construct the criterion (or, in other words, critical set W) with the maximum power $1 - \beta$. The solution of this problem is given by the well-known result of Neumann and Pearson that for continuous variables is formulated as follows. Let us introduce the function of the sample,

$$\varphi = \varphi(X_1, \ldots, X_n) = \frac{L(X_1, \ldots, X_n, \theta_1)}{L(X_1, \ldots, X_n, \theta_0)},$$

which is the ratio of the two likelihood functions: null and alternative. The optimal, or, for the given level of significance α, the maximally powerful, *Neyman–Pearson criterion* is constructed in the following way. The critical region W includes all those $(X_1, \ldots, X_n)$ for which the inequality

$$\varphi = \varphi(X_1, \ldots, X_n) \geq C$$

holds. Here C is a constant found from the condition

$$P\{\varphi \geq C | H_0\} = \alpha;$$

that is, the level α is guaranteed. In more detail, this condition can be written as

$$\int \underset{\varphi(X_1,\ldots,X_n)\leq C}{\cdots} \int L(x_1, \ldots, x_n, \theta_0) dx_1 \cdots dx_n = \alpha.$$

The Neyman–Pearson result remains analogous for discrete r.v.'s. The only difference is in a possible necessity of "randomization" of the optimal criterion (for details, see Lehman, 1959).

Example 1.26 Let us construct the Neyman–Pearson criterion for a normal distribution. Consider the test of two simple hypotheses

$$H_0\colon \mu = \mu_0 \quad \text{and} \quad H_1\colon \mu = \mu_1,$$

where μ is the mean, μ_0, μ_1 are some given values, $\mu_0 < \mu_1$, and the variance σ^2 is supposed to be known.

In this case the likelihood function is

$$L(X_1, \ldots, X_n) = \left(\frac{1}{\sigma\sqrt{2\pi}}\right)^n \exp\left[-\frac{1}{2\sigma^2} \sum_{1\le i\le n} (x_i - \mu)^2\right].$$

Thus, the likelihood ratio is

$$\varphi = \frac{L(X_1, \ldots, X_n, \mu_1)}{L(X_1, \ldots, X_n, \mu_0)}$$
$$= \exp\left[\frac{\mu_1 - \mu_0}{\sigma^2} \sum_{1\le i\le n} x_i\right] \exp\left[-\frac{n(\mu_1 - \mu_0)^2}{2\sigma^2}\right].$$

From here we see that the critical region W is given in this case as

$$\sum_{1\le i\le n} X_i \ge C, \tag{1.28}$$

where constant C is chosen from the condition of the given level of significance α:

$$P\left\{\sum_{1\le i\le n} X_i \ge C \middle| \mu = \mu_0\right\} = \alpha. \tag{1.29}$$

Random variable $\Sigma_{1\le i\le n} X_i$ has a normal distribution with mean $n\mu$ and variance $n\sigma^2$ (see Section 1.2). Thus the latter equality can be rewritten as

$$1 - \Phi\left(\frac{C - n\mu_0}{\sigma\sqrt{n}}\right) = \alpha, \tag{1.30}$$

where $\Phi(\cdot)$ is the function of the standard normal distribution (Laplace function). Thus,

$$\frac{C - n\mu_0}{\sigma}\sqrt{n} = U_{1-\alpha},$$

where $U_{1-\alpha}$ is the quantile of the standard normal distribution of the level $1 - \alpha$. Thus, constant C, which determines the critical region (1.28), can be found as

$$C = n\mu_0 + U_{1-\alpha}\sigma\sqrt{n}.$$

The error of type II is defined as

$$\beta = P\left\{\sum_{1\le i\le n} x_i < C|\mu = \mu_1\right\} = \Phi\left\{\frac{C - n\mu_1}{\sigma\sqrt{n}}\right\}. \tag{1.31}$$

By the Neyman–Pearson lemma, this value of β is the minimum possible for the specified value of α. □

Example 1.27 Consider the previous example for the condition $\mu_1 < \mu_0$. Following the steps in Example 1.26, we find that the optimal Neyman–Pearson criterion with level of significance α is given by the critical region

$$\sum_{1\le i\le n} x_i \le C,$$

where constant C is chosen from the condition

$$P\left\{\sum_{1\le i\le n} x_i \le C|\mu = \mu_0\right\} = \alpha$$

or

$$\Phi\left\{\frac{C - n\mu_0}{\sigma\sqrt{n}}\right\} = \alpha.$$

This gives

$$\frac{C - n\mu_0}{\sigma\sqrt{n}} = U_\alpha = -U_{1-\alpha}$$

and finally

$$C = n\mu_0 - U_{1-\alpha}\sigma\sqrt{n}. \ \square$$

Example 1.28 Let us construct the optimal Neyman–Pearson criterion for a parameter of the exponential distribution.

Consider two simple hypotheses

$$H_0\colon \lambda = \lambda_0 \quad \text{and} \quad H_1\colon \lambda = \lambda_1,$$

where λ_0 and λ_1 are the given levels for the distribution with density $f(x, \lambda) = \lambda e^{-\lambda x}$, $x > 0$. In this case the likelihood function is

$$L(X_1, \ldots, X_n) = \lambda^k \exp\left(-\lambda \sum_{1 \le i \le n} X_i\right).$$

The likelihood ratio has the form

$$\varphi = \left(\frac{\lambda_1}{\lambda_0}\right)^n \exp\left\{-(\lambda_1 - \lambda_0) \sum_{1 \le i \le n} X_i\right\}.$$

It follows that the critical region is defined by the inequality

$$\sum_{1 \le i \le n} X_i \le C,$$

where the constant C is chosen from the condition

$$\Pr\left\{\sum_{1 \le i \le n} x_i \le C | \lambda = \lambda_0\right\} = \alpha.$$

The latter condition can be rewritten in the form

$$P\left\{2\,\lambda_0 \sum_{1 \le i \le n} x_i \le 2\,\lambda_0\, C | \lambda = \lambda_0\right\} = \alpha.$$

Random variable $2\lambda_0 \sum_{1 \le i \le n} x_i$ for $\lambda = \lambda_0$ has a gamma distribution (see Section 1.2) with parameters $(\frac{1}{2}, n)$, which is equivalent to a χ^2 distribution with $2n$ degrees of freedom. Thus, the latter condition is equivalent to

$$H_{2n}(2\lambda_0 C) = \alpha,$$

where $H_{2n}(\cdot)$ is the cumulative function of the χ^2 distribution with $2n$ degrees of freedom. Finally, for constant C, we have

$$C = \frac{\chi^2_\alpha(2n)}{2\lambda_0},$$

where $\chi^2_\alpha(2n)$ denotes the quantile of the level α for a χ^2 distribution with $2n$ degrees of freedom. In this case the minimum type II error equals

$$\begin{aligned}\beta &= P\left\{\sum_{1 \le i \le n} X_i > C | \lambda = \lambda_1\right\} \\ &= 1 - H_{2n}(2\lambda_1 C) = 1 - H_{2n}\left[\chi^2_\alpha(2n)\left(\frac{\lambda_1}{\lambda_0}\right)\right].\end{aligned}$$

It follows from the Neyman–Pearson statement that this value cannot be decreased for a given type I error α. □

Example 1.29 Let us construct the optimal Neyman–Pearson criterion for a parameter of the binomial distribution.

Consider two simple hypotheses

$$H_0: p = p_0 \quad \text{and} \quad H_1: p = p_1,$$

where p is the probability of success for a sequence of Bernoulli trials, p_0 and p_1 are the given values, and $p_0 < p_1$. Let X_j denote the result of the jth trial. Here, $X_j = 1$ means success in the jth trial and $X_j = 0$ means failure, $\Pr\{X_j = 1\} = p$ and $\Pr\{X_j = 0\} = 1 - p$.

The likelihood function in this case is

$$L(X_1, \ldots, X_n, p) = \binom{n}{m} p^m (1 - p)^{n-m},$$

where $m = \Sigma_{1 \le j \le n} x_j$ is the total number of observed successes in the sequence of n trials. The likelihood ratio function is

$$\varphi = \frac{L(X_1, \ldots, X_n, p_1)}{L(x_1, \ldots, x_n, p_0)} = \left(\frac{p_1}{p_0}\right)^m \left(\frac{1 - p_0}{1 - p_1}\right)^{n-m}.$$

From the latter expression, we easily find that the critical region for the Neyman–Pearson criterion has the form

$$m \ge C. \tag{1.32}$$

The DeMoivre–Laplace limit theorem states that for large n the distribution of r.v. m is approximated by the normal distribution with mean $\mu = np$ and variance $\sigma^2 = np(1 - p)$. Using the normal approximation, constant C in (1.32) can be found from the condition

$$\Pr\{m \ge C | p = p_0\} \approx 1 - \Phi\left(\frac{C - np_0}{\sqrt{np_0(1 - p_0)}}\right) = \alpha,$$

and, consequently, we have

$$C = np_0 + U_{1-\alpha}\sqrt{np_0(1 - p_0)}. \tag{1.33}$$

The type II error is defined as

$$\beta = P\{m \le C | p = p_1\} \approx \Phi\left(\frac{C - np_1}{\sqrt{np_1(1 - p_1)}}\right). \qquad (1.34)\ \square$$

1.5.4 Sample Size

Let some values of type I and II errors (α and β) be specified so that $\alpha + \beta < 1$. Before the test we need to determine the needed sample size n^* to construct the criterion for testing two simple hypotheses H_0: $\theta = \theta_0$ and H_1: $\theta = \theta_1$ with given levels of α and β. The quantity of n^* is defined as a minimum integer n for which the inequalities

$$\begin{aligned} p\{\varphi(X_1, \ldots, X_n) \ge C | \theta = \theta_0\} &\le \alpha, \\ P\{\varphi(X_1, \ldots, X_n) < C | \theta = \theta_1\} &\le \beta \end{aligned} \qquad (1.35)$$

hold for some constant $C = C^*$. In this case the Neyman–Pearson criterion delivering given α and β has the critical region defined by the inequality

$$\varphi(X_1, \ldots, X_n) \ge C^*.$$

Example 1.30 Consider a sample from a normal distribution. For the situation considered in Example 1.26, from (1.30) and (1.31) we find that (1.35) in this case has the form

$$1 - \Phi\left(\frac{C - n\mu_0}{\sigma\sqrt{n}}\right) \le \alpha, \qquad \Phi\left(\frac{C - n\mu_1}{\sigma\sqrt{n}}\right) \le \beta.$$

From the above formula, we obtain that the minimum sample size n^* and the corresponding critical constant C^* for given α and β are determined from the system of equations

$$1 - \Phi\left(\frac{C - n\mu_0}{\sigma\sqrt{n}}\right) = \alpha, \qquad \Phi\left(\frac{C - n\mu_1}{\sigma\sqrt{n}}\right) = \beta.$$

Using quantiles of the standard normal distribution, we can write these equations in the form

$$\frac{C - n\mu_0}{\sigma\sqrt{n}} = U_{1-\alpha}, \qquad \frac{C - n\mu_1}{\sigma\sqrt{n}} = U_\beta = -U_{1-\beta}.$$

Thus, we obtain a sample size as

$$n^* = \frac{\sigma^2(U_{1-\alpha} + U_{1-\beta})^2}{(\mu_1 - \mu_2)^2}. \tag{1.36}$$

For example, let hypotheses

$$H_0: \mu = \mu_0 = 3.5 \quad \text{and} \quad H_1: \mu = \mu_1 = 3.8$$

be tested for known $\sigma = 0.8$ and given $\alpha = 0.05$ and $\beta = 0.1$. Using (1.36) and taking into account that $U_{1-\alpha} = U_{0.95} = 1.64$ and $U_{1-\beta} = U_{0.9} = 1.28$, we find that the sample size is $n^* = 61$. □

Example 1.31 Find the sample size for testing by the Bernoulli scheme. For the problem considered in Example 1.29, we obtain from (1.33) and (1.34) the following system of equations to determine n^* and C^*:

$$1 - \Phi\left(\frac{C - np_0}{\sqrt{np_0(1 - p_0)}}\right) = \alpha, \qquad \Phi\left(\frac{C - np_1}{\sqrt{np_1(1 - p_1)}}\right) = \beta,$$

or, equivalently,

$$\frac{C - np_0}{\sqrt{np_0(1 - p_0)}} = U_{1-\alpha},$$

$$\frac{C - np_1}{\sqrt{np_1(1 - p_1)}} = U_\beta = -U_{1-\beta},$$

and, finally,

$$n^* = \frac{[U_{1-\alpha}\sqrt{p_0(1 - p_0)} + U_{1-\beta}\sqrt{p_1(1 - p_1)}]^2}{(p_1 - p_0)^2}.$$

This expression gives the needed sample size to test simple hypotheses of type $H_0: p = p_0$, $H_1: p = p_1$ for a sequence of Bernoulli trials to satisfy the given probabilities of errors α and β. Since the quantity n^* is usually fractional, in practice we take the smallest integer larger than n^*. □

The sample size obtained with the use of the Neyman–Pearson criterion cannot be improved (decreased) if this sample size has been determined and fixed in advance. Nevertheless, the average sample size can be decreased for the same error probabilities α and β in sequential trials where the decision about stopping the test is made during testing depending on the obtained data (see details in Part III).

1.5.5 Composite Parametric Hypotheses

Let us test two composite hypotheses

$$H_0\colon \theta \in D_0 \quad \text{and} \quad H_1\colon \theta \in D_1, \tag{1.37}$$

where D_0 and D_1 are some nonintersected regions of domain of θ; for instance, these regions are $\theta \leq \theta_0$ and $\theta \geq \theta_1$, where θ_0 and θ_1 are given values, $\theta_0 < \theta_1$.

Parametric hypotheses of type (1.37) can be one-parametric (if parameter $\boldsymbol{\theta}$ is a scalar) or multi-parametric (if parameter θ is a vector). The criterion of testing of composite hypotheses is also defined via a critical set W of the sample $(X_1, \ldots, X_n)$. As before, if the sample $(X_1, \ldots, X_n)$ belongs to W, then the null hypothesis H_0 is rejected and the alternative hypothesis is accepted. If the sample does not belong to W, then the alternative hypothesis is rejected and the null hypothesis is accepted.

The error probabilities of types I and II have the same meaning as above and are defined as

$$\alpha(\theta) = P\{(X_1, \ldots, X_n) \in W|\theta\}, \quad \theta \in D_0,$$

$$\beta(\theta) = P\{(X_1, \ldots, X_n) \in \overline{W}|\theta\}, \quad \theta \in D_1,$$

where $P\{\cdot|\theta\}$ is the conditional probability under the condition that the true value of the parameter equals θ. Contrary to simple hypotheses, values $\alpha(\theta)$ and $\beta(\theta)$ are some functions of parameter θ. The maximum possible value of the type I error is

$$\alpha = \max_{\theta \in D_0} \alpha(\theta)$$

and it is called a *criterion scale* or a *level of significance.*

The function

$$E(\theta) = P\{(X_1, \ldots, X_n) \in W|\theta\} = \int \underset{W}{\cdots} \int L((x_1, \ldots, x_n, \theta)\, dx_1 \cdots dx_n,$$

which expresses the probability of rejection of hypothesis H_0 depending on parameter θ, is called the *function of criterion power* (*power function*). If there exists such a criterion that maximizes function $E(\theta)$ in all possible criteria simultaneously for all $\theta \neq D_1$ (for fixed α), this criterion is called uniformly most powerful. Such criteria exist only in some particular cases for simple hypotheses (see examples below).

Probabilities of errors are expressed via the power functions as follows:

$$\alpha(\theta) = E(\theta) \qquad \text{for } \theta \in D_0, \tag{1.38}$$

$$1 - \beta(\theta) = 1 - E(\theta) \quad \text{for } \theta \in D_1. \tag{1.39}$$

Thus the uniformly most powerful criterion (if it exists) minimizes the type II error (for fixed α) for all $\theta \neq D_1$.

Remark 1.4 Equalities (1.38) and (1.39) are true only for values of parameter θ indicated there. For different values of θ, the values of $\alpha(\theta)$ and $\beta(\theta)$ cannot be interpreted as probabilities of error. □

Sometimes, together with the power function, the *operative characteristic* of a criterion is used,

$$S(\theta) = P\{(X_1, \ldots, X_n) \in \overline{W}|\theta\} = \int \underset{W}{\cdots} \int L(x_1, \ldots, x_n, \theta)\, dx_1 \cdots dx_n,$$

that is, the probability to accept the null hypothesis H_0 if the true value of the parameter equals θ. Obviously, the operative characteristic and power function can be expressed via each other as $S(\theta) = 1 - E(\theta)$.

Let us construct the criteria for testing composite hypotheses on a normal distribution.

Example 1.32 Consider the test of a simple hypothesis H_0: $\mu = \mu_0$ versus the composite hypothesis H_1: $\mu > \mu_0$, where μ is the mean of the normal distribution with known variance σ^2.

For any $\mu_1 > \mu_0$, the critical region of the most powerful Neyman–Pearson criterion of significance α for simple hypotheses $\mu = \mu_0$ against $\mu = \mu_1$ has the form (1.28), where the constant C is chosen from (1.29) or (1.30) and, consequently, does not depend on μ_1. This means that the criterion constructed above for that simple hypothesis with the critical region

$$\sum_{1 \le i \le n} X_i \ge C = n\mu_0 + U_{1-\alpha}\sigma\sqrt{n} \tag{1.40}$$

is the uniformly most powerful criterion (with the significance level α) for the composite alternative hypothesis H_1: $\mu > \mu_0$. □

Example 1.33 Using the conditions in Example 1.32, consider the test of a simple hypothesis H_0: $\mu = \mu_0$ versus the composite hypothesis H_1: $\mu < \mu_0$.

In this case, using the results obtained in Example 1.27, we find that the uniformly most powerful criterion of significance α for this case is given by the critical region

$$\sum_{1 \le i \le n} X_i \le C = n\mu_0 - U_{1-\alpha}\sigma\sqrt{n}. \ \square$$

Example 1.34 For the conditions in Example 1.32, consider the test of two composite hypotheses

$$H_0\colon \mu \le \mu_0 \quad \text{and} \quad H_1\colon \mu \ge \mu_1, \tag{1.41}$$

where $\mu_0 < \mu_1$.

Notice that for the criterion with critical region (1.40), the probability of a type I error

$$\begin{aligned}\alpha(\mu) &= P\left\{\sum_{1\le i\le n} X_i \ge C|\mu\right\} \\ &= 1 - \Phi\left(\frac{C - n\mu}{\sigma\sqrt{n}}\right) = 1 - \Phi\left(U_{1-\alpha} + (\mu_0 - \mu)\frac{\sqrt{n}}{\sigma}\right)\end{aligned}$$

monotonically increases in μ. Thus, the maximum value of the probability of a type I error is determined as

$$\alpha = \max_{\mu \le \mu_0} \alpha(\mu)$$

and is achieved at the point $\mu = \mu_0$. It follows that this criterion used to test composite hypotheses (1.41) has the significance level $\alpha = \alpha(\mu_0)$. Following the same arguments as in Example 1.32, we find that the criterion with the critical region (1.40) is the uniformly most powerful criterion for the problem with composite hypotheses. $\square$

Example 1.35 Consider a test for two composite hypotheses

$$H_0\colon \mu = \mu_0 \quad \text{and} \quad H_1\colon \mu > \mu_1 \tag{1.42}$$

for the mean μ of the normal distribution with the unknown variance σ^2.

In contrast to Example 1.32, hypothesis H_0 in this case is also a composite. For $\mu = \mu_0$, the statistic

$$\left(\frac{\overline{X} - \mu_0}{S}\right)\sqrt{n-1} \tag{1.43}$$

has a Student distribution with $n - 1$ degrees of freedom (see Section 1.2). From here, we obtain that the criterion with the significance level α for hypotheses (1.42) is given by the critical region

$$\left(\frac{\overline{X} - \mu_0}{S}\right)\sqrt{n-1} \geq t_{1-\alpha}(n-1),$$

where $t_{1-\alpha}(n-1)$ is the quantile of the level $1-\alpha$ of the Student distribution with $n-1$ degrees of freedom.

Analogously, using the statistic (1.43), we construct the criterion for the test of composite hypotheses of the form

$$H_0\colon \mu = \mu_0 \quad \text{and} \quad H_1\colon \mu < \mu_0 \tag{1.44}$$

or

$$H_0\colon \mu = \mu_0 \quad \text{and} \quad H_1\colon \mu \neq \mu_0. \tag{1.45}$$

For the hypothesis (1.44) the criterion of significance α is given by the critical region

$$\left(\frac{\overline{X} - \mu_0}{S}\right)\sqrt{n-1} \leq t_{\alpha}(n-1)$$

and for the alternative hypothesis (1.45) by

$$\left(\frac{|\overline{X} - \mu_0|}{S}\right)\sqrt{n-1} \geq t_{1-\frac{\alpha}{2}}(n-1) \; \square$$

Example 1.36 Consider a test hypothesis about equality of the mathematical expectations of two normal distributions. Assume two independent samples $X_1, \ldots, X_n$ and $Y_1, \ldots, Y_m$ of sizes n and m from normal distributions with parameters μ_1, σ_1 and μ_2, σ_2, respectively. Consider the following cases of composite hypotheses where the variances σ_1^2 and σ_2^2 are known:

$$\begin{array}{lll} 1. & H_0\colon \mu_1 = \mu_2 \quad \text{and} & H_1\colon \mu_1 > \mu_2, \\ 2. & H_0\colon \mu_1 = \mu_2 \quad \text{and} & H_1\colon \mu_1 < \mu_2, \\ 3. & H_0\colon \mu_1 = \mu_2 \quad \text{and} & H_1\colon \mu_1 \neq \mu_2. \end{array} \tag{1.46}$$

The difference of the sample means $(\overline{X} - \overline{Y})$ has a normal distribution with mean $\mu_1 - \mu_2$ and variance $(\sigma_1^2/n) + (\sigma_2^2/m)$. It follows that if the null hypothesis is true, that is, if $\mu_1 = \mu_2$, the statistic

$$\frac{\overline{X} - \overline{Y}}{\sqrt{\sigma_1^2/n + \sigma_2^2/m}} \tag{1.47}$$

has a normal distribution with parameters (0, 1). Then the significance criterion α for the problems formulated above are given by

1. $(\bar{X} - \bar{Y})/\sqrt{\sigma_1^2/n + \sigma_2^2/m} \geq U_{1-\alpha}$,
2. $(\bar{X} - \bar{Y})/\sqrt{\sigma_1^2/n + \sigma_2^2/m} \leq U_{\alpha}$, and
3. $|\bar{X} - \bar{Y}|/\sqrt{\sigma_1^2/n + \sigma_2^2/m} \geq U_{1-\alpha/2}$,

where u_γ is the quantile of level γ of the standard normal distribution.

Let us now consider problems (1.46) for the case where the variances are unknown but equal; that is, $\sigma_1 = \sigma_2 = \sigma$. In this case statistics nS_1^2/σ^2 and nS_2^2/σ^2 have χ^2 distributions with $n - 1$ and $m - 1$ degrees of freedom, respectively. Thus, statistic $nS_2^2/\sigma^2 + nS_{21}^2/\sigma^2$ also has a χ^2 distribution with $n + m - 2$ degrees of freedom. Taking into account that statistic (1.47) has the standard normal distribution for $\mu_1 = \mu_2$, we conclude that statistic $(\bar{X} - \bar{Y})\sqrt{n + m - 2}/\sqrt{1/n + 1/m}\sqrt{nS_1^2 + mS_2^2}$ has a Student distribution with $n + m - 2$ degrees of freedom (see Section 1.2). Thus, we find that the criteria of significance α for the considered problems are

1. $(\bar{X} - \bar{Y})b_{nm}/\sqrt{nS_1^2 + mS_2^2} \geq t_{1-\alpha}(n + m - 2)$,
2. $(\bar{X} - \bar{Y})b_{nm}/\sqrt{nS_1^2 + mS_2^2} \geq t_{\alpha}(n + m - 2)$, and
3. $(\bar{X} - \bar{Y})b_{nm}/\sqrt{nS_1^2 + mS_2^2} \geq t_{1-\alpha/2}(n + m - 2)$,

where $b_{nm} = \sqrt{nm[(n + m - 2)/(n + m)]}$ and $t_\gamma(n + m - 2)$ is the quantile of level γ of the Student distribution with $n + m - 2$ degrees of freedom. □

PROBLEMS

1.1. Prove Theorem 1.1.

1.2. Prove that the estimator

$$\hat{\mu}_r = \frac{1}{n} \sum_{1 \leq i \leq n} x_i^r$$

for the sample moment of order r is unbiased for moment $\mu_r = E\{\xi^r\}$.

1.3. Find a system of equations for calculating the MLE of parameters λ, α of the Weibull–Gnedenko distribution with density $f(x, \lambda, \alpha) = \lambda\alpha x^{\alpha-1}e^{-\lambda x^\alpha}$, $x > 0$.

1.4. Find the MLE of parameters a and b of uniform distribution in interval (a, b).

1.5. Show that the total number of successes m observed in n independent Bernoulli trials with an unknown probability of success p represents the sufficient statistic.

1.6. Construct a confidence interval for parameter Θ of a uniform distribution on interval $[0, \Theta]$ if a sample $x_1, \ldots, x_n$ of size n from this distribution is given.

1.7. Construct a confidence interval for parameter Θ of a "shifted" exponential distribution with density $f(x, \Theta) = e^{-(x-\theta)}$, $x > 0$, if a sample $X_1, \ldots, X_n$ of size n from this distribution is given.

1.8. Construct a confidence interval for the difference $\mu_1 - \mu_2$ where each of these values is the mean of two distinct normal distributions with parameters (μ_1, σ_1) and (μ_2, σ_2) if there are given test results of two independent samples $x_1, \ldots, x_n$ and $y_1, \ldots, y_m$, respectively. (Variances σ_1^2 and σ_2^2 are assumed to be known.)

1.9. Solve the preceding problem under the assumption that the variances are unknown but equal, that is, $\sigma_1^2 = \sigma_2^2 = \sigma^2$.

1.10. Let $x_1, \ldots, x_n$ and $y_1, \ldots, y_m$ be independent random samples from two different normal distributions with parameters (μ_1, σ_1) and (μ_2, σ_2), respectively. Construct the confidence interval for the ratio of variances, $K = \sigma_1^2/\sigma_2^2$.

1.11. The following problem arises for confidence estimation of the availability coefficient of a repairable unit on the basis of the test results. One needs to construct a confidence interval for ratio λ/μ of parameters of two exponential distributions with densities $f(x) = \lambda e^{-\lambda x}$, $x \geq 0$, and $g(x) = \mu e^{-\mu x}$, $x \geq 0$, on the basis of two independent samples $x_1, \ldots, x_n$ and $y_1, \ldots, y_m$ taken from these respective distributions.

1.12. The following problem also often arises in reliability practice. Let some units be tested during sequential independent cycles. The probability of a unit's successful operation, p, is unknown. The test is performed until the first failure. Construct the confidence interval for parameter p if it is known that the first failure has occurred at the nth cycle.

1.13. For a sample of size $n = 9$, construct the Neyman–Pearson criterion for testing two simple hypotheses about the mean μ of the normal distribution,

$$H_0: \mu = \mu_0 = 53 \quad \text{and} \quad H_1: \mu = \mu_1 = 54,$$

with the given level of significance $\alpha = 0.1$ and the known variance $\sigma^2 = 16$. For the constructed criterion, find the probability of a type II error and the power of the criterion.

1.14. In the previous problem find the minimum size of a sample, n^*, that delivers $\alpha = 0.1$ and $\beta = 0.1$. Construct the corresponding Neyman–Pearson criterion for this case.

1.15. Parties of wire rolls are tested for strength. It is known that the value of strength (measured in some conditional units) x has a normal distribution with variance $\sigma^2 = 9$. The party of rolls is considered as acceptable if the mean of the strength of all rolls is $\mu \geq 14$ and inadmissible if $\mu \leq 10$. For each party, one chooses at random n items and observes values $x_1, \ldots, x_n$. We need to test the two hypotheses

$$H_0\colon \mu \geq 14 \quad \text{vs.} \quad H_1\colon \mu \leq 10,$$

with the given probabilities of errors $\alpha = 0.1$ and $\beta = 0.05$. For these conditions, solve the following problems:

a. Find a necessary sample size n_* for which the required α and β will be satisfied.

b. Construct the corresponding uniformly most powerful criterion for the found sample size.

c. For the constructed criterion, find the power function and operative characteristic.
The operative characteristic of the criterion $S(\mu) = 1 - E(\mu)$.

CHAPTER 2

PLANS OF TESTS WITH A SINGLE CENSORSHIP

In this chapter, we consider simple test plans with a single censorship, that is, tests performed for a duration of time T specified in advance or up to a fixed number of failures r. Statistical inferences for these test plans were almost completely developed for exponential distributions by Epstein and Sobel (1953), Epstein (1960a–c), and others. However, for many other important practical cases, there is no solution even for simple cases. Some new results for multiple censorship will be considered in the next chapter.

2.1 INTRODUCTION

2.1.1 Test Without Replacement

Consider a unit whose random TTF ξ has an unknown distribution $F(t) = P\{\xi \leq t\}$. Assume N identical units are to be tested at $t = 0$. First, consider a case where the test continues until all N units fail. In this case we observe the sequence of failure moments

$$0 < t_1 < t_2 < \cdots < t_r < \cdots < t_N, \tag{2.1}$$

where t_r is the moment of the rth failure, $1 \leq r \leq N$.

This type of test forms the standard complete sample of N independently distributed r.v.'s ξ. Test results (2.1) are ordered statistics.

In practice, we usually have no possibility to perform a test until all units fail. A test terminates either when the rth failure has occurred at time t_r, $r < N$, or when the fixed (in advance) duration T is exceeded.

Below we will use the following notation for test plans:

1. Plan $[N\ U\ r]$. Here the first symbol, N, denotes the number of identical units tested at $t = 0$. The second symbol, U, means that the tested units are unrepairable while testing. The third symbol specifies the stopping rule. In this particular case, a symbol r reflects that the test terminates at time t_r, $r \leq N$, that is, when the rth failure has happened. Thus the moment at which the test terminates is random. The results of the test are the failure moments

$$0 < t_1 < t_2 < \cdots < t_r.$$

(For remaining units the test is stopped at moment t_r.)

2. Plan $[N\ U\ T]$. Here the first two symbols have the same meaning as in the previous case, and symbol T reflects that the test continues up to the preassigned moment. The number of failures for this plan, d, is random. The results of the test are the moments of failure before time T,

$$0 < t_1 < t_2 < \cdots < t_d < T,$$

where $d \leq N$. Obviously, d is a random number.

3. Plan $[N\ U\ (r, T)]$. Here the first two symbols have the same sense as in the previous cases, and symbol (r, T) reflects that the test terminates at moment $\tau = \min(t_r, T)$, that is, either at the preassigned moment T or at the moment of the rth failure, depending on what happens earlier. The number of failures for this plan, d, is a r.v. restricted by r. The moment of test termination is also random and restricted from the right. The results of the test are the moments of failure before time T,

$$0 < t_1 < t_2 < \cdots < t_d < T,$$

where $d \leq r$.

The first test plan described by statistic (2.1) is plan $[N\ U\ r]$ with $r = N$; that is, plan $[N\ U\ N]$ corresponds to the complete (uncensored) sample of size N. Different test plans can be depicted on plane $(t, d(t))$, where $d(t)$ is the number of failures that occur at time t (Figures 2.1–2.3). The test terminates when the process $d(t)$ reaches the bound of the corresponding region.

Note that for all plans of type $[\cdot\ U\ \cdot]$ a normalized trajectory of the process $d(t)/N$ coincides with the graph of the empirical d.f. $\hat{F}_N(t)$ for all t for which this trajectory was observed. In other words, the graph of the type mentioned above represents an empirical d.f. on time interval $[0, \tau]$, where τ is the stopping moment.

Test termination before moment t_N is called *censorship*. The corresponding moment is called the *moment of censorship*. For each plan considered above,

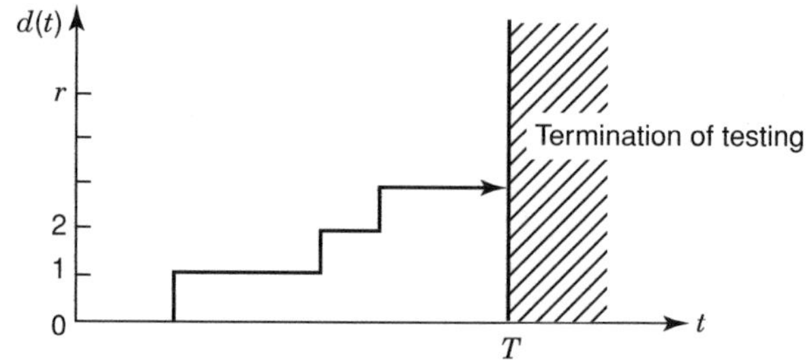

Figure 2.1 Sample of behavior of function $d(t)$ for plans $[N\ R\ T]$ and $[N\ W\ T]$. (*Source:* From I. A. Ushakov (Ed.), *Handbook of Reliability Engineering,* Wiley, New York, 1994, p. 289. Copyright © 1994 John Wiley & Sons, Inc. Reprinted by permission.)

the moment of censorship is unique. Such plans are called plans with a single censoring. (More complex plans are considered in Chapter 3.)

2.1.2 Test Plans with Renewable Units

In some test plans units are renewed after failure. It is assumed that replaced units are identical to the initial ones in a reliability sense. Thus, the number of tested units is constant during the entire testing period.

Test plans for which units are instantly replaced after a failure are denoted as $[\cdot\ R\ \cdot]$. All other test plan notation for renewable (replaceable) units remains the same:

1. Plan $[N\ R\ r]$. There are N units at the initial moment $t = 0$. All units are independent and after each failure a new unit completely identical to the initial one replaces the failed one. The test terminates at the moment t_r, $r \leq N$, when the rth failure has happened.

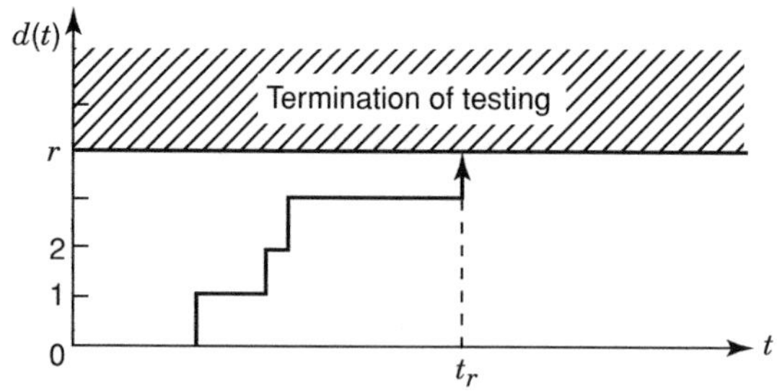

Figure 2.2 Sample of behavior of function $d(t)$ for plans $[N\ R\ r]$ and $[N\ W\ r]$. (*Source:* From I. A. Ushakov (Ed.), *Handbook of Reliability Engineering,* Wiley, New York, 1994, p. 289. Copyright © 1994 John Wiley & Sons, Inc. Reprinted by permission.)

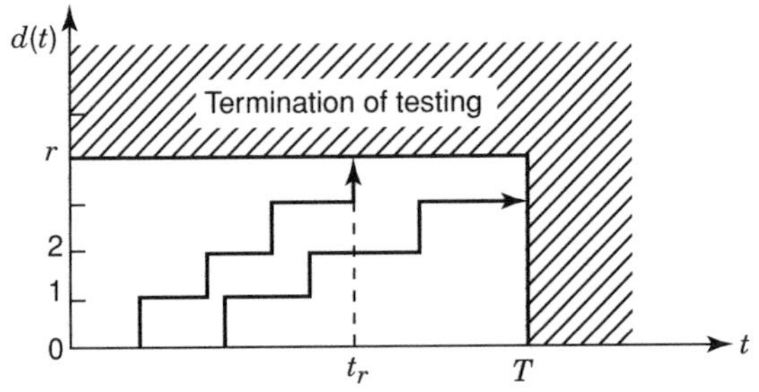

Figure 2.3 Sample of behavior of function $d(t)$ for plans [N R (r, T)] and [N W (r, T)]. (*Source:* From I. A. Ushakov (Ed.), *Handbook of Reliability Engineering,* Wiley, New York, 1994, p. 289. Copyright © 1994 John Wiley & Sons, Inc. Reprinted by permission.)

2. Plan [N R T]. This plan differs from the previous one: The test continues up to the preassigned moment T.

3. Plan [N R (r, T)]. Here the test terminates at moment $\tau = \min(t_r, T)$ which we have explained above.

Test plans without replacement will be called U-type, and test plans with replacement will be called R-type. Graphical illustration of the R-plans coincides with that of the U-plans, but in this case there is no analogy of the trajectory with an empirical d.f.

2.2 EXPONENTIAL DISTRIBUTION

Here we briefly present the main results for the simplest and better investigated case where the d.f. of the unit TTF is exponential, $F(t) = 1 - e^{-\lambda t}$, with unknown parameter λ, $\lambda > 0$ (Epstein and Sobel, 1953; Epstein, 1960a–c; Gnedenko et al., 1969). For this distribution all reliability indices of interest can be expressed through parameter λ; therefore, estimating this parameter is the main issue.

2.2.1 Test Plan [*N U r*]

In this case the test results are the moments of the first r failures,

$$t_1 < t_2 < \cdots < t_r. \tag{2.2}$$

Due to the memory property of the exponential distribution, r.v.'s t_1, $t_2 - t_1$, . . . , $t_r - t_{r-1}$ are independent. An r.v. $t_j - t_{j-1}$ has the exponential distribution with parameter $(N - j + 1)\lambda$, $1 \leq j \leq r$. (Obviously, $t_0 = 0$.) Hence the likelihood function, that is, the joint density function of test results (2.2), is

$$L = \prod_{1 \le j \le n} (N - j + 1)\lambda e^{-(N-j+1)\lambda(t_j - t_{j-1})},$$

or after simple transformations,

$$L = C\lambda^r e^{-\lambda S}, \tag{2.3}$$

where C is a normalizing constant, $C = N(n - 1) \cdots (N - r + 1)$, and the value of S equals the total testing time of all units, that is,

$$S = \sum_{1 \le j \le r} (N - j + 1)(t_j - t_{j-1}) = t_1 + \cdots + t_r + (n - r)t_r. \tag{2.4}$$

Thus, the likelihood function depends on test results via the value of S. It means (see Section 1.3.4) that statistic S is a sufficient statistic for this test plan.

***Point Estimate of Parameter* λ** Applying the maximum-likelihood method, we obtain from (2.3) that the maximum-likelihood equation has the form

$$\frac{\partial \ln L}{\partial \lambda} = \frac{r}{\lambda} - S = 0$$

and hence the maximum-likelihood estimate (MLE) for λ is

$$\hat{\lambda} = \frac{r}{S}. \tag{2.5}$$

This estimate is biased. We can prove this by finding the density function of the distribution of statistic S. The r.v.'s $\eta_j = (N - j + 1)(t_j - t_{j-1})$, $1 \le j \le r$, are independent and identical r.v.'s exponentially distributed with parameter λ. Therefore, as we can see from (2.4), the r.v. S is the sum of r independent exponentially distributed r.v.'s. From this, it follows (see Section 1.2) that statistic S has the gamma distribution with parameters (λ, r) and density function

$$f(x) = \frac{\lambda^r x^{r-1}}{(r - 1)!} e^{-\lambda x}. \tag{2.6}$$

From (2.6), we can easily find the mathematical expectation of estimate $\hat{\lambda}$:

$$E\{\hat{\lambda}\} = \int_0^\infty \frac{r}{x} \frac{\lambda^r x^{r-1}}{(r - 1)!} e^{-\lambda x}\, dx = \left(\frac{r}{r - 1}\right) \lambda.$$

Consequently, the MLE (2.5) is biased. Because of the bias, we usually use the estimate of the form

$$\hat{\lambda}' = \frac{r-1}{r}\hat{\lambda} = \frac{r-1}{S},$$

which is unbiased (for $r > 1$). Further, we can easily show on the basis of (2.6) that the variance of the unbiased estimate is (for $r > 2$)

$$\text{Var}\{\hat{\lambda}'\} = \frac{\lambda^2}{r-2}.$$

Confidence Limits of Parameter λ Applying a general approach (see Section 1.4), let us take an initial centered statistic in the form

$$W = \lambda S.$$

Expressing the density function of the distribution of S from formula (2.6), we can easily find that statistic W has the gamma distribution with parameters $(1, r)$. Let us denote the quantile of the level q for this distribution by $\Gamma_q(1, r)$. Then the inequalities

$$\Gamma_\alpha(1, r) \le \lambda S \le \Gamma_{1-\beta}(1, r)$$

hold with probability $\gamma = 1 - \alpha - \beta$. Hence the expressions for the lower and upper confidence limits with confidence probability $\gamma = 1 - \alpha - \beta$ (γ-confidence limits) for parameter λ are

$$\underline{\lambda} = \frac{1}{S}\Gamma_\alpha(1, r), \qquad \overline{\lambda} = \frac{1}{S}\Gamma_{1-\beta}(1, r).$$

Taking into account the relation between the gamma and χ^2 distributions (see Section 1.2.3), we can write

$$\underline{\lambda} = \frac{1}{2S}\chi^2_\alpha(2r), \qquad \overline{\lambda} = \frac{1}{2S}\chi^2_{1-\beta}(2r). \tag{2.7}$$

where $\chi^2_q(2r)$ is the quantile of level q of the χ^2 distribution with $2r$ degrees of freedom. Numerical tables for these quantiles are given in the Appendix (Table 2). Additionally, these confidence limits are often expressed as

$$\underline{\lambda} = \frac{1}{S}\Delta_\alpha(r-1), \qquad \overline{\lambda} = \frac{1}{S}\Delta_\beta(r-1), \tag{2.8}$$

where $\Delta_{1-\alpha}(d)$ and $\Delta_{\beta}(d)$ are the standard lower and upper γ-confidence limits, $\gamma = 1 - \alpha - \beta$, for parameter Λ of the Poisson d.f. These limits are constructed on the basis of the observed value of d (see Section 1.4.8). Numerical tables of $\Delta_{1-\alpha}(d)$ are given in the Appendix (Table 12.4).

Example 2.1 We tested nine units ($N = 9$) by plan $[N\ U\ r]$ until three failures ($r = 3$) occurred. The following failure moments were recorded (in hours): $t_1 = 144$, $t_2 = 182$, and $t_3 = 243$. We need to construct the 0.9-confidence interval for parameter λ. We are also interested in the corresponding γ-confidence limits for the MTTF, $\tau = 1/\lambda$, and the probability of failure-free operation (PFFO), $P = e^{-\lambda t_0}$ for $t_0 = 10$ hr.

In this case the total testing time equals

$$S = t_1 + \cdots + t_r + (N - r)t_r$$
$$= 144 + 182 + 243 + 6 \times 243 = 2027.$$

Taking $\alpha = \beta = \frac{1}{2}\gamma = 0.05$ and applying formula (2.7) and Table 2 from the Appendix, we obtain the following lower and upper γ-confidence limits:

$$\underline{\lambda} = \frac{\chi^2_{0.05}(6)}{2S} = \frac{1.635}{2 \cdot 2027} = 0.402 \times 10^{-3}\ \text{hr}^{-1},$$

$$\overline{\lambda} = \frac{\chi^2_{0.95}(6)}{2S} = \frac{12 \cdot 59}{2 \cdot 2027} = 3.09 \times 10^{-3}\ \text{hr}^{-1}.$$

The corresponding confidence limits for τ and P are

$$\underline{\tau} = \frac{1}{\overline{\lambda}} = 322\ \text{hr}, \qquad \overline{\tau} = \frac{1}{\underline{\lambda}} = 2480\ \text{hr},$$

and

$$\underline{P} = e^{-\overline{\lambda} t} = e^{-3.09\times 10^{-3}\times 10} = 0.970, \qquad \overline{P} = e^{-\underline{\lambda} t} = e^{-0.402\times 10^{-3}\times 10} = 0.996. \ \square$$

2.2.2 Test Plan [*N U T*]

The result of this test is represented by the set of sequential failure moments

$$t_1 < t_2 < \cdots < t_{d-1} < t_d < T,$$

where d is the number of failures that occurred during time T. For this plan, the number of failures, d, is a r.v., $d \le N$. As above, in the result of simple transforms, we obtain that the likelihood function in this case can be written as

$$L = C\lambda^d e^{-\lambda s}, \tag{2.9}$$

where $C = N(N - 1) \times \cdots \times (N - d + 1)$ is a normalizing constant and S is again the total testing time of all units. The value of S is calculated as

$$S = t_1 + \cdots + t_d + (N - d)T.$$

From (2.9) it follows that the sufficient statistic is the two-dimensional statistic (d, S). The maximum-likelihood equation in this case is

$$\frac{\partial \ln L}{\partial \lambda} = \frac{d}{\lambda} - S = 0$$

and, consequently, the MLE for λ is

$$\hat{\lambda} = \frac{d}{S}.$$

Construction of the confidence limits for parameter λ, based on the two-dimensional statistic (d, S), leads to many calculations. In addition, for the case of most practical interest, where units are highly reliable (i.e., where $\lambda T << 1$ and $d << N$), most of the information is contained in statistic d. Therefore, in practice, one often has to construct the confidence limits only on the basis of statistic d.

It is clear that the r.v. d is binomially distributed with parameter $p = 1 - e^{-\lambda T}$. Therefore the confidence limits for parameter λ can be easily obtained on the basis of the standard Clopper–Pearson γ-confidence limits for binomial trials (see Section 1.4.6). The lower and upper γ-confidence limits for binomial parameter p (if d failures have occurred while testing N units) obtained by the Clopper–Pearson equations are given by

$$\sum_{d \le j \le N} \binom{N}{j} \underline{p}^j (1 - \underline{p})^{N-j} = \tfrac{1}{2}\alpha, \quad \sum_{0 \le j \le d} \binom{N}{j} \overline{p}^j (1 - \overline{p})^{N-j} = \tfrac{1}{2}\alpha. \tag{2.10}$$

(Values of $\underline{p}$ and $\overline{p}$ for $\gamma = 1 - \alpha = 0.99$ and $\gamma = 1 - \alpha = 0.95$ are given in Tables E.14 and E.15.) Let us denote

$$\underline{\Delta}'_{1-\alpha}(d) = -\ln(1 - \underline{p}), \qquad \overline{\Delta}'_{1-\alpha}(d) = -\ln(1 - \overline{p}), \tag{2.11}$$

In accordance with (2.10) the lower and upper γ-confidence limits, $\gamma = 1 - \alpha$, for parameter λ can be calculated using the formulas

$$\underline{\lambda} = \frac{1}{T} \underline{\Delta}'_{1-\alpha}(d), \qquad \overline{\lambda} = \frac{1}{T} \overline{\Delta}'_{1-\alpha}(d). \tag{2.12}$$

Values of $\underline{\Delta}'_{1-\alpha}(d)$ and $\overline{\Delta}'_{1-\alpha}(d)$ for $\gamma = 1 - \alpha = 0.95$ are given in the Appendix in Table 3.

Example 2.2 Fifty units ($N = 50$) are tested during $t^* = 30$ hr by plan $[N\ U\ t^*]$. Six failures ($d = 6$) have occurred. We need to find the lower and upper γ-confidence limits, $\gamma = 0.95$, for the parameter λ, MTTF $\tau = 1/\lambda$, and probability of successful operation $P(t_0) = e^{-\lambda t_0}$ for $t_0 = 4$ hr.

Applying formulas (2.12) and Table 3, we obtain the following lower and upper γ-confidence limits for the parameter λ:

$$\underline{\lambda} = \frac{1}{T}\underline{\Delta}'_{0.95}(6) = \frac{0.046}{30} = 15.3 \times 10^{-4}\ \text{hr}^{-1},$$

$$\overline{\lambda} = \frac{1}{T}\overline{\Delta}'_{0.095}(6) = \frac{0.278}{30} = 92.7 \times 10^{-4}\ \text{hr}^{-1}.$$

The remaining lower and upper γ-confidence limits are

$$\underline{P}(t_0) = e^{-\overline{\lambda} t_0} = e^{-(92.7)\times 10^{-4}\times 4} = 0.963,$$

$$\overline{P}(t_0) = e^{-\underline{\lambda} t_0} = e^{-(15.3)\times 10^{-4}\times 4} = 0.994. \square$$

$$\underline{\tau} = \frac{1}{\overline{\lambda}} = 108 \text{ hours}$$

$$\overline{\tau} = \frac{1}{\underline{\lambda}} = 651 \text{ hours}$$

2.2.3 Test Plan [*N U* (*r*, *T*)]

Results of this plan are given by the set of sequential moments of failure

$$t_1 < t_2 < \cdots < t_d < T,$$

where d is the number of failures during time T, $d \le r$. The likelihood function has the form

$$L = C\,\lambda^d e^{-\lambda S},$$

where $C = N(N-1) \times \cdots \times (N - d + 1)$ is the normalizing constant, and

$$S = \begin{cases} t_1 + \cdots + t_d + (N - d)T & \text{if } d < r, \\ t_1 + \cdots + t_r + (N - r)t_r & \text{if } d = r. \end{cases}$$

The value of S, as stated previously, represents the total testing time for all units. The sufficient statistic is again a pair (d, S). The MLE for parameter λ is the same as for the previous test plans: $\hat{\lambda} = d/S$.

Applying the general approach given in Section 1.4, we can obtain the following confidence limits with confidence probability $\gamma = 1 - \alpha$ for parameter λ:

For $d < r$ the lower and upper γ-confidence limits are the same as for test plan $[N\ U\ T]$, that is,

$$\underline{\lambda} = \frac{1}{T}\underline{\Delta}'_{1-\alpha}(d), \qquad \overline{\lambda} = \frac{1}{T}\overline{\Delta}'_{1-\alpha}(d).$$

For $d = r$ (or for $t_r < t^*$) the confidence limits for λ can be found from equations

$$\varphi(\underline{\lambda}S, \underline{\lambda}T) = \tfrac{1}{2}\alpha, \tag{2.13}$$

$$\sum_{0 \le j \le r-1} \binom{N}{j} (1 - e^{-\overline{\lambda}T})e^{-(N-j)\overline{\lambda}T} + \varphi(\overline{\lambda}S, \overline{\lambda}T) = \tfrac{1}{2}\alpha,$$

where $\varphi(x, y) = P\{\lambda S < x, \lambda t_r < y\}$.

Finding the lower and upper γ-confidence limits from these equations needs cumbersome calculations and compiling special tables of the function $\varphi(x, y)$. For highly reliable units (i.e., for $\lambda T << 1$, $r << N$), which is the most interesting case for practical applications, we emphasize that all test plans of types $[N\ U\ r]$, $[N\ U\ T]$, and $[N\ U\ (r, T)]$ can be considered approximately equivalent to the R-plans $[N\ R\ r]$, $[N\ R\ T]$, and $[N\ R\ (r, T)]$. For this reason, in practice, one often uses simple formulas for test plan $[N\ R\ (r, T)]$ for construction of the confidence limits $\underline{\lambda}$, $\overline{\lambda}$ instead of the exact equations (2.13).

2.2.4 Test Plan [*N R r*]

Test results of this plan are given by the set of sequential moments of failure

$$t_1 < t_2 < \cdots < t_r. \tag{2.14}$$

Due to the memoryless property of the exponential distribution, r.v.'s t_1, $t_2 - t_1, \ldots, t_r - t_{r-1}$ are independent and identical r.v.'s exponentially distributed with parameter $N\lambda$. Thus the failure moments of (2.14) form the Poisson process with intensity $N\lambda$. Therefore the likelihood function has the form

$$L = N\lambda e^{-N\lambda t_1} N\lambda e^{-N\lambda(t_2 - t_1)} \times \cdots \times N\lambda e^{-N\lambda(t_r - t_{r-1})} = (N\lambda)^r e^{-\lambda S},$$

where $S = Nt_r$ is, as before, the total testing time of all units. Thus, for this case, the sufficient statistic is S (or the moment of the rth failure, t_r). The MLE is

$$\hat{\lambda} = \frac{r}{S} = \frac{r}{Nt_r}.$$

Notice that the value of S can be written for this test plan as

$$S = Nt_1 + N(t_2 - t_1) + \cdots + N(t_r - t_{r-1});$$

that is, S is the sum of r independent and identical r.v.'s distributed exponentially with parameter λ. So, S has the distribution coinciding with that for test plan $[N\ U\ r]$. It follows that the confidence limits for the considered test plan $[N\ R\ r]$ can be found by the analogous formulas

$$\underline{\lambda} = \frac{1}{S}\Delta_{1-\alpha}(r - 1), \qquad \overline{\lambda} = \frac{1}{S}\Delta_{\beta}(r - 1),$$

where $S = Nt_r$.

2.2.5 Test Plan [*N R T*]

Results of this plan are given by the set of sequential moments of failure

$$t_1 < t_2 < \cdots < t_d < T, \tag{2.15}$$

where d is the number of failures during time T. These moments of failure form the Poisson process with parameter $N\lambda$. Consequently, the likelihood function is determined by the expression

$$\begin{aligned} L &= N\lambda e^{-N\lambda t_1} N\lambda e^{-N\lambda(t_2-t_1)} \times \cdots \times N\lambda e^{-N\lambda(t_d-t_{d-1})} e^{-N\lambda(T-t_d)} \\ &= (N\lambda)^d e^{-N\lambda T} \end{aligned} \tag{2.16}$$

and the sufficient statistic is d, the number of failures having occurred during time T. From (2.16), we can obtain, as above, that the MLE of parameter λ has the form

$$\hat{\lambda} = \frac{d}{NT}. \tag{2.17}$$

Confidence Limits for λ The number of failures during time T, d, has a Poisson distribution with parameter $\Lambda = \lambda NT$. Consequently, the task is reduced to the construction of the lower and upper γ-confidence limits for the parameter of the Poisson distribution (see Section 1.4.8). The lower and upper γ-confidence limits, $\gamma = 1 - \alpha - \beta$, for parameter λ can be found by the formulas

$$\underline{\lambda} = \frac{1}{NT} \Delta_{1-\alpha}(d - 1), \qquad \overline{\lambda} = \frac{1}{NT} \Delta_{\beta}(d), \tag{2.18}$$

where $\Delta_{1-\alpha}(d - 1)$ and $\Delta_{\beta}(d)$ are the standard lower and upper γ-confidence limits, $\gamma = 1 - \alpha - \beta$, of the parameter of the Poisson distribution for d failures. The value of $\Delta_q(d)$ is the solution of the equation

$$e^{-\Delta} \sum_{0 \le j \le d} \frac{\Delta^j}{j!} = q.$$

Numerical tables of these values are given in the Appendix in Table 12.4. The lower and upper γ-confidence limits (2.17) can also be expressed via quantiles of the χ^2 distribution (see Section 1.4.8):

$$\underline{\lambda} = \frac{\chi^2_{\alpha}(2d)}{2NT}, \qquad \overline{\lambda} = \frac{\chi^2_{1-\beta}(2d + 2)}{2NT}.$$

Values $\underline{\lambda}$ and $\overline{\lambda}$ can also be used as one-side limits: $\underline{\lambda}$ is the one-side lower $(1 - \alpha)$-confidence limit and $\overline{\lambda}$ is the one-side upper $(1 - \beta)$-confidence limit for λ.

Example 2.3 Ten units ($N = 10$) were tested according to plan $[N\ R\ T]$ for 150 hr ($T = 150$). Six failures were registered, $d = 6$ failures.

Find the point estimate and lower and upper γ-confidence limits, $\gamma = 0.98$, for λ; the MTTF, $\tau = 1/\lambda$; and probability of successful operation $P(t_0) = e^{-\lambda t_0}$ for $t_0 = 1$ hr.

Using formula (2.17), the point estimate of parameter λ is given as

$$\hat{\lambda} = \frac{d}{NT} = \frac{6}{10 \cdot 150} = 4 \times 10^{-3}\ \text{hr}^{-1}.$$

Setting $\gamma = 1 - \alpha - \beta$, $\alpha = 0.001$, and $\beta = 0, 01$ and using formula (2.18) and Table 4 in the Appendix, the lower and upper γ-confidence limits for λ can be found as

$$\underline{\lambda} = \frac{\Delta_{0.99}(5)}{10 \cdot 150} = \frac{1.79}{10 \cdot 150} = 1.19 \times 10^{-3}\ \text{hr}^{-1},$$

$$\overline{\lambda} = \frac{\Delta_{0.01}(6)}{10 \cdot 150} = \frac{14.75}{10 \cdot 150} = 9.69 \times 10^{-3}\ \text{hr}^{-1}.$$

The corresponding point estimate and confidence limits for the MTTF and PFFO are

$$\hat{\tau} = \frac{1}{\hat{\lambda}} = 250 \text{ hr}, \qquad \underline{\tau} = \frac{1}{\overline{\lambda}} = 103 \text{ hr}, \qquad \overline{\tau} = \frac{1}{\underline{\lambda}} = 833 \text{ hr},$$

$$\hat{P}(t_0) = e^{-\hat{\lambda} t_0} = 0.996, \qquad \overline{P}(t_0) = e^{-\underline{\lambda} t_0} = 0.99. \ \square$$

2.2.6 Plan [*N R* (*r*,*T*)]

In this case the test results are the moments of failure

$$t_1 < t_2 < \cdots < t_d < T,$$

where d is the number of failures ($d \leq r$). As above we obtain that the likelihood function in this case has the form

$$L = (N\lambda)^d e^{-\lambda S}, \tag{2.19}$$

where S is the total testing time of all units. This value is determined by following formulas

$$S \begin{cases} NT & \text{if } d < r, \\ Nt_r & \text{if } d = r. \end{cases}$$

According to (2.19), the sufficient statistic for this plan is the two-dimensional statistic (d, S). The MLE of parameter λ has the same form as for the previous plans, that is, $\hat{\lambda} = d/S$.

The confidence limits for parameter λ for this plan are calculated exactly as for plans [N R T] and [N R r]. If $d < r$, we use the formulas for plan [N R T]:

$$\underline{\lambda} = \frac{\Delta_{1-\alpha}(d-1)}{NT}, \qquad \overline{\lambda} = \frac{\Delta_\beta(d)}{NT}.$$

If $d = r$ (or, in other words, $t_r < T$), the formulas for plan [N R r] are used:

$$\underline{\lambda} = \frac{\Delta_{1-\alpha}(r-1)}{S}, \qquad \overline{\lambda} = \frac{\Delta_\beta(r-1)}{S},$$

where $S = Nt_r$.

2.3 EXPONENTIAL METHODS FOR IFR DISTRIBUTIONS

The assumption that the distribution of time to failure, $F(t)$, is "exponential" is widely used in reliability theory and its applications. This is understandable (though not satisfactory or forgivable) because in this case analytical approaches are very simple and mathematically attractive. (The argument reminds us of searching for something that is lost under a street lamp only because it is a brightly lit place!) But for practical purposes, this assumption is sometimes far from realistic. Naturally, the question arises: How does the assumption of exponentiality affect statistical inferences if the real distribution is not exponential? In some sense, this is a question about the stability, or robustness, of statistical inferences in relation to the form of a distribution function $F(t)$. This problem and related ones were studied by Barlow and Proschan (1966, 1967), Zelen and Dannemiller (1961), Barlow and Gupta (1966), Pavlov (1974, 1977b), and others. In particular, these works show that some results for an exponential distribution can be extended to IFR distributions (distributions with increasing failure rate). For the sake of brevity, we will discuss the application of "exponential" methods (i.e., methods for estimation of the exponential distribution) for more general cases, in particular, estimation of the IFR class of distributions.

We will assume that a distribution function has a density: $f(t) = F'(t)$. Denote the reliability function by $P(t) = 1 - F(t)$ and the intensity function by $\lambda(t) = f(t)/P(t)$. Let us also introduce the function Λ, defined as the cumulative hazard function $\Lambda(t) = \int_0^t \lambda(x)\, dx$. This new function relates to the reliability function via the formula $P(t) = e^{-\Lambda t}$.

A distribution function $F(t)$ is IFR if $\lambda(t)$ is monotonically nondecreasing in $t \geq 0$. It is easy to see that for IFR distributions, function $\Lambda(t)$ is convex in t for all t, for which $P(t) > 0$.

The assumption that the function $\lambda(t)$ is nondecreasing coordinates with the common understanding of aging. Indeed, for the overwhelming majority of real physical objects, time leads to deterioration of materials and worsening of its reliability properties. This is the reason why IFR distributions are widely used in reliability analysis (Barlow and Proschan, 1975; Gnedenko et al., 1969; Ushakov, 1994; and others). It is necessary to emphasize that the class of IFR distributions includes as particular cases such standard parametrical distributions as exponential, normal, Weibull–Gnedenko (with form parameter $\alpha > 1$), gamma, and others.

2.3.1 Plan [*N U r*]

We need to construct the lower γ-confidence limit for the reliability function $P(t) = 1 - F(t)$ on the basis of the test results by plan $[N\ U\ r]$:

$$t_1 < t_2 < \cdots < t_r. \tag{2.20}$$

For an exponential function $F(t)$ this confidence limit has the form

$$\underline{P}(t) = e^{-\overline{\lambda}t}, \tag{2.21}$$

where $\overline{\lambda} = (1/S)\Delta_{1-\gamma}(r-1)$ is the upper γ-confidence limit for the parameter of the exponential distribution, and $S = t_1 + \cdots + t_r + (N-r)t_r$ is the total testing time of all units (see Section 2.2).

The lower γ-confidence limit for the IFR distribution is given by the expression

$$\underline{P}^*(t) = \begin{cases} \exp\left(-\frac{1}{S}\Delta_{1-\gamma}(r-1)t\right) & \text{if } t \le \frac{S}{N}, \\ 0 & \text{if } t > \frac{S}{N}. \end{cases} \tag{2.22}$$

This lower limit coincides with the corresponding limit for the exponential distribution (2.21) at the beginning of the time interval $0 \le t \le S/N$.

Another important reliability index is the time of failure-free operation, t_q, with probability q. This value is determined from the equation $P(t_q) = q$. It is clear that this value coincides with the quantile of level $1 - q$ of the distribution $F(t)$. For the exponential distribution, we can write $e^{-\lambda t_q} = q$, and, consequently, $t_q = -\ln q/\lambda$. The lower γ-confidence limit for t_q for the exponential distribution has the form

$$\underline{t}_q = -\frac{\ln q}{\overline{\lambda}} = \frac{(-\ln q)S}{\Delta_{1-\gamma}(r-1)}. \tag{2.23}$$

The analog of the limit (2.23) for t_q is

$$\underline{t}_q^* = \min\left\{\frac{S}{N}, \frac{(-\ln q)S}{\Delta_{1-\gamma}(r-1)}\right\}. \tag{2.24}$$

This value delivers the lower γ-confidence limit for t_q if $F(t)$ is IFR. The lower confidence limits in (2.22) and (2.24) for $P(t)$ and t_q are connected by the obvious relation

$$\underline{t}_q^* = \max\{t: \underline{P}^*(t) \ge q\}.$$

The lower confidence limit in (2.24) coincides with the analogous limit for the exponential distribution in (2.23) if the level q satisfies the inequality

$$q \ge e^{-\Delta_{1-\gamma}(r-1)/N} \approx e^{-r/N}. \tag{2.25}$$

The limit of type (2.24) for the class of IFR distributions was obtained by Barlow and Proschan (1966). The limit of type (2.22) was found by Pavlov (1977b). In the latter work, improved limits of types (2.22) and (2.24) were also obtained for IFR and IFRA distributions. More general test plans of type $[N\ U\ (r, T)]$ were also considered there. We avoid those results because they are too clumsy.

Example 2.4 On the basis of data given in the form (2.20), we need to test hypotheses for $P(t_0)$:

$$H_0\colon P(t_0) > b \quad \text{vs.} \quad H_1\colon P(t_0) \le b,$$

where b is some required level of the reliability index $P(t_0)$. Let us consider first an exponential distribution. In this case the following standard rule based on the lower γ-confidence limit of type (2.21) is used:

$$\begin{aligned}&\text{To accept } H_0 \text{ if } \underline{P}(t_0) > b,\\ &\text{To accept } H_1 \text{ if } \underline{P}(t_0) \le b.\end{aligned} \tag{2.26}$$

In this case the probability of a type II error is not larger than $\beta = 1 - \gamma$. Indeed, by the definition of the lower limit, for $P(t_0) \le b$, the probability of a type II error (i.e., the probability of accepting an item if it must be rejected) is

$$\begin{aligned}P\{\text{accept } H_0\} &= P\{\underline{P}(t_0) > b\} \le P\{\underline{P}(t_0) > P(t_0)\}\\ &= 1 - P\{\underline{P}(t_0) \le P(t_0)\} \le 1 - \gamma.\end{aligned}$$

Let us now assume that the distribution $F(t)$ is IFR rather than exponential. Then from expressions (2.21) and (2.22) for the lower confidence limits, the rule (2.26) delivers a type II error not larger than $\beta = 1 - \gamma$ if the critical level b satisfies the inequality

$$b \ge e^{(\Delta_{1-\gamma}(r-1)/N}. \tag{2.27}$$

In practice, the required level b is usually close to 1 and the right side of (2.27) is of order $\exp(-r/N) \approx 1 - r/N$. Hence condition (2.27) is not too restrictive. Thus, the decision rule (2.26) based on the exponentiality of $F(t)$ usually preserves the supplier risk β (but not the consumer risk) for distributions of the IFR class.

The analogous result can be obtained for the index of type t_q. □

Example 2.5 On the basis of data (2.20) let us test the hypotheses

$$H_0\colon t_q > C \quad \text{vs.} \quad H_1\colon t_q < C,$$

where C is a required level of t_q.

Consider the decision rule for the exponential distribution based on the lower γ-confidence limit (2.23):

$$\begin{aligned} &\text{To accept } H_0 \text{ if } \underline{t}_q > C, \\ &\text{To accept } H_1 \text{ if } \underline{t}_q < C. \end{aligned} \tag{2.28}$$

For the exponential distribution, this decision rule delivers a consumer risk (type II error) not larger than $\beta = 1 - \gamma$. The proof of this fact completely coincides with the previous case. The comparison of confidence limits (2.23) and (2.24) shows that the decision rule (2.28) for the exponential distribution preserves the consumer risk for IFR distributions if the value of q satisfies the inequality (2.25). In practice, we usually choose q close to 1. □

Thus, for indices of type $P(t_0)$ and t_q, the decision rules for the exponential distribution can be extended to IFR distributions with conditions that are not too restrictive, and the meaning of β is preserved. Let us emphasize once more that in general it is not a correct estimate of supplier risk (this is considered in detail in Part III).

The confidence limits (2.22) and (2.24) for the class of IFR distributions are not consistent. This deficiency, however, becomes significant only for large sample sizes. At the same time, the standard nonparametric confidence limits for $P(t_0)$ and t_q based on an empirical distribution function or order statistic, though consistent, gives worse confidence limits than (2.22) and (2.24) if

- the sample size is small or/and
- an item is highly reliable (either t_0 is small or q is close to 1).

Notice that the cases mentioned above are the most important in practice (see problems 2.3 and 2.4). The standard nonparametric confidence limits for t_q based on an order statistic have the following deficiency (see Walsh, 1962; Zacks, 1971; and others). For a specified confidence coefficient γ, the lower confidence limit for t_q can be constructed only for a large enough sample size, namely, for $N \geq N(\gamma, q)$, where $N(\gamma, q)$ is some threshold value. The standard nonparametric confidence limits for $P(t_0)$ based on an empirical distribution function (see Section 2.5) have an analogous deficiency: For a given sample size N, the lower γ-confidence limit for $P(t_0)$ does not exceed some threshold level $Q(N, \gamma) < 1$ for arbitrary small t. At the same time, for IFR distributions $P(t) \to 1$ if $t \to 0$.

Confidence Strip for Reliability Index By construction, (2.22) is the lower γ-confidence limit for $P(t)$ for any fixed moment t. It is possible to show a stronger statement: Inequality

$$\underline{P}^*(t) \leq P(t) \tag{2.29}$$

simultaneously holds for all $t \geq 0$ with probability not less than γ if $F(t)$ is IFR. Thus, (2.22) gives a γ-confidence strip for $P(t)$ in the case of IFR distributions. The proof of this statement follows from the fact that the confidence limit $\underline{P}^*(t)$ for different t depends on the same statistic S (for details, see Pavlov, 1977).

Let us denote the unit's MTTF by $\tau = \int_0^\infty P(t)\, dt$. By integration of the left side of (2.29), we obtain the following lower γ-confidence limit for the class of IFR distributions:

$$\begin{aligned} \underline{\tau}^* &= \int_0^\infty \underline{P}^*(t)\, dt = \int_0^{S/N} \exp\left(-\frac{\Delta_{1-\gamma}(r-1)}{S}\, t\right) dt \\ &= C_r \frac{S}{\Delta_{1-\gamma}(r-1)}, \end{aligned} \tag{2.30}$$

where the constant is

$$C_r = 1 - \exp\left(\frac{\Delta_{1-\gamma}(r-1)}{N}\right).$$

Let us compare this limit with the lower γ-confidence limit for τ for the exponential distribution

$$\underline{\tau} = \frac{S}{\Delta_{1-\gamma}(r-1)}. \tag{2.31}$$

The value of C_r is of the order $C_r \approx 1 - \exp(-r/N)$. It follows that in the case of no censorship (or with an insignificant one, i.e., where r is close to N) there is no practical difference between the two confidence limits $\underline{\tau}^*$ and $\underline{\tau}$. If $r << N$, the difference becomes significant. The constant C_r can be interpreted as the correction coefficient for the confidence limit for τ that should be used if exponential methods are applied to a more general IFR distribution.

Example 2.6 The test of four units by plan $[N\ U\ r]$ with $r = N = 4$ gave the following failure moments (in hours):

$$t_1 = 40, \qquad t_2 = 80, \qquad t_3 = 110, \qquad t_4 = 240.$$

Then for the exponential distribution the lower γ-confidence limit (2.31) with confidence coefficient $\gamma = 0.9$ equals

$$\underline{\tau} = \frac{S}{\Delta_{1-\gamma}(r-1)} = \frac{470}{\Delta_{0.1}(3)} = \frac{470}{6.68} = 70.4 \text{ hr.}$$

For the IFR distribution the lower γ-confidence limit (2.30) equals

$$\underline{\tau}^* = C_r\underline{\tau} = (1 - e^{-(\Delta_{0.1}(3)/4)})\underline{\tau} = 0.812\underline{\tau} = 57.3 \text{ hr.}$$

In this case the coefficient C_r equals 0.812. □

Assume that, under the assumption that $F(t)$ is IFR, we need to check on the basis of (2.20) the following standard hypotheses related to the MTTF:

$$H_0\colon \tau > b \quad \text{vs.} \quad H_1\colon \tau \le b,$$

where b is a specified level of the reliability index τ. Consider the following decision rule based on the lower γ-confidence limit (2.30):

$$\text{To accept } H_0 \text{ if } \underline{\tau}^* > b,$$

$$\text{To accept } H_1 \text{ if } \underline{\tau}^* \le b.$$

By definition of the lower confidence limit, this rule delivers the probability of a type II error (consumer risk) not higher than $\beta = 1 - \gamma$ for an IFR distribution. This decision rule can also be expressed via the confidence limit (2.31) for the exponential distribution as

$$\begin{aligned} &\text{To accept } H_0 \text{ if } \underline{\tau} > \frac{b}{C_r}, \\ &\text{To accept } H_1 \text{ if } \underline{\tau} < \frac{b}{C_r}, \end{aligned} \tag{2.32}$$

where coefficient C_r introduced above is less than 1. It means that to apply the exponential method the required level of reliability index equal to b must be increased $1/C_r$ times to keep the same type II error for the class of IFR distributions.

Example 2.7 Under the conditions of Example 2.6, where a test was made using a plan of type $[N\ U\ r]$ with $N = r = 4$, to obtain the needed level b, the required level of reliability must be increased $1/C_r = 1/0.812 = 1.22$ times. □

Remark 2.1 If we use the decision rule (2.32) for index τ for the exponential distribution with no modification (i.e., $C_r = 1$), consumer risk increases. Barlow and Proschan (1967) showed that producer risk might be preserved under some additional conditions on N, r, and γ. For reliability indices of types $P(t_0)$ and t_q, there is an inverse situation: Consumer risk is preserved and producer risk is increased. In this connection, the questions arise: How can reliability indices dealing with the class of IFR distributions be simultaneously guaranteed for both types of risk? Is it necessary to increase the sample size for testing? If so, by how much should the sample size be increased? These and related questions are considered in Part III. □

2.3.2 Test Plan [*N R T*]

Consider a test plan of type [N R T] for systems consisting of units with IFR distributions of TTF (Pavlov, 1974). Using plan [N R T], at any moment t we test N identical units. Each failed unit is immediately replaced by a new one that is completely identical. The test continues up to some moment T specified in advance. The results of the test are failure moments

$$t_1 < t_2 < \cdots < t_d < T, \tag{2.33}$$

where d is the number of failures occurring during the test.

The lower γ-confidence limit for the IFR reliability function $P(t)$ is found in the work mentioned above:

$$\underline{P}^*(t) = \begin{cases} \exp\left(-\dfrac{\Delta_{1-\gamma(d)}}{S}\, t\right) & \text{if } t \le \dfrac{NT}{N+D}, \\ 0 & \text{if } t > \dfrac{NT}{N+D}, \end{cases} \tag{2.34}$$

where $S = NT$ is the total testing time of all units. This limit coincides for $t \le NT/(N + d)$ with the standard lower γ-confidence limit for the exponential distribution for the same plan (see Section 2.2):

$$\underline{P}(t) = \exp\left(-\frac{\Delta_{1-\gamma}(d)}{S}\, t\right). \tag{2.35}$$

Notice that for plan [N R T] the total number of units tested and the number of failure-free intervals equals $N + d$. Thus, the value of $\tau^* = NT/(N + d)$ represents the MTTF of a unit tested. Therefore the lower confidence limit of the reliability function for the exponential d.f. remains correct for the class of IFR distributions if $t \le \tau^*$ where τ^* is a statistical estimate of the MTTF, τ. This fact can serve as statistical analogue of the well known lower bound of the IFR reliability function with the given MTTF (see Barlow and Proschan (1967)):

$$P(t) \geq \begin{cases} e^{-t/\tau} & \text{if } t \leq \tau, \\ 0 & \text{if } t > \tau. \end{cases}$$

The value of $\underline{P}^*(t)$ delivers the lower γ-confidence limit of $P(t)$ for each fixed moment of time $t > 0$. Besides, there exists even more strong statement, namely, that the system of inequalities

$$\underline{P}^*(t) \leq P(t) \text{ for all } t > 0$$

holds with probability not less than γ. This gives us an opportunity to get the corresponding confidence limits for such reliability indices as t_q and τ.

Let us define $\underline{\tau}_q^*$ as the moment when function $\underline{P}^*(t)$ crosses the level of q:

$$\underline{t}_q^* = \max\{t\colon \underline{P}^*(t) \geq q\},$$

from where

$$\underline{t}_q^* = \min\left\{\frac{(-\ln q)NT}{\Delta_{1-\gamma}(d)}, \frac{NT}{N+d}\right\}. \tag{2.36}$$

By integration (2.34), we obtain

$$\underline{\tau}^* = \int_0^\infty \underline{P}^*(t)dt = \int_0^{NT/(N+d)} \exp\left(-\frac{\Delta_{1-\gamma}(d)}{NT}\, t\right) dt = C_d \frac{NT}{\Delta_{1-\gamma}(d))}, \tag{2.37}$$

where

$$C_d = 1 - e^{-\Delta_{1-\gamma}(d)/(N+d)}.$$

Values of $\underline{t}_q^*$ and $\underline{\tau}^*$ deliver the lower γ-confidence limits for indices t_q and τ, respectively, for IFR distributions.

Example 2.8 On the basis of tests performed by plan $[N\ R\ T]$, we need to test hypotheses related to the probability of failure-free operation, $P(t_0)$, for some fixed moment t:

$$H_0\colon P(t_0) > b \quad \text{vs.} \quad H_1\colon P(t_0) \leq b,$$

where b is the required level of index $P(t_0)$.

Consider the decision rule based on the lower γ-confidence limit (2.34):

$$\begin{aligned} &\text{To accept } H_0 \text{ if } \underline{P}^*(t_0) > b, \\ &\text{To accept } H_1 \text{ if } \underline{P}^*(t_0) \leq b, \end{aligned} \tag{2.38}$$

By definition of the lower confidence limit, this rule delivers the probability of a type II error (consumer risk) not higher than $\beta = 1 - \gamma$ for IFR distributions. Compare this decision rule with the standard case of the exponential distribution $F(t)$,

$$\begin{aligned}&\text{To accept } H_0 \text{ if } \underline{P}(t_0) > b,\\ &\text{To accept } H_1 \text{ if } \underline{P}(t_0) \le b.\end{aligned} \tag{2.39}$$

where $\underline{P}(t_0)$ is the lower confidence limit (2.35) for the exponential distribution. This decision rule delivers consumer risk equal to $\beta = 1 - \gamma$ for the exponential d.f. □

From (2.34) and (2.35), it follows that decision rules (2.38) and (2.39) coincide if the required level of reliability, b, satisfies the inequality

$$b \ge q^*, \tag{2.40}$$

where

$$q^* = \exp\left(-\min_d \left\{\frac{\Delta_{1-\gamma}(d)}{N + d},\ d = 0, 1, \ldots\right\}\right).$$

Using simple transformation and taking into account that $\Delta_{1-\gamma}(0) = \ln[1/(1 - \gamma)]$, we obtain

$$q^* = \exp\left(-\min\left\{1, \frac{\Delta_{1-\gamma}(0)}{N}\right\}\right) = \max\left\{\frac{1}{e}, \sqrt[N]{1 - \gamma}\right\}. \tag{2.41}$$

To be correct, notice that in (2.40) we find the exact lower limit of the value $\Delta_{1-\gamma}(d)/(N + d)$ in d rather than the minimum, since this value exceeds 1 only as $d \to \infty$.

Thus, if the required level of reliability is sufficiently high, that is, condition (2.40) holds, then the decision rule (2.39) preserves consumer risk for IFR distributions. An analogous fact is true for the index of type t_q. We show this in the following example.

Example 2.9 On the basis of data obtained using plan $[N\ R\ T]$, let us test hypotheses

$$H_0\colon t_q > b \quad \text{vs.} \quad H_1\colon t_q \le b,$$

where b is the required level of t_q. Let us consider the decision rule based on the lower γ-confidence limit (2.36) for IFR distributions,

$$\begin{aligned} &\text{To accept } H_0 \text{ if } \underline{t}^*_q > b, \\ &\text{To accept } H_1 \text{ if } \underline{t}^*_q \le b, \end{aligned} \tag{2.42}$$

and the analogous decision rule

$$\begin{aligned} &\text{To accept } H_0 \text{ if } \underline{t}_q > b, \\ &\text{To accept } H_1 \text{ if } \underline{t}_q \le b, \end{aligned} \tag{2.43}$$

which is based on the standard lower γ-confidence limit for an exponential distribution:

$$\underline{t}_q = \frac{(-\ln q)NT}{\Delta_{1-\gamma}(d)}. \tag{2.44}$$

The confidence limits (2.44) and (2.36) coincide, and consequently, decision rules (2.42) and (2.43) coincide if the inequality

$$q \ge q^* \tag{2.45}$$

holds. (This inequality has the same sense as above.) □

In practice, the required level of reliability is usually high; it is almost always higher than 0.9. Therefore conditions (2.45) and (2.40) hold if, roughly speaking, $\sqrt[N]{1-\gamma} \le 0.9$, or

$$N \le \frac{\ln\,[1/(1-\gamma)]}{\ln(1/0.9)} \approx 10 \ln\left(\frac{1}{1-\gamma}\right).$$

For confidence coefficients 0.8–0.99, the left side of the above expression equals from 20 to 40. Thus, for test plans with replacement, exponential methods applied to the class of IFR distributions usually preserve consumer risk for indices of types $P(t_0)$ and t_q.

2.4 ESTIMATION OF UNIT RELIABILITY INDICES FOR GENERAL PARAMETRIC CASE

Let $F(t, \boldsymbol{\theta})$ be the d.f of a unit's TTF and depend on some parameter $\boldsymbol{\theta}$. This parameter may in general be a vector $\boldsymbol{\theta} = (\theta_1, \cdots, \theta_r)$. We denote the corresponding density function by $f(t, \boldsymbol{\theta}) = F'(t, \boldsymbol{\theta})$. The value of parameter $\boldsymbol{\theta}$ (or components of $\boldsymbol{\theta}$) is unknown. We wish to determine the point estimate and confidence limits for some reliability index R on the basis of the test data. This index is assumed to depend on the above-mentioned parameter

$$R = R(\boldsymbol{\theta}) = R(\theta_1, \ldots, \theta_r).$$

In practice we are usually interested in the following standard reliability indices of a unit:

1. the probability of unit failure-free operation (PFFO) during a required time t_0, $R(\boldsymbol{\theta}) = 1 - F(t_0, \boldsymbol{\theta})$;
2. the mean time to failure (MTTF), $R(\boldsymbol{\theta}) = \Sigma_0^\infty[1 - F(t, \theta)]\, dt$; and
3. the guaranteed time to failure (or q100th percent life), $R(\boldsymbol{\theta})$, which is determined from the equation $1 - F(r, \boldsymbol{\theta}) = q$.

Remark 2.2 In Section 2.4 we consider only units with no renewal, that is, only units working until its first failure. Renewable units will be considered in Chapter 3 and in Part II. □

2.4.1 Point Estimate of Reliability Indices

The point estimate of the reliability index $R = R(\theta)$ is usually calculated as

$$\hat{R} = R(\hat{\boldsymbol{\theta}}) = R(\hat{\theta}_1, \ldots, \hat{\theta}_r),$$

where $\hat{\boldsymbol{\theta}} = (\hat{\theta}_1, \ldots, \hat{\theta}_r)$ is a vector of point estimates of parameters that were found on the basis of test results by some standard method, usually by the maximum-likelihood method. The likelihood function is usually written for different test plans in the following ways.

1. For plan $[N\ U\ r]$ the likelihood function has the form

$$L = C \prod_{1 \le i \le r} f(t_i, \boldsymbol{\theta})\{\overline{F}((t_r, \boldsymbol{\theta})\}^{N-r}, \tag{2.46}$$

   where $t_1 < t_2 < \cdots < t_{r-1} < t_r$ are the observed failure moments, $\overline{F}(t, \boldsymbol{\theta}) = 1 - F(t, \boldsymbol{\vartheta})$ is a complementary function, and C is a norm constant: $C = N(N - 1) \times \cdots \times (N - r + 1)$. This formula for the likelihood function (as well as the one below) is obtained similar to the one for the exponential distribution (see Section 2.2, (2.16)).
2. For plan $[N\ U\ T]$ this formula is

$$L = C \prod_{1 \le i \le d} f(t_i, \boldsymbol{\theta})\{\overline{F}(T, \boldsymbol{\theta})\}^{N-d}, \tag{2.47}$$

   where d is the number of observed failures, $t_1 < t_2 < \cdots < t_d < T$ are the observed failure moments, and C is a norm constant (the meaning of it is insignificant).

3. For plan $[N\ U\ (r, T)]$ the likelihood function is determined by (2.46) if $d = r$ (i.e., if $t_r \le T$) or by (2.47) if $d < r$ (i.e., if $t_r > T$).
4. For plan $[N\ R\ r]$ the likelihood function is

$$L = C \prod_{1 \le i \le r} f(s_i, \boldsymbol{\theta}) \prod_{1 \le j \le N-1} \overline{F}(u_j, \boldsymbol{\theta}), \tag{2.48}$$

   where s_i are the intervals terminated by failure and u_j are the incomplete intervals where the test was interrupted before a failure occurrence.
5. For plans with replacement $[N\ R\ T]$ the likelihood function is

$$L = C \prod_{1 \le i \le d} f(s_i,\theta) \prod_{1 \le j \le N} \overline{F}(u_j,\theta) \tag{2.49}$$

   where the notation coincides with the previous case.
6. For plan $[N\ R\ (r, T)]$ the likelihood function is determined by (2.48) if $d = r$ (i.e., if $t_r \le T$) or by (2.49) if $d < r$ (i.e., if $t_r > T$).

It is easy to see that the likelihood function can be written for all these cases in the uniform type as

$$L = C \prod_{1 \le i \le d} f(s_i, \boldsymbol{\theta}) \prod_{1 \le j \le \nu} \overline{F}(u_j, \boldsymbol{\theta}), \tag{2.50}$$

where, in addition to the previous notation, d is the number of complete intervals (up to the failure occurrence) and ν is the number of incomplete intervals. The latter values might be random or specified in advance.

In accordance with the maximum-likelihood method, the estimates of parameters $\hat{\boldsymbol{\theta}} = (\hat{\theta}_1, \ldots, \hat{\theta}_r)$ are those values of the parameters $\boldsymbol{\theta} = (\theta_1, \ldots, \theta_r)$ that deliver the minimum of the likelihood function (2.50) for given fixed test results s_i, u_j, d, and v. As is well known, from a computational viewpoint it is better to consider a logarithm of the likelihood function.

As an illustration, consider the two-parameter Weibull–Gnedenko distribution often applied to practical reliability problems. This distribution has cumulative function and density

$$F(t, \boldsymbol{\theta}) = 1 - e^{-(t/\beta)^\alpha}, \qquad f(t, \boldsymbol{\theta}) = \frac{\alpha t^{\alpha-1}}{\beta^\alpha} e^{-(t/\beta)^\alpha},$$

where $\boldsymbol{\theta} = (\alpha, \beta)$ is the vector of parameters.

This distribution was applied by Weibull for reliability engineering problems in mechanics (Weibull, 1939, 1951). Almost simultaneously Gnedenko showed that this distribution is a particular case of the class of limit distributions of extremum values of a large number of independent r.v.'s (Gnedenko, 1941, 1943).

The likelihood function (2.50) for the Weibull–Gnedenko distribution has the form

$$L = C \prod_{1 \le i \le d} \alpha \left(\frac{s_i^{\alpha-1}}{\beta^\alpha} \right) e^{-(s_i/\beta)^\alpha} \prod_{1 \le j \le \nu} e^{-(u_j/\beta)^\alpha}.$$

Various ways of calculating the point estimates of parameters $\hat{\alpha}$ and $\hat{\beta}$ for the Weibull–Gnedenko distribution are considered by Kao (1956), Lieberman (1962), Mennon (1963), Cohen (1965), Mann (1968), Kudlaev (1986), and others.

If the point estimates $\hat{\alpha}$ and $\hat{\beta}$ are found, the PFFO during time t_0,

$$R = e^{-(t_0/\beta)^\alpha},$$

can be easily found as

$$\hat{R} = e^{-(t_0/\hat{\beta})^{\hat{\alpha}}}.$$

In an analogous manner, we can find estimates for other reliability indices.

2.4.2 Confidence Limits for Reliability Indices

The usual general approach for calculating approximate confidence limits for a reliability index $R = R(\boldsymbol{\theta})$ consists in finding the point estimate of this index, $\hat{R} = R(\hat{\boldsymbol{\theta}})$, and after this a normal approximation is used for a large sample size. In this case the lower and upper γ-confidence limits for R are obtained as

$$\underline{R} = \hat{R} - u_\gamma \sqrt{V(\hat{\boldsymbol{\theta}})}, \qquad \overline{R} = \hat{R} + u_\gamma \sqrt{V(\hat{\boldsymbol{\theta}})}, \tag{2.51}$$

where u_γ is the quantile of the level γ of the standard normal distribution, $V(\boldsymbol{\theta})$ is the variance of the estimate $\hat{R}$ for the given values of parameters $\boldsymbol{\theta} = (\theta_1, \ldots, \theta_r)$, and $\hat{\boldsymbol{\theta}} = (\hat{\theta}_1, \ldots, \hat{\theta}_r)$ is the vector of point estimates. (Particular cases of this approach were considered in Sections 1.4.5 and 1.4.6.) The interval $(\underline{R}, \overline{R})$ forms the approximate two-sided confidence interval with the confidence coefficient equal to $2\gamma - 1$. Besides, if the point estimates $\hat{\boldsymbol{\theta}} = (\hat{\theta}_1, \ldots, \hat{\theta}_r)$ are asymptotically unbiased and effective, then the confidence limits (2.51) are also asymptotically effective. (Notice that the MLEs possess such properties under conditions that are not too restrictive.)

Nevertheless, this method is essentially approximate and can be applied only if the sample size is very large. We should emphasize that the error is generated by the following two main factors:

1. the normal approximation is used instead of an exact distribution of estimate $\hat{R}$ and

2. in (2.51) we use the estimate $\hat{\boldsymbol{\theta}}$ instead of its exact value.

We will show below that such an approximation can lead to serious mistakes if the sample size is not large. In particular, this approximation might lead to a significant increase in the lower confidence limit of reliability indices (MTTF, or PFFO).

Example 2.10 For illustration, let us consider a simple example with the exponential distribution $F(t, \lambda) = 1 - e^{-\lambda t}$ for which the strict confidence limits are known. Consider plan $[N\ R\ T]$ with $N = 10$, $T = 150$ hr. During the test we observe $d = 6$ failures.

We are interested in the upper γ-confidence limit, $\gamma = 0.99$, for parameter λ. (This case corresponds to the lower estimation of a reliability index). In this case the MLE of parameter λ has the form

$$\hat{\lambda} = \frac{d}{NT}.$$

This estimate is unbiased with mathematical expectation $E\hat{\lambda} = \lambda$ and variance $V(\lambda) = E(\hat{\lambda} - \lambda)^2 = \lambda/NT$. Applying formula (2.51) and Table 1 in the Appendix for quantiles of the standard normal distribution, we obtain the following approximate upper γ-confidence limit for λ:

$$\bar{\lambda} = \hat{\lambda} + u_\gamma\sqrt{V(\hat{\lambda})} = \hat{\lambda} + u_\gamma\sqrt{\frac{\hat{\lambda}}{NT}} = \frac{d}{NT} + u_{0.99}\sqrt{\frac{d}{(NT)^2}}$$

$$= \frac{6}{10 \cdot 150} + 2.32\sqrt{\frac{6}{(10 \cdot 150)^2}} = 0.0078 \text{ hr}^{-1}.$$

We can find the same upper confidence limit by the strict formula (2.18). By this formula and Table 12.4 in the Appendix, we have

$$\bar{\lambda}' = \frac{\Delta_{1-\gamma}(d)}{NT} = \frac{\Delta_{0.001}(6)}{10 \cdot 150} = \frac{14.57}{1500} = 0.0097 \text{ hr}^{-1}.$$

This difference becomes more visible if we consider such a standard reliability index as MTTF $\tau = 1/\lambda$. In this case the approximate lower γ-confidence limit $\gamma = 0.99$ is given as

$$\underline{\tau} = \frac{1}{\bar{\lambda}} = \frac{1}{0.0078} = 128 \text{ hr}$$

and the strict limit as

$$\underline{\tau}' = \frac{1}{\overline{\lambda}'} = \frac{1}{0.0097} = 103 \text{ hr.}$$

Thus, the use of normal approximation (2.51) gives about 24% error in the undesirable direction. For practical use it is always better to have a conservative evaluation of a reliability index.

However, except for exponential distribution, strict confidence limits for reliability indices can be constructed for very few particular cases. Some of them are considered below. □

Confidence Limits for the Normal Distribution Consider a unit that has a normal distribution of TTF,

$$F(t, \boldsymbol{\theta}) = \Phi\left(\frac{t - \mu}{\sigma}\right),$$

where $\Phi(\cdot)$ is the function of the standard normal distribution (Laplace function) and μ and σ are the mean and standard deviation, respectively. In this case the two-dimensional parameter is $\boldsymbol{\theta} = (\mu, \sigma)$.

Consider a test by plan $[N\ U\ N]$ that corresponds to the complete sample. The results of this test are

$$t_1 < t_2 < \cdots < t_N. \tag{2.52}$$

We are interested in the confidence limits for the reliability function for fixed t_0, that is,

$$R = 1 - \Phi\left(\frac{t_0 - \mu}{\sigma}\right) = \Phi\left(\frac{\mu - t_0}{\sigma}\right). \tag{2.53}$$

Since $\Phi(\cdot)$ is a monotone increasing function, the problem is in finding the confidence limits for $(\mu - t_0)/\sigma$. The standard empirical mean and variance found by the test results (2.52) are

$$\overline{x} = \frac{1}{N} \sum_{1 \le i \le N} t_i \quad \text{and} \quad S^2 = \frac{1}{N} \sum_{1 \le i \le N} (t_i - \overline{x})^2.$$

Then the statistic

$$W = \left(\frac{\overline{x} - t_0}{S}\right) \sqrt{N - 1} \tag{2.54}$$

has a noncentered Student distribution with $N - 1$ degrees of freedom and parameter of noncentrality

$$\delta = \left(\frac{\mu - t_0}{S}\right)\sqrt{N}.$$

(For more details about the Student distribution see, e.g., Rao, 1965, Section 3a.) Now we can construct the desired confidence limits. Let us denote the noncentered Student distribution with $N - 1$ degrees of freedom and parameter of noncentrality by $G(y, \delta) = P(W < y)$. This function is decreasing in δ. Applying the general method of confidence limit construction [see formula (1.20)], we find that the lower and upper γ-confidence bounds for parameter δ follow from the equations

$$G(W, \overline{\delta}) = 1 - \gamma, \qquad G(W, \underline{\delta}) = \gamma,$$

where W is the value of statistic (2.54). After this the lower and upper γ-confidence limits for the reliability index (2.53) can be found by the formulas

$$\underline{R} = \Phi\left(\frac{\underline{\delta}}{\sqrt{N}}\right), \qquad \overline{R} = \Phi\left(\frac{\overline{\delta}}{\sqrt{N}}\right).$$

The interval $(\underline{R}, \overline{R})$ is the two-sided confidence interval for R with confidence coefficient $2\gamma - 1$.

Confidence Limits for the Weibull–Gnedenko Distribution Consider the test of a unit by plan $[N\ U\ r]$. Let the unit's TTF be distributed by the Weibull–Gnedenko function

$$F(t) = F(t, \alpha, \beta) = 1 - e^{-(t/\beta)^{\alpha}},$$

where $\alpha > 0$ and $\beta > 0$ are unknown parameters. Test results are presented by the moments of the first r failures,

$$t_1 < t_2 < \cdots < t_{r-1} < t_r. \tag{2.55}$$

The task is to construct the confidence limits for the reliability function for fixed t_0,

$$P(t_0) = 1 - F(t_0) = e^{-(t_0/\beta)^{\alpha}}, \tag{2.56}$$

on the basis of the test results.

In practice, the most important is to construct the lower confidence limit for (2.56). The solution to this task is obtained by Johns and Lieberman (1966).

Instead of test results in the form of (2.55), let us introduce new r.v.'s: $y_i = \ln(t_i/t_0)$, $i = 1, \ldots, r$. The distribution function of the r.v. $y = \ln(\xi/t_0)$ can be written in the form

$$G(u) = P(y \leq u) = P\left(\ln\frac{\xi}{t_0} \leq u\right) = P(\xi \leq t_0 e^u)$$
$$= F(t_0 e^u) = 1 - e^{-(t_0 e^u/\beta)^\alpha}. \tag{2.57}$$

We introduce the new parameters $\mu = \ln(\beta/t_0)$ and $\sigma = 1/\alpha$. Then the distribution function (2.57) can be written in the form

$$G(u) = 1 - \exp\{-e^{(u-\mu)/\sigma}\},$$

or alternatively,

$$G(u) = H\left(\frac{u - \mu}{\sigma}\right),$$

where $H(z) = 1 - \exp(-e^z)$. Thus, μ and σ represent parameters of bias and scale for the distribution of the r.v. y. Notice that the reliability index defined in (2.56) is expressed via μ and σ as

$$P(t_0) = \exp\left\{-\left(\frac{t_0}{\beta}\right)^\alpha\right\} = \exp\{-e^{-\mu/\sigma}\}. \tag{2.58}$$

So, the problem is reduced to the construction of confidence limits for the value μ/σ.

Let us introduce linear statistics

$$Z_a = \sum_{1\leq i\leq r} a_i y_i \qquad Z_b = \sum_{1\leq i\leq r} b_i y_i,$$

where a_i and b_i are some constants such that $\Sigma_{1\leq i\leq r}\, a_i = 1$ and $\Sigma_{1\leq i\leq r}\, b_i = 0$. Let us also introduce r.v.'s

$$V_a = \frac{Z_a - \mu}{\sigma}, \qquad V_b = \frac{Z_b}{\sigma}. \tag{2.59}$$

These r.v.'s can be represented in the form

$$V_a = \sum_{1\leq i\leq r} a_i \left(\frac{y_i - \mu}{\sigma}\right), \qquad V_b = \sum_{1\leq i\leq r} b_i \left(\frac{y_i - \mu}{\sigma}\right).$$

The distribution of r.v.'s $(y_i - \mu)/\sigma$ does not depend on parameters μ and σ. So, the two-dimensional r.v. (V_a, V_b) does not depend on μ and σ. This allows us to construct the confidence limits for μ/σ and $P(t_0)$. From (2.59), we obtain

$$\frac{\mu}{\sigma} = \left(\frac{Z_a}{Z_b}\right) V_b - V_a.$$

For each fixed t, consider the r.v. $tV_b - V_a$ and find the value of $L(t)$ from the condition

$$P\{tV_b - V_a \geq L(t)\} = \gamma, \tag{2.60}$$

where γ is the specified confidence coefficient. Then

$$\begin{aligned} P\left\{L\left(\frac{Z_a}{Z_b}\right) \leq \frac{\mu}{\sigma}\right\} &= P\left\{\frac{Z_a}{Z_b} \leq h\left(\frac{\mu}{\sigma}\right)\right\} = P\left\{\frac{\mu + \sigma V_a}{\sigma V_b} \leq h\left(\frac{\mu}{\sigma}\right)\right\} \\ &= P\left\{\frac{\mu}{\sigma} \leq h\left(\frac{\mu}{\sigma}\right) V_b - V_a\right\}, \end{aligned} \tag{2.61}$$

where $h(\cdot)$ is the inverse of the function L. Denoting $t = h(\mu/\sigma)$ and taking into account that $L(t) = L[h(\mu/\sigma)] = \mu/\sigma$, we obtain from (2.61) that

$$P\left\{L\left(\frac{Z_a}{Z_b}\right) \leq \frac{\mu}{\sigma}\right\} = P\{L(t) \leq tV_b - V_a\} = \gamma.$$

So $L(Z_a/Z_b)$ is the lower γ-confidence for r.v. μ/σ. It follows from (2.58) that the lower γ-confidence for the reliability index $P(t_0)$ has the form

$$\underline{P}(t_0) = \exp\{-e^{-L(Z_a/Z_b)}\}, \tag{2.62}$$

where $L(\cdot)$ can be calculated using Monte Carlo simulation.

The above procedure of confidence limit construction depends on the choice of constants a_i and b_i while statistics Z_a and Z_b are determined. Johns and Lieberman (1966) provide tables for constructing optimal confidence limits that are asymptotically effective. They also give tables of confidence limits (2.62) for various values of N, r, and γ.

The next two examples compare the confidence limits obtained by the method of Johns and Lieberman with the confidence limits for an IFR distribution given in Section 2.3.

Example 2.11 Ten units ($N = 10$) were tested by plan $[N\ U\ r]$. The test was terminated with occurrence of the fifth failure ($r = 5$). The moments of failure were (measured in hours)

$$t_i = 50, \quad t_i = 75, \quad t_3 = 125, \quad t_4 = 250, \quad t_5 = 300.$$

Assume that a unit has TTF with the Weibull–Gnedenko distribution. We need to construct the lower γ-confidence limit, $\gamma = 0.9$, for the PFFO, $P(t_0)$, with the fixed required time of failure-free operation, $t_0 = 40$ hr.

For this example, in Johns and Lieberman (1966) the lower confidence limit (2.62) was found to be

$$\underline{P}(t_0) = 0.796. \square \tag{2.63}$$

Example 2.12 For the previous example, assume that the unit's TTF distribution is IFR.

The total testing time of all units in this case equals

$$s = 50 + 75 + 125 + 250 + 300 + 5 \cdot 300 = 2300 \text{ hr}.$$

The time interval t_0 for which we estimate the unit reliability index satisfies the inequality $40 = t_0 < S/N = 230$. Therefore the lower confidence limit can be constructed by exponential methods (see Section 2.3). Applying formula (2.22) from the previous section and Table 12.4 in the Appendix, we find, for the same confidence coefficient $\gamma = 0.9$,

$$\begin{aligned} \underline{P}^*(t_0) &= \exp\left\{-\frac{\Delta_{1-\gamma}(r-1)}{S}\, t_0\right\} = \exp\left\{-\frac{\Delta_{0.1}(4)}{2300} \times 40\right\} \\ &= \exp\left\{-\frac{7.99}{2300} \times 40\right\} = 0.869. \square \end{aligned} \tag{2.64}$$

The lower confidence limit (2.64) for the class of IFR distributions is essentially higher than the lower confidence limit (2.63) found for the Weibull–Gnedenko distribution. Notice that the class of IFR distributions includes a family of Weibull–Gnedenko distributions (with the shape parameter $\alpha \geq 1$). In our opinion, such a significant difference can be explained by the fact that the $\alpha < 1$ family of Weibull–Gnedenko distributions also contains a subclass of DFR distributions. So, the general approach can give a gain in comparison with the strict results obtained for specific cases.

2.5 NONPARAMETRIC CONFIDENCE LIMITS FOR DISTRIBUTION FUNCTION

At the beginning of this section let us consider the test plan $[N\ U\ N]$. The test results in this case are moments $t_1 < t_2 < \cdots < t_N$ that form a complete sample of size N. We assume that the d.f. of the unit's TTF, $F(t)$, is continuous. Let $\hat{F}_N(t)$ be the empirical distribution function constructed on the basis of the test results and

$$D_N = \max_{0 \le t < \infty} |\hat{F}_N(t) - F(t)| \tag{2.65}$$

be the maximum deviation of the empirical d.f. from the theoretical one. It is important that the d.f. or the r.v. D_N does not depend on the d.f. $F(t)$. To convince the reader, let us introduce a new r.v. $\eta = F(\xi)$ that, as is well known, has the uniform distribution on the interval [0, 1]. Then the statement above follows from the equality

$$\max_{0 \le t \le \infty} |\hat{F}_N(t) - F(t)| = \max_{0 \le z \le 1} |\hat{G}_N(z) - z|,$$

where $\hat{G}_N(z)$ is the empirical d.f. constructed on the basis of the sample $\eta_i = F(t_i)$, $i = 1, \ldots, N$ taken from the uniform distribution.

Let the d.f. of the r.v. (2.65) be denoted by

$$H_N(u) = P\{D_N \le u\}. \tag{2.66}$$

Distribution (2.66) is the distribution of r.v. (2.65) that is defined for any distribution of initial r.v ξ. So, it is enough to find (2.66) for any distribution, for instance, for the uniform distribution of an r.v. ξ. Let us denote the quantile of the level $1 - q$ for the distribution (2.66) by an $A(N, q)$. We obtain that the inequalities

$$\hat{F}_N(t) - A(N, 1 - \gamma) \le F(t) \le \hat{F}_n(t) + A(N, 1 - \gamma) \tag{2.67}$$

simultaneously hold for all $0 \le t < \infty$ with probability γ. Thus, these inequalities give a γ-confidence strip for the reliability function $P(t) = 1 - F(t)$. In accordance with (2.67), this strip has the form

$$\hat{P}_N(t) - A(N, 1 - \gamma) \le P(t) \le \hat{P}_N(t) + A(N, 1 - \gamma) \tag{2.68}$$

for all $0 \le t < \infty$. Here $\hat{P}_N(t) = 1 - \hat{F}_N(t)$ is the empirical reliability function. The numerical values of $A(N, q)$ are given in Table E.11 in the Appendix. For $N \to \infty$, the asymptotic distribution of $\sqrt{N}D_N$ is given by the Kolmogorov distribution (see Section 1.5.6). Therefore for large sample size N, the γ-

confidence strip of the reliability function can be approximately determined by the inequalities

$$\hat{P}_N(t) - \frac{y_{1-\gamma}}{\sqrt{N}} \leq P(t) \leq \hat{P}_N(t) + \frac{y_{1-\gamma}}{\sqrt{N}},$$

where y_q is the quantile of the level $1 - q$ of the Kolmogorov distribution (for the numerical tables see in Table E.12 in the Appendix).

***Plans* [N U r], [N U T], *and* [N U (r, T)]** For the censored plans listed above, the considered confidence strips for distributions and reliability functions remain valid but only for the time interval $0 \leq t \leq \tau$, where τ is the moment of censorship. This moment for the plans mentioned above are determined as $\tau = t_r$, $\tau = T$, and $\tau = \min(t_r, T)$. For $t > \omega\tau$ in the nonparametrical case, that is, where there is no assumption about a distribution belonging to any distribution family, it is impossible to declare anything except the trivial inequalities

$$0 \leq P(t) \leq \hat{P}_N(\tau) + A(N, 1 - \gamma), \qquad t > \tau.$$

2.6 BOOTSTRAP METHOD OF APPROXIMATE CONFIDENCE LIMITS

The bootstrap method will be demonstrated using the following simple example. Assume that we use plan $[N\ U\ r]$. A unit's TTF has the distribution $F(t, \boldsymbol{\theta})$ with an unknown vector of parameters $\boldsymbol{\theta} = (\theta_1, \ldots, \theta_r)$. The results of the test are presented by the random vector

$$\mathbf{x} = (t_1, \ldots, t_r),$$

where $t_1 < t_2 < \cdots < t_r$ are the moments of sequential failures.

We need to construct the confidence limits for some unit's reliability index $R = R(\boldsymbol{\theta})$. For instance, if we are interested in the unit's MTTF, then

$$R(\boldsymbol{\theta}) = \int_0^\infty [1 - F(t, \boldsymbol{\theta})]\, dt;$$

if this reliability index is PFFO, then

$$R(\boldsymbol{\theta}) = 1 - F(t_0, \boldsymbol{\theta});$$

and so on. Let $\hat{\theta} = \hat{\theta}(\mathbf{x})$ and $\hat{R} = \hat{R}(\mathbf{x})$ be point estimates of $\boldsymbol{\theta}$ and $R(\boldsymbol{\theta})$, respectively, which are obtained on the basis of the test results $\mathbf{x}$. Let us denote the d.f. of estimate $\hat{R}$ for a given value of $\boldsymbol{\theta}$ by

$$H(t, \boldsymbol{\theta}) = P_{\theta}(\hat{R} \le t). \tag{2.69}$$

For the sake of simplicity, we assume that this function is continuous and strictly monotone increasing in t.

The lower and upper confidence limits $\underline{R}$ and $\overline{R}$, with the confidence coefficient $\gamma = 1 - \alpha - \beta$ for the reliability index R, are determined from the conditions

$$H(\underline{R}, \boldsymbol{\theta}) = \alpha, \qquad H(\overline{R}, \boldsymbol{\theta}) = 1 - \beta; \tag{2.70}$$

that is, they are the corresponding quantiles of the distribution $H(t, \boldsymbol{\theta})$ of the estimate $\hat{R}$. However, the function $H(t, \boldsymbol{\theta})$ depends on the parameter θ, which is unknown by the assumption. Let us set $\boldsymbol{\theta} = \hat{\theta}$, where $\hat{\theta}$ is the estimate of $\boldsymbol{\theta}$ found on the basis of the test results. It gives us the estimate of $H(t, \boldsymbol{\theta})$ in the form of $H(t, \hat{\theta})$. The final form of the confidence limits for $\underline{R}$ and $\overline{R}$ is

$$\begin{aligned} H(\underline{R}, \hat{\theta}) &= \alpha \\ H(\overline{R}, \hat{\theta}) &= 1 - \alpha. \end{aligned} \tag{2.71}$$

An analytical form of the d.f. $H(t, \boldsymbol{\theta})$ is usually unknown, and if known, then it is rather complex. Therefore in most cases the function and confidence limits $\underline{R}$ and $\overline{R}$ are approximately found by Monte Carlo simulation by setting $\boldsymbol{\theta} = \hat{\theta}$.

Thus, the procedure of this method is as follows:

1. On the basis of real test results $\mathbf{x}$, we calculate the point estimate of the vector of unknown parameters $\hat{\theta} = \hat{\theta}(\mathbf{x})$.
2. Then with the help of computer modeling, we find M independent realizations of the process of testing with $\boldsymbol{\theta} = \hat{\theta}$:

   $$(x_1, \hat{R}_1), (x_2, \hat{R}_2), \ldots, (x_M, \hat{R}_M),$$

   where x_i is the ith realization of modeling and $\hat{R}_i = \hat{R}_i(x_i)$ is the estimate of the reliability index obtained in the ith realization, $1 \le i \le M$.
3. On the basis of the obtained values, we construct by a common way the corresponding empirical d.f. $\hat{H}_M(t)$. Finally, the quantiles of levels α (from the left) and $1 - \beta$ (from the right) are chosen. They are the confidence limits $\underline{R}$ and $\overline{R}$, respectively. For instance, if $\alpha = \beta = 0.05$ and the number of realizations $N = 1000$, we take $\underline{R} = \rho_{(50)}$ and $\overline{R} = \rho_{(950)}$, where $\rho_{(j)}$ is the jth order statistic.

Remark 2.3 This approach can be extended in an obvious way to a nonparametric case if we consider $\boldsymbol{\theta}$ as the d.f. of a unit's TTF, $F(t)$, and $\hat{\theta}$ as its empirical d.f., $\hat{F}_N(t)$ (for plan $[N\ U\ N]$). The interval $(\underline{R}, \overline{R})$ constructed in such a way is not, generally speaking, the confidence interval for R in a precise "frequency" sense. Indeed, it does not follow from the construction of the interval that this interval satisfies the condition

$$P_\theta\{\underline{R} \le R(\boldsymbol{\theta}) \le \overline{R}\} \ge \gamma$$

for all $\boldsymbol{\theta}$. (Here $\gamma = 1 - \alpha - \beta$.) Therefore the bootstrap limits $(\underline{R}, \overline{R})$ can be interpreted as approximate confidence limits. These limits are sometimes close to the confidence limits and sometimes weaker (see Example 2.15). □

Example 2.13 A unit with the normal of d.f of TTF is tested by plan $[N\ U\ N]$. The mean of this normal d.f. equals μ and the variance equals σ^2; assume that $\mu >> \sigma^2$. The result of this test has the form of the complete sample of size N: $t_1, t_2, \ldots, t_n$.

Let us construct the boot strap limits for the parameter μ under the assumption that the variance σ^2 is known. The standard point estimate of the mean μ is its empirical average $\hat{\mu} = (1/N)\ \Sigma_{1 \le i \le N}\ t_i$. For the given value of μ, the estimate $\hat{\mu}$ has the normal d.f. with the mean μ and variance σ^2, that is,

$$H(t, \mu) = P_\mu(\hat{\mu} \le t) = \Phi\left(\frac{t - \mu}{\sigma/\sqrt{N}}\right),$$

where $\Phi(u)$ is the function of the standard normal distribution with parameters $\mu = 0$ and $\sigma = 1$. Applying (2.71), we obtain the lower and upper bootstrap limits $\underline{\mu}$ and $\overline{\mu}$ for μ from the equations

$$\Phi\left(\frac{\underline{\mu} - \hat{\mu}}{\sigma/\sqrt{N}}\right) = \alpha, \qquad \Phi\left(\frac{\overline{\mu} - \hat{\mu}}{\sigma/\sqrt{N}}\right) = 1 - \beta.$$

From here it follows that

$$\underline{\mu} = \hat{\mu} - u_{1-\alpha}\frac{\sigma}{\sqrt{N}}, \qquad \overline{\mu} = \hat{\mu} + u_{1-\beta}\frac{\sigma}{\sqrt{N}},$$

where u_q is the quantile of level q of the standard normal distribution. Thus in this case the bootstrap limits coincide with the standard exact confidence limits for μ. □

Example 2.14 Consider the conditions of the previous example with the following difference: The variance σ^2 is unknown. In this case $\boldsymbol{\theta} = (\mu, \sigma)$ is the two-dimensional parameter and the reliability index of interest is $R = R(\mu, \sigma) \equiv \mu$. The estimate of the vector of parameters θ is $\hat{\theta}, \hat{\sigma})$, where

$$\hat{\mu} = \frac{1}{N} \sum_{1 \le i \le N} t_i \quad \text{and} \quad \hat{\sigma} = \sqrt{\frac{1}{N-1} \sum_{1 \le i \le N} (t_i - \hat{\mu})^2}$$

are the empirical average and STD, respectively. For the given $\theta = (\mu, \sigma)$, the distribution of the estimate $\hat{\mu}$ has the same form as in the previous case:

$$H(t, \mu, \sigma) = P_{\mu,\sigma}(\hat{\mu} \le t) = \Phi\left(\frac{t - \mu}{\sigma/\sqrt{N}}\right).$$

Applying Equation (2.71), we find that the lower and upper bootstrap limits $\underline{\mu}$ and $\overline{\mu}$ can be found from the equations

$$\Phi\left(\frac{\underline{\mu} - \hat{\mu}}{\hat{\sigma}/\sqrt{N}}\right) = \alpha, \qquad \Phi\left(\frac{\overline{\mu} - \hat{\mu}}{\hat{\sigma}/\sqrt{N}}\right) = 1 - \beta,$$

and from here it follows that

$$\underline{\mu} = \hat{\mu} - u_{1-\alpha} \frac{\hat{\sigma}}{\sqrt{N}}, \qquad \overline{\mu} = \hat{\mu} + u_{1-\beta} \frac{\hat{\sigma}}{\sqrt{N}}.$$

In this case the bootstrap limits do not coincide with the exact confidence limits for parameter μ based on the Student distribution. Thus these limits can only be considered approximate. □

Example 2.15 Consider now the exponential distribution $F(t, \theta) = 1 - \exp(-t/\boldsymbol{\theta})$. In this case, parameter $\boldsymbol{\theta}$ represents the MTTF. Applying the approach described above, construct the bootstrap limits for the parameter $\boldsymbol{\theta}$ for plan $[N\ U\ r]$. The standard point estimate for $\boldsymbol{\theta}$ is $\hat{\theta} = S/r$ (see Section 2.3), where $S = t_1 + \cdots + t_r + (N - r)t_r$ is the total testing time of all units during the entire test. The d.f. of the estimate $\hat{\theta}$ for a specified value of $\boldsymbol{\theta}$ is given by the expression

$$H(t, \boldsymbol{\theta}) = P_\theta(\hat{\theta} \le t) = P_\theta\left(\frac{S}{r} \le t\right) = P_\theta(S \le rt).$$

Taking into account that r.v. S has the gamma distribution with parameters λ and r, where $\lambda = 1/\theta$ (see Section 2.3), we obtain

$$H(t, \boldsymbol{\theta}) = \Gamma\left(\frac{rt}{\boldsymbol{\theta}}\right), \tag{2.72}$$

where

$$\Gamma(u) = 1 - e^{-u} \sum_{0 \le j \le r-1} \frac{u^j}{j!}$$

is the gamma d.f. □

Thus, for this example, the d.f. of estimate $\hat{\theta}$ can be easily found in analytical form. It allows us to obtain analytical expressions for the bootstrap limits $\underline{\theta}$ and $\overline{\theta}$. In correspondence to (2.71), these limits can be found from the equations

$$H(\underline{\theta}, \hat{\theta}) = \Gamma\left(\frac{r\underline{\theta}}{\hat{\theta}}\right) = \alpha, \qquad H(\overline{\theta}, \hat{\theta}) = \Gamma\left(\frac{r\overline{\theta}}{\hat{\theta}}\right) = 1 - \beta.$$

Finally, we have

$$\begin{aligned} \underline{\theta} &= \frac{\Delta_{1-\alpha}(r-1)}{r}\hat{\theta} = \frac{\Delta_{1-\alpha}(r-1)}{r^2}S, \\ \overline{\theta} &= \frac{\Delta_{\beta}(r-1)}{r}\hat{\theta} = \frac{\Delta_{\beta}(r-1)}{r^2}S, \end{aligned} \tag{2.73}$$

where values $\Delta_{1-\alpha}(r-1)$ and $\Delta_{\beta}(r-1)$ were introduced in Section 2.3.1. At the same time, the exact confidence limits for parameter θ for the exponential distribution and for test plan $[N\ U\ r]$ has the form (see Section 2.3.1)

$$\underline{\theta}' = \frac{S}{\Delta_{\alpha}(r-1)}, \qquad \overline{\theta}' = \frac{S}{\Delta_{1-\beta}(r-1)}. \tag{2.74}$$

From the expressions obtained above, we have that the ratios of the bootstrap limits to the confidence limits and vice versa are written as

$$\frac{\underline{\theta}}{\underline{\theta}'} = \frac{\Delta_{1-\alpha}(r-1)\Delta_{\alpha}(r-1)}{r^2}, \qquad \frac{\overline{\theta}}{\overline{\theta}'} = \frac{\Delta_{1-\beta}(r-1)\Delta_{\beta}(r-1)}{r^2}.$$

By direct calculation with the help of tables for $\Delta_q(r-1)$ (see Table 12.4 in the Appendix), it is easy to check that both ratios are less than 1. It means that the lower and upper bootstrap limits are biased to the left with respect to the exact confidence limits; that is, $\underline{\theta} < \underline{\theta}'$ and $\overline{\theta} < \overline{\theta}'$.

Example 2.16 The test was performed by plan $[N\ U\ r]$ until the occurrence of two failures, $d = 2$. The total unit testing time was $S = 100$ hr. Then for $\gamma = 0.9$ and $\alpha = \beta = 0.05$ using Table 12.4, we find that

$$\underline{\theta}' = \frac{S}{\Delta_{\alpha}(r-1)} = \frac{100}{\Delta_{0.05}(1)} = \frac{100}{4.74} = 21.1 \text{ hr.}$$

$$\overline{\theta}' = \frac{S}{\Delta_{1-\beta}(r-1)} = \frac{100}{\Delta_{0.95}(1)} = \frac{100}{40.35} = 286 \text{ hr.}$$

$$\underline{\theta} = \frac{\Delta_{1-\alpha}(r-1)}{r^2} S = \frac{0.35}{2^2} \times 100 = 8.7 \text{ hr.}$$

$$\overline{\theta} = \frac{\Delta_{\beta}(r-1)}{r^2} S = \frac{4.74}{2^2} \times 100 = 118 \text{ hr.}$$

Using the bootstrap method in this case is not sensible since the exact confidence limits are known. However, this example clearly shows that the bootstrap method can produce significant bias. Thus this method should be used with caution.

Nevertheless, this method has the obvious advantages of being simple, understandable, and universal and is thus often used in practice. This approach and some of its improvements were considered by Efron (1981, 1985) and others. The bootstrap method is useful for the situations where exact confidence limits are not calculable. In particular, this method is very effective in estimating reliability indices of complex systems from the results of unit tests (see Part II).

PROBLEMS

2.1. Eight units ($N = 8$) were tested by plan $[N\ U\ r]$. The test was terminated with the occurrence of the second failure ($r = 2$). There were two recorded failures, at $t_1 = 60$ hr and $t_2 = 110$ hr. A unit has the exponential distribution $F(t) = 1 - e^{-\lambda t}$.

Construct γ-confidence limit $\gamma = 0.9$ for the parameter λ, for the PFFO $P(t_0)$ if $t_0 = 5$ hr, and for the MTTF τ.

2.2. The test in problem 2.1 was performed by plan $[N\ U\ T]$. Ten units were tested ($N = 10$). The test was terminated at the predetermined moment, $T = 100$ hr. Two failures are recorded, at $t_1 = 15$ hr and at $t_2 = 72$ hr. The unit has the exponential distribution $F(t) = 1 - e^{-\lambda t}$.

Construct the point estimates and lower confidence limits with confidence coefficient $\gamma = 0.975$ for the parameter λ, for the PFFO $P(t_0)$ if $t_0 = 10$ hr, and for the MTTF τ.

2.3. Seven units ($N = 7$) were tested by plan $[N\ U\ r]$. The test was terminated at the moment of occurrence of the third failure ($r = 3$). Failures occurred at $t_1 = 150$ hr, $t_2 = 250$ hr, and $t_3 = 400$ hr. The unit's TTF distribution $F(t)$ is assumed to be IFR.

Construct the lower confidence limit with confidence coefficient $\gamma = 0.975$ for the PFFO $P(t_0) = 1 - F(t_0)$ if $t_0 = 20$ hr.

2.4. Using the conditions in problem 2.3, find the nonparametrical lower confidence limit for $P(t_0)$ with the same confidence coefficient, $\gamma = 0.975$. Compare the result with the case where $F(t)$ is IFR.

2.5. Assume that $F(t)$ is IFR. Using the conditions in problem 2.3, find the lower confidence limit for MTTF τ with confidence coefficient $\gamma = 0.975$. Compare the results with the result for the exponential distribution.

2.6. Four units ($N = 4$) were tested by plan $[N\ R\ T]$. Failed units were immediately replaced by new ones. The test was terminated at a predetermined time, $T = 100$. Six failures were recorded ($d = 6$). We assume that the unit's TTF distribution $F(t)$ is IFR.

Construct the lower confidence limit with confidence coefficient $\gamma = 0.95$ for $P(t_0)$ if $t_0 = 1$ hr.

2.7. Using the conditions in problem 2.6, find the lower confidence limit with confidence coefficient $\gamma = 0.95$ for the guarantee unit's TTF t_q (with the guarantee level $q = 0.9$).

2.8. Using the conditions in problem 2.5, find the lower confidence limit with confidence coefficient $\gamma = 0.95$ for the MTTF τ.

2.9. Ten units ($N = 10$) were tested by plan $[N\ U\ r]$. The test was terminated after the fifth failure ($r = 5$). Failures occurred at $t_1 = 50$ hr, $t_2 = 75$ hr, $t_3 = 125$ hr, $t_4 = 250$ hr, and $t_5 = 300$ hr. It is known that the unit's TTF distribution $F(t)$ is IFR. Find the nonparametric lower confidence limit with confidence coefficient $\gamma = 0.9$ for $P(t_0)$ if $t_0 = 40$ hr. Compare this result with that obtained for the Weibull–Gnedenko distribution (see Example 2.11) and the IFR distribution (see Example 2.12).

CHAPTER 3

CENSORED SAMPLES

3.1 INTRODUCTION

Censored samples appear in practice if, for some reason, we stop testing before all tested units have failed. The procedure of test interruption may be different: The test of an entire sample might be stopped at some moment of time (chosen in advance or spontaneously); the test of some group of tested units might be stopped at some specified moments; or the test of some individual units might be terminated for some reason. At the end, we know several moments of unit failures and also know that some units have their time to failure (TTF) t_j larger than the moment of their test stopping, τ_j.

Sometimes we have information that the moment of failure belongs to the time interval $a_j \leq t_j \leq b_j$. Such a situation appears if we perform only periodical inspections. At moment b_j we find that the unit j has failed though at the previous inspection at moment a_j it was operational. These test plans cover most but not all possible types of censorship.

3.2 INDEPENDENT RANDOM CENSORSHIP

One of the most known and best investigated problems of censored testing is the following. We have a unit with unknown d.f. $F(t) = P\{\xi \leq t\}$ of a random TTF ξ. We are testing N identical independent units. Testing of each unit is terminated at a random time η if it has not failed. Random variables ξ and η are independent. The d.f. of η is denoted by $G(t) = P\{\eta \leq t\}$. Thus unit j is tested up to the moment

$$z_j = \min(\xi_j, \eta_j)$$

where ξ_j is the moment of the jth unit failure and η_j is the moment of the jth unit test termination, $j = 1, \ldots, N$. We assume that all r.v.'s $\xi_1, \ldots, \xi_N$, $\eta_1, \ldots, \eta_N$ are mutually independent. Thus, the test results are N pairs of r.v.'s

$$(z_j, I_j), \qquad j = 1, \ldots, N, \tag{3.1}$$

where $I_j = 1$ if $\xi_j \le \eta_j$, that is, the jth unit has failed during the test, and $I_j = 0$ if $\xi_j > \eta_j$, that is, the test was terminated before a failure.

Using the test results (3.1), we need to estimate this d.f. These results can be represented in more convenient form by introducing an additional notation,

$$s_j = z_j \quad \text{if } I_j = 1,$$
$$u_j = z_j \quad \text{if } I_j = 0.$$

Values s_j are called *complete intervals* and values u_j are called *censored* ones. In this notation (3.1) can be rewritten as $(s_1, \ldots, s_d)$, $(u_1, \ldots, u_{N-d})$, where d is the number of failed units and $N - d$ is the number of units whose test was terminated.

This model is called the model of *independent random censorship.* Sometimes this model is also called the model of "*concurrent risks,*" in which moments of test termination are interpreted as failures of some other types.

Example 3.1 Consider a series system consisting of two independent units. Their TTFs are denoted by ξ and η, and their d.f.'s by $F(t) = P\{\xi \le t\}$ and $G(t) = P\{\eta \le t\}$, respectively. The system TTF is defined as

$$z = \min(\xi, \eta).$$

Assume that we wish to estimate the d.f. of the first unit, that is, $F(t)$, under the following restriction: This unit can be tested only within the test of the entire system. Assume that N systems are being tested and that every time the test is continued until system failure. We have test results in the form (z_j, I_j), $j = 1, \ldots, N$, where $z_j = \min(\xi_j, \eta_j)$ is the moment of system failure, ξ_j and η_j are the moments of unit failures within the jth system, and

$$I_j = \begin{cases} 1 & \text{if } \xi_j \le \eta_j, \\ 0 & \text{if } \xi_j > \eta_j. \end{cases}$$

It is clear that this is a particular case of the model of (independent random censorship (concurrent risks). □

Example 3.2 Consider a series system consisting of n units of the first type and m units of the second type with the d.f.'s $F(t) = \Pr\{\xi \le t\}$ and $G(t) = \Pr\{\eta \le t\}$, respectively. All system units are independent. The moment of system failure is determined as

$$z = \min(\xi_1, \ldots, \xi_n, \eta_1, \ldots, \eta_m),$$

where ξ is the moment of failure of the kth unit of the first type, $k = 1, \ldots, n$, and η_r is the moment of failure of the rth unit of the second type, $r = 1, \ldots, m$. We are testing N identical systems until failure. Let us again estimate $F(t)$. The test results are represented by the set of Nn TTF intervals of units of the first type: d complete and $Nn - d$ censored, where d is the number of system failures due to units of the first type.

In this case, the test model is not based on independent random censorship; after a failure of a unit of the first type, testing of all remaining units of this type is terminated. It leads to the fact that the moment of test termination of units of the first type might depend on the failure moments of units of the same type. □

Remark 3.1 These examples lead to the problem of estimation of unit reliability on the basis of system testing. The "inverse problems" concerning estimation of system reliability on the basis of censored testing of its units are less investigated (see Part II). □

Remark 3.2 (Terminology) Let t_j be the moment of the jth unit failure and τ be the moment of censorship. Then if $\tau < t_j$ *censorship is from the left.* (Examples above are of this type.) If $t_j < T$ where T is some moment, *censorship is from the right.* Finally, if $a < t_j < b$, *it is interval censorship.* □

Example 3.3 Units are tested by the plan $[N\ U\ T]$, that is, N units are tested starting at $t = 0$ with no replacement of failed units and the test continues until a prior specified moment T. During the test the moments of failure are not recorded.

Thus, the only information is that d units have failed up to the moment T: $t_j \le T$, $j = 1, \ldots, d$ (censorship from the left). All the remaining $N - d$ units do not fail, that is, $t_k > T$, $k = 1, \ldots, N - d$ (censorship from the right). □

Example 3.4 Consider the test plan $[N\ U\ T]$. There is a possibility to check the unit states only at some specified moments $0 < T_1 < \cdots < T_n < T$. We have observed d failures for each of which it is known that the moment of failure t_j belongs to an interval (T_1, T_{1+1}). Thus, for d units we have interval censorship and for the remaining $N - d$ units $t_k > T$, that is, there is censorship from the right. □

We see that many various situations lead to test censorship. Our main focus will be on models with censorship from the right.

3.3 MARKOV MODEL OF CENSORED TESTING WITHOUT RENEWAL

The following testing model with censorship was considered elsewhere (Ushakov, 1980; Pavlov and Ushakov, 1984). It is a natural generalization of the model of independent random censorship (concurrent risks). A close model was considered by Belyaev (1984). Let us present an informal description of the problem (a formal description is given in Section 3.8.2).

Let a test of N identical and independent units with unknown d.f. begin at moment $t = 0$. At some moment τ_1 we terminate the test of n_1 units (operational up to this moment); then at some moment τ_2, $\tau_2 > \tau_1$, we terminate the test of n_2 units; and so on. The sequential moments of time $\tau_1, \tau_2, \ldots, \tau_k, \ldots$ and corresponding numbers $n_1, n_2, \ldots, n_k, \ldots$ can be chosen in advance or in the process of testing (even depending on the result of the test). All of these values might be deterministic or random. The moments at which we terminate testing of some number of units $\tau_1 < \tau_2 < \cdots < \tau_k < \cdots$ form a monotone increasing sequence of Markov random moments (see Section 3.8.1). The number of units n_1 whose testing is terminated at the moment τ_1 is random and might depend on the prehistory of test evolution up to the moment τ_1 but does not depend on the future testing process trajectory for $t > \tau_1$. Analogously, the number of units n_k, $k > 1$, whose testing is terminated at the moment τ_k is random and might arbitrarily depend on the prehistory but does not depend on the future for $t > \tau_k$.

The moment of termination of the entire test, ν, might also be random. The total number of inner termination points might be random (although naturally restricted by N). So, it is sufficient to consider a finite sequence of N moments $\tau_1 < \tau_2 \cdots < \tau_k < \cdots < \tau_N \leq \nu$. For a formal description, it is convenient to assume that some n_k might be equal to zero. The "real" number of termination r is defined as the number of such moments τ_k for which $n_k > 0$. Thus the model is given by the set

$$M = \{(\tau_1, n_1), (\tau_2, n_2), \ldots, (\tau_N, n_N), \nu\}. \tag{3.2}$$

We call this model of testing a Markov model. We repeat once more that the intervention in the testing process does not depend on the future.

This model generalizes the model of independent random censorship (concurrent risks), standard testing plans without replacement of types [N U T], [N U r], and [N U (r, T)] considered in Chapter 2, and factually all other testing plans without replacement. For instance, it includes the case where a sample is tested simultaneously by several different plans of class U (some by plan [N_1 U r_1], another by plan [N_2 U T_2], etc.).

The main characterization of the plan (3.2) is an absence of replacement of failed units. We will call this the Markov model of type U and use the abbreviation [MMU]. More general models including tests with replacement (renewal) are considered below in Section 3.6.

As above, an interval is called complete if it terminates by unit failure and is censored otherwise. Let $N(t)$ be the number under testing at moment t. A graph of function $N(t)$ is a convenient form of presenting the process. The graph can be constructed by putting ordered intervals one over another. The staircase function $N(t)$ is represented by the envelope of this set of intervals (see Figure 3.1). The results might also be presented by the set of failure moments $t_1, t_2, \ldots, t_j, \ldots$, moments of termination $\tau_1, \tau_2, \ldots, \tau_k, \ldots$, and numbers of units whose test is terminated at those moments, $n_1, n_2, \ldots, n_k, \ldots$. The same information might also be presented by the set of complete and censored intervals $s_1, s_2, \ldots, s_d$; $u_1, u_2, \ldots, u_{N-d}$, where d is the total number of failures. Notice that some u_k may coincide if $n_k > 1$.

3.4 NONPARAMETRIC ESTIMATES OF RELIABILITY FUNCTION

3.4.1 Kaplan–Meier Estimate

Let us consider the problem of constructing the point estimate of a unit reliability function, $P(t) = 1 - F(t)$. The d.f. $F(t)$ is assumed continuous. Es-

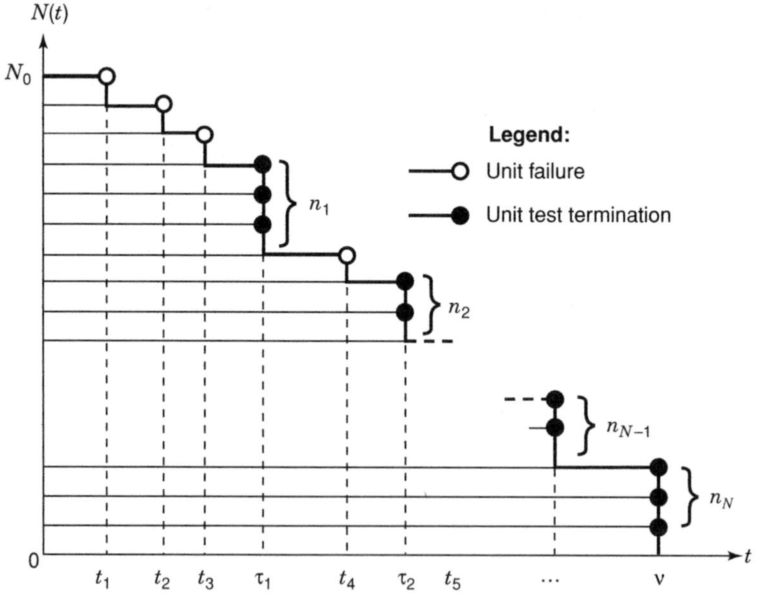

Figure 3.1 Sample of a staircase function $N(t)$.

timate (3.3) given below was found by Kaplan and Meier (1958). This estimate is often called a *product limit estimator.* It can be easily extended on the general Markov model described above.

For convenience, let us introduce the following notation:

For convenience, let us introduce the following notation: $D(t)$ is the number of failures on interval $(0, t]$, or in other words, the number of failure moments t_j such that $t_j \leq t$;

$L(t)$ is the number of units whose test has been terminated on interval $(0, t]$, or in other words,

$$L(t) = \sum_{k:\tau_k \leq t} n_k;$$

$N(t)$ is the number of units under test at the moment t (i.e., not failed or terminated),

$$N(t) = N - D(t) - L(t).$$

All random functions $D(t)$, $L(t)$, and $N(t)$ are continuous in t from the right. Let us also denote

$$D(t^-) = \lim_{\varepsilon>0,\varepsilon\to 0} D(t - \varepsilon), \qquad N(t^-) = \lim_{\varepsilon>0,\varepsilon\to 0} N(t - \varepsilon)$$

as the limits of these functions from the left at time t.

Let $0 < \tau_1 < \tau_2 < \cdots < \tau_r$ be the moments of test termination for groups of units and 1 is the total number of such termination. Then the estimate $\hat{P}(t)$ of the reliability function $P(t)$ at the moment t is constructed as

$$\hat{P}(t) = \prod_{0 \leq k \leq m-1} \left(1 - \frac{d_k}{N_k}\right)\left(1 - \frac{d_m(t)}{N_m}\right) \tag{3.3}$$

for $\tau_m < t \leq \tau_{m+1}$, $m = 0, 1, \ldots, r$, where $d_k = D(\tau_{k+1}) - D(\tau_k)$ is the number of failures on the kth interval $(\tau_k, \tau_{k+1}]$ between two sequential termination points; $N_k = N(\tau_k)$ is the number of units testing at the beginning of the kth interval $(\tau_k, \tau_{k+1}]$, or in other words, the number of units under the test after current test termination for a group of units at moment τ_k, $k = 0, 1, \ldots, r$; and $d_m(t) = D(t) - D(\tau_m)$ is the number of failures on the interval $(\tau_m, t]$. Here we assume that $\tau_0 = 0$, $N_0 = N(0) = N$, $\tau_{l+1} = \infty$.

It is easy to see that (3.3) can be rewritten in the form

$$\hat{P}(t) = \prod_{1 \leq j \leq D(t)} \left(1 - \frac{\Delta D(t_j)}{N(t_j^-)}\right), \tag{3.4}$$

where t_j is the moment of the jth failure; $N(t_j^-) = N(t_j - 0)$ is the number of units under the test just before the moment of the jth failure; and $\Delta D(t_j) = D(t_j) - D(t_j^-)$ is the number of failures at the moment t_j. If the d.f. $F(t)$ is continuous, then $\Delta D(t_j) = 1$ with probability 1. It means that (3.3) and (3.4) can be rewritten as

$$\hat{P}(t) = \prod_{1 \le j \le D(t)} \left(1 - \frac{1}{N(t_j^-)}\right). \tag{3.5}$$

Sometimes it is more convenient to write (3.4) and (3.5) in the form of an infinite product as

$$\hat{P}(t) = \prod_{u \le t} \left(1 - \frac{\Delta D(u)}{N(u^-)}\right),$$

where $N(u^-) = N(u - 0)$ and $\Delta D(u) = D(u) - D(u^-)$.

Remark 3.3 Expressions (3.3)–(3.5) are valid for all situations except the only special case where $m = 1$, $\tau_1 < t$, and $N_1 = N(\tau_1) = 0$; that is, the test of all units has been terminated at some moment τ_1 before t. (For illustration, in Figure 3.2 we depict the case with $r = 4$.) In this case one needs to reject

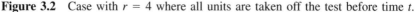

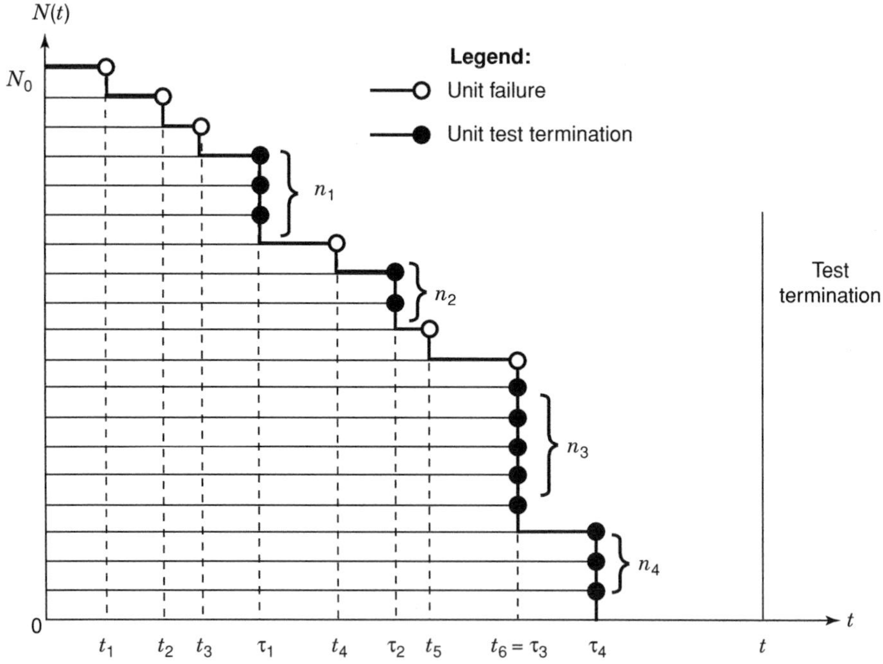

Figure 3.2 Case with $r = 4$ where all units are taken off the test before time t.

attempts at constructing the estimate, or formally set $\hat{P}(t) = 0$. (In the latter case, we should use a conditional agreement that the ratio $0/0 = 1$.) □

Example 3.5 Nineteen units have been tested ($n = 19$). The test was stopped at moment t. The test results are presented in Tables 3.1 and 3.2, where t_j are the moments of failure (in hours) and τ_k are the moments of test termination for a group of n_k units. The graph for this case is depicted in Figure 3.3.

We need to construct the estimate $\hat{P}(t)$ for the reliability function $P(t)$ for the specified moment $t = 300$ hr. In this case the total number of test terminations $r = 4$. The moment, for which the reliability function is estimated, is $t = 300$ and satisfies the inequalities $\tau_1 < \tau_2 < t < \tau_3 < \tau_4$, that is, $m = 2$. The numbers of failures between two sequential test termination moments are

$$d_0 = D(\tau_1) - D(0) = D(\tau_1) = 3,$$

$$d_1 = D(\tau_2) - D(\tau_1) = 4 - 3 = 1.$$

The number of failures on the interval $(\tau_2, t]$ is given as

$$d_2(t) = D(t) - D(\tau_2) = 5 - 4 = 1.$$

The values of N_k at the beginning of these intervals are

$$N_0 = N(0) = 19,$$

$$N_1 = N(\tau_1) = N_0 - d_0 - n_1 = 13,$$

$$N_2 = N(\tau_2) = N_1 - d_1 - n_2 = 10.$$

Thus, the estimate (3.3) in this case is

$$\hat{P}(t = 300) = \left(1 - \frac{d_0}{N_0}\right)\left(1 - \frac{d_1}{N_1}\right)\left(1 - \frac{d_2(t)}{N_2}\right)$$

$$= (1 - \tfrac{3}{19})(1 - \tfrac{1}{13})(1 - \tfrac{1}{10}) = 0.698.$$

Calculation with the help of formula (3.5) gives the same result:

Table 3.1 Failure Moments for Example 3.1

j	1	2	3	4	5	6
t_j	34	79	107	177	257	344

Table 3.2 Termination Moments τ_k and Group Size n_k

k	1	2	3	4
τ_k	150	240	344	400
n_k	3	2	5	3

$$\hat{P}(t = 300) = \prod_{1 \le j \le 5} \left(1 - \frac{1}{(t_j^-)}\right)$$

$$= (1 - \tfrac{3}{19})(1 - \tfrac{1}{18})(1 - \tfrac{1}{17})(1 - \tfrac{1}{13})(1 - \tfrac{1}{10}) = 0.698.$$

Notice that (3.5) usually leads to simpler calculations because it does not need any ordering of intervals $(\tau_k, \tau_{k+1}]$ on the time axis. □

Example 3.6 (Moment of Test Termination of Group of Units Coincides with Moment of One Failure) Using the conditions of Example 3.5 let us calculate the estimate of the reliability function $P(t)$ for the moment $t = 375$. In this case, one of the termination moments, $\tau_3 = 344$, coincides with the

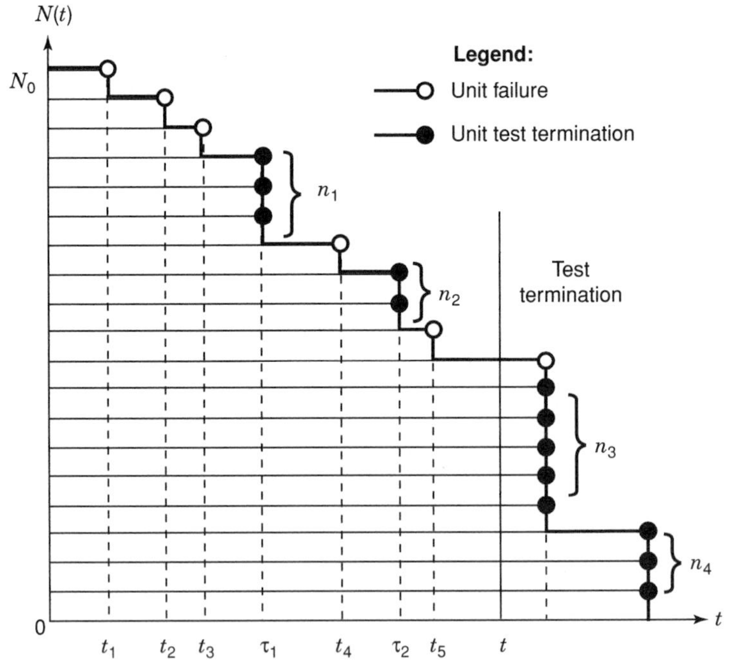

Figure 3.3 Case where the test was stopped at moment t before all units have failed or been taken off.

moment of failure, $t_6 = 344$. The calculation is analogous to the previous case. In this case the moment $t = 375$ satisfies the inequalities

$$\tau_1 < \tau_2 < \tau_3 < t < \tau_4,$$

that is, $m = 3$. The values of d_k and N_k are given as

$$d_0 = D(\tau_1) - D(0) = 3, \qquad d_1 = D(\tau_2) - D(\tau_1) = 1,$$
$$d_2 = D(\tau_3) - D(\tau_2) = 2, \qquad d_3(t) = D(t) - D(\tau_3) = 0;$$
$$N_0 = N(0) = 19, \quad N_1 = N(\tau_1) = 13, \quad N_2 = N(\tau_2) = 10, \quad N_3 = N(\tau_3) = 3.$$

The estimate (3.3) in this case is

$$\begin{aligned}\hat{P}(t = 375) &= \left(1 - \frac{d_0}{N_0}\right)\left(1 - \frac{d_1}{N_1}\right)\left(1 - \frac{d_2}{N_2}\right)\left(1 - \frac{d_3(t)}{N_3}\right) \\ &= (1 - \tfrac{3}{19})(1 - \tfrac{1}{13})(1 - \tfrac{2}{10}) = 0.621.\end{aligned}$$

Calculation with the help of formula (3.5) gives the same result:

$$\begin{aligned}\hat{P}(t = 375) &= \prod_{1 \le j \le 6} \left(1 - \frac{1}{N(t_j^-)}\right) \\ &= (1 - \tfrac{1}{19})(1 - \tfrac{1}{18})(1 - \tfrac{1}{17})(1 - \tfrac{1}{13})(1 - \tfrac{1}{10})(1 - \tfrac{1}{9}) \\ &= 0.621. \ \square\end{aligned}$$

3.4.2 Discrete Scheme of Testing

Let us consider a simple approximate method of obtaining the estimates (3.3)–(3.5). Let us divide the interval $(0, t]$ into M equal parts, each of the length $h = t/M$. Consider the case where test termination moments of groups of units, τ_k, can take only discrete values $h, 2h, \ldots, jh, \ldots$, where the size of a step h is fixed but might be infinitesimally small. Consider the conditional probability of unit failure on the interval $(jh, jh + h)$ under the condition that the unit has not failed before the moment jh or has not been removed from the test. Denote this probability by

$$q_j = \frac{P(jh) - P(jh + h)}{P(jh)}. \tag{3.6}$$

Then $P(t)$ might be written as

$$P(t) = \prod_{0 \le j \le M-1} (1 - q_j). \tag{3.7}$$

Formula (3.7) can easily be obtained from the definition of the reliability function and the conditional probabilities (3.6). (See problem 3.1.)

Consider the testing process on the interval $(jh, jh + h)$. At the moment jh there are some random number of units $N(jh)$ that have not failed or whose test has not been terminated. On the interval $(jh, jh + h)$ these units are operating without external "intervention" because the next termination might occur only at the moment $jh + h$. So, on this interval we deal with a simple binomial scheme of testing. Namely, we have $N(jh)$ independent trials with the probability q_j to fail at each trial. The standard estimate for the binomial parameter q_j is the value

$$\hat{q}_j = \frac{d_j}{N(jh)}, \tag{3.8}$$

where $d_j = D(jh + h) - D(jh)$ is the number of failed units onto the interval $(jh, jh + h)$. Substituting (3.8) into (3.7) gives the estimate of the reliability function

$$\hat{P}(t) = \prod_{0 \le j \le M-1} \left(1 - \frac{d_j}{N(jh)}\right). \tag{3.9}$$

It is easy to see that this is just what was given above in (3.3)–(3.5), the Kaplan–Meier estimate, where the test termination moments τ_k are jh, $j = 1, 2, \ldots$. (The estimate for the general case can be obtained by the limit for $h \to 0$.)

Remark 3.4 It can be seen from the arguments above that the discrete scheme of censored tests is a sequence of Bernoulli trials with varied sample size from step to step. For parameter q_j, the following approximate formula for small h is valid:

$$q_j = \frac{P(jh) - P(jh + h)}{P(jh)} = 1 - \exp\left[-\int_{jh}^{jh+h} \lambda(t)\, dt\right] \approx \lambda(jh)h,$$

where $\lambda(t)$ is the failure rate of the unit. Thus, in the general case where $h \to 0$, the process of testing can be interpreted as a sequence of binomial trials with a randomly changed size of a sample $N(t)$ and infinitesimally small binomial parameter equal to the probability of failure $q \approx \lambda(t)h$. On the basis of this experiment, one needs to estimate some reliability index that depends on $\lambda(t)$, for instance, the reliability function

$$P(t) = \exp\left[-\int_0^t \lambda(u)\, du\right]. \ \square$$

Remark 3.5 Let us mention an interesting analogy with the problem of system reliability estimation on the basis of the results of unit tests. In the discrete model of censored tests considered above, the estimate (point or interval) of $P(t)$ is reduced to the estimate (3.7) of the binomial parameters q_j on the basis of corresponding binomial tests,

$$\{N(jh), d_j\}, \qquad j = 0, 1, \ldots, M-1, \tag{3.10}$$

where $N(jh)$ is the number of tests of the parameter q_j and d_j is the number of observed failures. But this problem coincides with the problem of reliability estimation of the series system, consisting of M different units, based on the results of binomial trials (3.10) for each unit individually (see Part II). The difference is that, in the problem considered above, the sample size $N(jh)$ for each jth parameter q_j is random and depends on other parameter test results. In system reliability estimation for different parameters q_j, sample sizes N_j are specified in advance and fixed. $\square$

Limit for $h \to 0$ shows us that, on the basis of censored tests, the estimate of the unit reliability function $P(t)$ can be reduced to the following asymptotic problem: to construct an estimate (point or interval) for the reliability function of a series system consisting of M different units on the basis of binomial trials for each unit if the number of system units $M \to \infty$ and the probability of failure of each jth unit $q_j \to 0$, $j = 1, \ldots, M$.

Ushakov (1980) suggested another recurrent procedure of the calculation of estimate $\hat{P}(t)$ that produces the estimate coinciding with the Kaplan–Meier estimate (3.3)–(3.5). (See problems 3.2 and 3.3.) Pavlov and Ushakov (1984), using a Markov model of censored tests, obtained conditions whereby this estimate and estimates (3.3)–(3.5) are unbiased.

Let us describe this recurrent procedure. As before, m is the index that satisfies the inequalities

$$0 = \tau_0 < \tau_1 < \cdots < \tau_m < t \le \tau_{m+1} < \cdots < \tau_r < \tau_{r+1} = \infty,$$

where τ_k, $k = 1, \ldots, r$, is the moment of test termination for a group of units and r is the total number of such moments (in this particular test). Let us introduce values $r_m, r_{m-1}, \ldots, r_0$, which are calculated by the recurrent procedure

$$\begin{aligned} r_m &= d_m(t)\left(1 + \frac{n_m}{N_m}\right), \\ r_k &= (r_{k+1} + d_k)\left(1 - \frac{n_k}{N_k}\right), \qquad k = m-1, m-2, \ldots, 0, \end{aligned} \tag{3.11}$$

where values $d_m(t)$, N_k, d_k have the same meanings as in (3.3) and n_k is the number of units whose test has been terminated at moment τ_k, $k = 0, 1, \ldots, m - 1$; $n_0 = 0$. This procedure has the following simple meaning. Based on the number of failures $d_m(t)$ observed on the interval $(\tau_m, t]$, we can extrapolate the number of failures r_m in a natural way, which may occur within this interval if n_m units, whose test was terminated at the moment τ_m, would continue to be tested. In an analogous way, based on the observed number of failures $d_k, d_{k+1}, \ldots, d_m$, the value of r_k is the extrapolation of the number of failures that could occur on the interval $(\tau_k, t]$ if n_k units whose test has been terminated at the moment τ_k would continue to be tested. After finding all values described above, the estimate of the reliability function for the moment t can be found as

$$\hat{P}(t) = 1 - \frac{r_0}{N}, \tag{3.12}$$

where $N = N(0)$ is the initial number of units tested. The recurrent procedure is defined for all situations except $m = r$, $\tau_1 < t$, and $N_1 = N(\tau_1) = 0$, that is, when at some moment τ_1 before the moment t the test of all units has been terminated (see also Remark 3.1 and Figure 3.2). In this situation, as above, one should suggest no estimate, or formally set $\hat{P}(t) = 0$.

3.4.3 Unbiased Estimator

Let us introduce the condition

$$P\{\tau_r < t, N(\tau_r) = 0\} = 0. \tag{3.13}$$

This condition means that not all unit tests have been terminated before moment t. An even stronger condition is

$$P\{N(\tau_r) = 0\} = 0. \tag{3.14}$$

The value of τ_r is the test termination of the last group of units. Notice that if there was no termination, that is, it was a complete sample, then we assume $r = 0$, $\tau_1 = \tau_0 = 0$, $n_0 = 0$. The condition (3.14) means that (3.13) holds for all t. In other words, some units always remain under testing.

If condition (3.13) holds, the equality

$$E\{r_0\} = NF(t)$$

is valid; that is, estimate (3.12) and, consequently, (3.3)–(3.5) are unbiased estimates of the reliability function $P(t) = 1 - F(t)$ at the moment t. (The proof of this statement is given in Section 3.8.3.) If (3.14) holds, then the estimate mentioned above is unbiased for all t.

Example 3.7 As a simple example consider test plan $[N\ U\ T]$. In this case, obviously, condition (3.14) does not hold, and condition (3.13) holds for $t \leq T$. So, the estimates considered above give an unbiased estimate for $P(t)$ for all $t \leq T$. □

Example 3.8 Consider test plan $[NB\ U\ r]$, where $r < N$. In this case, for any $t > 0$, tests of all units might be terminated before t with the positive probability at the rth failure moment t_r. The probability of this event equals

$$\sum_{r \leq d \leq N} \binom{N}{d} [F(t)]^d [1 - F(t)]^{N-d}.$$

Condition (3.13) does not hold for $t > 0$. □

Example 3.9 Consider the model of independent random censoring (concurrent risks). For this model the probability of the left side of (3.13) also differs from 0 for any $t > 0$ if $G(t) > 0$. It means, as in the previous example, that condition (3.13) does not hold, and consequently, it is impossible to declare that the estimates considered above are unbiased. □

Example 3.10 Assume that the test is performed by the following plan. At the moment $t = 0$ ten units are tested. After the first failure, test of $n_1 = 2$ units is terminated. After the second failure, test of $n_2 = 4$ units is terminated. After this the test is continued until two remaining units fail. In this case condition (3.14) holds, or in other words, condition (3.13) holds for all $t > 0$. So, the estimates considered above are unbiased for all $t > 0$. □

Example 3.11 Let two tests be performed simultaneously: by plan $[N_1\ U\ r]$, where $r < N_1$, and by plan $[N_2\ U\ T]$. In this case condition (3.13) holds, and consequently, the estimates considered above are unbiased for all $t \leq T$. □

Example 3.12 Let the units be tested simultaneously by several plans of type $[N_j\ U\ r_j]$, $r_j < N_j$, $j = 1, \ldots, k$, and several plans of type $[N_j'\ U\ T_j]$, $j = 1, \ldots, m$. In this case condition (3.13) holds; consequently, the estimates considered above are unbiased for any $t \leq \max_{1 \leq j \leq m} T_j$. □

3.4.4 Nonparametric Estimator of Resource Function

In reliability characterization, one often uses a "*resource function*" $\Lambda(t)$ that relates to the reliability function $P(t)$ via the well-known expression

$$\Lambda(t) = -\ln P(t) = \int_0^t \lambda(u)\, du, \tag{3.15}$$

where $\lambda(u)$ is the hazard function. [In this chapter we assume that the function $\lambda(u)$ is continuous in $t \geq 0$.]

Consider the Markov testing model [MMU] corresponding to censorship without replacement of failed units. Consider an approximate derivation of the point estimate of $\hat{\Lambda}(t)$ for the resource function. We turn our attention again to a discrete testing scheme (see Section 3.3.2) where test termination moments τ_k for groups of units might take only discrete meanings h, $2h$, . . . , jh, . . . with some step h.

From (3.6) and (3.7) we obtain

$$\Lambda(t) = -\ln P(t) = -\sum_{0 \leq j \leq M-1} \ln(1 - q_j), \tag{3.16}$$

where

$$q_j = 1 - \exp\left[-\int_{jh}^{jh+h} \lambda(u)\, du\right].$$

It is clear that if $h \to 0$ values $q_j \to 0$. So, for small h from (3.16) the approximate equality follows

$$\Lambda(t) \approx \sum_{0 \leq j \leq M-1} q_j. \tag{3.17}$$

More precisely, it can be shown that (see Section 3.8.4)

$$\Lambda(t) = \sum_{0 \leq j \leq M-1} q_j + \delta \tag{3.18}$$

where $|\delta| < Ch$ and C is some constant. Thus, for a discrete testing scheme estimation of the resource function $\Lambda(t)$ is reduced (with some error δ that can be chosen arbitrarily small by an appropriate choice of h) to estimation of the sum of the binomial parameters

$$\Lambda(t) = \sum_{0 \leq j \leq M-1} q_j.$$

The latter value is estimated in an obvious way. One needs to substitute parameter estimates $\hat{q}_j$ from (3.8) to obtain

$$\hat{\Lambda}(t) = \sum_{0 \le j \le M-1} \hat{q}_j = \sum_{0 \le j \le M-1} \frac{d_j}{N(jh)}. \tag{3.19}$$

The limit in $h \to 0$ gives

$$\hat{\Lambda}(t) = \sum_{0 \le j \le D(t)} \frac{1}{N(t_j^-)}, \tag{3.20}$$

where t_j is the moment of the jth failure, $D(t)$ is the total number of failures on the interval $(0, t]$, and $N(t_j^-) = N(t_j - 0)$ is the number of units tested just before the moment of the jth failure, t_j. Indeed, from the graph of the function $N(t)$ in Figure 3.1 we can see that right sides of (3.19) and (3.20) begin to coincide for $h < \varepsilon$, where $\varepsilon > 0$ is the minimum length of an interval between two neighboring failures t_j and termination moments τ_k.

Note that the estimate (3.20) is defined for $N(t^-) > 0$ or for $\nu \ge t$, where ν is the moment of termination of the entire test, which means that at least one unit is still tested at moment t. If $N(t^-) = 0$ (or $\nu < t$), that is, all units have failed or their testing has been terminated up to the moment t, the estimate is undefined. In this case one might reject the estimation completely or set, for instance, either $\hat{\Lambda}(t) = \hat{\Lambda}(\nu)$ or $\hat{\Lambda}(t) = \infty$.

The estimate $\hat{\Lambda}(t)$ for plan [MMU] constructed in such a way is called an empirical resource function. First estimates of type (3.20) were obtained by Nelson (1969, 1972). (For details, see also Belyaev, 1984, 1987; Zamyatin, 1986; and others.) Conditions for the consistency of estimates of the reliability function $\hat{P}(t)$ and resource function $\hat{\Lambda}(t)$ for the initial number of tested units $N = N(0) \to \infty$ are given in Section 3.8.6.

Below we give some examples of calculations of estimates of the reliability function and resource function on the basis of formula (3.20).

Example 3.13 Using the conditions of Example 3.1, let us find estimates of the reliability function $P(t)$ and resource function $\Lambda(t)$ for $t = 200$ and $t = 300$. On the basis of formula (3.20), we obtain

$$\begin{aligned}\hat{\Lambda}(200) &= \sum_{1 \le j \le 4} \frac{1}{N(t_j^-)} \\ &= \frac{1}{N(34^-)} + \frac{1}{N(79^-)} + \frac{1}{N(107^-)} + \frac{1}{N(177^-)} = \tfrac{1}{19} + \tfrac{1}{18} + \tfrac{1}{17} + \tfrac{1}{13} \\ &= 0.244,\end{aligned}$$

$$\begin{aligned}\hat{\Lambda}(300) &= \sum_{1\le j\le 5} \frac{1}{N(t_j^-)} \\ &= \frac{1}{N(34^-)} + \frac{1}{N(79^-)} + \frac{1}{N(107^-)} + \frac{1}{N(177^-)} + \frac{1}{N(257^-)} \\ &= \tfrac{1}{19} + \tfrac{1}{18} + \tfrac{1}{17} + \tfrac{1}{13} + \tfrac{1}{10} \\ &= 0.344.\end{aligned}$$

Corresponding estimates of the reliability function for $t = 200$ and $t = 300$ are

$$\hat{P}(200) = e^{-\hat{\Lambda}(200)} = e^{-0.244} = 0.783,$$
$$\hat{P}(300) = e^{-\hat{\Lambda}(300)} = e^{-0.344} = 0.709.$$

The latter estimate has insignificant deviation from the estimate $\hat{P}(300) = 0.698$ found in Example 3.1 using the Kaplan–Meier formula. □

3.5 ASYMPTOTIC CONFIDENCE LIMITS

The problem of construction of confidence limits for the reliability function and resource function in the case of censored samples is more complex for a complete sample. It explains why most often for large samples these confidence limits are constructed as asymptotic. The model of independent random censorship (concurrent risks) is most often developed. This problem was considered by Breslow and Crowley (1974), Hall and Wellner (1980), Gill (1983), Belyaev (1984, 1985), and others. For test plans with renewal of type $[N\ R\ T]$ analogous problems were considered by Gill (1981), Belyaev (1987), and Zamyatin (1986). (See also Section 3.6.)

Consider the construction of estimates of the reliability function $P(t)$ and resource function $\Lambda(t)$ for a fixed time t. For the model of independent random censorship, let the number of tested units be large, $N \to \infty$, and for the chosen t let the condition $[1 - F(t)]\{1 - G(t)] > 0$ hold. From the works mentioned above, it follows that the lower and upper γ-confidence bounds for $\Lambda(t)$ has the form

$$\underline{\Lambda}(t) = \hat{\Lambda}(t) - u_{1-\alpha}\sqrt{\hat{V}(t)},\ \overline{\Lambda}(t) = \hat{\Lambda}(t) + u_{1-\alpha}\sqrt{\hat{V}(t)}, \tag{3.21}$$

where $\alpha = \frac{1}{2}(1 - \gamma)$, $u_{1-\alpha}$ is the quantile of the level $1 - \alpha$ of the standard normal distribution and $\hat{V}(t)$ is the estimate of the variance of the value $\hat{\Lambda}(t)$ determined by the expression

$$\hat{V}(t) = \sum_{1 \le j \le D(t)} \frac{1}{[N(t_j^-)]^2}. \tag{3.22}$$

Corresponding lower and upper confidence limits for the reliability function $P(t)$ are

$$\underline{P}(t) = e^{-\overline{\Lambda}(t)} \quad \text{and} \quad \overline{P}(t) = e^{-\underline{\Lambda}(t)}. \tag{3.23}$$

The confidence limits (3.21) and (3.23) are approximate and its confidence coefficient is close to the specified value γ only if the sample size N and the number of failures up to the moment t, $D(t)$, are sufficiently large.

Example 3.14 In the condition of Example 3.1, let us construct an approximate γ-confidence limit for $\gamma = 0.9$ for the reliability function $P(t)$ and resource function $\Lambda(t)$ for $t = 300$.

Using formula (3.22), we obtain the estimate of the variance

$$\begin{aligned}\hat{V}(300) &= \sum_{1 \le j \le 5} \frac{1}{[N(t_j^-)]^2} \\ &= \frac{1}{[N(34^-)]^2} + \frac{1}{[N(790^-)]^2} + \frac{1}{[N(107^-)]^2} + \frac{1}{[N(177^-)]^2} + \frac{1}{[N(257^-)]^2} \\ &= \frac{1}{19^2} + \frac{1}{18^2} + \frac{1}{17^2} + \frac{1}{13^2} + \frac{1}{10^2} = 0.0250.\end{aligned}$$

Applying formula (3.21) and using the value of the estimate $\hat{\Lambda}(300) = 0.344$ obtained in Example 3.13, we have the following lower and upper confidence limits:

$$\begin{aligned}\underline{\Lambda}(300) &= \hat{\Lambda}(300) - u_{0.95}\sqrt{\hat{V}(300)} \\ &= 0.344 - 1.64\sqrt{0.025} = 0.085, \\ \overline{\Lambda}(300) &= \hat{\Lambda}(300) + u_{0.95}\sqrt{\hat{V}(300)} \\ &= 0.344 + 1.64\sqrt{0.025} = 0.603.\end{aligned}$$

Corresponding confidence limits for the reliability function $P(t)$ for $t = 300$ are

$$\begin{aligned}\underline{P}(300) &= e^{-\overline{\Lambda}(300)} = e^{-0.603} = 0.547, \\ \overline{P}(300) &= e^{-\underline{\Lambda}(300)} = e^{-0.085} = 0.918. \ \square\end{aligned}$$

3.6 ACCURATE CONFIDENCE LIMITS

In practice we often meet situations where sample size is a moderate quantity. The use of approximate formulas might lead to significant errors. In particular, the lower confidence limit of the reliability function is higher than its real value. We propose an approach (see Pavlov (1982a, 1983a,b, 1995) and others) that allows for models of censored tests to construct nonparametric confidence limits that are accurate; that is, they guarantee the specified level of confidence coefficient for any sample size.

Let us first consider the Markov model of censored tests without renewal considered in Section 3.2. Construction of the confidence limits and corresponding proofs are given in Section 3.8.5 in more detail. Here we give only results and illustrate them with examples.

Assume that the hazard function $\lambda(t)$ is continuous. Let $A(t)$ be some function of time that is chosen by a researcher during the testing $\{N(t), D(t)\}$ in such a way that the following conditions hold:

Condition 1: The inequality $A(t) \leq 1$ for all $t \geq 0$.

Condition 2: Function $A(t)$ is piecewise continuous in t and continuous on the right and has finite limits on the left at any $t > 0$.

Function $A(t)$ is called the *control function.* The actual function can be arbitrarily chosen by the researcher at any current moment t depending on all information of testing behavior on the interval $(0, t]$. (A more detailed description of the control function is given in Section 3.8.5.)

The confidential strip for the hazard rate $\lambda(u)$ with the confidence coefficient γ is constructed at any current moment of time during the testing. This strip includes all functions $\lambda(u)$, $u > 0$, that satisfy the system of inequalities

$$\int_0^s A(u)N(u)\lambda(u)\,du + \sum_{1\leq j\leq D(s)} \ln[1 - A(t_j^-)] \leq |\ln(1 - \gamma)| \qquad (3.24)$$

for all $s \leq t$, where $A(t_j^-) = A(t_j - 0)$ is the limit on the left of the function $A(t)$ at the moment of the jth failure, t_j.

It is easy to see that the confidence strip (3.24) becomes tighter with increasing t. It follows from the fact that new inequalities add to (3.24) as t increases. For $t = \nu$, where ν is the termination of the entire test, the formula (3.24) gives the final confidence strip for $\lambda(u)$ with the confidence coefficient not less than γ by the system of inequalities

$$\int_0^s A(u)N(u)\lambda(u)\,du + \sum_{1\leq j\leq D(s)} \ln[1 - A(t_j^-)] \leq |\ln(1 - \gamma)| \qquad (3.25)$$

for all $s \leq \nu$.

Then by appropriate choice of the control function $A(t)$ from (3.25) one can obtain the confidential strips for the resource function $\Lambda(t)$, reliability function $P(t)$, and confidence limits for other reliability indices, for instance, the MTTF. Particularly, choosing $\nu = 1 - \alpha$ and $A(t) = -a/N(t)$, where $a > 0$ is some constant, we obtain the γ-confidence strip for $\lambda(u)$ as the inequality

$$a \int_0^s \lambda(u)\, du \geq |\ln \alpha| + \sum_{1 \leq j \leq D(s)} \ln \left[1 + \frac{a}{N(t_j^-)} \right] \tag{3.26}$$

for all $s \leq \nu$. For $s > \nu$ this inequality also holds, and for $s \geq \nu$ the right side remains constant.

Analogously, assuming in (3.25) that $\gamma = 1 - \beta$ and $A(t) = b/N(t)$, where b is a positive constant, $0 < b \leq N(0)$, we obtain the confidence strip for $\lambda(u)$ with confidence coefficient not less than $1 - \beta$ as the inequality

$$b \int_0^s \lambda(u)\, du \geq |\ln \beta| - \sum_{1 \leq j \leq D(s)} \ln \left[1 - \frac{b}{N(t_j^-)} \right] \tag{3.27}$$

for all $s \leq \nu$.

Notice that (3.27) is valid for $s \leq \nu' = \min(\nu, \nu_b)$, where $\nu_b = \min\{t: N(t) \leq b\}$ is the moment of crossing the level b by the function $N(t)$. For $s > \nu'$ in a nonparametric case, the right side of (3.27) should be assumed to be equal to ∞. For this purpose, we can set $\ln z = -\infty$ for $z \leq 0$.

Formulas (3.26) and (3.27) give the following confidence strip for the resource function with confidence coefficient not less than $1 - \alpha - \beta$:

$$P\{\underline{\Lambda}(t) \leq \Lambda(t) \leq \overline{\Lambda}(t) \text{ for all } t \geq 0\} \geq -\alpha - \beta,$$

where the lower and upper strips have the form

$$\underline{\Lambda}(t) = -\frac{|\ln \alpha|}{a} + \frac{1}{a} \sum_{1 \leq j \leq D(t)} \ln \left[1 + \frac{a}{N(t_j^-)} \right], \tag{3.28}$$

$$\overline{\Lambda}(t) = \frac{|\ln \beta|}{b} - \frac{1}{b} \sum_{1 \leq j \leq D(t)} \ln \left[1 - \frac{b}{N(t_j^-)} \right]. \tag{3.29}$$

The corresponding confidence strips for the reliability function with confidence coefficient not less than $1 - \alpha - \beta$ are given as

$$\underline{P}(t) \leq P(t) \leq \overline{P}(t), \qquad t \geq 0, \tag{3.30}$$

where the lower and upper limits are determined by the formulas

$$\underline{P}(t) = e^{-\overline{\Lambda}(t)}, \qquad \overline{P}(t) = e^{-\underline{\Lambda}(t)}.$$

Section 3.8.6 gives conditions that allow us to obtain consistent confidence limits for the reliability function and resource function if the initial number of tested units is $N = N(0) \to \infty$.

From the expressions given above, one can obtain the nonparametric lower limit for the MTTF, $\underline{\mu}$, with confidence coefficient not less than $1 - \beta$. Indeed, using the formula for the MTTF expressed via the reliability function, we have

$$\underline{\mu} = \int_0^\infty \underline{P}(t)\, dt = \int_0^\infty e^{-\overline{\Lambda}(t)}\, dt.$$

In an analogous way, on the basis of confidence strip (3.30), we can construct nonparametric limits for guaranteed time t_q with level of guarantee q. In this case lower and upper limits with confidence coefficient not less than $1 - \alpha - \beta$ are given by the formulas

$$\underline{t}_q = \inf\{t\colon \underline{P}(t) \le q\}, \qquad \overline{t}_q = \sup\{t\colon \overline{P}(t) \ge q\}.$$

In other words, $\underline{t}_q$ and $\overline{t}_q$ are defined as the moments of crossing the level q by the graphs of the lower and upper confidence limits $\underline{P}(t)$ and $\overline{P}(t)$.

Example 3.15 (Case of No Failure Test) Let us choose the control function as

$$A(t) \equiv 1, \qquad t \ge 0. \tag{3.31}$$

Let there be no failures or test termination up to moment t. Then from (3.24) it follows that

$$\int_0^s N(u)\lambda(u)\, du \le |\ln(1 - \gamma)| \quad \text{for all } s \le t.$$

Since in this case $N(u) \equiv N = N(0)$ for all $u \le t$, it follows that

$$\int_0^s \lambda(u)\, du \le \frac{|\ln(1 - \gamma)|}{N} \quad \text{for all } s \le t. \tag{3.32}$$

From (3.32) we obtain the following upper confidence limit for the resource function with the confidence coefficient not less than γ:

$$\Lambda(s) \le \overline{\Lambda} = \frac{|\ln(1 - \gamma)|}{N} \quad \text{for all } s \le t,$$

which gives the following lower γ-confidence limit for the reliability function:

$$P(s) \geq \underline{P}(t) = e^{-\overline{\Lambda}} = \sqrt[N]{1 - \gamma} \quad \text{for all } s \leq t.$$

Thus, choosing the control function $A(t)$ from condition (3.31), on the interval $[0, t]$ with no failures, we obtain the lower γ-*confidence limit* $\underline{P}$, which coincides with the standard Clopper–Pearson limit for the case of testing N units with no failures. □

Example 3.16 (Testing Until the First Failure) Let the test be performed until the first failure; that is, the termination of the entire test is $\nu = t_1$. No censorship is available. Choosing again the control function as $A(t) \equiv 1$ for all $t \geq 0$, from (3.24) and (3.25) we obtain

$$\int_0^s N(u)\lambda(u)\, du \leq |\ln(1 - \gamma)| \quad \text{for all } s < \nu$$

or

$$\int_0^s \lambda(u)\, du \leq \frac{|\ln(1 - \gamma)|}{N} \quad \text{for all } s < \nu,$$

from which the γ-confidence limit for the reliability function follows:

$$\Lambda(s) \leq \overline{\Lambda} = \frac{|\ln(1 - \gamma)|}{N} \quad \text{for all } s < \nu.$$

The corresponding lower γ-confidence limit for the reliability function is

$$P(s) \geq \underline{P} = e^{-\overline{\Lambda}} = \sqrt[N]{1 - \gamma} \quad \text{for all } s < \nu. \ \square$$

Example 3.17 Using the conditions of Example 3.1, let us construct the lower τ-*confidence limit for the reliability function* $P(t = 160)$ with confidence coefficient not less than $\gamma = 0.95$. Setting the constant $b = 10$ into (3.29), we obtain the following upper confidence limit for the resource function:

$$\begin{aligned}\underline{\Lambda}(160) &= \frac{|\ln 0.05|}{10} - \frac{1}{10} \sum_{1 \leq j \leq 3} \ln \left[1 - \frac{10}{N(t_j^-)}\right] \\ &= \frac{2.99}{10} - \frac{1}{10}\left[\ln\left(1 - \frac{10}{19}\right) + \ln\left(1 - \frac{10}{18}\right) + \ln\left(1 - \frac{10}{17}\right)\right] \\ &= 0.515.\end{aligned}$$

So, the lower 0.95-confidence limit for the reliability function $P(160)$ is

$$\underline{P}(160) = e^{-\overline{\Lambda}(160)} = e^{-0.515} = 0.597. \ \square$$

Notice that on the basis of the confidence limits constructed above one can solve the problem of testing statistical hypotheses relating different reliability indices, for instance,

$$H_0\colon P(t) \geq P_0 \quad \text{vs.} \quad H_1\colon P(t) \leq P_1,$$

where $P_0 > P_1$ are specified critical levels of the reliability index $P(t)$, and so on.

An interesting and still not appropriately solved question is the problem of choice of an optimal (in some sense) control function $A(t)$. [The conditions on $A(t)$ for which the confidence limits constructed above are consistent for the initial number of tested units $N = N(0) \to \infty$ are given in Section 3.8.6.]

We considered above only the model [*MMU*], but this approach can be easily extended to more general models, including models with renewal (some are considered below).

3.7 MORE GENERAL MODELS OF CENSORED TESTS: MARKOV MODEL WITH RENEWAL

Consider a test model that differs from [*MMU*] by the possibility of replacing failed units by identical new ones or to add new units to previously tested ones. Let us call such a model [*MMR*], *R* stands for "renewal." At the initial moment $t = 0$ there are $N = N(0)$ identical and independent units. At some time $\sigma_1 > 0$ we add m_2 new units identical to the initial ones, at the moment $\sigma_2 > \sigma_1$ we add m_1 units, and so on. We assume that all units fail independently. So, new units are tested at moments

$$0 < \sigma_1 < \sigma_2 < \cdots < \sigma_i < \cdots,$$

which form Markov and "independent of the future" sequences of time. In other words, the moments and corresponding numbers of added units might be appointed arbitrarily, depending on any information known at the moment t (see Section 3.8.1). So, at moment σ_i we add m_i new units, where m_i is a discrete random number that might arbitrarily depend on the current results of the test process on the interval $[0, \sigma_i]$ but does not depend on the future development of the process for $t > \sigma_i$. (More accurate formal definitions of σ_i and m_i are given in Section 3.8.7.)

Thus the test model is given by the set of data $M = \{(\tau_1, n_1), \ldots, (\tau_k, n_k), \ldots; (\sigma_1, m_1), \ldots, (\sigma_i, m_i), \ldots; \nu\}$, where ν is the termination of the entire test. This moment is defined by some rule and in general might be a Markov random moment. Values (τ_k, n_k) are defined as above: τ_k is the moment of the kth test termination for a group of units and n_k is the number of units of the group, $k = 1, 2, \ldots$. If in the model [MMU] the number of test terminations was restricted by the value of $N = N(0)$, in the model [MMR] the number of test termination moments τ_k and unit additions σ_i are not restricted. The sequential moments of unit failures are denoted as above:

$$0 < t_1 < \cdots < t_j < t_{j+1} < \cdots .$$

We also keep the notation $D(t)$ for the number of moments t_j such that $t_j \leq t$ (i.e., the number of failures) and $N(t)$ for the number of units tested at moment t:

$$N(t) = N(0) + B(t) - D(t) - L(t),$$

where

$$B(t) = \sum_{i:\ \sigma_i \leq t} m_i, \qquad L(t) = \sum_{k:\ \tau_k \leq t} n_k;$$

that is, $B(t)$ is the number of units that were put to the test before t, and $L(t)$ is the number of units whose test has been terminated up to moment t. All functions $N(t)$, $D(t)$, $B(t)$, and $L(t)$ are continuous from the right in t. A graphical illustration of a possible track of the function $N(t)$ is depicted in Figure 3.4.

Remark 3.6 In the previous model [MMU] all $N(t)$ tested units at moment t have the same operational time equal to t. In this model, which is more general, the "ages" of tested units are different. Denote the age of unit e at moment t by $S_e(t)$. Obviously, this value is

$$S_e(t) = 1 - \sigma_i,$$

where σ_i is the moment when this unit was added to the set of tested units. Denote the set of subscripts of all units tested at moment t by $E(t)$. This set

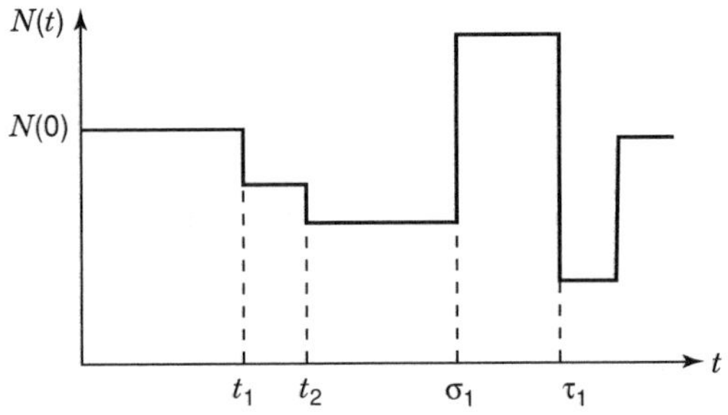

Figure 3.4 Sample of realization of the function $N(t)$ for test with renewal.

includes $N(t)$ different subscripts. So, units at their testing termination moment τ_k have different operational time (age): $S_e(\tau_k)$, $e \in E(\tau_k^-)$, where $E(\tau_k^-) = E(\tau_k - 0)$ is the set of subscripts of units under test just before moment τ_k. Thus, for the complete description of the model, it is necessary to point out not only the number of units whose test has been terminated but their age: $S_e(\tau_k)$, $e \in E(\tau_k)$, where $E(\tau_k)$ is the set of subscripts of the units whose test has been terminated at the moment τ_k, $E(\tau_k) \subset E(\tau_k^-)$. It is clear that this condition is redundant only for the exponential distribution $F(t)$ for which the distribution of residual time coincides with the initial one.

This model includes the previously described model [MMU] as a particular case. This model also covers all standard plans of type R (with unit renewal or replacement after failure). □

Example 3.18 (Test Plan of Type [1 R T]) Consider a standard renewal process, or in our terminology the test plan [1 R T]. (See Chapter 2.) At the moment $t = 0$, a single unit is in testing, $N(0) = 1$. At the failure moment, a failed unit is replaced by a new one; then at the moment of the second failure, a failed unit is again replaced by a new one; and so on. It is clear that this plan is a particular case of plan [MMR] with $N(0) = 1$; $n_k = 0$ for all $k = 1, 2, \ldots$; with moments of new unit additions, $\sigma_i = t_i$, $i = 1, 2, \ldots$; and $m_i = 1$, $i = 1, 2, \ldots$. Obviously, $N(t) \equiv 1$ for all $t > 0$. The entire test termination moment is $\nu = T$, where T is a specified constant. □

Example 3.19 (Test Plan of Type [N R . . .]) Let us consider a test consisting of N simultaneous observations of the processes discussed in Example 3.18. We call this plan [N R . . .]. This model can be considered as a particular case of the model [MMR] with $N(0) = n$; $n_k = 0$, $k = 1, 2, \ldots$; $\sigma_i = t_i$; $m_i = 1$, $i = 1, 2, \ldots$. At any time $t \geq 0$ there are $N(t) \equiv N$ tested units. Moments of test termination depend on the particular test plan. For the plan [N R T], it is $\nu = T$; for the plan [N R r], it is $\nu = t_r$; and for the plan [N R (r, T)], it is $v = \min(T, t_r)$. □

The method of constructing confidence limits for the model [MMU] extends to the more general model [MMR] in the following way. Each unit with subscript e is set in correspondence with some "control function" $A(e, t)$ that satisfies the same conditions as $A(t)$ (see Section 3.5). For each unit $e \in E(t)$, the meaning of $A(e, t)$ might be chosen arbitrarily by taking into account any previous information about the current test on the interval $[0, t]$. [Of course, we need to maintain conditions 1 and 1 formulated for the control function $A(t)$.] As above, the confidence strip for the hazard rate, $\lambda(u)$, with confidence coefficient not less than the specified value γ is constructed for any t during the testing. At moment t this confidence strip includes all continuous functions $\lambda(u)$, $u \geq 0$, that satisfy the system of inequalities

$$\int_0^s \sum_{e\in E(u)} A(e, u)\lambda[S_e(u)]\, du + \sum_{1\le j\le D(t)} \ln[1 - A(e_j, t_j^-)] \le |\ln(1 - \gamma)| \quad (3.33)$$

for all $s \le t$, where $S_e(u)$ is the "testing age" of unit e at the moment u; $A(e, t_j^-) = A(e, t_j - 0)$ is the limit from the left for the control function $A(e, t)$ at the moment of the jth failure, t_j; e_j is the subscript of the unit that fails at t_j; and $E(u)$ is the set of unit indices tested at moment u.

From the definition that follows, with an increase in time t the confidence strip is monotone narrowing for any trajectory of the testing process. It follows again from the fact that, with increasing t, more inequalities are included in (3.33). At $t = \nu$, where ν is the moment of the entire test termination, system (3.33) gives the final confidence strip for the hazard rate $\lambda(u)$:

$$\int_0^s \sum_{e\in E(u)} A(e, u)\lambda[S_e(u)]\, du + \sum_{1\le j\le D(t)} \ln[1 - A(e_j, t_j^-)] \le |\ln(1 - \gamma)| \quad (3.34)$$

for all $s \le \nu$.

The proof of the confidence strips (3.33) and (3.34) for the model [MMR] is analogous to that for the model [MMU] (see below in Section 3.8.5).

As we did in the previous section, the nonparametric confidence strips can be constructed on the basis of (3.33) and (3.34) for the reliability function $P(t)$ and resource function $\Lambda(t)$ as well as the confidence limits for other reliability indices.

3.8 APPENDICES

3.8.1 Markov Moments of Time

Let $(\Omega, \mathfrak{L}, P)$ be the main probabilistic space where the process of testing, $x_t, t \ge 0$, is defined: $\mathfrak{L}_t \subset \mathfrak{L}, t \ge 0$, be a nondecreasing family of σ-subalgebras; and the stochastic process x_t correspond to the family of $\mathfrak{L}_t$, $t \ge 0$.

A random moment $\tau = \tau(\omega) \ge 0$ is called a Markov moment with respect to the system of $\mathfrak{L}_t$, $t \ge 0$, if the event $\{\omega: \tau(\omega) \le t\} \in \mathfrak{L}_t$ for any $t \ge 0$ where ω is an element of the set Ω. Here the σ-algebra $\mathfrak{L}_t$ represents a set of all events whose occurrences or nonoccurrences up to the moment t are uniquely predictable. In other words, the sense of this statement is the following: A random moment τ is Markovian if, for any t, the answer about the occurrence of an event $\{\tau \le t\}$, based on the prehistory of the process before the moment t, is unique.

Each Markov moment τ corresponds to a σ-subalgebra $\mathfrak{L}_\tau \subset \mathfrak{L}$ that is a set of all event occurrences that are uniquely predictable at the moment τ. This σ-subalgebra $\mathfrak{L}_\tau$ includes all events $A \in \mathfrak{L}$ such that the event $A \cap \{\tau \le t\}$

$\in \mathfrak{L}_\tau$ for any $t \geq 0$. If some r.v. $\eta = \eta(\omega)$ is measurable with respect to the σ-algebra $\mathfrak{L}_\tau$, it means that at the moment t the meaning of this r.v. is known. [More detail may be found in Doob (1953) or Shiryaev (1980).]

3.8.2 Markov Censored Test Model Without Renewal

Assume that, at the moment $t = 0$, N identical and independent units enter testing. Each unit has a continuous d.f. $F(t)$ of a random TTF ξ. At some moment $\tau_1 > 0$, a group of n_1 units terminate the test, then at $\tau_2 > \tau_1$ a group of n_2 units exit the test, and so on. Any exit moment can be arbitrarily chosen depending on the information about the "prehistory" of the testing at the current moment t but does not depend on the "future trajectory" of the testing process, that is, for $t > \tau_k$. The number of units in a group, n_k, whose tests are terminated at the moment τ_k is also random and can be chosen on the basis of the same information.

Let $D(t)$ be the number of failures on the interval $[0, t]$, or, in other words, the number of moments t_j such that $t_j \leq t$, where t_j is the moment of the jth failure. Let $L(t) = \Sigma_{k:\tau_k \leq t} n_k$ be the number of units whose tests have been terminated on the interval $[0, t]$ and $N(t) = N - D(t) - L(t)$ be the number of units testing at the moment t. Let $(\Omega, \mathfrak{L}, P)$ be the main probabilistic space where the stochastic process $x_t = \{N(t), D(t)\}$, $t \geq 0$, is defined in correspondence to the nondecreasing family of σ-algebras $\mathfrak{L}_t \subset \mathfrak{L}$, $t \geq 0$. The stochastic functions $N(t)$ and $D(t)$ are continuous from the left in t. Assume that the test termination moments of the group of units,

$$\tau_1 < \tau_2 < \cdots < \tau_k < \cdots,$$

are Markovian with respect to the family $\mathfrak{L}_t$, $t \geq 0$, and a r.v. $n_k = 0, 1, 2, \ldots, N$ is measurable with respect to σ-algebra $\mathfrak{L}_{\tau k} \subset \mathfrak{L}$ related to the Markov moment τ_k, $k = 1, 2, \ldots$. The moment of the entire test termination is denoted by ν. We assume that this moment also is Markovian, related to the family $\mathfrak{L}_t$, $t \geq 0$. Since the initial number of units is $N = N(0)$, we might say that the total number of termination moments τ_k does not exceed N. Thus in this model it is enough to consider the finite sequence of N Markov moments

$$\tau_1 < \tau_2 < \cdots < \tau_N \leq \nu.$$

The number of "real" termination moments, that is, such moments for which $n_k > 0$, might be fewer than N. So, r is a r.v., $r = 1, 2, \ldots, N$. Thus the test model is defined by the set

$$M = \{(\tau_1, n_1), (\tau_2, n_2), \ldots, (\tau_N, n_N), \nu\}. \tag{3.35}$$

We call this test model Markovian to emphasize that any intervention into the process (test termination of a group of units or entire test) bears the Markovian character.

This model does not assume additional switching on new units for testing. More general models that take this factor into account will be considered in Section 3.7.7.

3.8.3 Unbiasedness of Estimate (3.12)

Let us show that if condition (3.13) holds for the model [MMU], then

$$E\{r_0\} = NF(t); \tag{3.36}$$

that is, the estimate (3.12) is an unbiased estimate of the reliability function $P(t) = 1 - F(t)$ for fixed $t > 0$. This model is defined by (3.35). Notice that since the initial number of units equals N, and n_k might take the value of zero, we can say that the last moment of test termination of a group of units, τ, coincides with the moment of the complete termination of the entire test, that is, $\tau_n \nu$. Therefore the model is defined by the set

$$M = \{(\tau_1, n_1), (\tau_2, n_2), \ldots, (\tau_N, n_N)\}.$$

Consider first the case where the moments of test termination, τ_k, take discrete values t_k:

$$0 = t_0 < t_1 < \cdots < t_m < t_{m+1} = t < t_{m+2} < \cdots .$$

For the sake of simplicity, let t, the moment for which the reliability function $P(t)$ is estimated, coincide with t_{m+1}. Introduce the test model

$$M^* = \{(\tau_1^*, n_1^*), (\tau_2^*, n_2^*), \ldots, (\tau_N^*, n_N^*), \ldots\},$$

where $\tau_i^* \equiv t_i$ and

$$n_i^* = \sum_{1 \le k \le N} n_k I(\tau_k = t_i), \qquad i = 1, 2, \ldots,$$

where $I(A)$ is the indicator for event A. In discrete cases models M and M^* are equivalent. The model M^* is another definition of the model M where all moments τ_i^* are deterministic and exhaust all possible discrete sets of $\{t_i\}$ and the model is defined by the numbers of units for which the test has been terminated at these moments, n_i^*, $i = 1, 2, \ldots$. Since the r.v. n_k is measurable with respect to the σ-algebra $\mathfrak{L}_{ti} \subset \mathfrak{L}$ corresponding to the Markov moment τ_k $(k = 1, 2, \ldots, N)$, n_i^*, is measurable with respect to the σ-algebra $\mathfrak{L}_{ti} \subset \mathfrak{L}$, $i = 1, 2, \ldots$.

Let us introduce the following notation for the model M^*: $N_i = N(t_i)$, $N_i' = N_i + n_i^*$, and $d_i = N_i - N_{i+1}' = D(t_{i+1}) - D(t_i)$, $i = 0, 1, 2, \ldots$. The recurrent procedure (3.11) for the discrete model M^* has the form

$$r_m = \begin{cases} N_m' \left(\dfrac{d_m}{N_m}\right) & \text{if } N_m' > 0, \\ 0 & \text{if } N_m' = N_m = d_m = 0, \end{cases}$$

$$r_i = N_i' \left(\frac{d_i + r_{i+1}}{N_i}\right) \quad \text{if } N_i' > 0,$$

$$r_0 = 0 \quad \text{if } N_i' = N_i = d_i = r_{i+1} = 0,$$

where $0 \le i \le m - 1$. Let us introduce the quantities

$$q_i = \frac{P(t_i) - P(t_{i+1})}{P(t_i)}, \qquad Q_i = \frac{P(t_i) - P(t)}{P(t_i)},$$

equal to the conditional probabilities of a unit's failure on intervals (t_i, t_{i+1}) and (t_i, t), respectively, under the condition that the unit has not failed before moment t_i. These quantities are related to each other as

$$Q_i = q_i + (1 - q_i)Q_{i+1}, \qquad i = 0, 1, \ldots, m, \tag{3.37}$$

where $Q_0 = F(t)$ and $Q_{m+1} = 0$.

In the discrete model m^*, the group of unit test terminations might occur only at moments t_i, $i = 1, 2, \cdots$. Thus on each interval (t_i, t_{i+1}) there is no "intervention" in the testing process (no termination). It means that on this interval of time the conditional distribution of the number of failures, d_i, is a standard binomial with parameters N_i (number of tested units) and q_i (probability of failure). It follows that the conditional mathematical expectation can be written as

$$E\{d_i | \mathfrak{L}_{t_i}\} = N_i q_i, \qquad i = 0, 1, \ldots, m. \tag{3.38}$$

Here we omit the standard remark that the equality for the conditional mathematical expectation is correct with excluding sets with null measure P. Notice also that the probabilistic measure P and corresponding mathematical expectation by this measure on $(\Omega, \mathcal{L})$ essentially depend on the estimated function $F = F(t)$; that is, $P = P_F$ and $E = E_F$. Subscript F is omitted for brevity. In the discussions to follow we will use a shortened notation: $\mathcal{L}_i = \mathcal{L}_{t_i}$.

Let us show that the conditional mathematical expectation is

$$E\{r_m|\mathcal{L}_m\} = N'_m q_m = N'_m Q_m.$$

For $N'_m = 0$ this equality directly follows from the definition of the r.v. r_m. Let $N'_m > 0$. Then by condition (3.13) the inequality $N_m > 0$ holds. Since quantities N'_m and N_m are $\mathcal{L}_m$-measurable,

$$E\{r_m|\mathcal{L}_m\} = E\left\{N'_m \frac{d_m}{N_m} |\mathcal{L}_m\right\} = \frac{N'_m}{N_m} E\{d_m|\mathcal{L}_m\},$$

and taking into account (3.38), it follows that

$$E\{r_m|\mathcal{L}_m\} = N'_m q_m = N'_m Q_m. \tag{3.39}$$

Let us now show that the equality

$$E\{r_i|\mathcal{L}_i\} = N'_i Q_i \tag{3.40}$$

follows from the equality

$$E\{r_{i+1}|\mathcal{L}_{i+1}\} = N'_{i+1} Q_{i+1}.$$

(Here $i = m - 1, m - 2, \ldots, 1, 0$.) Since $N'_0 = N_0 = N$, and taking into account (3.39), one has the proof of the equality (3.36) for the discrete model M^*.

Let (3.40) hold. Then for $N'_i = 0$ the equality (3.41) directly follows from the definition of the r.v. r_i. Let $N'_i > 0$. From condition (3.13) the equality $N_i > 0$ also holds. Since the r.v. N'_i and N_i are $\mathcal{L}_i$-measurable, using (3.38), we have

$$E\{r_i|\mathcal{L}_i\} = E\left\{N'_i \frac{d_i + r_{i+1}}{N_i} |\mathcal{L}_i\right\} = \frac{N'_i}{N_i} E\{d_i + r_{i+1}|\mathcal{L}_i\}$$

$$= \frac{N'_i}{N_i} E\{d_i|\mathcal{L}_i\} + \frac{N'_i}{N_i} E\{r_{i+1}|\mathcal{L}_i\} = N'_i q_i + \frac{N'_i}{N_i} E\{r_{i+1}|\mathcal{L}_i\}. \tag{3.41}$$

Since $\mathcal{L}_i \subset \mathcal{L}_{i+1}$, the following equality holds for the conditional mathematical expectation:

$$E\{r_{i+1}|\mathcal{L}_i\} = E\{E\{r_{i+1}|\mathcal{L}_{i+1}\}|\mathcal{L}_i\} = E\{N'_{i+1}Q_{i+1}|\mathcal{L}_i\}$$
$$= Q_{i+1}E\{N_i - d_i|\mathcal{L}_i\} = Q_{i+1}(N_i - N_i q_i).$$

Taking (3.37) into account, we have

$$E\{r_i|\mathcal{L}_i\} = N'_i[q_i + Q_{i+1}(1 - q_i)] = N'_i Q_i,$$

where $i = m - 1, m - 2, \ldots, 1, 0$. This proves (3.36) for the discrete model.

In the general case, consider a sequence of divisions of the time axis by discrete moments:

$$S_j = \{0 = t_{0j} < t_{1j} < \cdots < t_{ij} < \cdots\},$$

where $t_{ij} = i(h/2^j)$, $i = 0, 1, \ldots$; $h = t/L$; L is an arbitrary positive integer; and $S_j \subset S_{j+1}$, $j = 1, 2, \cdots$. For the jth division

$$\tau_{kj} = \left[\frac{\tau_k 2^j}{h}\right]^* \left(\frac{h}{2^j}\right),$$

where $[z]^*$ denotes the closest integer that is more than (or equal to) z. Consider the process $y(t)$, which denotes the total number of units being tested (including those for which testing has been terminated) and that have not failed up to moment t. By definition, the trajectory of this process is continuous in t on the right. Let us introduce the event

$$B_{kj} = \{y(\tau_k) = y(\tau_{kj})\},$$

which corresponds to the fact that there is no failure on interval $(\tau_k, \tau_{kj}]$ including those for which tests have been terminated. Since the trajectory of the process $y(t)$ is monotone decreasing,

$$B_{kj} \subset B_{k,j+1}, \qquad j = 1, 2, \ldots.$$

Then let us set

$$n_{kj} = n_k I(B_{kj}), \qquad k = 1, \ldots, N; j = 1, 2, \ldots.$$

The r.v. n_{kj} is measurable relative to a σ-algebra $\mathcal{L}_{\tau_{kj}}$ that is tied to the Markov moment τ_{kj} because $\tau_k \le \tau_{kj}$ and $\mathcal{L}_{\tau_k} \subset \mathcal{L}_{\tau_{kj}}$.

Now introduce the Markov test model

$$M_j^* = \{(\tau_{kj}, n_{kj}), k = 1, \ldots, N\}.$$

For this model moments of intermediate terminations τ_{kj} might take only discrete values t_{ij}, $i = 1, 2, \ldots$, which corresponds to the jth division of S_j. Let $1 - (r_{0j}/N)$ be the estimate of the reliability function $P(t)$ for model M_j^*. Then, as was proved before, for any $j = 1, 2, \ldots$, the equality

$$E\{r_{0j}\} = NF(t)$$

holds. The sequence $r_{0j} \to r_0$ as $j \to \infty$ everywhere on the set

$$B = \bigcap_{1 \le j < \infty} \bigcap_{1 \le j < \infty} \bigcap_{1 \le k \le N} B_{kl} = \{y(\tau_k) = y(\tau_k + 0),\ k = 1, \ldots, N\}.$$

Because of the continuity of trajectories of the process $y(t)$ on the left in t, the set B has the probability 1. Taking into account the restrictiveness of the sequence τ_{0j}: $0 \le r_{0j} \le N$, it follows that

$$E\{r_0\} = NF(t).$$

3.8.4 Proof of Formula (3.18)

Parameter q_j can be written in the form $q_j = 1 - e^{-z_j}$, where

$$z_j = \int_{jh}^{jh+h} \lambda(u)\, du, \qquad j = 0, 1, \ldots, M - 1. \tag{3.42}$$

From the continuity of function $\lambda(u)$ its bounds on interval $[0, t]$ follows: $\lambda(u) \le C < \infty$ for all $0 \le u \le t$. It gives, taking into account (3.42), that $z_j \le Ch$ for all $j = 0, 1, \ldots, M$; $h = t/M$, $M = 1, 2, \ldots$. It follows that

$$q_j = z_j + \varepsilon_j,$$

where $|\varepsilon| \le C'h^2$, $C' = \frac{1}{2}C^2$ for all $j = 0, 1, \ldots, M - 1$, and $h = t/M$, $M = 1, 2, \ldots$. Summation in j of the left and right sides of (3.42) gives

$$\Lambda(t) = \sum_{0 \le j \le M-1} z_j = \sum_{0 \le j \le M-1} q_j + \delta,$$

where

$$|\delta| \le \sum_{0 \le j \le M-1} |\varepsilon_j| \le C'Mh^2 = C''h,$$

where, in turn, $C'' = C't$, which proves (3.18).

3.8.5 Nonparametric Confidence Limits for Test Markov Model [MMU]

We further assume that $\lambda(t)$ is continuous in $t \geq 0$. The class of all such functions is denoted by W. Let us also denote λ: $\lambda = \{\lambda(t), t \geq 0\}$. The probabilistic measure on $(\Omega, \mathfrak{L})$ for given $\lambda \in W$ is denoted by P_λ and the mathematical expectation (by measure P_λ) by E_λ.

Let

$$x^t = \{x_u, 0 \leq u \leq t\} = \{[N(u), D(u)], 0 \leq u \leq t\}$$

be a collection of all observed statistical data at the moment t, that is, the trajectory of the process $x_u = [N(u, D(u)]$ on interval $[0, t]$. For the test model x^t, it is a set of all failure moments t_j, intermediate terminations, τ_k, and respective numbers n_k that are observed up to moment t.

Assume that at any current moment we construct the confidence set on the basis of information x^t,

$$H_t = H_t(x^t) \subset W \tag{3.43}$$

for $\lambda \in W$. A collection of all sets (3.43) is called a system of γ-confidence sets for λ if the following conditions hold:

1. For any trajectory of the testing process x_t, $0 \leq t < \infty$, and any moment $u \leq t$ the relation

$$H_t(x^t) \subset H_u(x^u)$$

holds; that is, the confidence set H_t becomes monotonically narrower with t increasing for any trajectory of observation.

2. For any Markovian (with respect to the family $\mathfrak{L}_t$, $t \geq 0$) termination moment ν such that $P_\lambda(\nu < \infty) = 1$ for all $\lambda \in W$, the inequality

$$P_\lambda(\lambda \in H_\nu) \geq \gamma$$

holds for all $\lambda \in W$, where $H_\nu = H_\nu(x^\nu)$ is the confidence set constructed up to the moment ν. The sense of this condition is in the fact that we can terminate the entire test at any moment ν and the set H_ν constructed up to this moment is γ-confident for λ.

Now we give the procedure for constructing the system of sequential confidence sets for a model of censored tests (see Pavlov, 1982b, 1983a,b). Let $A(t)$ be a stochastic function related to the family $\mathfrak{L}_t$, $t \geq 0$, and satisfying the following conditions:

1. $A(t) \leq 1$ for all $t \geq 0$;
2. $A(t)$ is a linear-spline function continuous on the right and having finite limits on the left at any point $t > 0$; and
3. on any finite interval $0 \leq u \leq t$, function $A(u)$ is restricted from below:

$$P_\lambda\{A(u) \geq C > -\infty \text{ for all } 0 \leq u \leq t\} = 1$$

for all $\lambda \in W$.

Let γ be the given level of the confidence coefficient, $0 < \gamma < 1$. Let each $\lambda \in W$ correspond to a set of events $p_\lambda^t \in \mathcal{L}_t$, $t > 0$, where events p_λ^t are defined as

$$p_\lambda^t = \left\{ \int_0^t A(u)N(u)\lambda(u)\,du \right. \tag{3.44}$$

$$\left. + \sum_{12 \leq j \leq D(s)} \ln[1 - A(t_j^-)] \leq |\ln(1 - \gamma)| \text{ for all } s \leq t \right\}.$$

Introduce subsets

$$H_t \subset W, \qquad t > 0, \tag{3.45}$$

which we define on the basis of events $p_{\lambda t}$, $t > 0$, as

$$H_t = \{\lambda \in W\colon I(p_\lambda^t) = 1\},$$

where $I(p_\lambda^t)$ is the indicator of event p_λ^t. So, H_t is the set of all $\lambda \in W$ such that event $p_\lambda^t \in \mathcal{L}_t$ for this trajectory of the testing process takes place. In accordance with this definition, the set H_t is written as

$$H_t = \left\{ \lambda\colon \int_0^s A(u)N(u)\lambda(u)\,du \right.$$

$$\left. + \sum_{1 \leq j \leq D(s)} \ln[1 - A(t_j^-)] \leq |\ln(1 - \gamma)| \text{ for all } s \leq t \right\}.$$

Let us now show that the system of sets (3.45) constructed in this way is the system of sequential γ-confidence sets for $\lambda \in W$. For this it is enough to show that

$$P_\lambda(\lambda \in H_t) = P_\lambda, \qquad (p_\lambda^t) \geq \gamma, \tag{3.46}$$

for any $\lambda \in W$ and any fixed finite $t > 0$. Indeed, the events $\wp_\lambda^t$ and $\lambda \in H_t$ are equivalent by construction. Then for some $\lambda \in W$ let the inequality (3.46) hold for any $t > 0$. Let us, then, show that

$$P_\lambda(\lambda \in H_\nu) \geq \gamma$$

for any Markovian moment of the total termination ν such that

$$P_\lambda(\nu < \infty) = 1. \tag{3.47}$$

Introduce event $\wp_\lambda^t \in \mathcal{L}$, which is defined as

$$\wp_\lambda = \lim_{t \to \infty} \downarrow \wp_\lambda^t = \bigcap_{t>0} \wp_\lambda^t$$

From (3.46) it follows that

$$P(\wp_\lambda) = \lim_{t \to \infty} P_\lambda(\wp_\lambda^t) \geq \gamma.$$

For any finite $t > 0$, the relation $\wp_\lambda \subset \wp_\lambda^t$ holds, and it follows that

$$\wp_\lambda \bigcap (\nu < \infty) \subset \wp_\lambda^\nu \bigcap (\nu < \infty).$$

Besides, due to equivalency of events $\wp_\lambda^t$ and $\lambda \in H_t$, the relation

$$(\lambda \in H_\nu) = \bigcap (\nu < \infty) = \wp_\lambda^\nu \bigcap (\nu < \infty)$$

holds. Taking into account (3.47), we have

$$\begin{aligned} P_\lambda(\lambda \in H_\nu) &= P_\lambda\{(\lambda \in H_\nu) \bigcap (\nu < \infty) \\ &= P_\lambda\{\wp_\lambda^\nu \bigcap (\nu < \infty)\} \\ &\geq P_\lambda\{\wp_\lambda \bigcap (\nu < \infty)\} = P_\lambda\{\wp_\lambda\} \geq \gamma. \end{aligned}$$

Notice that, in the considered model, the condition $P_\lambda(\nu < \infty) = 1$ automatically holds if $F(\infty) = 1$ because any total test termination does not last beyond the failure of the last unit.

Thus, it is enough to show that inequality (3.46) holds for any fixed, finite $t > 0$ and for any $\lambda \in W$. Let us first prove this inequality for a discrete model M_j^*, introduced in Section 3.8.3. For this model, the intermediate termination moments τ_k might take only discrete values t_{ij}, $i = 1, 2, \ldots, j = 1, 2, \ldots, n$. We introduce the following additional notation for this model:

The term $N_j(t)$ is the number of units on test up to moment t.

The term d_{ij} is the number of failures on the interval $(t_{ij}, t_{i+1,j})$.

The term

$$\delta_{ij} = \begin{cases} 1 & \text{if } d_{ij} > 0, \\ 0 & \text{if } d_{ij} = 0 \end{cases}$$

is an indicator showing that at least one failure has occurred on this interval.

The probability

$$q_{ij} = \frac{P(t_{ij}) - P(t_{i+1,j})}{P(t_{ij})} = 1 - \exp\left[-\int_{t_{ij}}^{t_{i+1,j}} \lambda(u)\, du\right] \tag{3.48}$$

is the conditional probability of the unit failure on the interval $(t_{ij}, t_{i+1,j})$ under the condition that until moment t_{ij} the unit has not failed.

The probability

$$\pi_{ij} = P_\lambda(\delta_{ij} = 1 | \mathfrak{L}_{t_{ij}}) = E_\lambda(\delta_{ij} | \mathfrak{L}_{t_{ij}}) \tag{3.49}$$

is the conditional probability of at least one failure on the interval $(t_{ij}, t_{i+1,j})$ under the condition that there is information about the entire previous trajectory of the process on interval $[0, t_{ij}]$.

In discrete model M_j^*, the testing process develops without external interventions (intermediate test terminations) on each interval $(t_{ij}, t_{i+1,j})$. It means that the conditional distribution of the number of failures d_{ij} on this interval (under the condition that all previous information about the process on interval $[0, t_{ij}]$ is known) has the standard binomial distribution with two parameters: number of trials $N_j(t_{ij})$ and probability of failure q_{ij}. Thus the conditional probability (3.49) is defined as

$$\pi_{ij} = 1 - (1 - q_{ij})^{N_{ij}(t_{ij})} = 1 - \exp\left(-N_j(t_{ij}) \int_{t_{ij}}^{t_{i+l+1}} \lambda(u)\, du\right). \tag{3.50}$$

A random sequence for the discrete model M_j^* is given as

$$\xi_{nj} = \prod_{1 \le i \le n-1} [1 + A(t_{ij})(\pi_{ij} - \delta_{ij})], \qquad n = 1, 2, \ldots \tag{3.51}$$

with $\xi_{1j} = 1$. In accordance with (3.49) the equality

$$E_\lambda(\xi_{n+1,j} | \mathfrak{L}_{t_{nj}}) = \xi_{nj}, \qquad n = 0, 1, \ldots,$$

holds. Thus the sequence ξ_{nj}, $n = 1, 2, \ldots$, forms a martingale with respect to the family $\mathfrak{L}_{t_{nj}}$, $n = 1, 2, \ldots$, with the mathematical expectation $E_\lambda \xi_{nj} = 1$, $n = 1, 2, \ldots$. Because of continuity of $\lambda(u)$, this function is restricted on each finite interval for $0 \leq u \leq t$ and it follows that $\pi_{ij} \leq C(h/2^j)$ for all $i = 0, 1, \ldots, r_j$, where $r_j = L \cdot 2^j$ and C is a constant. Taking into account that $A(t)$ is restricted and $\delta_{ij} \leq 1$, it follows that the martingale (3.51) is nonnegative beginning from some subscript j_1: $\delta_{ij} \geq 0$, $j \geq j_1$ for all $n = 1, 2, \ldots, r_j$. Using the known Doob–Kolmogorov inequality (see, e.g., Doob, 1953, or Shiryaev, 1980), we find, for $j \geq j_1$, that

$$P_\lambda \left\{ \max_{1 \leq n \leq l_j} \xi_{nj} \geq b \right\} \leq \frac{1}{b},$$

where level $b > 0$. Taking $b = 1/(1 - \gamma)$ in this inequality, we obtain

$$P_\lambda \left\{ \max_{1 \leq n \leq r_j} \xi_{nj} \leq \frac{1}{1 - \gamma} \right\} \geq \gamma. \tag{3.52}$$

Let the event

$$\wp^t_{\lambda, j} = \left\{ \max_{1 \leq n \leq r_j} \xi_{nj} \leq \frac{1}{1 - \gamma} \right\}$$

and $\bar{\wp}^t_\lambda$ and $\bar{\wp}^t_{\lambda, j}$ be events complementary to $\wp^t_\lambda$ and $\wp^t_{\lambda, j}$, respectively. Let the event $\bar{\wp}^t_\lambda$ take place, that is,

$$I(\bar{\wp}^t_\lambda) = 1. \tag{3.53}$$

The latter equality is equivalent to the condition that for some $s \leq t$ the inequality

$$\int_0^s A(u)N(u)\lambda(u)\, du + \sum_{1 \leq j \leq D(s)} \ln[1 - A(t_j^-)] > |\ln(1 - \gamma)| \tag{3.54}$$

holds.

Let us now show that beginning from some j_2 the event $\bar{\wp}^t_{\lambda, j}$ takes place:

$$I(\bar{\wp}^t_{\lambda, j}) = 1, \qquad j \geq j_2. \tag{3.55}$$

In other words, we need to show that (3.53) implies (3.55). It will follow that, for $j \to \infty$,

$$\overline{p_\lambda^t} \subset \liminf \overline{p_{\lambda,j}^t}$$

or

$$p_\lambda^t \supset \limsup p_{\lambda,j}^t,$$

which, with (3.53), proves the inequality (3.46).

Introduce a value

$$n(j) = \left[\frac{s2^j}{h}\right]^*,$$

where s is defined in (3.54). It is enough to show that if (3.54) holds, then, for some j_3, so does the inequality

$$\xi_{n(j),j} > (1 - \gamma)^{-1}, \qquad j \geq j_3. \tag{3.56}$$

This inequality can be rewritten in the form

$$\Sigma_j' + \Sigma_j'' > |\ln(1 - \gamma)|, \tag{3.57}$$

where the first sigma equals

$$\Sigma_j' = \sum_{\substack{1 \leq i \leq n(j)-1 \\ \delta_{ij}=0}} \ln[1 + A(t_{ij})\pi_{ij}],$$

where the sum is taken over all indices $i = 1, \ldots, n(j) - 1$ for which $\delta_{ij} = 0$, and the second sigma equals

$$\Sigma_j'' = \sum_{\substack{1 \leq i \leq n(j)-1 \\ \delta_{ij}=1}} \ln[1 + A(t_{ij})(\pi_{ij} - 1)],$$

where the sum is taken over all indices $i = 1, \ldots, n(j) - 1$ for which $\delta_{ij} = 1$.

Let us now show that for $j \to \infty$ the left side of (3.57) converges to the left side of (3.54) everywhere on the set B introduced in Section 3.8.3. Since this set has probability equal to 1, the following statement follows. From (3.50), function $\lambda(u)$ is restricted on the interval $0 \leq u \leq t$ and values $N(u)$ and $N_j(u)$ are restricted by constant $N(0)$, and the equality

$$\pi_{ij} = N_j(t_{ij}) \int_{t_{ij}}^{t_{i+1,j}} \lambda(u)\, du + \varphi \tag{3.58}$$

holds, where $|\varphi| < C\varepsilon^2$, C being some constant and $\varepsilon = t_{i+1,j} - t_{ij} = h/2^j$. So, the first sigma, with the condition that $A(u)$ is restricted, can be written as

$$\Sigma_j' = \int_0^s A(u)N_j(u)\lambda(u)\,du + \varphi_1,$$

where $|\varphi_1| < C_1\varepsilon$. Taking into account that, on set B, the value of $N_j(u)$ differs from $N(u)$ beginning from some j only on some finite number of intervals of length ε, we can write

$$\Sigma_j' = \int_0^s A(u)N_j(u)\lambda(u)\,du + \varphi_2$$

for $j \geq j_4$, where $|\varphi_2| < C_2\varepsilon$. Thus, for $j \to \infty$, the first sigma,

$$\Sigma_j' \to \int_0^s A(u)N_j(u)\lambda(u)\,du,$$

converges everywhere on the set B.

Further, from (3.58) it follows that $\pi_{ij} \leq C_3\varepsilon$ for all $i = 1, \ldots, r_j$. Taking into account the restrictiveness of $A(u)$ on the interval $0 \leq u \leq \underline{t}$, we have

$$\Sigma_j'' = \sum_{1 \leq i \leq n(j)-1;\delta_{ij}=1} \ln[1 - A(t_{ij})] + \varphi_3$$

for $j \geq j_5$, where $|\varphi_3| < C_4\varepsilon$. Thus, for $j \to \infty$ on the set B the second sigma converges,

$$\Sigma_j'' \to \sum_{1 \leq j \leq D(s)} \ln[1 - A(t_j^-)].$$

This proves the initial statement.

Example 3.7 The main condition on which the above proof is based is formula (3.50) for the conditional probability (3.49). In more descriptive form, it means the following: We know that the intermediate termination cannot be performed on the time interval $(t, t + h)$. Then the conditional probability of unit failure on this interval under the condition that it has not failed on the interval $(0, t]$ equals

$$q(t, t + h) = \frac{P(t) - P(t + h)}{P(t)} = 1 - \exp\left[-\int_t^{t+h} \lambda(u)\,du\right]$$

independent of the "prehistory" of the testing process up to the moment t. In other words, there is no dependence on the intermediate termination or failure of other units on the interval $(0, t]$. This and the assumption of the independence of tested units give us that the conditional distribution of the number of failures $d = D(t + h) - D(t)$ on this interval is binomial with parameters $N = N(t)$, the number of independent trials, and $q = q(t, t + h)$, the probability of failure. The same condition [formula (3.38)] was the key for the proof of unbiasedness of the estimate $\hat{P}(t)$ in Section 3.8.3. □

Remark 3.8 For the more general test model, [MMR], where new tested units might appear at moments $\sigma_1, \sigma_2, \ldots$ (see Section 3.6), the analogous statement is as follows. Let it be known that on the time interval $(t, t + h)$ no external interventions are possible (no intermediate termination, no addition of new units). Then the conditional probability of unit failure under the condition that it has not failed or its testing has not been terminated up to this moment equals

$$q = \frac{P[S(t)] - P[S(t) + h]}{P[S(t)]} = 1 - \exp\left[-\int_{S(t)}^{S(t)+h} \lambda(u)\, du\right], \qquad (3.59)$$

where $S(t)$ is the time for which the unit was operating up to the moment t, that is, $S(t) = t - \sigma$, where σ is the moment when this unit began to be tested. On the basis of this condition, the proof of the confidence limits (3.33) and (3.34) for the model [MMR] can be performed in a manner completely analogous to [MMU]. The only additional condition for this case is $E_\lambda B(t) < \infty$, $\lambda \in W$, for all $t > 0$, where $B(t)$ is the number of new units that began to be tested on the time interval $[0, t]$. □

3.8.6 Consistency of Confidence Limits (3.24) and (3.25)

Let $t > 0$ be some fixed moment. Consider a sequence of Markov test models $[\text{MMU}]_r$, $r = 1, 2, \ldots$. The characteristic for the rth model is denoted by subscript r: $N_r(t)$, $D_r(t)$, $\cdots$. Let the initial number of tested units be

$$N_r = N_r(0) \to \infty$$

for $r \to \infty$, with the condition (for fixed t)

$$\frac{N_r\sqrt{N_l}}{N_r^2(t^-)} \underset{P}{\to} 0. \qquad (3.60)$$

Assume that the intermediate termination moments can take only discrete values $v_j = ih$, $i = 1, 2, \ldots$ with some fixed infinitesimally small $h > 0$. For the sake of simplicity, assume that $h = t/K$, where K is an integer. Set $\gamma = 1 - \alpha$ and choose the function $A(t)$ as

$$A(t) = -\frac{a\sqrt{N_r}}{N_r(t)}, \qquad t \geq 0, \tag{3.61}$$

where a is a positive constant. From (3.24) we have the following lower $(1 - \alpha)$-confidence limit for the resource function:

$$\underline{\Lambda}_l(t) = -\frac{|\ln \alpha|}{a\sqrt{N_r}} + \frac{1}{a\sqrt{N_r}} \sum_{1 \leq j \leq D_l(t)} \ln \left[1 + \frac{a\sqrt{N_r}}{N_r(t_j^-)} \right]. \tag{3.62}$$

In an analogous way, setting $\gamma = 1 - \beta$ and

$$A(t) = \frac{b\sqrt{N_r}}{N_r(t)}, \qquad t \geq 0, \tag{3.63}$$

where b is a positive constant, we obtain the following upper $(1 - \beta)$-confidence limit for the resource function:

$$\overline{\Lambda}_r(t) = -\frac{|\ln \beta|}{b\sqrt{N_r}} - \frac{1}{b\sqrt{N_r}} \sum_{1 \leq j \leq D_l(t)} \ln \left[1 - \frac{b\sqrt{N_r}}{N_r(t_j^-)} \right]. \tag{3.64}$$

Let us show that for $r \to \infty$ there is the convergence

$$\underline{\Lambda}_r(t) \xrightarrow[p]{} \Lambda(t), \tag{3.65}$$

$$\overline{\Lambda}_r(t) \xrightarrow[p]{} \Lambda(t). \tag{3.66}$$

First let us prove (3.65). The first term in (3.62) is a constant converging to zero. The second term after some transformations can be rewritten as

$$\frac{1}{a\sqrt{N_r}} \sum_{1 \leq j \leq D_l(t)} \ln \left[1 + \frac{a\sqrt{N_r}}{N_r(t_j^-)} \right] = \hat{\Lambda}_r(t) + \delta,$$

where

$$\hat{\Lambda}_r \sum_{1 \leq j \leq D_l(t)} \frac{1}{N_r(t_j^-)}, \qquad |\delta| \leq \frac{a\sqrt{N_r}}{2} \sum_{1 \leq j \leq D_l(t)} \frac{1}{N_r^2(t_j^-)}. \tag{3.67}$$

From the latter inequality, taking into account that $N_r(t_j^-) \geq N_r(t^-)$ for all $j = 1, \ldots, D_r(t)$, we have

$$|\delta| \leq \frac{a\sqrt{N_r}}{2} \frac{D_r(t)}{N_r^2(t^-)}.$$

Thus now it is enough to show that

$$\frac{D_r(t)\sqrt{N_l}}{N_r^2(t^-)} \underset{p}{\rightarrow} 0, \tag{3.68}$$

$$\hat{\Lambda}_r(t) \underset{p}{\rightarrow} \Lambda(t). \tag{3.69}$$

Convergence of (3.68) directly follows from condition (3.60) if one takes into account that $D_r(t) \leq N_r$. Let us show the convergence of (3.69). The value $\hat{\Lambda}_r(t)$ can be written in the form

$$\hat{\Lambda}(t) = \sum_{1 \leq i \leq K} \eta_i, \tag{3.70}$$

where

$$\eta_i = \sum_{j \in J_i} \frac{1}{N_r(t_j^-)}$$

and the sum is taken over the set of subscripts $J_i = \{j\colon \nu_i - h < t_j \leq \nu_i\}$, $i = 1, \ldots, K$. So, it is enough to show that

$$\eta_i \underset{p}{\rightarrow} \Lambda(\nu_i) - \Lambda(\nu_i - h)$$

for all $i = 1, \ldots, K$. Since on the interval $(v_i - h, v_i)$ there are no intermediate terminations,

$$\eta_i = \sum_{N_i^- < \leq n \leq N_i^+} \frac{1}{n},$$

where $N_i^- = N(v_i)$ and $N_i^+ = N(v_i - h)$. For this value are given the limits

$$\int_{N_j^-}^{N_i^+ + 1} \frac{dx}{x} \leq \eta_i \leq \int_{N_i^- - 1}^{N_i^+} \frac{dx}{x}.$$

Denoting the number of failures on the interval $(v_i - h, v_i)$ by $d_i = N_i^+ - N_i^-$, we have

$$\ln\left(1 - \frac{d_i}{N_i^+}\right)^{-1} \le \eta_i \le \ln\left(1 - \frac{d_i}{N_i^+}\right)^{-1} + \ln\left(1 - \frac{1}{N_i^-}\right)^{-1}. \quad (3.71)$$

For assumed conditions $N_i^- \underset{P}{\to} \infty$ and $N_i^+ \underset{P}{\to} \infty$, $i = 1, \ldots, K$, it follows that

$$\frac{1}{N_i^-} \underset{P}{\to} 0, \qquad \frac{d_i}{N_i^+} \underset{P}{\to} q_i, \quad (3.72)$$

where $q_i = 1 - \exp[\Lambda(v_i - h) - \Lambda(v_i)]$ is the conditional probability of unit failure on the interval $(v_i - h, v_i)$ under the condition that up to that interval the unit has not failed. From (3.71) and (3.72) the convergence of (3.69) and (3.65) follows.

The proof of the convergence of (3.66) is completely analogous.

Remark 3.9 The main condition (3.60) for the consistency of confidence limits means that the number of units for which testing has been terminated up to the moment t must not be too large, or, in other words, $N_r(t^-)$ must grow to infinity faster than $N_r^{(3/4)}$. This condition automatically holds for the model of independent random censoring (concurrent risks) for all t such that $[1 - F(t)][1 - G(t)] > 0$, because in this case

$$\frac{N_r(t^-)}{N_r} \underset{P}{\to} [1 - F(t)][1 - G(t)],$$

from which it follows that

$$\frac{N_r\sqrt{N_r}}{N_r^2(t^-)} = \frac{1}{\sqrt{N_r}}\left[\frac{N_r}{N_r(t^-)}\right]^2 \underset{P}{\to} 0.$$

Notice that for the conditions mentioned above estimates of the resource function $\hat{\Lambda}(t)$ and reliability function $\hat{P}(t)$ are consistent for $N_r \to \infty$, though for consistency a weaker condition is satisfactory: $N_r(t^-) \underset{P}{\to} \infty$, or, in other words, $P_\lambda\{N_r(t^-) < C\} \to 0$ for any finite C. □

3.8.7 Markov Test Model [MMR] with Censorship and Renewal

Let us add the following two sequences to the model [MMU]:

$$\sigma_1 < \sigma_2 < \cdots < \sigma_i < \sigma_{i+1} < \cdots,$$

$$m_1, m_2, \cdots, m_i, m_{i+1}, \cdots,$$

where σ_i is the sequence of monotonically increasing Markov moments (with respect to the family $\mathfrak{L}_t$, $t \geq 0$) and $m_i = 0, 1, 2, \ldots$ is a $\mathfrak{L}_{\sigma_i}$-measurable r.v., $i = 1, 2, \ldots$. Here σ_i is the moment at which m_i new identical units are added to the tested units.

The total test termination moment ν is also Markovian with respect to the family $\mathfrak{L}_t$, $t \geq 0$, such that $P_\lambda(\nu < \infty) = 1$, $\lambda \in W$. Thus, this model is defined by the set

$$M = \{(\tau_1, n_1), \ldots, (\tau_k, n_k), \ldots; (\sigma_1, m_1), \ldots, (\sigma_i, m_i), \ldots; \nu\}, \quad (3.73)$$

where τ_k is the moment of the testing termination for a group of n_k units, $k = 1, 2, \ldots$ (see Section 3.8.2).

One more difference in this model from the model [MMU] is in the fact that the sequence (τ_k, n_k), as well as the sequence (σ_i, m_i), is, in general, infinite.

It is clear that the model [MMR] includes the model [MMU] as a particular case for $m_i = 0$, $i = 1, 2, \ldots$.

Notice also that in the model [MMU] all operating units at any moment t have the same "age" (operational time) equal to t. In the more general model [MMR] $N(t)$ units currently testing at time t have different age $S_e(t)$, $e \in E(t)$, where e is the subscript of the unit, $S_e(t)$ is the age of the eth unit at time t, and $E(t)$ is the set of subscripts of the units testing at moment t. The set $E(t)$ consists of $N(t)$ subscripts, so to define the test model, the set (3.73) must be widened: For each intermediate termination moment τ_k, we need to specify the set of the unit's subscripts rather than just their quantity, n_k: $\{e_r(\tau_k), r = 1, \ldots, n_k\}$, where $e_r(\tau_k)$ is the subscript of the rth unit whose test has been terminated at the moment τ_k, $e_r(\tau_k) \in E)\tau_k^-)$. Of course, it is possible to show the age of all units that are still being tested after moment τ_k: $S_e(\tau_k)$, $e \in E(\tau_k) \subset E(\tau_k^-)$.

PROBLEMS

3.1. Prove the formula for the reliability function,

$$P(t) = \prod_{0 \leq j \leq M-1} (1 - q_j), \quad (3.74)$$

Table 3.3 Failure Moments t_j (hr)

j	1	2	3	4	5
t_j	41	87	104	146	198

Table 3.4 Intermediate Termination Moments τ_k (hr) and Corresponding Number of Units in a Group, n_k

k	1	2	3	4
τ_k	100	146	180	350
n_k	2	1	4	3

where $t = Mh$,

$$q_j = P(jh < \xi \leq jh + h | \xi > jh) = \frac{P(hj) - P(hj + h)}{P(jh)},$$

$j = 0, 1, \ldots, M - 1$; that is, q_j is the conditional probability of unit failure on the interval $(jh, jh + h]$ under the condition that up to the moment jh the unit has not failed. (Here ξ denotes the unit's TTF.)

3.2. Using the conditions in Example 3.1, calculate the estimate of the reliability function $P(t)$ for $t = 300$ with the help of the recurrent procedure (3.11).

3.3. Using the conditions in Example 3.1, calculate the estimate of the reliability function $P(t)$ for $t = 375$ with the help of the recurrent procedure (3.11).

3.4. Consider the Markov test model [MMU] with censorship and without renewal. At moment $t = 0$ fifteen identical units are tested ($N = 15$). The test results are represented in Tables 3.3 and 3.4. To find the estimate of the reliability function $P(t = 200)$.

3.5. Using the conditions in problem 3.4, find the estimate for the reliability function $P(t = 120)$.

3.6. Using the conditions in problem 3.4, find the estimates for the reliability function $P(t = 150)$ and resource function $\Lambda(t = 150)$.

3.7. Using the conditions in Example 3.1 and an asymptotically normal approximation (see Section 3.4), construct the lower confidence limit with confidence coefficient $\gamma = 0.95$ for the reliability function $P(t = 160)$.

3.8. Using the conditions in problem 3.4 and the method described in Section 3.5, find the lower confidence limit with the confidence coefficient $\gamma = 0.9$ for the reliability function $P(t = 150)$ and resource function $\Lambda(t = 150)$.

CHAPTER 4

BAYES METHODS OF RELIABILITY ESTIMATION

4.1 INTRODUCTION

To illustrate the Bayes method, we consider the following standard case. We have a unit with d.f of TTF $F(t, \boldsymbol{\theta})$ and density $f(t, \boldsymbol{\theta})$, where $\boldsymbol{\theta} = (\boldsymbol{\theta}_1, \ldots, \boldsymbol{\theta}_m)$ is a parameter (in the general case, a vector) from some space Θ. Let $\mathbf{z}$ be a vector of test results obtained on the basis of some test plan and $L(\mathbf{z}|\boldsymbol{\theta})$ be a likelihood function or, in other words, the density of distribution of test results $\mathbf{z}$ for a given value of parameter $\boldsymbol{\theta}$. For instance, if $\mathbf{z}$ is a complete sample of size n, that s, $\mathbf{z} = (x_1, \ldots, x_n)$, where x_i are independent values of observed TTF, then the likelihood function has the form

$$L(\mathbf{z}|\boldsymbol{\theta}) = \prod_{1 \le i \le n} f(x_i, \boldsymbol{\theta}).$$

Following Bayes's approach, we consider parameter $\boldsymbol{\theta}$ a r.v. with some density $h(\boldsymbol{\theta})$, which is called the prior density of distribution. This density, roughly speaking, reflects our prior (before testing) knowledge about the parameter $\boldsymbol{\theta}$. The Bayes approach consists mainly in finding the conditional density of the distribution $h(\boldsymbol{\theta}|\mathbf{z})$ under the condition that the test result is $\mathbf{z}$:

$$h(\boldsymbol{\theta}|z) = \frac{h(\boldsymbol{\theta})L(z|\boldsymbol{\theta})}{\varphi(z)}, \tag{4.1}$$

where $\varphi(\mathbf{z})$ is the density of the distribution of $\mathbf{z}$ taking into account a prior distribution $h(\boldsymbol{\theta})$:

$$\varphi(z) = \int_{\Theta} h(\boldsymbol{\theta})L(z|\boldsymbol{\theta})\, d\boldsymbol{\theta}.$$

In formula (4.1) the function $h(\boldsymbol{\theta})L(z|\boldsymbol{\theta})$ is a joint density of the distribution of a pair $(\boldsymbol{\theta}, \mathbf{z})$ and the function $\varphi(\mathbf{z})$ is a marginal density of the distribution of $\mathbf{z}$. For given fixed test results $\mathbf{z}$, the value of function $h(\boldsymbol{\theta}|\mathbf{z})$ is proportional to the product of the prior density and likelihood function $h(\boldsymbol{\theta})L(z|\boldsymbol{\theta})$. The value of $\varphi(\mathbf{z})$ is a norm multiplier chosen in such a way that the integral of function $h(\boldsymbol{\theta}|\mathbf{z})$ by parameter $\boldsymbol{\theta}$ is equal to 1.

Relation (4.1) is the basis for the Bayes method. Function $h(\boldsymbol{\theta}|\mathbf{z})$ is called a posterior density of distribution of parameter $\boldsymbol{\theta}$ for given test results $\mathbf{z}$. This density reflects our knowledge about parameter $\boldsymbol{\theta}$ after the test. For instance, if D is some area in parameter space, then value $\int_D h(\boldsymbol{\theta}|z)\, d\boldsymbol{\theta}$ can be interpreted as a measure of our assurance in the fact that parameter $\boldsymbol{\theta}$ belongs to the area D if the test result is $\mathbf{z}$.

A posterior density (4.1) is the basis for obtaining statistical decisions about parameter $\boldsymbol{\theta}$. We will consider the construction of Bayes estimates and confidence limits.

4.2 POINT ESTIMATES AND CONFIDENCE LIMITS

4.2.1 Bayes Point Estimates

Let $R = R(\boldsymbol{\theta})$ be a unit reliability index that is a function of vector parameter $\boldsymbol{\theta}$. Let $\hat{R} = \hat{R}(\mathbf{z})$ be a point estimate for R. A natural measure of the quality of estimate $\hat{R}$ can be chosen to be the mathematical expectation of the square of the deviation of the estimate $\hat{R}$ from R, that is,

$$E(\hat{R} - R)^2 = \int_{\Theta} h(\boldsymbol{\theta})\, d\boldsymbol{\theta} \int_{Z} [\hat{R}(\mathbf{z}) - R(\boldsymbol{\theta})]^2 L(\mathbf{z}|\boldsymbol{\theta})\, d\mathbf{z}, \qquad (4.2)$$

where Z is a set of possible test results $\mathbf{z}$. The problem is to find the function $\hat{R}(\mathbf{z})$ for which (4.2) is a minimum.

Taking into account the definition of the posterior density (4.1), we can present (4.2) in the form

$$E(\hat{R} - R)^2 = \int_{Z} \varphi(\mathbf{z})\, d\mathbf{z} \int_{\Theta} [\hat{R}(\mathbf{z}) - R(\boldsymbol{\theta})]^2 h(\boldsymbol{\theta}|\mathbf{z})\, d\boldsymbol{\theta}.$$

Hence, to solve the problem formulated above, we need to find, for each fixed test results $\mathbf{z}$, a minimum in $\hat{R}$ of the function

$$G(\hat{R}) = \int_{\Theta} [\hat{R} - R(\boldsymbol{\theta})]^2 h(\boldsymbol{\theta}|\mathbf{z})\, d\boldsymbol{\theta}.$$

This function can be written in the form

$$G(\hat{R}) = \hat{R}^2 - 2\hat{R} \int_{\Theta} R(\boldsymbol{\theta}) h(\boldsymbol{\theta}|\mathbf{z})\, d\boldsymbol{\theta} + \int_{\Theta} R^2(\boldsymbol{\theta}) h(\boldsymbol{\theta}|\mathbf{z})\, d\boldsymbol{\theta}.$$

Calculating the derivative $G'(\hat{R})$ and setting it equal to zero, we see that the minimum of function $G(\hat{R})$ is reached at

$$\hat{R} = \int_{\Theta} R(\boldsymbol{\theta}) h(\boldsymbol{\theta}|\mathbf{z})\, d\boldsymbol{\theta}. \tag{4.3}$$

The right side of this equation is a posterior mathematical expectation for $R(\boldsymbol{\theta})$. Thus the best Bayes point estimate of reliability index $R = R(\boldsymbol{\theta})$, which minimizes the standard deviation (4.2) of $R(\boldsymbol{\theta})$ for given test results $\mathbf{z}$.

Notice that the choice of the type of measure of quality of estimate $\hat{R}$ essentially influences the type of "best" estimate. For instance, if we choose the mathematical expectation of the module of deviation of the estimate $\hat{R}$ from R as the measure of quality, that is,

$$E|\hat{R} - R| = \int_{\Theta} h(\boldsymbol{\theta})\, d\boldsymbol{\theta} \int_{Z} |\hat{R}(\mathbf{z}) - R(\boldsymbol{\theta})| L(\mathbf{z}|\boldsymbol{\theta})\, d\mathbf{z},$$

then we can show that the best Bayes point estimate of $\hat{R} = \hat{R}(\mathbf{z})$ is the median of the posterior distribution of $R(\boldsymbol{\theta})$ for fixed test results $\mathbf{z}$ (see problem 4.1).

4.2.2 Sufficient Statistics

Let $T = T(\mathbf{z})$ be a sufficient statistic; then (see Section 1.3.4) the likelihood function can be written as a product of two multipliers:

$$L(\mathbf{z}|\boldsymbol{\theta}) = C(\mathbf{z}) g[T(\mathbf{z}), \boldsymbol{\theta}],$$

where the function $C(\mathbf{z})$ depends only on the test results $\mathbf{z}$ but does not depend on parameter $\boldsymbol{\theta}$ and the function $g[T(\mathbf{z}), \boldsymbol{\theta}]$ depends on parameter $\boldsymbol{\theta}$ and test results $\mathbf{z}$ via statistic $T(\mathbf{z})$. Substituting this expression into (4.1), we have the following formula for the posterior density of distribution:

$$h(\boldsymbol{\theta}|\mathbf{z}) = \frac{h(\boldsymbol{\theta}) \cdot C(\mathbf{z}) \cdot g[T(\mathbf{z}), \boldsymbol{\theta}]}{C(\mathbf{z}) \int_{\Theta} h(\boldsymbol{\theta}) g[T(\mathbf{z}), \boldsymbol{\theta}]\, d\boldsymbol{\theta}} = \frac{h(\boldsymbol{\theta}) \cdot g[T(\mathbf{z}), \boldsymbol{\theta}]}{\int_{\Theta} h(\boldsymbol{\theta}) g[T(\mathbf{z}), \boldsymbol{\theta}]\, d\boldsymbol{\theta}}.$$

From here it follows that the posterior distribution density depends on test results $\mathbf{z}$ via a sufficient statistic $T = T(\mathbf{z})$. In other words, if different test results z_1 and z_1 are such that $T(\mathbf{z}_1) = T(\mathbf{z}_1)$, then the posterior distribution densities $h(\boldsymbol{\theta}|z_1)$ and $h(\boldsymbol{\theta}|z_2)$ coincide. Thus all Bayes statistical inferences (point estimates, confidence limits, etc.) depend only on a sufficient statistic $T = T(\mathbf{z})$. For instance (see Section 4.4), in the "exponential model," the two-dimensional statistic $T = (d, S)$ is sufficient. (Here d is the number of failures during the test and S is the total test time.)

4.2.3 Bayes Confidence Limits

Let $R = R(\boldsymbol{\theta})$ be a reliability index. Introduce the function

$$\Phi(t|\mathbf{z}) = P\{R(\boldsymbol{\theta}) \leq t|\mathbf{z}\} = \int_{R(\boldsymbol{\theta})\leq t} h(\boldsymbol{\theta}|\mathbf{z})\, d\boldsymbol{\theta} \tag{4.4}$$

which represents a posterior d.f. of reliability index $R(\boldsymbol{\theta})$ for a given test result $\mathbf{z}$. Bayes confidence limits for $R(\boldsymbol{\theta})$ are determined as corresponding quantiles of d.f. (4.4). Namely, Bayes lower confidence level (LCL) $\underline{R} = \underline{R}(\mathbf{z})$ with confidence coefficient $1 - \alpha$ for $R(\boldsymbol{\theta})$ can be found from the solution of the equality

$$\Phi(\underline{R}|\mathbf{z}) = \alpha. \tag{4.5}$$

The Bayes upper confidence level (UCL) $\overline{R} = \overline{R}(\mathbf{z})$ with confidence coefficient $1 - \beta$ for $R(\boldsymbol{\theta})$ is found from

$$\Phi(\overline{R}|\mathbf{z}) = 1 - \beta. \tag{4.6}$$

From (4.5) and (4.6) it also follows that the following equation is true:

$$P\{\underline{R} \leq R(\boldsymbol{\theta}) \leq \overline{R}|\mathbf{z}\} = \int_{\underline{R}\leq R(\boldsymbol{\theta})\leq \overline{R}} h(\boldsymbol{\theta})|\mathbf{z})\, d\boldsymbol{\theta} = 1 - \alpha - \beta;$$

that is, interval $(\underline{R}, \overline{R})$ is the Bayes confidence limit for $R(\boldsymbol{\theta})$ with confidence coefficient $\gamma = 1 - \alpha - \beta$.

Let us now consider some applications of the Bayes method.

4.3 BINOMIAL MODEL

The number of failures in N independent tests is found to be $z = d$. Parameter $\boldsymbol{\theta} = q$ is the probability of failure in a single test. In this case the likelihood function, that is, the probability of observing of d failures in a series of N trials, has the form

$$L(\mathbf{z}|\boldsymbol{\theta}) = L(d|q) = \binom{d}{N} q^d(1 - q)^{N-d}.$$

Let a prior distribution density $h(q)$ of parameter q be given. By formula (4.1) the posterior distribution density of parameter q for fixed d is determined as

$$h(q|d) = \frac{h(q)q^d(1 - q)^{N-d}}{\int_0^1 h(q)q^d(1 - q)^{N-d}\, dq}, \tag{4.7}$$

where $0 < q < 1$. Let a prior distribution $h(q)$ have the form of the standard beta distribution with parameters (a, b), that is,

$$h(q) = \frac{q^{a-1}(1 - q)^{b-1}}{B(a, b)}, \tag{4.8}$$

where $B(a, b) = \int_0^1 t^{a-1}(1 - t)^{b-1}\, dt$ is the beta function.

Substituting (4.8) into (4.7), we obtain

$$h(q|d) = \frac{q^{a+d-1}(1 - q)^{b+N-d-1}}{B(a + d, b + N - d)}; \tag{4.9}$$

that is, the posterior distribution density (4.9) is also a density of the beta distribution but with new parameters $(a + d, b + N - d)$. It shows in a very transparent form how test results (N, d) influence the prior distribution (4.8).

Remark 4.1 If prior and posterior distributions belong to the same family of distributions, such a family is called adjoint. For the "binomial model" a family of beta distributions is adjoint. In the "exponential model" such an adjoint distribution family is gamma (see Section 4.4). □

Applying (4.3), we obtain Bayes point estimate $\hat{q}$, a posterior mean of parameter q:

$$\hat{q} = \int_0^1 qh(q|d)\, dq = \frac{a + d}{a + b + M}. \tag{4.10}$$

The formula for the variance of the posterior distribution of parameter q can also be easily obtained:

$$E[(q - \hat{q})^2|d] = \int_0^1 (q - \hat{q})^2 h(q|d)\, dq = \frac{(a + d)(b + N - d)}{(a + b + N)^2(a + b + N + 1)}.$$

Using (4.5) and (4.6), the Bayes LCL $\underline{q}$ with confidence coefficient $1 - \alpha$ and the Bayes UCL $\overline{R}$ with confidence coefficient $1 - \beta$ for parameter q can be found from

$$I_{\underline{q}}(a + d, b + N - d) = \alpha, \tag{4.11}$$

$$I_{\overline{q}}(a + d, b + N - d) = 1 - \beta, \tag{4.12}$$

where $I_q(a, b)$ is an incomplete beta distribution with parameters a and b:

$$I_q(a, b) = \frac{1}{B(a, b)} \int_0^q t^{a-1}(1 - t)^{b-1}\, dt. \tag{4.13}$$

Notice that standard Clopper-Pearson confidence limits for the binomial parameter q are found from (see Section 1.4.6)

$$\sum_{d \le j \le N} \binom{j}{N} \underline{q}^j (1 - \underline{q})^{N-j} = \alpha, \qquad \sum_{1 \le j \le d} \binom{j}{N} \overline{q}^j (1 - \overline{q})^{N-j} = \beta.$$

Taking into account

$$\sum_{0 \le j \le d} \binom{j}{N} q^j (1 - q)^{N-j} = 1 - I_q(d + 1, N - d),$$

we can express Clopper–Pearson equations via the incomplete beta function in the form

$$I_{\underline{q}}(d, N - d + 1) = \alpha, \tag{4.14}$$

$$I_{\overline{q}}(d + 1, N - d) = 1 - \beta \tag{4.15}$$

Solutions of these equations calculated on the basis of statistic (d, N) are denoted by $\underline{q}_{1-\alpha}(N, d)$ and $\overline{q}_{1-\beta}(N, d)$ for the Clopper–Pearson LCL (with confidence coefficient $1 - \alpha$) and UCL (with confidence coefficient $1 - \beta$), respectively. Notice that the left sides of these equations are defined for any (not only for integer) values of N and d.

Comparing (4.11) and (4.12) with (4.14) and (4.15), we see that Bayes confidence limits $\underline{q}$ and $\overline{q}$ relate to the Clopper–Pearson confidence limits as follows:

$$\underline{q} = \underline{q}_{1-\alpha}(N + a + b - 1, d + a), \tag{4.16}$$

$$\overline{q} = \overline{q}_{1-\beta}(N + a + b - 1, d + a - 1). \tag{4.17}$$

These equations allow us to calculate Bayes confidence limits with the help of standard tables of binomial distribution (or tables of beta functions).

Formulas (4.61) and (4.17) can be interpreted in the following way. A knowledge of the prior distribution (4.8) of parameter q, roughly speaking, is equivalent to a situation where we have information about $a + b - 1$ additional trials with a observed failures. In the calculation of the upper confidence level $\overline{q}$, we have information about an additional $a + b - 1$ during which $a - 1$ failures have occurred. In an analogous way, the point Bayes estimate (4.10) can be interpreted.

Example 4.1 Let the parameters of the prior beta distribution be $a = 2$ and $b = 9$ and the test results be $N = 8$ and $d = 1$. Using (4.10), we obtain that the Bayes point estimate $\hat{q}$ for reliability index q (the failure probability) is

$$\hat{q} = \frac{d + a}{N + a + b} = \frac{3}{19} = 0.158.$$

Using formulas (4.16) and (4.17) and table E.14 in the Appendix, we found that 0.99 confidence limits $(\underline{q}, \overline{q})$ for parameter q (for $\alpha = \beta = 0.005$) are

$$\underline{q} = \underline{q}_{1-\alpha}(N + a + b - 1, d + a) = \underline{q}_{0.995}(18, 3) = 0.020,$$

$$\overline{q} = \overline{q}_{1-\beta}(N + a + b - 1, d + a - 1) = \overline{q}_{0.995}(18, 2) = 0.422. \ \square$$

Remark 4.2 The examples above illustrate the advantages and disadvantages of the Bayes approach. Indeed, if we could choose a prior distribution (4.8) so that ratio $a/(a + b)$ is close to the real value of parameter q, then we would obtain, roughly speaking, an equivalent number of trials $N + (a + b - 1)$ instead of the initial value of N. However, if this choice is not felicitous, then we need a large number of trials N to correct the results due to the wrong choice. □

4.4 EXPONENTIAL MODEL

Let unit TTF have exponential d.f. $F(t, \lambda) = 1 - e^{-\lambda t}$. In this case $\boldsymbol{\theta} = \lambda$. Assume that the test was performed by plan $[N\ U\ r]$, that is, with the number of failures fixed in advance (see Section 2.1). In this case test results are expressed as $z = S$, where S is the total test time determined by formula (2.4) and the likelihood function is

$$L\left(\frac{S}{\lambda}\right) = C\lambda^r e^{-\lambda s},$$

where C is a norm constant (see Section 2.2.1). Assume that a prior distribution density $h(\lambda)$ of parameter λ is given. Then in accordance with (4.1), the posterior distribution density of λ has the form

$$h(\lambda|S) = \frac{h(\lambda)\lambda^r e^{-\lambda S}}{\int_0^\infty h(\lambda)\lambda^r e^{-\lambda S}\, d\lambda}. \tag{4.18}$$

Take the standard gamma distribution (see Section 1.2 and Table 1.1) with parameters (U, a) as a prior distribution $h(\lambda)$, that is,

$$h(\lambda) = \frac{u^a \lambda^{a-1}}{\Gamma(a)} e^{-u\lambda}, \tag{4.19}$$

where $\Gamma(a) = \int_0^\infty t^{a-1} e^{-t}\, dt$ is the gamma function. Then, from (4.18), we obtain that the posterior distribution density is

$$h(\lambda|S) = \frac{(u + S)^{a+r} \lambda^{a+r-1} e^{-(u+S)\lambda}}{\Gamma(a + r)},$$

from which we see that it is also gamma distributed with parameters $(u + S, a + r)$. Thus, the exponential family is adjoint for the exponential model considered above.

Using (4.3), we obtain that the Bayes point estimate (the posterior mean) for parameter λ has the form

$$\hat{\lambda} = \int_0^\infty \lambda h(\lambda|S)\, d\lambda = \frac{a + r}{u + S}. \tag{4.20}$$

From (4.5) and (4.6) we find that the Bayes LCL $\underline{\lambda}$ (with confidence coefficient $1 - \alpha$) and the Bayes UCL $\overline{\lambda}$ (with confidence coefficient $1 - \beta$) for parameter λ can be found as

$$\underline{\lambda} = \frac{\Gamma_\alpha(1, a + r)}{u + S}, \qquad \overline{\lambda} = \frac{\Gamma_{1-\beta}(1, a + r)}{u + S},$$

where $\Gamma_\varepsilon(1, a + r)$ is the quantile of level ε for a gamma distribution with parameters $(1, a + r)$. These confidence limits can also be expressed via quantiles of a χ^2 distribution (see Section 1.2.3):

$$\underline{\lambda} = \frac{\chi^2_a(2a + 2r)}{2(u + S)}, \qquad \overline{\lambda} = \frac{\chi^2_{1-\beta}(2a + 2r)}{2(u + S)}, \tag{4.21}$$

where $\chi^2_\varepsilon(m)$ is the quantile of level ε for a χ^2 distribution with m degrees of freedom.

Comparing these formulas with the analogous ones in Section 2.2.1, we see that using the prior distribution (4.19) with parameters (u, a) corresponds to additional tests with total test time u and number of failure a.

Example 4.2 Let the parameters of the prior distribution be $u = 150$ hr and $a = 1$ and the test results be $S = 295$ hr and $r = 1$. Using (4.20) and (4.21), we obtain that the Bayes point estimate (the posterior mean) for parameter λ is equal to

$$\hat{\lambda} = \frac{a + r}{u + S} = \frac{1 + 1}{150 + 295} = 45 \times 10^{-4}\ \text{hr}^{-1}.$$

The Bayes 0.9 LCL $\underline{\lambda}$ and the 0.9 UCL $\overline{\lambda}$ for parameter λ are

$$\underline{\lambda} = \frac{\chi^2_{0.1}(4)}{2(150 + 295)} = \frac{1.064}{890} = 12 \times 10^{-4}\ \text{hr}^{-1},$$

$$\overline{\lambda} = \frac{\chi^2_{0.9}(4)}{2(150 + 295)} = \frac{7.78}{890} = 87.5 \times 10^{-4}\ \text{hr}^{-1}. \ \square$$

Test Plan $[N\ U\ T]$ Consider now plan $[N\ U\ T]$, that is, a test with time T fixed in advance. In this case test results are given by a pair $z = (S, d)$, where, as before, S is the total test time and d is the number of failures. Likelihood function is given by the expression

$$L\left(S, \frac{d}{\lambda}\right) = C\lambda^d e^{-\lambda S}, \tag{4.22}$$

where $C = C(d) = N(n - 1) \times \cdots \times (N - d + 1)$ depends on the test results (d) but does not depend on parameter λ. Consequently, the pair (S, D) is a sufficient statistic (see Section 2.2.2).

Let $h(\lambda)$ be a prior distribution density of parameter λ. Then from (4.1) and (4.22) it follows that the posterior distribution density has the form

$$h(\lambda | S, d) = \frac{h(\lambda)\lambda^d e^{-\lambda S}}{\int_0^\infty h(\lambda)\lambda^d e^{-\lambda S}\, d\lambda}. \tag{4.23}$$

It is easy to see that posterior densities (4.18) and (4.23) coincide if $d = r$.

It means that all results for plan [N U r] considered above can be extended to plan [N U T] for $d = r$.

Other Test Plans The same conclusions can be obtained for other test plans of type U (no renewal) and type R (with renewal). Indeed, for these plans the likelihood function has the form

$$L\left(S, \frac{d}{\lambda}\right) = C\lambda^d e^{-S\lambda}, \tag{4.24}$$

where C is a constant or a function depending on the test results but that does not depend on parameter λ (see Section 2.2). Notice that for plans [N U r] and [N R r] tests continue until the number of failures equals r, which is fixed in advance. We assume $d = r$. From (4.1) and (4.24) it follows that for all these test plans a posterior distribution density of parameter λ has the same form (4.23). The conclusion made on the basis of the posterior distribution density, in particular, formulas (4.20) and (4.21) for Bayes point estimate and confidence limits for test plan [N U r], can be expanded on the test plans mentioned above for $d = r$.

This important property of the Bayes approach ("insensitivity" with respect to the test plans) is expanded onto a more general case where a distribution of TTF is not necessarily exponential.

4.5 GENERAL PARAMETRIC MODEL

Consider a general case where TTF has d.f. $F(t, \boldsymbol{\theta})$ with density $f(t, \boldsymbol{\theta})$, where $\boldsymbol{\theta}$ is some (in the general case, a vector) parameter taking its values from set Θ. Let us denote $\overline{F}(t, \boldsymbol{\theta}) = 1 - F(t, \boldsymbol{\theta})$ a complementary distribution function (in other terms, a reliability function). For all test plans of type U (no renewal) and type R (with renewal), considered above, the likelihood function has the form (see Section 2.4)

$$L(z|\boldsymbol{\theta}) = C \prod_{1 \le i \le d} f(s_i, \boldsymbol{\theta}) \prod_{1 \le j \le v} \overline{F}(u_j, \boldsymbol{\theta}), \tag{4.25}$$

where $\mathbf{z} = (d, s_1, \ldots, s_d, v, u_1, \ldots, u_v)$ is a set of all test results:

- s_i are complete intervals of observations (i.e., those terminated by a failure), $1 \le i \le d$;
- u_j are incomplete intervals of observations (i.e., those terminated before failure has occurred), $1 \le j \le v$;

d is the number of failures (coincides with the number of complete intervals of observations);

ν is the number of units whose test was terminated before failure (coincides) with the number of censored test intervals); and

C is a function that can depend on the test results but does not depend on parameter $\boldsymbol{\theta}$.

Numbers d and υ might be determined in advance or randomly depending on the test plan (see Section 2.4).

Assume that a prior distribution density $h(\boldsymbol{\theta})$ of parameter $\boldsymbol{\theta}$ is given. From (4.1) and (4.25) we can obtain that, for all test plans, the posterior distribution density $h(\boldsymbol{\theta}|\mathbf{z})$ of parameter $\boldsymbol{\theta}$ for a given test result $\mathbf{z}$ has the same form:

$$h(\boldsymbol{\theta}|\mathbf{z}) = \frac{\mathrm{u}h(\boldsymbol{\theta})}{\varphi(\mathbf{z})} \prod_{1 \le i \le d} f(s_i, \boldsymbol{\theta}) \prod_{1 \le j \le \nu} \overline{F}(u_j, \boldsymbol{\theta}), \tag{4.26}$$

where

$$\varphi(\mathbf{z}) = \int_{\Theta} h(\boldsymbol{\theta}) \prod_{1 \le i \le d} f(s_i, \boldsymbol{\theta}) \prod_{1 \le j \le \nu} \overline{F}(u_j, \boldsymbol{\theta})\, d\boldsymbol{\theta}.$$

Thus, for fixed test results $\mathbf{z}$ the posterior distribution density (4.26) and all statistical inferences do not depend on the test plan. So, the Bayes approach is insensitive to test plans.

In conclusion, notice that the Bayes approach possesses some essential advantages: simplicity, universality, and inner logic. However, this method has its weak side, determined by an arbitrary choice of prior distributions.

Since the choice of the prior distribution is completely in the hand of the researcher, there is a possibility of "adapting" the final numbers to almost any "needed" results. This dependence on the subjective viewpoint of the researcher is a serious disadvantage of the method. At the same time, this method is very effective for "aggregation" of statistical data obtained from different sources. The problem of the choice of a prior distribution is discussed in detail by Cox and Hinkley (1978).

The Bayes approach has been widely discussed. Probably, the first intensive wave of works dedicated to the application of the Bayes approach to reliability problems began in the late 1960s and the 1970s: Springer and Thompson (1964, 1966, 1967a–c), Ferguson (1973), Cole (1975), Mastran (1976), Mastran and Singpurwalla (1978), and Smith and Springer (1976). Among recent works are Barlow (1985), Martz and Waller (1982, 1990), and Martz et al. (1988). Some monographs on the theme are Mann et al. (1974), Belyaev (1982), and Savchuk (1989).

PROBLEMS

4.1 Find the Bayes point estimate $\hat{R} = \hat{R}(\mathbf{z})$ for $R = R(\boldsymbol{\theta})$ that minimizes the posterior mathematical expectation of module $|\hat{R} - R|$.

4.2 Let a prior distribution parameter q (failure probability) in a "binomial model" be uniform in interval (0, 1) (see Section 4.3). For this case find the Bayes posterior distribution, point estimate (posterior mean), and confidence limits for parameter q.

4.3 Assume that in a "binomial model" we know in advance that failure probability q satisfies inequality $q \le q^*$, where $q^* \le 1$. A prior distribution of parameter q is chosen uniform in the interval $(0, q^*)$. Find the Bayes point estimate and confidence limits for parameter q.

4.4 A prior distribution parameter q (failure probability) in a binomial model is uniform in the interval (0, 1) and the test results are $d = 1$ (number of failures) and $N = 10$ (number of tested units). Find the Bayes point estimate and confidence limits with confidence coefficient 0.95 for parameter q.

4.5 Solve the previous problem for the same test results if it is known in advance that the failure probability q satisfies the inequality $q \le q^* = \frac{1}{2}$ and a prior distribution of parameter q is chosen uniform in the interval $(0, q^*)$.

4.6 Let a prior distribution of parameter λ (failure rate) in an exponential model be gamma (4.19) with parameters $u = 200$ hr and $a = 2$. The test results are $S = 400$ hr and $d = 1$. Find the posterior distribution density, the Bayes point estimate, and confidence limits with confidence coefficient 0.9 for parameter λ.

CHAPTER 5

ACCELERATED TESTING

5.1 INTRODUCTION

The true level of modern hardware reliability can be estimated only on the basis of tests under specified conditions or on the basis of real data. To estimate the reliability of a large population of items, one needs to test a number of items and to treat the data thus obtained using statistical methods. This seemingly straightforward way of reliability estimation is often difficult in practice due to monetary and time restrictions. Let us assume that an item has the exponential TTF distribution with parameter $l = 10^{-9}\ \text{hr}^{-1}$. For satisfactory statistical estimation, one needs to observe, say, 10 failures. To obtain 10 failures, approximately, 10^{10} item-hours are required. Thus, one unit should be tested for approximately 10 million years (!); or one million items should be tested for 10 years (!). Neither approach makes any practical sense.

As another example, consider an item with an extremely stable performance parameter that determines the item's reliability. (For instance, it can be a quartz timer in a synchronized system.) To estimate a time when the examined parameter deviates from its tolerance, one also needs neither many items tested or enormous testing time.

Such situations lead to the need for development of some special methods for reliability estimation. One of the most important practical methods used in these situations is the so-called accelerated life testing (ALT).

There are two main ways to accelerate tests:

This chapter was prepared by Dr. Mark Kaminsky.

- to put the test items under conditions much more severe than the nominal (operational) ones, which will result in faster failure occurrence; or/and
- to choose much more rigid tolerances and consider crossing these limits as a degradation (conditional) failure.

These approaches are based on the natural assumption that increasing stress (as well as rigid tolerances) results in decreasing the time of failure-free operation. Both approaches need credible techniques for extrapolation from the "accelerated domain" into the "normal domain" and they are based on the hypothesis of similarity of the item behavior under accelerated and normal (use) stress conditions. In mechanics, for instance, there exists a special approach, the so-called similarity theory, which is dedicated to such theoretical constructions and has many practical applications. Unfortunately, a hidden danger of erroneous conclusions always exists if one makes "too brave" an extrapolation.

Indeed, increasing the stress factors (shock, vibration, temperature, humidity) we can completely change the failure mechanisms, so that all of our predictions might turn out to be useless.

For the case of rigid tolerances we meet analogous situations. A parameter of interest can be very stable within a narrow band around the nominal level, and, at the same time, there might be a catastrophic failure mode because of various influences of positive feedback.

In this chapter we will formally consider the first above-mentioned case, though some results might be naturally extended to the second case.

5.2 BASIC NOTIONS AND PROBABILISTIC MODEL

5.2.1 TTF Distributions and Accelerated Life Reliability Model

A *reliability model* in accelerated life testing (*AL reliability model*) is usually defined as a relationship between the time to failure (TTF) distribution of a device and *stress factors,* such as load, cycling rate, temperature, humidity, and voltage. The AL reliability models are based on the considerations of physics of failure. From the mathematical point of view, an AL reliability model can be considered as a deterministic transformation of TTF.

5.2.2 Stress Severity in Terms of TTF Distribution

Let $F_1(t; z_1)$ and $F_2(t; z_2)$ be TTF cumulative distribution functions (CDFs) of the item under the constant stress conditions z_1 and z_2, respectively.[1] The stress condition z_2 is more severe than z_1 if, for all values of t,

[1]Stress condition z in the general case is a vector of the stress factors.

$$F_2(t;\ z_2) > F_1(t;\ z_1). \tag{5.1}$$

This inequality means that a more severe stress condition accelerates the TTF, so that the reliability of the item under stress condition z_2 is less than the reliability under stress condition z_1.

5.2.3 Time Transformation Function for the Case of Constant Stress

For monotonically increasing CDFs $F_1(t;\ z_1)$ and $F_2(t;\ z_2)$, if constant stress condition z_1 is less severe than z_2 and t_1 and t_2 are the times at which $F_1(t_1;\ z_1) = F_2(t_2;\ z_2)$, there exists a function g (for all t_1 and t_2) such that $t_1 = g(t_2)$ and

$$F_2(t_2;\ z_2) = F_1(g(t_2),\ z_1). \tag{5.2}$$

Because $F_1(t;\ z) < F_2(t;\ z)$, $g(t)$ must be an increasing function with $g(0) = 0$ and $\lim_{t\to\infty} g(t) = \infty$. The function $g(t)$ is called the *acceleration* or the *time transformation function.*

As mentioned above, an AL reliability model is a deterministic transformation of TTF. Two main time transformations are considered in life data analysis. These transformations are known as the *accelerated life (AL) model* and the *proportional-hazards (PH) model.*

5.2.4 Accelerated Life Model (Linear Time Transformation Function)

Without loss of generality, one may assume that $z = 0$ for the normal (use) stress condition. Denote a failure time CDF under a normal stress condition by $F_0(\cdot)$. The AL time transformation is given in terms of $F(t;\ z)$ and $F_0(\cdot)$ by the relationship (Cox and Oaks, 1984)

$$F(t;\ z) = F_0[(t\psi(z,\ A)], \tag{5.3}$$

where $y(z,\ A)$ is a positive function connecting TTF with a vector of stress factors z; and A is a vector of unknown parameters; for $z = 0$, $y(z,\ A)$ is assumed equal to 1.

The corresponding relationship for the probability density function is given by

$$f(t;\ z) = f_0[t\psi(z,\ A)]\psi(z,\ A). \tag{5.3'}$$

Relationship (5.3) is a scale transformation. It means that a change in stress does not result in a change in the shape of the distribution function but changes its scale only. Relationship (5.3) can be written in terms of the acceleration function as

$$g(t) = \psi(z, A)t. \tag{5.4}$$

In other words, relationship (5.3) is equivalent to the linear one with a time acceleration function.

The distributions considered are geometrically *similar* to each other. They are called belonging to the class of TTF distribution functions that is closed with respect to scale (Leemis, 1995). The similarity property is extremely useful in physics and engineering. Because it is difficult to imagine that any change of failure modes or mechanism would not result in a change in the shape of the failure time distribution, the relationship (5.3) can also be considered as a principle of failure mechanism conservation or a *similarity* principle. Analysis of some sets of real-life data using the statistical procedures described below (*criteria of linearity of time transformation function*) often shows that the similarity of TTF distributions really exists, so that a violation of the similarity can identify a change in a failure mechanism.

The relationship for the 100*p*th percentile of TTF $t_p(z)$ can be obtained from (5.3) as

$$t_p(z, B) = \frac{t_p^0}{\psi(z, A)}, \tag{5.5}$$

where t_p^0 is the 100*p*th percentile for the normal stress condition $z = 0$.

The relationship (5.4) is the *percentile AL reliability model,* usually written in the form

$$t_p(z, B) = \eta(z, B). \tag{5.6}$$

The AL reliability model is related to the relationship for percentiles, (5.5), as

$$\eta(z, B) = \frac{t_p^0}{\psi(z, A)}. \tag{5.7}$$

The corresponding relationship for failure rate can also be obtained from (5.3) as

$$\lambda(t; z) = \psi(z, A)\lambda^0[t\psi(z, A)]. \tag{5.8}$$

It is easy to see that the relationship for percentiles is the simplest; that is why it is being used in the following sections on AL data analysis.

5.2.5 Cumulative Damage Models and Accelerated Life Model

Some known cumulative damage models can result in the similarity of TTF distributions under quite reasonable restrictions. As an example, consider the

Barlow and Proschan model resulting in an aging TTF distribution. They consider an item subjected to shocks appearing randomly in time. Let these shocks arrive according to the Poisson process with constant intensity l. Each shock causes a random amount x_i of damage, where $x_1, x_2, \ldots$ are independently distributed with a common CDF $F(x)$ (damage distribution function). The item fails when accumulated damage exceeds a threshold X. The TTF CDF is given by

$$H(t) = \sum_{k=0}^{\infty} e^{-\lambda t} F^{(k)}(X) \frac{(\lambda t)^k}{k!},$$

where $F^{(k)}$ is k times convolution of $F(x)$. It was shown by Barlow and Proschan that for *any* damage distribution function $F(x)$ the TTF distribution function has increasing failure rate on the average (IFRA).

Now consider an item under the stress conditions characterized by different stress intensities λ_i and different damage distribution functions $F_i(x)$. It can also be shown that the similarity of the corresponding TTF distribution functions $H_i(t)$ [or the failure mechanism conservation (Equation (5.3)] will hold for all these stress conditions $(\lambda_i, F_i(x))$ if they have the same damage CDF $F_i(x) = F(x)$, $i = 1, 2, \ldots$. A similar example from fracture mechanics is considered elsewhere (Crowder et al., 1991).

5.2.6 Proportional-Hazards Model

For the PH model the basic relationship analogous to (5.3) is given by

$$F(t; z) = 1 - [1 - F_0(t)]^{\psi(z,A)}, \tag{5.9}$$

or, in terms of the reliability function R, as

$$R(t; z) = R_0(t)^{\psi(z,A)}. \tag{5.9$'$}$$

The proper PH model is known as the relationship for hazard rate, which can be obtained from (5.9) as

$$\lambda(t; z) = \psi(z, A)\lambda^0(t), \tag{5.10}$$

where $y(z, A)$ is usually chosen as a loglinear function.

The PH model does not normally retain the shape of the CDF, and the function $\psi(z)$ no longer has a simple relationship to the acceleration function or a clear physical sense. That is why the PH model is not as popular in reliability applications as the AL model.

Nevertheless it can be shown (Cox and Oaks, 1984) that for the Weibull distribution (and only for the Weibull distribution) the PH model coincides with the AL model.

It should be noted that the AL model time transformation is more popular for reliability applications, while the PH model is widely used in biomedical life data analysis.

5.2.7 Some Popular AL (Reliability) Models for Percentiles

Most commonly used models for the percentiles (including median) are log-linear models. Two such models are the power rule model and the Arrhenius reaction model. The power rule model for the 100*p*th percentile is given by

$$t_p(x) = \frac{a}{x^c}, \qquad a > 0,\ c > 0,\ x > 0, \tag{5.11}$$

where x is a mechanical or electrical stress and a and c are constants. In reliability of electrical insulation and capacitors, x is usually applied voltage. In estimating fatigue life, the model is used as the analytical representation of the S–N or *Wöhler curve,* where S is a stress amplitude and N is life in cycles to failure, that is,

$$N = kS^{-b}, \tag{5.11'}$$

where b and k are material parameters estimated from test data. Because of the probabilistic nature of fatigue life at any given stress level, one has to deal with not one S–N curve but a family of S–N curves, so that each curve is related to a probability of failure as a parameter. These curves are called S-P curves, or curves of constant probability of failure on a stress-versus-life plot. It should be noted that relationship (5.11′) is only empirical (Sobczyk and Spencer, 1992).

The Arrhenius reaction rate model is the following relationship between life and absolute temperature T:

$$t_p(T) = a \exp\left(\frac{E_a}{T}\right), \tag{5.12}$$

where E_a is activation energy. This model is the most widely used one to examine the effect of temperature on reliability. Originally, it was introduced as a chemical reaction rate model.

The model combining the above models is given by

$$t_p(x, T) = ax^{-c} \exp\left(\frac{E_a}{T}\right). \tag{5.13}$$

In the fracture mechanics of polymers this model is known as Bruhanova's and Bartenev's model. It is also used as a model for the electromigration failures in aluminum thin films of integrated circuits; in this case x is current density.

Another popular AL reliability model is Jurkov's model:

$$t_p(x, T) = t_0 \exp\left(\frac{E_a - \gamma x}{T}\right). \tag{5.14}$$

This model is considered as an empirical relationship reflecting the thermal fluctuation character of long-term strength (Regel, et al., 1974; Goldman, 1994). For mechanical long-term strength, parameter t_0 is a constant that is numerically close to the period of thermal atomic oscillations (10^{-11}–10^{-13} sec), E_a is the effective activation energy numerically close to vaporization energy for metals and to chemical bond energies for polymers, and γ is a structural coefficient. The model is widely used for reliability problems of mechanical and electrical long-term strength.

The a priori choice of a model (or some competing models) is being made using physical considerations. Meanwhile, statistical data analysis of ALT results and failure mode and effects analysis (FMEA) afford to check the adequacy of the model chosen or to discriminate the best model among the competing ones.

5.2.8 Accelerated Life Model for Time-Dependent Stress

Models considered in the previous section are for stress constant in time. The case of time-dependent stress is not only more general but also of more practical importance because applications in engineering reliability are not limited to ALT problems. As an example, consider the time-dependent stress analog of the model (5.11′). The stress amplitude S experienced by a structural element often varies during its service life, so that the straightforward use of Equation (5.11′) is not possible. In such situations the so-called Palmgren-Miner rule is widely used to estimate fatigue life. The rule treats the fatigue fracture as a result of a *linear accumulation* of partial fatigue damage fractions. According to the rule, the damage fraction Δ_i at any stress level S_i is proportional to the ratio n_i/N_i, where n_i is the number of cycles of operation under stress level S_i and N_i is the total number of cycles to failure (life) under the constant stress level S_i, that is,

$$\Delta_i = \frac{n_i}{N_i}, \qquad n_i \le N_i.$$

Total accumulated damage D under different stress levels S_i ($i = 1, 2, \ldots$) is defined as

$$D = \sum_i \Delta_i = \sum_i \frac{n_i}{N_i}.$$

It is assumed that failure occurs if $D \geq 1$.

Accelerated life tests with time-dependent stress such as step-stress and ramp tests are also of a great importance. For example, one of the most common reliability tests of thin silicon dioxide films in metal-oxide-semiconductor integrated circuits is the so-called ramp-voltage test. In this test, the oxide film is stressed to a breakdown by a voltage that increases linearly with time (Chan, 1990).

Let $z(t)$ be a time-dependent stress vector such that $z(t)$ s integrable. In this case, the relationship (5.3) can be written in the form

$$F\{t;\ [z(\cdot)]\} = F_0[\Psi(t)], \tag{5.15}$$

where

$$\Psi(t^{(z)}) = \int_0^{t(z)} \psi[z(s), A]\, ds$$

and $t^{(z)}$ is the time related to an item under the stress condition $z(t)$.

The relationship (5.15) was given by Cox and Oaks (1984). Based on this relationship, Cox also obtained the analogous relationships for the probability density function and hazard rate function.

The corresponding relationship for the 100pth percentile of TTF $t_p[z(t)]$ can be obtained from (5.15) as

$$t_p^0 = \int_0^{t_p[z(t)]} \psi[z(s), A]\, ds.$$

Using (5.6) and (5.7), the last relationship can be rewritten as

$$1 = \int_0^{t_p[z(t)]} \frac{1}{t_p^0\{\psi[z(s, A]\}^{-1}}\, ds \equiv \int_0^{t_p[z(t)]} \frac{1}{\eta[z(s), B\}}\, ds. \tag{5.16}$$

5.2.9 AL Reliability Model for Time-Dependent Stress and Miner's Rule

It should be noted that relationship (5.16) is an exact nonparametric probabilistic continuous form of the Palmgren–Miner (PM) rule mentioned in the previous section. So, the problem of using AL tests with time-dependent stress is identical to the problem of cumulative damage addressed by the PM rule. Moreover, there exists a useful analogy between mechanical damage accumulation and electrical breakdown; for example, Jurkov's model is used as the relationship for mechanical as well as for long-term electrical strength.

In the theory of cumulative damage, a certain *damage measure* $D(t)$ is introduced $[0 \leq D(t) \leq 1]$.

Assuming that $D(t)$ depends on its value at some initial time t_0 and on an external action $Q(t)$, the following general equation for $D(t)$ is being postulated (Sobczyk and Spencer, 1992):

$$\frac{dD(t)}{dt} = f[D(t), Qt)], \tag{5.17}$$

where $f(D, Q)$ is a nonnegative function that satisfies the conditions ensuring the existence and uniqueness of the solution of Equation (5.17). The equation is regarded as a *kinetic equation* for damage evaluation (Bolotin, 1989).

If the right side of Equation (5.17) is independent of $D(t)$, the solution of the equation with the initial condition $D(0) = 0$ is the *linear damage accumulation* model given by

$$D(t) = \int_0^t f[Q(\tau)]\, d\tau. \tag{5.18}$$

The time T at which the damage reaches its critical value corresponds to the condition $D(t) = 1$. Using the notation $t(q) = 1/f(Q)$, one gets

$$\int_0^T \frac{d\tau}{t[Q(\tau)]} = 1. \tag{5.19}$$

Equation (5.19) formally coincides with Equation (5.16), so it is clear that the AL model is the linear damage accumulation model. Nevertheless Equation (5.19) is deterministic and, from an engineering point of view, it is also not clear how to measure or estimate the *external action* $Q(t)$, what the function $f[Q(t)]$ is, and how to validate the correctness of Equation (5.19).

On the contrary, Equation (5.16), depicting the general case of the time-dependent stress AL model, is expressed in terms of quantiles of TTF and usual (constant stress) AL reliability models. The correctness of the equation can be tested using the statistical procedures considered in Section 5.2.

There could be two main kinds of application of Equation (5.16): fitting an AL reliability model [estimating the vector or parameters, B, of the percentile reliability model, $\eta(z, B)$, on the basis of AL tests with time-dependent stress] and reliability (percentiles of time to failure) estimation (when the model is known) for the given time-dependent stress, in the stress domain, where the similarity of TTF distributions exists (conservation of failure mechanisms holds).

5.3 ACCELERATED LIFE TEST DATA ANALYSIS

5.3.1 Exploratory Data Analysis (Criteria of Linearity of Time Transformation Function for Constant Stress)

The possibility to verify the correctness of relationship (5.3) experimentally not only is important for failure mechanism study but also has a great practical importance, because almost all the statistical procedures for AL test planning and data analysis (for both the constant and time-dependent stress) are based on the assumption (5.3).

Several techniques can be used for verification of the linearity of the time transformation function or verification of the PM rule. Let us start with the historically first criterion that show similarity of TTF distribution functions and the linear damage accumulation model. This criterion requires two special tests (Gugushvili et al., 1975). During the first test a sample is tested at z_1 constant stress level for a time t_1 at which z_1 is changed to a constant stress z_2 for a time t_2. During the second test another sample is first tested under z_2 for t_2 and then it is tested under the stress level z_1 for time t_1. The time transformation function will be linear in time if the reliability functions of the items after the first and the second tests are equal; that is, a change of loading order does not change the cumulative damage. We will come back to this criterion while considering exploratory data analysis for the case of time-dependent stress, because this test is based on the time-dependent loading.

The second criterion is associated with the variation coefficient (i.e., standard deviation to average ratio, s/m) of TTF. It is easy to show that if the time transformation function is linear for the constant stress levels z_1, z_2, . . . , z_k, the variation coefficient of TTF will be the same for all these stress levels. Let x be TTF under stress condition z_i and y be TTF under stress condition z_j. Under the basic AL model assumption (5.3) these random variables are related to each other as $y = kx$, where k is a constant. The corresponding mean values $E(x)$ and $E(y)$ are related to each other as $E(y) = kE(x)$. The analogous relationship for the standard deviations $s(x)$ and $s(y)$ is, obviously, $s(y) = ks(x)$, so that the variation coefficients $s(y)/E(y)$ and $s(x)/E(x)$ are equal.

Thus, the analysis of the variation coefficient not only provides information about aging of the TTF distribution but also helps one to understand if the conservation of failure mechanisms holds true. (Recall that $s/m < 1$ for IFR and IFRA TTF distributions, $s/m = 1$ for the exponential distribution, and $s/m > 1$ for DFR and DFRA distributions). Analogously, it can be shown that under the same assumption the variance of the logarithm of TTFs will be the same for these stress levels. Consider the TTFs x (under stress condition z_i) and y (under stress condition z_j). Under AL model assumption (5.3), the logarithms of these random variables are related to each other as

$$\log y = \log x + \log k,$$

where k is the same constant. Taking variances of both sides of the equation above, one gets

$$\mathrm{Var}(\log y) = \mathrm{Var}(\log x).$$

If TTF is lognormal, Bartlett's and Cochran's tests can be used for checking if the variances are constant.

The third criterion is based on the use of quantile-quantile plots. The quantile-quantile plot is a curve such that the coordinates of every point are the TTF quantiles (percentiles) for investigated pairs of stress conditions. If the time transformation function is linear in time [i.e., (5.3) holds], the quantile-quantile plot will be a straight line going through the origin. The corresponding data analysis is realized in the following way. All sample quantiles of a given constant stress condition are plotted on one axis and the sample quantiles of another stress condition are plotted on the other axis. Using the points obtained (a pair of quantiles of the same level gives a point), the straight regression line can be fitted. The time transformation function will be considered as linear, provided one gets linear dependence between the sample quantilies and the hypothesis that the intercept of the fitted line is equal to zero is not rejected.

Example 5.1 (*Fatique Life Data*) Consider the Birnbaum-Saunders et al. data (Bogdanoff and Kozin, 1985, p. 53):

> The test specimens were 6061-T6 aluminum strips, 0.061 in. thick, 4.5 in. long, and 0.5 in. wide. The specimens were cut parallel to the direction of rolling of the sheet stock. The specimens were mounted in simple supported bearings and deflected at the center with a Teflon clamp in reverse bending. The center was deflected 18 times per second and three stress amplitudes were used; these amplitudes were 21, 26, and 31 Kpsi. There were 101 specimens at 21 Kpsi, 102 specimens at 26 Kpsi, and 101 specimens at 31 Kpsi. Specimens were tested to failure. Life in 10^3 units were recorded.

The sample means m, the sample variations s^2, and the sample variation coefficients s/m are summarized in Table 5.1.

Note that all the sample variation coefficients are less than 1, so the TTF distributions for all the stress levels might be considered as aging (IFR or IFRA). The variation coefficients for the amplitude 26 and 311 Kpsi are approximately equal, so these stress conditions might be considered as leading to the same failure mechanisms. This conclusion can be supported by quantile-quantile plot analysis: All the quantile-quantile plots show strong linear dependence (all the sample correlation coefficients are about 0.99), but zero (insignificant) intercept gives the quantile-quantile plot for the amplitudes 26 and 31 Kpsi only. Thus the TTF distribution functions for these amplitudes

Table 5.1 Parameters for Example 5.1

Stress Level				
Kpsi	MPa	$m/10^3$, cycles	$s^2/10^6$, cycles^2	$(S^2)^{1/2}/m$
± 21	144.79	1,400	152,881	0.28
± 26	179.26	396	3,881	0.16
± 31	213.74	134	502	0.17

are similar (AL model is applicable) and the failure mechanism conservation takes place in the amplitude range 26–31 Kpsi. □

5.3.2 Statistical Methods of Reliability Prediction on the Basis of AL Tests with Constant Stress

Statistical methods of reliability prediction on the basis of AL tests can be divided into parametric and nonparametric ones. In the first case TTF distribution is related to a particular parametric distribution—Normal, exponential, Weibull; in the second case the only assumption is the assumption about a particular class of TTF distributions—continuous, IFR, IFRA.

The most commonly used parametric methods are parametric regression (normal and lognormal, exponential, Weibull, and extreme value) and the least-squares method and maximum-likelihood methods (Lawless, 1982; Nelson, 1990; Leemis, 1995), which are briefly discussed below. We also consider the nonparametric regression procedure for the percentile AL reliability model fitting for the cases of constant stress and time-dependent stress (Kaminskiy, 1994).

Maximum-Likelihood Approach Consider the maximum-likelihood approach to statistical analysis of AL models based on the Weibull TTF distribution as a typical example.

The reliability function for the Weibull distribution is given by

$$R(t) = \exp\left[-\left(\frac{t}{\alpha}\right)^{\beta}\right], \tag{5.20}$$

where α is a scale parameter and β is a shape parameter.

Under the AL model assumption, the scale parameter is considered as a function of stress factors, z. Consider the case when this function (reliability model) is loglinear, that is,

$$\log \alpha(Z, B) = ZB, \tag{5.21}$$

where $Z = (z_0, z_1, \ldots, z_p)$ is a vector of stress factors and $B = (b_0, b_1, \ldots, b_p)^T$ is a vector of model parameters, $z_0 \equiv 1$.

For the following discussion it is better to deal with the logarithm of TTF. Denote $y = \log t$. It is easy to show that y has the type I (Gumbel) extreme value distribution for the minimum. The probability density function (PDF) of the distribution is given by

$$f(y) = \frac{1}{\sigma} \exp\left[\frac{y - \log \alpha}{1/\beta} - \exp\left(\frac{y - \log \alpha}{1/\beta}\right)\right], \qquad -\infty < y < \infty. \tag{5.22}$$

Using the AL model in the form of Equation (5.3′), the stress-dependent PDF can be written as

$$f(y - z) = \frac{1}{\sigma} \exp\left[\frac{y - ZB}{\sigma} - \exp\left(\frac{y - ZB}{\sigma}\right)\right], \qquad \sigma = \frac{1}{\beta}. \tag{5.23}$$

Let y_i be either a logarithm of TTF or a logarithm of censoring time associated with a stress condition Z_i. Denote the sets of observations for which y_i is a logarithm of TTF and a logarithm of right censoring time by U and C, respectively. The likelihood function for the given observations and PDF (5.23) is (Lawless, 1982)

$$\begin{aligned} L(B, \sigma) = {} & \prod_{i \in U} \frac{1}{\sigma} \exp\left[\frac{y_i - Z_i B}{\sigma}\right. \\ & \left. - \exp\left(\frac{y_i - Z_i B}{\sigma}\right)\right] \prod_{i \in C} \exp\left[-\exp\left(\frac{y_i - Z_i B}{\sigma}\right)\right], \end{aligned} \tag{5.24}$$

$$\begin{aligned} \log L(B, \sigma) = {} & -r \log \sigma + \sum_{i \in U} \frac{y_i - Z_i B}{\sigma} \\ & - \sum_{i=1}^{n} \exp\left(\frac{y_i - Z_i B}{\sigma}\right). \end{aligned}$$

The maximum likelihood equations are

$$\begin{aligned} \frac{\partial \log L}{\partial b_l} &= -\frac{1}{\sigma} \sum_{i \in U} Z_{il} + \frac{1}{\sigma} \sum_{i=1}^{n} z_{il} e^{x_i}, \qquad l = 0, 1, \ldots, p, \\ \frac{\partial \log L}{\partial \sigma} &= -\frac{r}{\sigma} - \frac{1}{\sigma} \sum_{i \in U} x_i + \frac{1}{\sigma} \sum_{i=1}^{n} x_i e^{x_i}, \qquad x_i = \frac{1}{\sigma}(y_i - z_i, B). \end{aligned} \tag{5.25}$$

The likelihood function can also be used for discriminating between two competing models (Crowder et al., 1991). Let us have a model M_1 with p_1 parameters and a model M_2 with p_2 parameters, and let $p_2 > p_1$. Let L_1 and L_2 be the maximized values of log L for models M_1 and M_2, respectively. The *likelihood ratio* statistic $W = 2\ (\log L_1 - \log L_2)$ has an approximate χ^2

distribution with $p_2 - p_1$ degrees of freedom. Large values of W provide evidence against the null hypothesis that both models supply the same goodness of fit.

To get the observed variance-covariance matrix (or the observed information matrix) after solving the system (5.25), one needs the second derivatives of the log-likelihood function considered, which are given by

$$\frac{\partial^2 \log L}{\partial b_1\, \partial b_1} = \frac{1}{\sigma^2} \sum_{i\in U} z_{il} z_{is} e^{x_i}, \qquad l, s = 0, 1, \ldots, p,$$

$$\frac{\partial^2 \log L}{\partial \sigma^2} = \frac{r}{\sigma^2} - \frac{2}{\sigma^2} \sum_{i\in U} x_i - \frac{2}{\sigma^2} \sum_{i=1}^{n} x_i e^{x_i} - \frac{1}{\sigma^2} \sum_{i=1}^{n} x_i^2 e^{x_i}, \tag{5.26}$$

$$\frac{\partial^2 \log L}{\partial B_1\, \partial \sigma} = \frac{1}{\sigma^2} \sum_{i\in U} z_{il} \frac{1}{\sigma^2} \sum_{i=1}^{n} z_{il} e^{x_i} - \frac{1}{\sigma^2} \sum_{i=1}^{n} z_{il} x_i e^{x_i}.$$

The system (5.25) can only be solved numerically. The Newton-Raphson method is usually recommended. It is not as simple a problem as it is sometimes stated (Lawless, 1982). Moreover, using different software realizations of the same optimization method, one usually gets different solutions. In the following section the least-squares method as a shortcut procedure for AL data analysis is discussed. This method is not as effective as the maximum-likelihood one, but it is robust in the sense that it is not associated with numerical optimization, so using different software realizations, one always gets the same results. Moreover, even if a good software tool for the maximum-likelihood method is available, it is very important to have good values; the least-squares estimates can be used as such starting values.

Least-Squares Estimation The relationship (5.5) from Section 5.1.4 can be written in terms of random variables as

$$T = \frac{T_0}{\psi(z)},$$

where T_0 has TTF CDF $F_0(\cdot)$. Denote the expectation of $\log T_0$ by μ_0. Using the equation above, one can write

$$\log T = \mu_0 - \log \psi(z) + \varepsilon, \tag{5.27}$$

where ε is a random error of zero mean with a distribution not depending on z.

If $\log \sigma(z)$ is, again, a linear function (the case of the loglinear reliability model), that is,

$$\log \psi(Z, B) = ZB, \tag{5.28}$$

Equation (5.27) can be written as

$$\log T = \mu_0 - ZB + \varepsilon, \tag{5.29}$$

which is a linear model satisfying the conditions of the Gauss-Markov theorem.

When TTF samples are uncensored, the regression equation for observations T_i, Z_i $(i = 1, 2, \ldots, n)$ is

$$\log T_i = \mu_0 - Z_i B + \varepsilon_i, \tag{5.30}$$

where for any TTF distribution the e_i $(i = 1, 2, \ldots, n)$ are independent and identically distributed with an unknown variance and known distribution (if the distribution of TTF is known). Thus, on the one hand, the least-squares technique for AL data analysis can be used as a nonparametric one and, on the other hand, if TTF distribution is known, one can try a parametric approach. The lognormal TTF distribution is an example of the last case, which is reduced to standard normal regression, which makes clear the popularity of the lognormal distribution in AL practice.

Example 5.2 (Class H Insulation Data) The data are hours to failure of 40 units divided in four samples of equal size (10) (Nelson, 1990). The samples were tested at 190, 220, 240, and 260°C. The test purpose was to estimate the median life at the design temperature 180°C. The test results are given in Table 5.2.

In Table 5.3 some statistics useful for the exploratory data analysis are summarized. It should be noted that the data considered are not distinct TTFs but grouped (inspection) data; nevertheless, the test results under 260°C look

Table 5.2 Class H Insulation Test Results: Hours to Failure

190°C	220°C	240°C	260°C
7228	1764	1175	600
7228	2436	1175	744
7228	2436	1521	744
8448	2436	1569	744
9167	2436	1617	912
9167	2436	1665	1128
9167	3108	1665	1320
9167	3108	1713	1464
10511	3108	1761	1608
10511	3108	1953	1896

Table 5.3 Some Basic Statistics for Class H Insulation Data

T (°C)	Sample Mean	Sample Standard Deviation	Sample Variance of log (TTF)
190	8782	1244	0.021
220	2638	143	0.033
240	1581	244	0.027
260	1116	439	0.154

suspicious. If the data were treated as having the lognormal distribution, Bartlett's test would immediately show that this stress level (260°C) results in significantly greater variance compared with other stress levels. Analysis of failure modes for this insulation show that failure modes at 260°C are different from failure modes at lower temperatures (Nelson, 1990).

The Arrhenius model for the logarithms of TTF can be easily fitted using any software tool having a linear least-squares or regression procedure. Using the four temperature levels, the fitting results in the following estimates of the model parameters: $a = 6.90 \times 10^{-4}$ hr and activation energy $E_a = 7531$ K >> 0.65 eV, which give the prediction for median life at the design temperature 180°C, $t_{50\%}$ (180°C) >> 11,500 hr. The proportion of the variance of the logarithm of TTF explained by the model (adjusted R^2) is 91%. Deleting the sample obtained under 260°C (which is quite reasonable) results in the following estimates of the model parameters: $a = 1.50 \times 10^{-4}$ hr and activation energy $E_a = 8260$ K >> 0.71 eV, which give a little more optimistic prediction for the median life at the design temperature 180°C, $t_{50\%}$ (180°C) >> 12500 hr, and better proportion of the variance explained by the model (adjusted R^2) is 95%. □

Simple Percentile Regression The procedure is based on the following formal assumptions:

1. Accelerated life time transformation (5.3) is true and the AL reliability model $\eta(\mathbf{z}, \mathbf{B})$ for a quantile t_p (100pth percentile) is a given function of the stress factors $\mathbf{z}$ with an unknown vector of parameters $\mathbf{B}$ [Equation (5.6)].
2. The TTF distributions for all stress conditions z_i ($i = 1, \ldots, k$) are IFRA distributions having continuous density functions $f(t; z_i)$.
3. The test results are type II right-censored samples, with number of uncensored failure times r_i ($i = 1, \ldots, k$) and sample sizes n_i large enough to estimate the t_p as the sample percentile $\hat{t}_p$:

$$\hat{t}_p = \begin{cases} t_{([n_i p])} & \text{if } n_i p \text{ is not integer,} \\ \text{any value from interval } [t_{(n_i p)}, t_{(n_i p+1)}] & \text{if } n_i p \text{ is integer,} \end{cases}$$

where $t_{(x)}$ is the failure time (order statistic) and $[x]$ means *the greatest integer that does not exceed x.*

4. The sample sizes are large enough that the asymptotic normal distribution of the above estimate can be used. This normal distribution has mean equal to t_p and variance equal to $p(1 - p)/nf^2(t_p)$, where n is the sample size.

The goal is to estimate vector **B** of the parameters of model (5.6) and to predict the percentile at the normal (or any given) stress condition on the basis of AL tests at different constant stress conditions $z_1, \ldots, z_k$, where k is greater than the dimension of the vector **B**, that is, ($k >$ dim **B**).

Based on the preceding assumptions, the statistical model corresponding to Equation (5.6) can be written in a typical regression form as

$$\hat{t}_p(z_i, \mathbf{B}) = \eta(z_i, \mathbf{B}) + \varepsilon_i, \tag{5.31}$$

where ε_i are normally distributed with mean zero and variance $p(1 - p)/n_i f^2(t_p)$, that is, using standard notation $N(0, p(1 - p)/n_i f^2(t_p))$. Note that the distribution of error ε is dependent [through TTF probability density function $f(t_p)$] on the particular TTF distribution; that is, the model is not distribution free.

Rewrite the model (5.31) in the multiplicative form

$$\hat{t}_p(z_i, \mathbf{B}) = \eta(z_i, \mathbf{B}) \cdot \varepsilon, \tag{5.32}$$

where ε has distribution $N(1, p(1 - p)/n_i \eta^2(z_i, \mathbf{B}) f^2[\eta(z_i, \mathbf{B})])$.

Now we try to transform the multiplicative model (5.32) to the model with normally distributed additive error, that is, to the standard normal regression.

Taking the logarithm,[2] model (5.21) can be written as

$$\log \hat{t}_p(z_i, \mathbf{B}) = \log \eta(z_i, \mathbf{B}) + \log(1 + \varepsilon_1), \tag{5.33}$$

where ε_1 has distribution

$$N\left(0, \frac{p(1 - p)}{n_i \eta^2(z_i, \mathbf{B}) f^2[\eta(z_i, \mathbf{B})]}\right). \tag{5.34}$$

Using AL model relationships for the probability density function (5.3′) and for percentiles (5.5), it is easy to show that $\eta(\mathbf{z}, \mathbf{B}) f[\eta(\mathbf{z}, \mathbf{B})]$ is a constant,

[2]Percentile reliability models $\eta(z, B)$ usually are not linear but loglinear.

so, if the sample sizes n_i $(i = 1, 2, \ldots, k)$ are equal, the variances in (5.34) are also equal.

To avoid the distribution dependence on the variance in (5.34), let us find the nonparametric nonrandom lower bound for $f(t_p)$. The bound is given by the following:

Corollary 5.1 If $F(t)$ is an IFRA cumulative distribution function with probability density $f(t)$ and $F(t_p) = p$, then

$$f(t_p) \geq -\frac{(1-p)\log(1-p)}{t_p}. \quad \square$$

Proof Consider the function

$$G(t) = 1 - \exp(\alpha t),$$

where

$$\alpha = \frac{-\log(1-p)}{t_p}.$$

According to Theorem 6.1, p. 110, in Barlow and Proschan (1975), the function $F(t) - G(t)$ does not have more than one change of sign. If this change occurs, the plus will be changed to a minus. Since $F(t)$ and $G(t)$ have the same percentile t_p, this change of sign can occur at the point t_p only. The result follows. Now demand the satisfying of the inequality $\varepsilon_1 << 1$. The inequality will be satisfied if the standard deviation of ε_1 is much less than unity, that is, for each n_i $(i = 1, 2, \ldots, k)$,

$$\left(\frac{p}{n_i(1-p)\log^2(1-p)}\right)^{1/2} << 1. \tag{5.35}$$

In this case model (5.32) can be written as

$$\log \hat{t}_p(z, B) = \log \eta(z, B) + \varepsilon_2 \tag{5.36}$$

where ε_2 has distribution $N(0, \sigma^2)$ and σ^2 is an unknown constant, if the sample sizes n_i $(i = 1, 2, \ldots, k)$ are equal; otherwise the weights proportional to the values of n_i should be used.

Thus the multiplicative model (5.32) can be transformed to the model with additive normally distributed error (i.e., to the standard normal regression model) if inequality (5.35) is satisfied. This inequality is the restriction superimposed on the test sample size. As far as the procedure considered is already based on the asymptotic properties of sample percentile distribution, this restriction should be easily satisfied. So, the problem of the reliability prediction is reduced to the estimation of the parameters of a normal regression followed by point and interval predicting, for example, ranging the stress factors according to their influence on reliability using stepwise regression methods. Using this approach, the standard regression experiment design can be applied to AL test planning. □

Example 5.3 (*Capacitor Breakdown Data*) Certain capacitors were tested by using voltage in conjunction with temperature as accelerating stress factors. One hundred capacitors were tested at each voltage-temperature combination and each test was terminated after not less than 11 failures had been observed. The purpose of the test was to predict the 10th percentile of TTF distribution under the stress condition 63 V and 100°C. The test plan and results are given in Table 5.4.

Model (5.14) was used as the relationship between the 10th percentile and the stress factors (voltage and temperature). The estimation of the model parameters resulted in the following relationship

$$t_p(V, T) = a \exp\left(\frac{E_a - BV}{T}\right),$$

where $a = 7.735 \times 10^{-15}\ \text{hr}^{-1}$, $E_a = 16{,}099$ K (or 1.39 eV), $B = 4.6256$ K /V, V is voltage in volts, and T is temperature in kelvin.

The results of prediction obtained on the basis of this model are given in the right column of Table 5.4. The multiple correlation coefficient for this model is greater than 0.99. To estimate the influence of the voltage on the capacitor's life, the Arrhenius model was fitted for the same data set. The correlation coefficient for the fitted Arrhenius model turned out to be 0.633 only. So, it is easy to conclude that both temperature and voltage are significant. The same conclusion could be drawn using the F ratio for residual variances for these models. This example is a simple illustration of stepwise regression. □

Table 5.4 Capacitor Test Plan and Results

Temperature (°C)	Voltage (V)	Tenth Percentile, Estimate Based on Test Results (hr)	Tenth Percentile, Predicted by Model (hr)
125	150	512.2	500
85	300	5,643	5,435.8
100	300	991.4	1,041.3
100	63	19,371.7	19,257
90	63	—	63,244

5.3.3 Exploratory Data Analysis for Time-Dependent Stress

The first criterion considered in Section 5.2.1 is the criterion for the particular time-dependent stress. In the general case, the value of the integral in Equation (5.16) does not change when a stress history $z(s)$, $t_p \geq s \geq 0$, has been changed for $z(t_p - s)$, $t_p \geq s \geq 0$; it means that time is reversible under the AL model. Based on this property, it is not very difficult to realize the verification of the AL model. For example, each sample that is going to be tested under time-dependent stress can be divided in two equal parts, so that the first subsample could be tested under forward stress history, meanwhile the second subsample is tested under backward stress history.

5.3.4 Statistical Estimation of AL Reliability Models on the Basis of AL Tests with Time-Dependent Stress

Using Equation (5.16), the time-dependent analog of the model (5.32) can be written as

$$t_p^0 = \int_0^{\hat{t}_p[z(t)]} \psi[z(s), \mathbf{A}]\, ds, \tag{5.37}$$

where $\hat{t}_p[z(t)]$ is the sample percentile for an item under the stress condition (loading history) $z(t)$.

The problem of estimating the vector $\mathbf{A}$ and t_p^0 in this case cannot be reduced to parameter estimation for a standard regression model as in the previous case of constant stress.

Consider k different time-dependent stress conditions (loading histories) $z_i(t)$, $i = 1, 2, \ldots, k$, $[k > (\dim \mathbf{A}) + 1]$, under which the test results are (as in the previous case) type II censored samples and the number of uncensored failure times and the sample sizes are large enough to estimate the t_p as the sample percentile $\hat{t}_p$. In this situation the parameter estimates for the AL reliability model (of the vector $\mathbf{A}$ and t_p^0) can be obtained using a least-squares method solution of the following system of integral equations:

$$t_p^0 = \int_0^{\hat{t}_p[z_i(t)]} \psi[z_i(s), \mathbf{A}]\, ds, \qquad i = 1, 2, \ldots, k. \tag{5.38}$$

Example 5.4 Assume a model (5.13) for the 10th percentile of TTF $t_{0.1}$ of a ceramic capacitor in the form

$$t_{0.1}(U, T) = aU^{-c} \exp\left(\frac{E_a}{T}\right), \tag{5.39}$$

where U is applied voltage and T is absolute temperature.

Table 5.5 Ceramic Capacitors Test Results

Temperature (K)	Voltage U_0 (V)	TTF Percentile Estimate (hr)
398	100	347.9
358	150	1688.5
373	100	989.6
373	63	1078.6

Consider a time–step stress AL test plan using step-stress voltage in conjunction with constant temperature as accelerating stress factors. A test sample starts at a specified low voltage U_0 and is tested for a specified time Δt. Then the voltage is increased by ΔU, and the sample is tested at $U_0 + \Delta U$ during Δt, that is,

$$U(t) = U_0 + \Delta U \text{ En}\left(\frac{t}{\Delta t}\right),$$

where $\mathrm{E_n}(x)$ means "nearest integer not greater than x." The test will be terminated after the portion $p \geq 0.1$ of items fails. So, the test results are the sample percentiles at each voltage-temperature combination. The test plan and simulated results with $\Delta U = 10$ V, $\Delta t = 24$ hr are given in Table 5.5.

For the example considered the system of integral equations (5.38) takes the form

$$a = \int_0^{t_{0.1i}} \exp\left(-\frac{E_a}{T_i}\right) [U(s_i)]^c \, ds, \qquad i = 1, 2, 3, 4.$$

Solving this system for the data above yields the following estimates for the model (5.39):

$a = 2.23 \times 10^{-8}$ [hr $\mathrm{V}^{1.88}$], $E_a = 1.32 \times 10^4$ K, $c = 1.88$, which are not bad compared with the following values of the parameters used to simulate the data: $a = 2.43 \times 10^{-8}$ [hr $\mathrm{V}^{1.87}$], $E_a = 1.32 \times 10^4$ K, $c = 1.87$. □

PART II

SYSTEMS RELIABILITY ESTIMATION

CHAPTER 6

TESTING WITH NO FAILURES

6.1 INTRODUCTION

One of the main problems in reliability engineering is estimating the system's reliability in the early phases of design. At this stage a designer only knows the results of reliability tests on separate components of the system. This problem is crucial when dealing with a continuously developing system such as telecommunication or power networks. Its parts (subsystems, interfaces, terminals, etc.) are operating within an existing system and reliability data are collected. New items can be specially tested in advance. Thus different statistical data concerning units are available. The problem is to figure out how the system reliability will change if the system changes its configuration or incorporates some new units.

This chapter investigates highly reliable systems or equipment. In this case, one knows the total testing time for each system part and that there were no failures for any of the parts being considered. A significant factor in this case is the possibility of obtaining best lower confidence limits for a wide class of complex structures.

6.2 SERIES SYSTEM

One of the most commonly used and theoretically well investigated test plans is the *binomial test plan*. This plan is described in the following way. Assume that a system has m different units and p_i is the probability of failure-free operation (PFFO) of unit i during the specified time period t. A system reliability index R can be defined as a function $R = R(\mathbf{p})$, which depends on unit reliability values p_i, where $\mathbf{p} = (p_1, \ldots, p_m)$. Values of p_i's are a priori unknown, but we know the results of tests: N_i units of type i were tested and d_i failures were observed. We are interested in the estimation of an unknown index R by test results $\mathbf{d} = (d_1, \ldots, d_m)$.

A value $\varphi = \varphi(\mathbf{d})$, which depends on test results, is called a point unbiased estimate of $R = R(\mathbf{p})$ if, for all possible $\mathbf{p}$,

$$E_p\{\varphi\} = R(\mathbf{p}),$$

where $E_p\{\varphi\}$ is the mean of the estimate φ for a given $\mathbf{p}$. Usually, one uses the variance of an unbiased estimate of φ,

$$D_p\{\varphi\} = E_p\{[\varphi - R(\mathbf{p})]^2\},$$

as a measure of its effectiveness. The variance characterizes a deviation of the estimate around a real value of R (which is unknown).

One often uses confidence intervals also. Remember that an interval $[\underline{\varphi}(\mathbf{d}), \overline{\varphi}(\mathbf{d})]$ is called the confidence interval with the confidence level γ for $R(\mathbf{p})$ if, for all $\mathbf{p}$,

$$P_p\{\underline{\varphi}(\mathbf{d}) \le R(\mathbf{p}) \le \overline{\varphi}(\mathbf{d})\} \ge \gamma.$$

Sometimes the main interest is only in the one-sided limit of the confidence interval that guarantees the value of the reliability index. So, for many systems, it is important to be sure that the PFFO is not lower than some specified level. Another example is the *coefficient of unavailability*, which is expected to be not larger than some given level. Value $\underline{\varphi}(\mathbf{d})$ is called a one-sided lower γ-confidence limit for $R(\mathbf{p})$ if, for all $\mathbf{p}$,

$$P_p\{\underline{\varphi}(\mathbf{d}) \le R(\mathbf{p})\} \ge \gamma.$$

In an analogous way, value $\overline{\varphi}(\mathbf{d})$ is called a one-sided upper γ-confidence limit for $R(\mathbf{p})$ if, for all $\mathbf{p}$,

$$P_p\{R(\mathbf{p}) \le \overline{\varphi}(\mathbf{d}) \ge \gamma.$$

As a sample, let us consider one of the simplest system structures—a series connection of units (see Figure 6.1). Such a system fails if any of its units have failed, that is, if a system consists of independent units

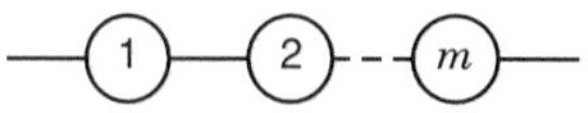

Figure 6.1 System with series structure.

$$R(\mathbf{p}) = \prod_{1 \le i \le m} p_i.$$

The most effective unbiased estimate for the PFFO of unit i, p_i, is a value

$$\hat{p}_i = 1 - \frac{d_i}{N_i}.$$

An unbiased estimate with the minimal variance for the system PFFO is defined as

$$\hat{P} = \prod_{1 \le i \le m} \left(1 - \frac{d_i}{N_i}\right).$$

In practice, a system with highly reliable units is typical, and consequently, the numbers of failures d_i are small. This leads to large values of the variance of the point estimate, which means that the resulting point estimate is very unstable: Values may significantly change from test to test. Notice that a value of a point estimate has no information about a confidence of the obtained result. For instance, the same unbiased estimate will be obtained for two different cases: (a) 1 failure is observed per 10 tested units and (b) 10 failures are observed per 100 tested units. However, even simple common sense tells us that the second case delivers more confidence in the results.

This leads to the necessity of characterization by confidence limits in addition to unbiased estimates. But, if constructing unbiased estimates is a standard task, then the construction of confidence limits is more sophisticated.

In this chapter we are dealing with a no-failure test case: $\mathbf{d} = 0$, that is, each $d_i = 0$. At the same time, the number of tested units of different types, where N_i, are different.

Probably, the first strong solution to this problem was obtained by Mirnyi and Solovyev (1964). They constructed the lower confidence limit for a series system (see Figure 6.1) for no-failure tests. In this case, the lower confidence limit $R*$ with confidence level γ for the system's PFFO is defined as

$$\underline{R} = \min_i \underline{p}_i, \tag{6.1}$$

where

$$\underline{p}_i = (1 - \gamma)^{1/N_i}$$

is the standard lower γ-confident Clopper–Pearson limit for a single unit i (see the beginning paragraphs of Section 6.7). It is clear that this produces

$$\underline{R} = (1 - \gamma)^{1/N^*},$$

where

$$N^* = \min_i N_i$$

is the minimal number of tested units.

At a first glance (6.1) seems paradoxical. For an explanation of this result on an intuitive level, imagine the following situation. We are testing N_i units each of m types, $1 \le i \le m$, and no failures have occurred. Now, we assemble series systems so that each system consists of units of different types. We are able to assemble only N^* complete series systems; that is, the number of such completed systems, consisting of required units, cannot exceed the minimum number of units. Imagine that we tested these units assembled in the system instead of as separate units. Then none of the systems would have failed during tests. (We suppose that assembling the units into a system does not worsen unit reliability.) All of the remaining units, which cannot be assembled into a complete system, give us no additional information about possible system behavior. Thus, the above described no-failure test of $\mathbf{N} = (N_1, N_2, \ldots, N_m)$ units is equivalent to the no-failure test of N^* series systems. These arguments explain the Mirnyi–Solovyev result.

Now consider the construction of a lower confidence limit of the PFFO for more complex systems when no unit failures are observed.

6.3 GENERAL EXPRESSION FOR BEST LOWER CONFIDENCE LIMIT

6.3.1 Systems with Monotone Structure

The state of unit i of a system at moment t, $t > 0$, can be described with the help of a Boolean variable $X_i(t)$ such that $X_i(t) = 0$ if the unit has failed and $X_i(t) = 1$ if the unit is operational. A system state can be described with a vector $\mathbf{X} = (X_1, \ldots, X_m)$, defined in space Ω consisting of 2^m discrete points. For independent units, the probability that the system is in state $\mathbf{X}$ is defined as

$$P\{\mathbf{X}\} = \prod_{1 \le i \le m} p_i^{x_i}(1 - p_i)^{1-X_i},$$

where $p_i = P\{X_i = 1\} = E\{X_i\}$ is the ith unit's PFFO and $E\{X_i\}$ is the mean of X_i. Introduce now the structural function of a system $\Psi(\mathbf{X})$ such the $\Psi(\mathbf{X}) = 1$ if a system is operational and $\Psi(\mathbf{X}) = 0$ otherwise. Of course, such a description of a system is possible only if the system's failure criterion is strictly defined. In this case, the space Ω can be divided into two disjoint subspaces G and $\overline{G}$ such that

$$\Psi(\mathbf{X}) = \begin{cases} 1 & \text{if } \mathbf{X} \in G, \\ 0 & \text{if } \mathbf{X} \in \overline{G}. \end{cases}$$

Such a description of complex system reliability is used by many researchers. For example, see Barlow, Proschan (1975, 1981); Gnedenko, Ushakov (1995); Ushakov, ed. (1994). In Barlow, Proschan (1975, 1981) a system structure is called *monotone* if it satisfies the following conditions:

1.
$$\Psi(\mathbf{X}) \geq \Psi(\mathbf{Y}) \tag{6.2}$$

if $X_i \geq Y_i$ for all $1 \leq i \leq m$, that is, a unit failure cannot lead to a system state improvement, and

2.
$$\Psi(\mathbf{1}) = 1, \quad \Psi(\mathbf{0}) = 0$$

where $\mathbf{1} = (1, \ldots, 1)$ and $\mathbf{0} = (0, \ldots, 0)$. In other words, if all of the system units are operational, the system itself is operational; if all of the system units have failed, the system itself has failed.

The system's PFFO is determined by the vector $\mathbf{p} = (p_1, \ldots, p_m)$ and can be expressed with the help of the structural function as

$$R(\mathbf{p}) = \Pr\{\mathbf{X} \in G\} = E\Psi(\mathbf{X}) = \sum_{\mathbf{X} \in \Omega} \Psi(\mathbf{X}) \prod_{1 \leq i \leq m} p_i^{X_i}(1 - p_i)^{1-X_i}. \tag{6.3}$$

It is clear that condition (6.2) leads to the monotonicity of the system's PFFO $R(\mathbf{p})$ by each p_i. This fact can be shown in the following way. Fix all parameters except p_1. Expression (6.3) can be presented in the form

$$R(\mathbf{p}) = p_1[C_1(p_2, \ldots, p_m) - C_0(p_2, \ldots, p_m)] + C_0(p_2, \ldots, p_m),$$

where

$$C_0(p_2, \ldots, p_m) = \sum_{X_2, \ldots, X_m} \Psi(0, x_2, \ldots, x_m) \prod_{2 \leq i \leq m} p_i^{X_i}(1 - p_i)^{1-X_i}$$

is the reliability function under the condition that the first unit is absolutely unreliable and

$$C_1(p_2, \ldots, p_m) = \sum_{X_2, \ldots, X_m} \Psi(1, x_2, \ldots, x_m) \prod_{2 \leq i \leq m} p_i^{X_i}(1 - p_i)^{1-X_i}$$

is the PFFO under the condition that unit i is absolutely reliable. By condition (6.2), $C_1 \geq C_0$ and, consequently, $R(\mathbf{p})$ monotonically increases when p_i increases.

Examples of a system with a monotone structure are considered below.

6.3.2 Best Lower Confidence Limit for No-Failure Test

Assume again that N_i units of type i, $i = 1, 2, \ldots, m$, where tested and no failures were observed for any units, that is, $d_i = 0$, $1 \le i \le m$. The lower γ-confidence limit $(\underline{R})$ for the system's PFFO can be found as the solution of the following optimization problem (see details in Section 2.2):

$$\min_{\mathbf{p} \in H_0} R(\mathbf{p}) = \underline{R}, \tag{6.4}$$

where the minimum is taken over the set H_0 of points $\mathbf{p} = (p_1, \ldots, p_m)$:

$$\prod_{1 \le i \le m} p_i^{N_i} \ge 1 - \gamma, \tag{6.5}$$

$$0 \le p_i \le 1, \qquad 1 \le i \le m. \tag{6.6}$$

In agreement with inequality (6.5), the set H_0, where we search for the minimum of $R(\mathbf{p})$, is such that for all values of the parameters included in the set H_0 the probability of observing an event of type $d_1 = \cdots = d_m = 0$ is not less than the level of significance $\varepsilon = 1 - \gamma$. Inequalities in (6.6) are obvious and follow from the definition of parameters p_i. As the lower confidence limit for the system PFFO, we take a minimum possible value $R(\mathbf{p})$ in set H_0.

Assume that function $R(\mathbf{p})$ is monotone in each p_i. One can show (see Section 6.7.1) that $\underline{R}$ found in (6.4) is the best γ-confidence lower limit for $R(\mathbf{p})$ under the condition that $d_1 = \cdots = d_m = 0$.

Monotone increments in the system's structural function by each parameter means that, in principle, the best lower confidence limit can be found as the solution of the above-mentioned nonlinear optimization problem (6.4). It is interesting to note that the only upper confidence limit for PFFO is trivial: $\overline{R} = 1$.

6.4 STRUCTURES WITH CONVEX CUMULATIVE HAZARD FUNCTION

To find the maximum limit in (6.4)–(6.6), it is convenient to introduce variables

$$p_i = e^{-z_i}, \qquad 1 \le i \le m. \tag{6.7}$$

For systems with PFFO equal to $R(\mathbf{p})$, we introduce the function

$$f(\mathbf{z}) = f(z_1, \ldots, z_m) = -\ln R(e_l^{-z}, \ldots, e^{-z_m}), \tag{6.8}$$

which is called a *cumulative hazard function.* Transformation (6.8) is used in Pavlov (1982a) for the analysis of systems with complex monotone structures. A cumulative hazard function is increased in each z_i, if PFFO $R(\mathbf{p})$ is increased in each p_i.

Transformations (6.7) and (6.8) have the following meaning. Let all of the system's units have an exponential distribution of time to failure (TTF):

$$p_i(t) = e^{-\lambda_i t},$$

where λ_i is the failure rate of unit i, $1 \le i \le m$. In this case z_i's coincide (with accuracy to the coefficient t) with parameters λ_i's. Now express the system's PFFO with the help of parameters λ_i:

$$R = e^{-f(\lambda_1 t, \ldots, \lambda_m t)}.$$

From the latter expression, one can see that the cumulative hazard function refers to a system's failure rate, which is expressed via unit parameters. Thus, for a series system, we can write

$$f(\mathbf{z}) = z_1 + \cdots + z_m, \tag{6.9}$$

which corresponds to the well-known fact that the failure rate of a series system equals the sum of the unit's failure rates.

If units have a distribution of TTF that is not exponential, the meaning of the system's cumulative hazard function remains the same if, instead of parameters λ_i, one uses parameters of the type

$$\Lambda_i = \frac{1}{t}\int_0^t \lambda_i(u)\, du,$$

where $\lambda_i(t)$ is the failure rate of unit i. A value of Λ_i refers to the average failure rate of unit i on time interval $(0, t)$.

The optimization problem (6.4)–(6.6) is quite easy to solve if the system's cumulative hazard function is convex in $\mathbf{z} = (z_1, \ldots, z_m)$. With the use of new variables $\mathbf{z}$, problem (6.4) can be written as follows:

$$\text{To find } \underline{R} = \min e^{-f(\mathbf{z})} \tag{6.10}$$

under the linear restrictions

$$\sum_{1\le i\le m} N_i z_i \le A, \qquad z_i > 0, \tag{6.11}$$

where $A = -\ln(1 - \gamma)$. As soon as function e_{-f} monotonically decreases by

f, computation of the minimum in (6.10) is equivalent to the problem of finding

$$\bar{f} = \max f(\mathbf{z}) \tag{6.12}$$

with the same restrictions. Then the lower confidence limit $\underline{R}$ can be found as $\underline{R} = \exp(-\bar{f})$. A region given by restrictions (6.11) is convex. If, in addition, the cumulative hazard function $f(\mathbf{z})$ is also convex, then in correspondence with the well-known results of convex programmming (see Section 6.7.3), the maximum of (6.12) under the restrictions of (6.11) is located in one of m "corner" points of type

$$z^i = \Bigl(\underbrace{0, \ldots, 0}_{i-1}, \frac{A}{N_i}, \underbrace{0, \ldots ,0}_{n-i}\Bigr), \qquad 1 \leq i \leq m,$$

where all coordinates except one are zeros. Thus, for systems with a convex cumulative hazard function, the solution is given by the simple expression

$$\bar{f} = \max_i f(\underbrace{0, \ldots, 0}_{i-1}), \frac{A}{N_i}, (\underbrace{0, \ldots, 0}_{n-i}),$$

from which we obtain

$$\underline{R} = \min_i R(\underbrace{1, \ldots, 1}_{i-1}), \underline{p}_i, (\underbrace{1, \ldots, 1}_{n-i}), \tag{6.13}$$

where

$$\underline{p}_i = e^{-A/N_i} = (1 - \gamma)^{1/N_i}$$

is the γ-confidence Clopper–Pearson limit for p_i. The results of the previous section are as follows: Limit (6.13) is the best γ-confidence Clopper–Pearson limit for the system's PFFO $R(\mathbf{p})$.

By its nature, (6.13) is similar to the Mirnyi and Solovyev result for a series system and includes the latter as a particular case. The procedure can be described in the following way: First, one computes m estimates for the system under the condition that unit i, $1 \leq i \leq m$, has the PFFO equal to its lower γ-confidence limit $\underline{p}_i$; then the lowest value is considered as the lower γ-confidence limit for the system's PFFO.

Now consider the main cases when the cumulative hazard function of the system is convex. We have already mentioned that the Mirnyi–Solovyev result is valid for a series system for which the reliability function is $R(\mathbf{p}) = \Pi_{1\leq i\leq m} p_i$.

6.4.1 Series Connection of Groups of Identical Units

Let a system consist of several groups of units in series. Different groups consist of different units but each group includes identical units in series: There are n_i units of type i each with PFFO equal to p_i. In this case

$$R(\mathbf{p}) = \prod_{1 \leq i \leq m} p_i^{n_i},$$

where m is the number of different system units. The cumulative hazard function of the system,

$$f(\mathbf{z}) = \sum_{1 \leq i \leq m} n_i z_i,$$

is linear, as in the previous case. For this case, (6.13) gives $\underline{R} = \min_{1 \leq i \leq m} (1 - \gamma)^{n_i/N_i}$.

6.4.2 Series–Parallel System with Identical Redundant Units

Consider a system consisting of m series redundant groups. Each groups i, $1 \leq i \leq m$, consists of n_i parallel identical units with PFFO equal to p_i (see Figure 6.2). The system fails if at least one redundant group has failed. A redundant group fails if all of its units have failed. In this case

$$R(\mathbf{p}) = \prod_{1 \leq i \leq m} [1 - (1 - p_i)^{n_i}], \qquad f(\mathbf{z}) = \sum_{1 \leq i \leq m} \varphi_i(z_i),$$

where

$$\varphi_i(z_i) = -\ln[1 - (1 - e^{-z_i})^{n_i}].$$

By direct differentiation, one can show that $\varphi_i^*(z_i) \geq 0$, $1 \leq i \leq m$, and, consequently, the cumulative hazard function $f(\mathbf{z})$ is convex because it is a sum of convex functions. In this case, the general expression (6.13) gives the best lower γ-confidence limit for the PFFO of a series–parallel system:

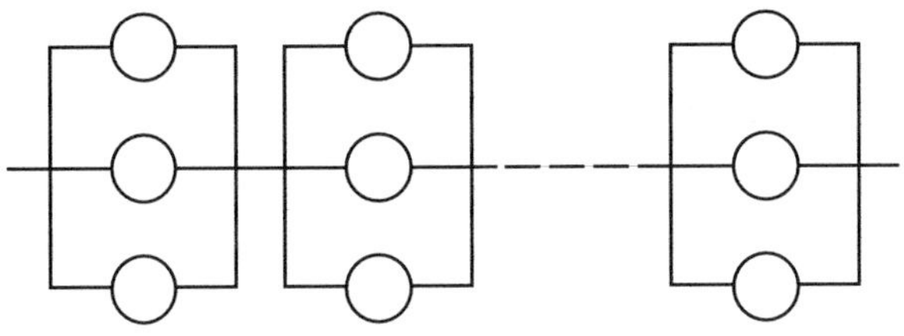

Figure 6.2 System with series–parallel structure.

$$\underline{R} = \min_{1 \le i \le m} \{1 - [1 - (1 - \gamma)^{1/N_i}]^{n_i}\}$$

This result was obtained in Sudakov (1974).

6.4.3 Series Connection of *K* out of *N* Subsystems

In comparison with the previous one, a more general case concerns a system in which the redundant group i fails if at least K_i of its n_i units have failed. The previous structures are specific examples of this general case: $K_i = 1$ in the first case and $K_i = n_i$ in the second case, $1 \le i \le m$. In this general case

$$R(\mathbf{p}) = \prod_{1 \le i \le m} \sum_{0 \le j \le K_i - 1} \binom{n_i}{j} (1 - p_i)^j \, p_i^{n_i - j},$$

$$f(\mathbf{z}) = \sum_{1 \le i \le m} \varphi_i(z_i),$$

$$\varphi_i(z_i) = -\ln \sum_{0 \le j \le K_i - 1} \binom{n_i}{j} (1 - e^{z_i})^j \, e^{-n_i z_i + j z_i}.$$

As previously, by a direct differentiation, one can show that $\underline{\varphi}_i^*(z_i) \ge 0$ for all $1 \le i \le m$, and, consequently, the cumulative hazard function $f(\mathbf{z})$ is convex, being a sum of convex functions. Expression (6.13) in this case gives

$$\underline{R} = \min_{1 \le i \le m} \sum_{0 \le j \le K_i - 1} \binom{n_i}{j} (1 - \underline{p}_i)^j \, \underline{p}_i^{n_i - j}.$$

When the system's cumulative hazard function is not convex, the problem becomes more difficult.

6.5 SERIES–PARALLEL STRUCTURE WITH DIFFERENT REDUNDANT UNITS

A system consists of m series connections of redundant groups (see Figure 6.2). Group i consists of n_i units each with its own PFFO: $p_1^{n_1}, p_2^{n_2}, \ldots, p_m^{n_m}$. The failure criteria for a redundant group and for the system remain the same as before. As one can see, here we rejected an assumption about the units' identity within a redundant group. The PFFO and the cumulative hazard function of the system are

$$R(\mathbf{p}) = \prod_{1 \le i \le m} \left[1 - \prod_{1 \le j \le n_i} (1 - p_i^j)\right], \qquad (6.14)$$

$$f(\mathbf{z}) = - \sum_{1 \le i \le m} \ln\left[1 - \prod_{1 \le j \le n_i} (1 - e^{-z^{ij}})\right].$$

The problem reduces to the computation of the maximum of the cumulative hazard function of (6.14) under restrictions

$$\sum_{1\le i\le m} \sum_{1\le j\le n_i} N_{ij} z_{ij} \le -\ln(1-\gamma),\ z_{ij} \ge 0, \qquad 1 \le j \le n_i, \qquad 1 \le i \le m, \tag{6.15}$$

where N_{ij} and p_i^j are the number of tested units and the PFFO for unit i within redundant group j, respectively; $z_{ij} = -\ln p_i^j$, $\mathbf{p} = \{p_i^j\}$, and $\mathbf{z} = \{z_{ij}\}$.

We use an auxiliary problem: Find

$$\varphi_i(x_i) = \max \prod_{1\le j\le n_i} (1 + p_i^j) \tag{6.16}$$

under the restrictions

$$\prod_{1\le j\le n_i} (p_i^j)^{N_{ij}} \ge 1 - e^{-x_i},\ 0 \le p_i^j \le 1, \qquad 1 \le j \le n_i. \tag{6.17}$$

An equivalent problem in variables z_{ij} is as follows: Find

$$\varphi_i(x_i) = \max \prod_{1\le j\le n_i} (1 - e^{-z_{ij}}) \tag{6.18}$$

under the restrictions

$$\sum_{1\le j\le n_i} N_{ij} z_{ij} = x_i,\ z_{ij} \ge 0, \qquad 1 \le j \le n_i. \tag{6.19}$$

From the definition of the auxiliary problem (6.16)–(6.19), it follows directly that the desired maximum of function (6.14) under restrictions (6.15) can be computed as

$$\max_{\mathbf{z}} f(\mathbf{z}) = \max_{\mathbf{x}} \left\{ - \sum_{1\le i\le m} \ln[1 - \varphi_i(x_i)] \right\}, \tag{6.20}$$

where a maximum is found under the restrictions

$$\sum_{1\le i\le m} x_i \le \ln(1-\gamma),\ x_i \ge 0, \qquad 1 \le i \le m. \tag{6.21}$$

For brevity the following statements are formulated in the form of theorems.

Theorem 6.1 Solution of auxiliary problem (6.18)–(6.19) has the form

$$\varphi_i(x_i) = \prod_{1 \le j \le n_i} \frac{t(x_i)}{t(x_i) + N_{ij}}, \tag{6.22}$$

where $t(x_i)$ is the solution of the equation

$$\sum_{1 \le j \le n_i} N_{ij} \ln\left(1 + \frac{T}{N_{ij}}\right) = x_i \tag{6.23}$$

relative to $t > 0$. (For the proof of the theorem, see Section 6.7.4.) □

Theorem 6.2 Function $\Psi_i(x_i) = -\ln[1 - \varphi_i(x_i)]$ is monotone increasing and convex for $x_i \ge 0$, $1 \le i \le m$. (For the proof of the theorem, see Section 6.7.5.) □

It is easy to see that Theorem 6.2 provides the solution of problem (6.14)–(6.15). Indeed, by the theorem, the function on the right side of (6.20) is convex in $\mathbf{x} = (x_i, \ldots, x_m)$, being a sum of convex functions and, consequently, the maximum of (6.20) is reached at one of m "corner" points of the type $(0, \ldots, 0 - \ln(1 - \gamma), \ldots, 0, \ldots, 0)$. Thus,

$$\max_{\mathbf{z}} f(\mathbf{z}) = -\ln\left(\min_{1 \le i \le m} \{1 - \varphi_i[-\ln(1 - \gamma)]\}\right)$$

and the final form of the best lower γ-confidence limit for the system's PFFO follows from this:

$$\underline{R} = \min_{1 \le i \le m} \{1 - \varphi_i[-\ln(1 - \gamma)]\}, \tag{6.24}$$

where $\varphi_i(\cdot)$ is defined from (6.22) and (6.23). Since the left side of (6.23) is monotone increasing for $t \ge 0$, numerical computation of (6.23) and the further computation of φ_i is not a difficult task.

6.5.1 Parallel Connection

Separately consider an important particular case of a parallel system that represents the above-considered system for $m = 11$ (see Figure 6.3). The system's PFFO of n units connected in parallel is defined by

$$r(\mathbf{p}) = 1 - \prod_{1 \le j \le n} (1 - p_j),$$

where p_j is the jth unit PFFO. From (6.22)–(6.24) for $m = 1$, $n_i = n$, and $N_{ij} = N_j$, we obtain that the best γ-confidence limit for $r(\mathbf{p})$ is determined as

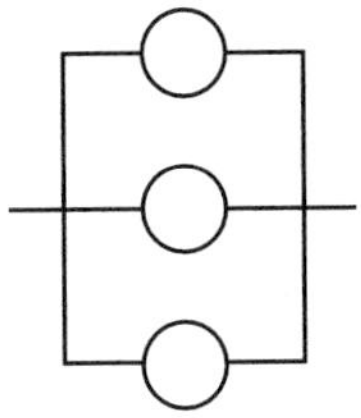

Figure 6.3 System with parallel structure.

$$r_* = 1 - [-\ln(1 - \gamma)] = 1 - \prod_{1 \le j \le n} \frac{t}{t + N_j}, \tag{6.25}$$

where N_j is the number of tested units of type j and t is determined from the equation

$$\sum_{1 \le j \le n} N_j \ln\left(1 + \frac{t}{N_j}\right) = -\ln(1 - \gamma) \tag{6.26}$$

with the left side monotone increasing in t.

In a specific case of equal numbers of tested units, $N_j = N$, $1 \le j \le n$, the result obtained from (6.25)–(6.26) in Tyoskin, Kursky (1986) is as follows:

$$r_* = 1 - [1 - (1 - \gamma)^{1/nN}]^n \tag{6.27}$$

for the lower γ-confidence limit of the PFFO of a parallel system under the condition of equality of all N_i.

Note that the value of $r_* = 1 - \varphi_i[-\ln(1 - \gamma)]$ in (6.24) gives the best lower γ-confidence limit for the PFFO of each separately considered redundant group i. Thus, (6.24) has the following meaning. For computation of the best lower γ-confidence limit for the PFFO of a series–parallel system, one needs to compute the best γ-confidence limit of the type (6.25)–(6.26) for each redundant group and then to take the minimum among them. In this sense, the procedure remains similar to the Mirnyi–Solovyev procedure for a series system, including the latter as a particular case for $n_i = 1$, $1 \le i \le m$.

Example 6.1 A system consists of three series redundant groups. The number of redundant units in these groups are $n_1 = 2$, $n_2 = 3$, and $n_3 = 3$. The system as a whole was tested $N = 6$ times and no failures were observed. The lower 90& confidence limits of the PFFO of different redundant groups by (6.27),

$$r_r * = 1 - [1 - (1 - \gamma)^{1/Nn_i}]^{n_i}, \qquad 1 \le i \le 3,$$

are 0.970, 0.998, and 0.998, respectively. The smallest number gives us the best lower 90% confidence limit for the system's PFFO, that is, $\underline{R} = 0.970$.

Now consider a case with all units within the group identical: $p_{11} = p_{12} = p_1$, $p_{21} = p_{22} = p_{23} = p_2$, $p_{31} = p_{32} = p_{33} = p_3$. The best lower γ-confidence limit for the system's PFFO is computed by (6.14):

$$\underline{R} = \min_{1 \le i \le m} \{1 - [1 - (1 - \gamma)^{1/N_i}]^{n_i}\},$$

where one needs to take $N_i = Nn_i$ because this is the total number of tested units of type i. The limit obtained in this case obviously coincides with the value of $\underline{R} = 0.97$ computed above. This means that in this case an assumption about identity of redundant units did not improve the confidence estimate of the system's PFFO. □

6.5.2 Parallel–Series System

Consider a system consisting of m series subsystems connected in parallel (see Figure 6.4). Subsystem i consists of n_i units. The system's PFFO can be determined by

$$R(\mathbf{p}) = 1 - \prod_{1 \le i \le m} \left[1 - \prod_{1 \le j \le n_i} p_{ij}\right],$$

where p_{ij} is the PFFO of unit j within series subsystem i, $1 \le i \le m$, $1 \le j \le n_i$.

The problem is to find the minimum of the function $R(\mathbf{p})$ under the restrictions of type (6.5) and (6.6) or variables of type $z_{ij} = -\ln p_i^j$,

$$\max \prod_{1 \le i \le m} \left[1 - \exp\left(- \sum_{1 \le j \le n_i} z_{ij}\right)\right], \tag{6.28}$$

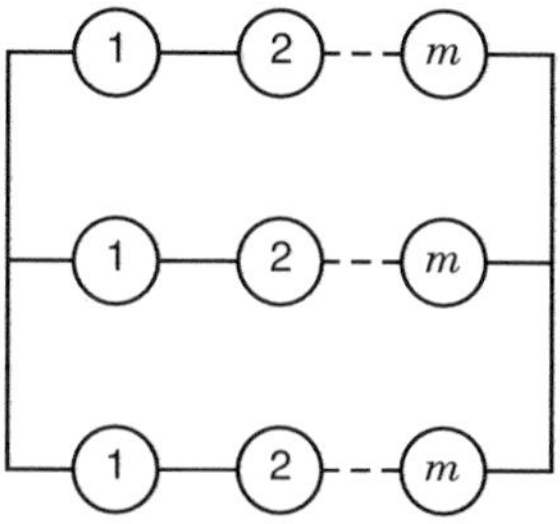

Figure 6.4 System with parallel–series structure.

with restrictions (6.15), where N_{ij} is the number of tested units with subscript ij. Denote

$$g_i(x_i) = \max \sum_{1 \le j \le n_i} z_{ij}, \tag{6.29}$$

where the maximum is taken under the restrictions

$$\sum_{1 \le j \le n_i} N_{ij}\, z_{ij} \le N_i^- x_i,\ z_{ij} \ge 0, \qquad 1 \le j \le n_i, \tag{6.30}$$

where

$$N_i^- = \min_{1 \le j \le n_i} N_{ij}$$

is the minimal number of tested units within series subsystem i.

Computation of the maximum in (6.28) under restrictions (6.15) is reduced to the following problem: Find

$$\max \prod_{1 \le i \le m} [1 - e^{-g_i(x_i)}]$$

under the restrictions

$$\sum_{1 \le i \le m} N_i^- x_i \le -\ln(1 - \gamma),\ x_i \ge 0, \qquad 1 \le i \le m, \tag{6.31}$$

In correspondence with (6.29) and (6.30) and $g_i(x_i) = x_i$, the problem is reduced to the computation of

$$\max \sum_{1 \le i \le m} [1 - e^{-x_i}] \tag{6.32}$$

under restrictions (6.31). This problem was solved in Theorem 6.1. Hence, the lower confidence limit of the system's PFFO can be written in the form

$$\underline{R} = 1 - [-\ln(1 - \gamma)], \tag{6.33}$$

where

$$\varphi(u) = \prod_{1 \le i \le m} \frac{t(u)}{t(u) + N_i^-}$$

and $t(u)$ is the solution of the equation

$$\sum_{1 \le i \le m} N_i^- \ln\left(1 + \frac{t}{N_i^-}\right) = u.$$

Thus the best lower confidence limit for the PFFO of a parallel–series system is found as the limit of type (6.25)–(6.26) for a parallel system when a number of tested units of type i equals the minimum number of tested units within series subsystem i of the original system. Particular cases for $m = 1$ (a series system) and for $n_1 = \cdots = n_m = 1$ (a parallel system) completely coincide with literature results (Mirnyi, Solovyev (1964), Sudakov (1974)).

6.6 SYSTEMS WITH COMPLEX STRUCTURE (IDENTICAL TESTS)

The structures considered above can be easily analyzed because of the specific nature of their reliability functions. In general, for systems with an arbitrary monotone structure there is no solution. But, nevertheless, a solution exists for a wide class of complex structures K' (see Definition 6.4) for a very important practical case where the number of tested units of each type is the same ($N_i = \cdots = N_m = N$). Note that this case coincides with N tests of the entire system. Of course, a system can be tested in an incomplete structure (not all units can be installed). It can occur if the system develops in time and enlarges from stage to stage during the practical utilization (e.g., telecommunication networks and power systems, which are continuously developing). Later, new units that will be installed in the system are separately tested and the statistical data obtained are incorporated with previously available data.

Let m equal the total number of the system's units and $S = (i_1, \ldots, i_n)$. There ae some subsets of the system's units, $n \le m$. Denote a vector of a system state $\mathbf{e}(i_i, \ldots, i_n)$ if units $i_1, \ldots, i_n$ are operational and the remaining have failed and a vector of a system state $\bar{\mathbf{e}}(i_1, \ldots, i_n)$ if units $i_1, \ldots, i_n$ have failed and the remaining are operational. (In some sense these are "mirror" states.) The following definitions are well known (Barlow, Proschan (1975, 1981); Gnedenko, Ushakov (1995)).

Definition 6.1 A subset of units $A = (i_1, \ldots, i_n)$ is called a path of a two-pole graph if $\Psi[\mathbf{e}(i_1, \ldots, i_n)] = 1$; that is, a system is operational if all units of the path are operational. □

Definition 6.2 A subset of units $B = (i_1, \ldots, i_n)$ is called a cut of a two-pole graph if $\Psi[\bar{\mathbf{e}}(i_1, \ldots, i_n)] = 0$; that is, a system is failed if all units of the cut have failed. □

Let us select the system cut that contains the smallest number of units. We call this cut a main cut S' and denote a number of its units as b. Consider a structure obtained as a parallel connection of all units of the main cut. This structure corresponds to the initial system under the assumption that all of its remaining units are absolutely reliable. It is clear that because of the monotonicity property, the new structure is more reliable than the original one:

$$R(\mathbf{p}) \leq R'(\mathbf{p}) = 1 - \prod_{i \in S'} p_i \tag{6.34}$$

for any values of unit reliability parameters $\mathbf{p} = (p_1, \ldots, p_m)$.

We now find a substructure of the original system structure that delivers the lower limit.

Definition 6.3 Paths $A_1, \ldots, A_k$ are called independent (nonoverlapping) if they do not contain common units, that is,

$$A_i \cap A_j = \emptyset$$

for any $i \neq j$, $(i, j) \in \{1, 2, \ldots, k\}$. □

Among all of the sets of independent paths we select one that contains a maximal number of independent paths. We denote this number by a. There may be several such sets of independent paths. We denote a set of all such sets of independent paths by Δ. We denote set k of independent paths by $A_1^k, \ldots, A_a^k$, where k belongs to Δ. Now consider a structure that is obtained as a parallel connection of these independent paths. This structure can be obtained from the original one by the assumption that all of the remaining units of the original structure are absolutely unreliable. (The same assumption can be obtained if these units are deleted from the original structure.) Again, it follows from the definition of system monotonicity that we obtain the lower limit,

$$R'_{(k)}(\mathbf{p}) = 1 - \prod_{1 \leq i \leq a} [1 - \prod_{j \in A_i^k} p_j \leq R(\mathbf{p}). \tag{6.35}$$

Inequalities (6.34) and (6.35) are true for all values of reliability parameters $\mathbf{p} = (p_1, \ldots, p_m)$; thus it follows that

$$\max_{j \in \Delta} \min_{H_0} R''_{(k)}(\mathbf{p}) \leq \min_{H_0} R(\mathbf{p}) \leq \min_{H_0} R'(\mathbf{p}),$$

where H_0 is a set given by restrictions (6.5) and (6.6). Consequently,

$$\underline{R}'' \leq \underline{R} \leq \underline{R}',$$

where $\underline{R}''$, $\underline{R}'$, and $\underline{R}$ are the best γ-confidence limits of the two majorant structures and the original one, respectively. Note that values of $\underline{R}'$ and $\underline{R}''$ can be easily computed with the use of previously obtained results. Using (6.27) and (6.33), we can finally obtain

$$1 - [1 - (1 - \gamma)^{1/Na}]^a \leq \underline{R} \leq 1 - [1 - (1 - \gamma)^{1/Nb}]^b, \qquad (6.36)$$

where the left and right sides are the lower γ-confidence limits for a parallel system consisting of a and b units, respectively.

Obviously, a number of units in the main cut coincides with a maximal number of independent paths.

Definition 6.4 Let us call a system structure K' if

$$a = b. \qquad (6.37)$$

For a structure of class K', inequalities (6.36) give the best lower γ-confidence limit for the system's PFFO:

$$\underline{R} = 1 - [1 - (1 - \gamma)^{1/Na}]^a, \qquad (6.38)$$

which coincides with a similar limit for the main cut of a system. □

It is easy to find that class K' includes all of the above-considered structures of series–parallel and parallel–series types. Condition (6.37) is not true for all monotone structures. For instance, in Section 6.2 we considered k out of n structures. For a structure with $n = 3$ and $k = 2$, we have $a = 1$ and $b = 2$. But for most monotone structures this condition is true.

Example 6.2 Consider the system represented in Figure 6.5. One can find from this figure that $a = b = 3$. The main cut of a system is represented in the figure by shadowed units. □

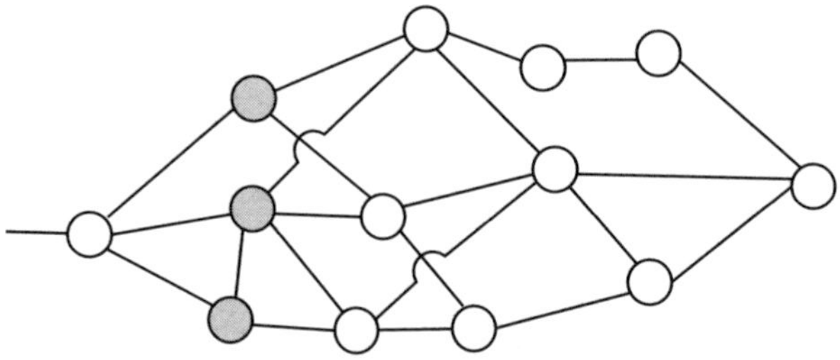

Figure 6.5 System with network-type structure.

Example 6.3 A radial-ring structure is represented in Figure 6.6. Consider the probability of a successful transmission of a signal between the central unit and the shadowed peripheral unit as a system reliability index. In this case $a = b = 4$ if transmission is possible in both directions and $a = b = 3$ if transmission is possible only from a peripheral unit to the central zone. □

Example 6.4 A lattice-type structure is represented in Figure 6.7. Consider the probability of a successsful transmission of a signal between two shadowed units as a system's reliability index. In this case $a = b = 2$. □

All of these nontrivial structures allow us to use a simple expression (6.38) to obtain the best lower confidence limit. Notice that, in these cases, direct attempts to solve the optimization problem (6.4)–(6.6) lead to huge calculations and usually are unsuccessful.

6.6.1 Computation of Confidence Limit for System Bases on a Known Limit for Another System

Assume that the same set of units with reliability parameters $\mathbf{p} = (p_1, \ldots, p_m)$ is used to build two different structures with reliability functions $R(\mathbf{p})$ and $R'(\mathbf{p})$. Call these structures main and auxiliary, respectively. Assume that for the PFFO of the auxiliary system we know the lower γ-confidence limit

$$\underline{R}'(\mathbf{d}) = \underline{R}'(d_1, \ldots, d_m).$$

This limit can be found by any known method.

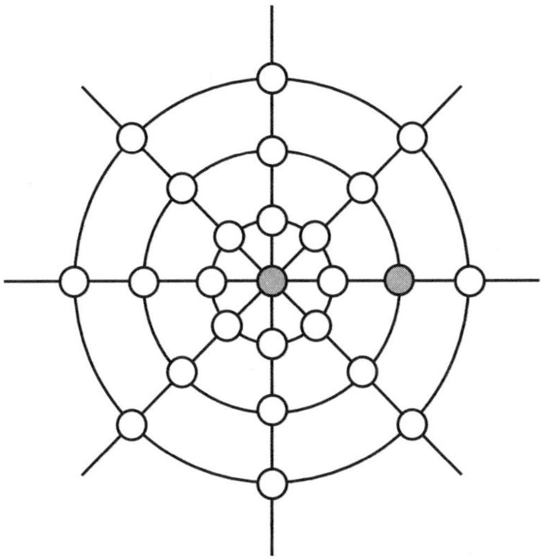

Figure 6.6 System with ring-radial structure.

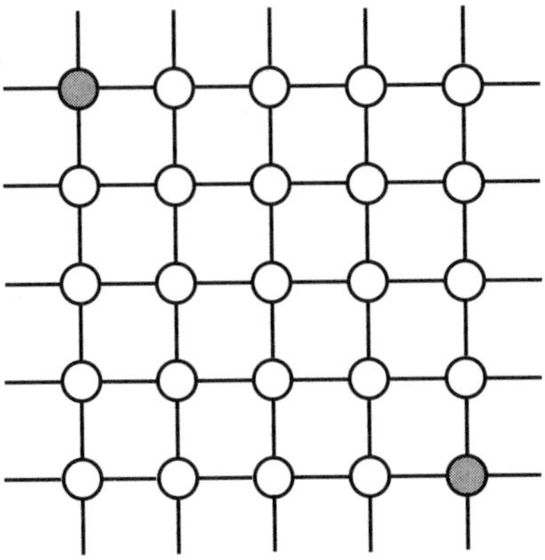

Figure 6.7 System with lattice structure.

We will consider the following problem: to find the lower γ-confidence limit R_* for the PFFO of the main system if the limit for the auxiliary system is known.

Let us introduce a system of sets

$$H_d = \{\mathbf{p}: \underline{R}'(\mathbf{p}) \geq \underline{R}'(\mathbf{d})\}. \tag{6.39}$$

By definition of a γ-confidence limit $\underline{R}'(\mathbf{d})$, the following inequality is true for all $\mathbf{p} = (p_1, \ldots, p_m)$:

$$P_p\{\mathbf{p} \in H_d\} = P_p\{\underline{R}'(\mathbf{d}) \leq R'(\mathbf{p})\} \geq \gamma, \tag{6.40}$$

from where it follows that sets (6.39) form a system of γ-confidence sets, and

$$\underline{R}(\mathbf{d}) = \min_{H_d} R(\mathbf{p}) \tag{6.41}$$

gives the lower γ-confidence limit for the main system's PFFO. Notice that all of these arguments are correct not only for binomial test plans and results in the form $\mathbf{d} = (d_1, \ldots, d_m)$ but also for arbitrary plans with results $\mathbf{x} \in X$ with substitution in (6.39)–(6.41) of $\mathbf{d}$ for $\mathbf{x}$.

Consider a case when the auxiliary structure is in series with the reliability function

$$R'(\mathbf{p}) = \prod_{1 \leq i \leq m} p_i.$$

By (6.39)–(6.41), the problem of construction of the lower γ-confidence limit for the main system's PFFO is equivalent to the problem of finding the minimum

$$\underline{R} = \min_{\mathbf{p}} R(\mathbf{p})$$

under the restrictions

$$\prod_{1 \le i \le m} p_i \ge \underline{R}', \; 0 \le p_i \le 1, \qquad 1 \le i \le m.$$

Thus, from a formal viewpoint, this problem is completely equivalent to problem (6.4)–(6.6) of computing the best lower confidence limit for $R(\mathbf{p})$ under binomial tests with no observed failure. Using the previous results, we can write expressions for the lower γ-confidence limit for the main system's PFFO if this system is series, parallel, series–parallel, or parallel series or has a structure belonging to class K'. For the most general class, namely, K', we have, from (6.38),

$$\underline{R} = 1 - [1 - (\underline{R}')^{1/b}]^b,$$

where $\underline{R}'$ is the known lower γ-confidence limit for the auxiliary system and b is a number of units in the main cut of the main system. (For a parallel structure, b is the number of the system's units; for a parallel–series structure b is the number of parallel connected series subsystems; for a series–parallel structure b is the number of units in the smallest redundant group.)

6.7 APPENDIX

6.7.1 Confidence Clopper–Pearson Limits for Parameter of Binomial Distribution

Let us observe d failures in sequence of N independent tests. Then the lower γ-confidence limit $\underline{p}$ for the PFFO can be found from the equation

$$\sum_{0 \le k \le d} \binom{N}{k} (1 - \underline{p})^k \, \underline{p}^{N-k} = 1 - \gamma.$$

The upper γ-confidence limit $\overline{p}$ for the same conditions can be found from the equation

$$\sum_{0 \le k \le N-d} \binom{N}{k} (1 - \overline{p})^k \, (\overline{p})^{N-k} = 1 - \gamma.$$

In particular, the lower γ-confidence limit for case $d = 0$ is found from the equation

$$\underline{p}^N = 1 - \gamma.$$

Tables for $\underline{p}$ and $\overline{p}$ are given in the Appendix of the book.

6.7.2 Best Lower Confidence Limit for PFFO in Case of No-Failure Test

Let us consider all possible lower limits for PFFO $R(p)$ with the confidence level of not less than γ, that is, possible functions of observed data $\varphi(\mathbf{d}) = \varphi(d_1, \ldots, d_m)$ such that

$$P_p\{\varphi(\mathbf{d}) \le P(\mathbf{p})\} \ge \gamma$$

for all of $\mathbf{p} \in \Theta$, where $\Theta = \{\mathbf{p}\colon 0 \le p_i \le 1,\ 1 < \underline{i} < m\}$ is a set of all possible values of parameters $\mathbf{p} = (p_1, \ldots, p_m)$.

Theorem 6.3 Let the function $R(\mathbf{p}) = R(p_1, \ldots, p_m)$ be monotone nondecreasing for each parameter and continuous for $\mathbf{p} \in \Theta$. Then any lower confidence limit $\varphi(\mathbf{d})$ with confidence level of not less than γ for $R(\mathbf{p})$ at the point $\mathbf{d} = (0, 0, \ldots, 0)$ satisfies the inequality

$$\varphi(0, 0, \ldots, 0) \le \underline{R},$$

where $\underline{R}$ is the lower confidence limit (6.4). □

Proof Assume the opposite: $\varphi(\mathbf{0}) > \underline{R}$. From continuity of the function $R(\mathbf{p})$ in H_0, the minimum in (6.4) is reached at some point

$$\tilde{\mathbf{p}} = (\tilde{p}_1, \ldots, \tilde{p}_m) \in H_0.$$

Because of the monotonicity of $R(\mathbf{p})$, the minimum in (6.4) is reached on the border of the area H_0, and

$$\prod_{1 \le i \le m} \tilde{p}_i^{N_i} = 1 - \gamma.$$

Consider an interval between points $\tilde{p}$ and $\mathbf{p}' = (1, 1, \ldots, 1)$ that can be given in a parametric form as a set of points

$$p = \tilde{p} + t(p' - \tilde{p}), \tag{6.42}$$

where $0 \le t \le 1$. Because of the convexity of region H_0, this interval belongs to this region. Consider the function $R(\mathbf{p})$ on interval (6.42). Let us introduce the function

$$g(t) = R\{\tilde{p} + t(p' + \tilde{p})\}, \qquad 0 \le t \le 1.$$

Under our assumptions, function $g(t)$ is continuous and monotone nondecreasing in t. Note that from

$$g(0) = R(\tilde{p}) = \underline{R}(0) < \varphi(0) \le R(1, 1, \ldots, 1) = g(1)$$

it follows that a point t', $0 < t \le 1$, exists such that

$$g(t') = \varphi(0), \; g(t) < \varphi(0, 0, \ldots, 0), \qquad 0 \le t < t. \tag{6.43}$$

Now consider the probability $P_p\{\varphi(\mathbf{d}) \le R(p)\}$, where p belongs to interval (6.42) and $0 \le t < t'$. Because of the monotonicity of (6.43), we have

$$\begin{aligned} g(0) = R(\tilde{p}) = \underline{R}(0) < \varphi(0) \le R(1, 1, \ldots, 1) = g(1), \; P_p\{\varphi(\mathbf{d}) \le R(\mathbf{p})\} \\ \le 1 - P_p\{d_1 = \cdots = d_m = 0\} = 1 - \prod_{1 \le i \le m} p_i^{N_i}. \end{aligned}$$

From the strong monotone increments of the function

$$h(t) = \prod_{1 \le i \le m} [\tilde{p}_i + t(1 - \tilde{p}_i)]^{N_i}$$

in t and $h(0) = 1 - \gamma$, it follows that the probability

$$P_p\{\varphi(\mathbf{d}) \le R(\mathbf{p})\} < \gamma$$

in all internal points p of interval (6.42) for parameter $0 < t < t$. Thus, the confidence limit $\varphi(\mathbf{d})$ has confidence level strictly less than γ, which contradicts the theorem's conditions. Consequently,

$$\varphi(0) \le \underline{R}.$$

6.7.3 Maximum of Convex Function on Convex Set

An area Γ in m-dimensional Euclidean space R_m is convex if, for each of its two points $\mathbf{x} = (x_1, \ldots, x_m)$ and $\mathbf{y} = (y_1, \ldots, y_m)$, this area also contains an entire interval between these points. In other words, any point of a type $\mathbf{z} = \alpha\mathbf{x} + (1 - \alpha)\mathbf{y}$, where $0 \le \alpha \le 1$ belongs to this area. A function of m variables $f(\mathbf{x}) = f(x_1, \ldots, x_m)$ is called convex (strictly convex) if

$$f[\alpha\mathbf{x} + (1 - \alpha)\mathbf{y}] \leq (<) \alpha f(\mathbf{x}) + (1 - \alpha)f(\mathbf{y})$$

for any $\mathbf{x}$, $\mathbf{y}$, $0 \leq \alpha \leq 1$.

A point x is called an inner point of a convex area Ω if it belongs to an interval that lies totally inside the area. The "surface" point is any point of the area Ω that is not inner. Let Ω be a closed convex region and $f(x)$ be a continuous and strictly convex function. Then $\max_\Gamma f(x)$ is reached at a surface point of the area Ω, and the point at which maximum is reached is unique. To prove it, assume the opposite: The maximum is reached at the inner point $z \in \Omega$. Then there exists such points x and y belonging to Ω and such $0 < \alpha < 1$ that $z = \alpha x + (1 - \alpha)y$, and, consequently, $f(z) < \alpha f(x) + (1 - \alpha) f(y)$, and from this $f(z) < \max[f(x), f(y)]$. But the latter contradicts the statement that max $f(x)$ is reached at point z. If function $f(x)$ is convex (not necessarily strictly convex), then $\max_\Gamma f(x)$ is reached on a surface point of Ω, but such a point might not be unique. (See details in [Pavlov (1977b)].)

Consider, for example, the problem of finding the maximum in (6.12) under the restrictions of (6.11). Because function $f(\mathbf{z})$ monotonically increases in each of its variables, its maximum value is reached on the points at which the first inequality in (6.11) turns into an equality, and, consequently, we can choose restrictions of the type

$$\sum_{1 \leq i \leq m} N_i z_i = A, \; z_i \geq 0, \qquad 1 \leq i \leq m. \tag{6.44}$$

Surface points (the corner) of the convex area Ω determined by restrictions (6.44) are points of the type $z^i = (0, \ldots, 0, A/N_i, 0, \ldots, 0)$, where all coordinates except one are zeros. The simplest way to check this is to turn to Figure 6.8, where (for the two-dimensional case) we marked in bold the two unique corner points.

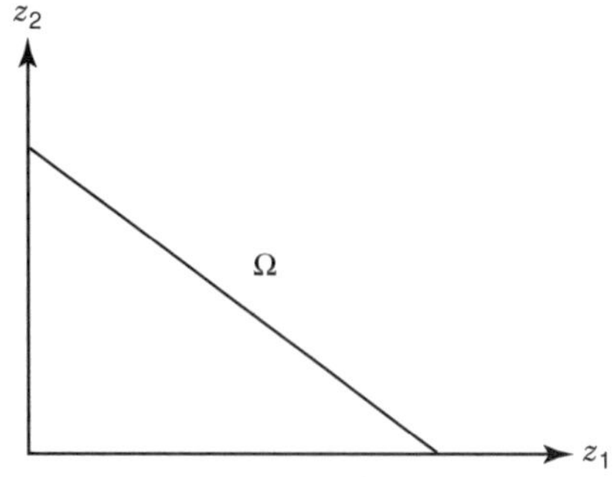

Figure 6.8 Area Ω for two-dimensional case.

6.7.4 Proof of Theorem 6.1

Maximization of function (6.18) is equivalent to maximization of the function

$$h(z) = \sum_{1 \le j \le n_i} \ln(1 - e^{-z_{ij}}) \tag{6.45}$$

under restrictions (6.19). Function (6.45) is monotonically increased by each of its variables and is strictly concave. It follows that inside the area

$$\left\{ z_{ij}: \sum_{1 \le j \le n_i} N_{ij} z_{ij} = x_i,\ x_{ij} \ge 0 \right\} \tag{6.46}$$

there is a unique conditional local minimum that, at the same time, is the global one and is determined by a system of Lagrange equations

$$\frac{\partial h}{\partial x_{ij}} = \frac{1}{e^{z_{ij}} - 1} = \alpha N_{ij}, \qquad 1 \le j \le n_i, \tag{6.47}$$

where α is the Lagrange multiplier. [It is not necessary here to investigate if the maximum belongs to the border of area (6.46) or not because $h(z) = -\infty$ for $z_{ij} = 0$.] Expressing z_{ij} from (6.47) through α and substituting this expression into condition (6.46), we obtain the statement of the theorem. (For the sake of convenience, we use the notation $t = 1/\alpha$.)

6.7.5 Proof of Theorem 6.2

From (6.22) and (6.23), after simple transformations, we obtain the following expression for the derivative:

$$\Psi_i'(x_i) = \frac{t^{n_i - 1}(x_i)}{\prod_{1 \le j \le n_i} [N_{ij} + t(x_i)] - t^{n_i}(x_i)}, \tag{6.48}$$

where $t(x_i)$ is determined by (6.23). Since the denominator of (6.48) is a polynomial of $t(x_i)$ with the power $n_i - 1$ and positive coefficients, (6.48) is monotone increasing in x_i. Function $t(x_i)$, in turn, is increasing in x_i, and it follows that $\Psi_i'(x_i)$ is increasing in x_i $\Psi_i(x_i)$ is convex.

PROBLEMS

6.1. Consider the system depicted in Figure 6.9. This system consists of two groups of redundant units connected in series. The first group consists

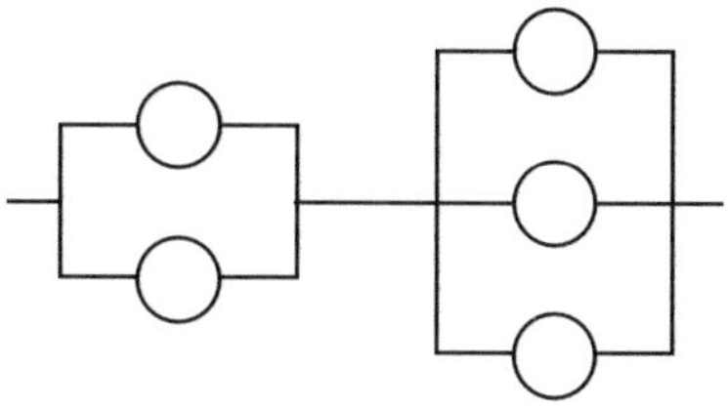

Figure 6.9 Structure of system considered in problem 6.1.

of two parallel units with parameters p_{11} and p_{12}, and the second consists of three parallel units with parameters p_{21}, p_{22}, and p_{23}. These parameters are unknown. Units within each group are not assumed identical. The system's probability of failure-free operation is expressed as

$$R = [1 - (1 - p_{11})(1 - p_{12})]\,[1 - (1 - p_{21})(1 - p_{22})(1 - p_{23})]. \quad (6.49)$$

During eight tests of the system ($N = 8$) there was no failure.

Construct the lower 90% confidence limit for the reliability function (6.49).

6.2. The problem almost completely coincides with the previous one. The difference is in the fact that the units within each group are identical:

$$p_{11} = p_{12} = p_1 \quad \text{and} \quad p_{21} = p_{22}\, p_{23} = p_2. \quad (6.50)$$

Construct the lower 90% confidence limit for the reliability function

$$R = [1 - (1 - p_1)^2)]\,[1 - (1 - p_2)^3]. \quad (6.51)$$

6.3. Solve the previous problem if it is know that

$$p_1 \geq p_2. \quad (6.52)$$

Construct the lower 90% confidence limit for the reliability function (6.51).

6.4. A series–parallel system has a structure of type $(n_1, n_2, \ldots, n_m)$ if it consists of m groups of redundant groups and the ith one consists of n_i identical units, each with parameter p_i. In other words, this system has the reliability function

$$R = \prod_{1 \leq i \leq m} [1 - (1 - p_i)^{n_i}].$$

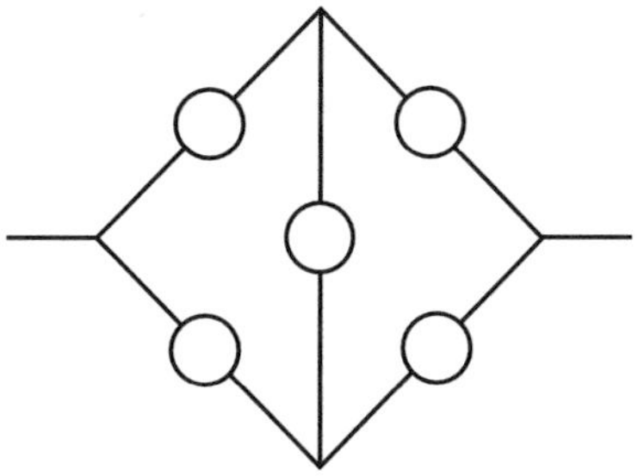

Figure 6.10 System with "bridge" structure for problem 6.4.

Assume the system with the structure (2, 3) depicted in Figure 6.10 was tested eight times ($N = 8$) with no failures.

Construct the lower 90% confidence limit for the reliability function of the system with structure (3, 2). In other words, we need to estimate the system reliability of one system on the basis of the test results of another system consisting of similar units.

6.5. Under the condition of the previous exercise, find the structure (n_1, n_2) with the maximum lower 90% confidence limit if the number of units are subjected to the restrictions

$$n_1 + n_2 \leq 5, \qquad n_1 \geq 1, n_2 \geq 1. \tag{6.53}$$

This task represents one of the variants of the optimal redundancy problem where we should choose the optimal structure (n_1, n_2) with restrictions (6.53) on the basis of a test of the system ($n_1 = 2$, $n_2 = 3$). Notice that this problem cannot be solved using point estimates because in this case all variants of the system have a trivial estimate equal to 1.

6.6. A "bridge system" (see Figure 6.10) was tested four times ($N = 4$) with no failures. All units are not assumed identical.

Construct the lower 95% confidence limit for the system reliability function.

CHAPTER 7

SYSTEM CONFIDENCE LIMITS BASED ON UNIT TEST RESULTS

7.1 INTRODUCTION

7.1.1 Practical Applications

In real-world situations we often need to estimate the system reliability *before* it has been designed. We are thus forced to predict system behavior based on statistical information obtained from unit (component or subsystem) testing. The goal is to use this information in such a way that the result will be adequate to what may be obtained from testing the system as a whole. The same situation arises when testing the entire system, for some reason, is very difficult or even impossible at the developmental phase. An analogous circumstance appears if we are only able to test a system in a truncated configuration, perhaps on some "pilot model" of the system. We wish to make a realistic confidence prediction of the prospective system on the basis of testing a truncated configuration and results of testing its separate units. At the other extreme, units and truncated configurations are all we have for a continuously developing system.

Sometimes unit tests are more effective than a test on the entire system. Assume that the system has a high order of redundancy. Such a system can be highly reliable and is protected against the failure of single units. In this case the total test volume, measured as *testing hours* times *number of tested units,* can be less for unit tests than for system tests. Of course, this situation can be transformed: We can test the system but collect all information about unit failures and use the full set of results for system reliability estimation.

Notice that in all these cases a new problem of aggregating statistical data arises: Different units can be tested differently, that is, by different testing plans (e.g., different numbers of tested units, different time of testing, truncation of tests). Thus, the problem of constructing confidence limits for the system on the basis of unit test data is of great practical interest.

7.1.2 Formulation of the Problem

We showed in the previous chapter that the problem of finding the best confidence limits for the system from unit data of a no-failure test can be easily solved analytically. In other cases, solving this problem is rather complicated.

In general, the problem of constructing confidence limits for the PFFO of a complex system based on unit testing can be formulated in the following way. Let m be the number of different types of units and $\boldsymbol{\theta} = (\theta_1, \theta_2, \ldots, \theta_m)$ be a vector of reliability parameters of these units. (Here θ_i is a reliability parameter of a unit of the ith type.) A set of all possible values of the vector $\boldsymbol{\theta}$ is denoted Θ. Let R be the system PFFO that depends on the unit parameters:

$$R = R(\boldsymbol{\theta}) = R(\theta_1, \theta_2, \ldots, \theta_m). \tag{7.1}$$

We assume that dependence (7.1) is known but the unit parameters *are unknown* though we possess the test results $\mathbf{x}_i$ for each unit of the ith type, $i = 1, \ldots, m$. Test results $\mathbf{x}_i$ for each unit can be obtained by either of two methods: (1) individual testing of units or (2) individual registration of unit failures during the test of the system as a whole. A set of all available test data is denoted by $\mathbf{x} = (x_1, x_2, \ldots, x_m)$. This vector is random and its distribution $P_\theta\{\mathbf{x}\}$ depends on the set of unknown parameters $\boldsymbol{\theta}$.

Let $\underline{R} = \underline{R}(\mathbf{x})$ and $\overline{R} = \overline{R}(\mathbf{x})$ be the lower and upper confidence limits for each test outcome $\mathbf{x}$. Both $\underline{R}$ and $\overline{R}$ depend on random outcomes. Consequently, they are random variables themselves. The interval $[\underline{R}, \overline{R}]$ is said to have the confidence probability γ for the unknown value $R = R(\boldsymbol{\theta})$ if

$$P_\theta\{\underline{R} \le R(\boldsymbol{\theta}) \le \overline{R}\} \ge \gamma \tag{7.2}$$

for all $\boldsymbol{\theta} \in \Theta$. The maximum possible value of γ, satisfying (7.2) for all $\boldsymbol{\theta}$, is usually called the confidence coefficient. In an analogous way, a function $\varphi(\mathbf{x})$ of test outcomes is called the lower (upper) confidence limit with the confidence coefficient γ for $R(\boldsymbol{\theta})$ if, for all $\boldsymbol{\theta} \in \Theta$,

$$P_\theta\{\varphi(\mathbf{x}) \le (\ge)\ R(\boldsymbol{\theta})R\} \ge \gamma. \tag{7.3}$$

Consider the requirement for validation of inequalities (7.2) and (7.3) for all possible $\boldsymbol{\theta}$ in more detail. As mentioned above, the distribution of outcomes $\mathbf{x}$ depends on $\boldsymbol{\theta}$. Therefore the confidence probability in (7.2) and (7.3) may also depend on $\boldsymbol{\theta}$. Thus, (7.2) guarantees that the confidence interval is valid for any $\boldsymbol{\theta}$. However, if we know that $\boldsymbol{\theta}$ belongs to a narrower subset Θ_0, $\Theta_0 \in \Theta$, then the confidence interval will also be narrower.

7.2 CALCULATION BY DIRECT SUBSTITUTION

7.2.1 Point Estimates for Units

Assume that test results x_i for different units are independent and for each parameter we can find a point estimate (e.g., a maximum likelihood) $\hat{\theta}_i = \hat{\theta}_i(x_i)$. The point estimate $\hat{R}$ for the system PFFO R is most often obtained by substitution of the point estimates of parameters $\hat{\theta}_i$ into the function (7.1), that is,

$$\hat{R} = R(\hat{\theta}_1, \hat{\theta}_2, \ldots, \hat{\theta}_m). \tag{7.4}$$

In some cases such a procedure delivers an unbiased estimate of R if the estimates $\hat{\theta}_i$ were unbiased themselves (see Example 7.1). However, even if the estimate (7.4) is biased, it usually possesses asymptotically optimal properties (unbiased and efficient) if estimates $\hat{\theta}_i$ possess these properties. The latter statement is illustrated in Example 7.2.

Example 7.1 (Series System, Binomial Test) Consider a series system of m different units. Assume that we apply the binomial testing plan. The system PFFO is expressed as

$$R = \prod_{1 \le i \le m} p_i,$$

where p_i is the PFFO of the ith unit. Each unit is assumed to be tested separately. The number of tested units, N_i, and the number of failures, d_i, for units of each type are known.

In this case $\theta_i = p_i$ and $x_i = d_i$. The standard unbiased point estimate for parameter p_i is

$$\hat{p}_i = 1 - \frac{d_i}{N_i}.$$

The corresponding estimate for a system PFFO such as (7.4) is

$$\hat{R} = \prod_{1 \le i \le M} \hat{p}_i = \prod_{1 \le i \le m} \left(1 - \frac{d_i}{N_i}\right).$$

Since the test results d_i were assumed to be independent, this estimate is unbiased for R. □

Example 7.2 (Series–Parallel System, Binomial Test) Consider a series–parallel system of m redundant groups. The ith group consists of n_i redundant units connected in parallel (loaded regime of redundant units). We again assume that the binomial testing plan is applied.

The system PFFO has the form

$$R = \prod_{1 \le i \le m} [1 - (1 - p_i)^{n_i}].$$

Suppose N_i units of each type were tested and d_i failures were observed. The number of tested units does not depend on the size of the redundant group.

In this case we generally use the following point estimate for the system PFFO:

$$\hat{R} = \prod_{1 \le i \le M} [1 - (1 - \hat{p}_i)^{n_i}] = \prod_{1 \le i \le m} \left[1 - \left(\frac{d_i}{N_i}\right)^{n_i}\right].$$

This estimate is biased for $n_i > 1$. This is because

$$E\left(\frac{d_i}{N_i}\right)^{n_i} \ne (1 - p_i)^{n_i}. \ \square$$

7.2.2 Confidence Limits

The problem of system point estimator construction can be easily solved by direct substitution of point estimates of unit parameters into the system reliability function (7.1). Finding the confidence limits for point estimate (7.4) is a more complex problem. To illustrate this, let us try to construct the lower limit for the system PFFO defined in (7.1) by direct use of the lower confidence limits of unit parameters θ_i. Assume that the function $R(\boldsymbol{\theta}) = R(\theta_1, \theta_2, \ldots, \theta_m)$ is monotonically increasing in each θ_i. This assumption translates to a natural condition that the system PFFO improves with the growth of the units' reliability. Assume further that the lower γ-confidence limit $\underline{\theta}_i = \underline{\theta}_i(x_i)$ is constructed for each parameter θ_i on the basis of test results x_i, that is,

$$P_\theta(\underline{\theta}_i \le \theta_i) \ge \gamma, \qquad i = \overline{1, m}. \tag{7.5}$$

Let us now find the lower confidence limit $\underline{R}$ for the system PFFO R by direct substitution of confidence limits $\underline{\theta}_i$ into (7.1),

$$\underline{R} = R(\underline{\theta}_1, \underline{\theta}_2, \ldots, \underline{\theta}_m). \tag{7.6}$$

Monotonicity of function $R(\boldsymbol{\theta})$ implies

$$\bigcap_{1\le i\le m} \{\underline{\theta}_i, \le \theta_i\} \subset \{R(\underline{\theta}_1, \underline{\theta}_2, \ldots, \underline{\theta}_m) \le R(\theta_1, \theta_2, \ldots, \theta_m)\}.$$

From this, Equation (7.5), and the independence of test results, we have

$$P_\theta\{\underline{R} \le R(\boldsymbol{\theta})\} \ge P_\theta\left\{\bigcap_{1\le i\le m} (\underline{\theta}_i \le \theta_i)\right\} = \prod_{1\le i\le m} P_\theta(\underline{\theta}_i \le \theta_i) \ge \gamma^m.$$

Thus, the only statement we can make is that the lower confidence limit (7.6) possesses the confidence coefficient *not less* than γ^m. However, we do not know to what real value of γ it corresponds. Moreover, this value decreases very fast with the growth of the number of redundant groups. This confidence interval is too conservative and ineffective even for small m. For instance, for $m = 10$ and $\gamma = 0.9$, $\gamma^m \approx 0.35$. It means, in particular, that if we need to construct the confidence limit for the system PFFO with confidence coefficient not less than 0.9, we should increase individual unit confidence limits up to $\gamma \approx 0.99$. Thus, the more the number of redundant groups in the system, the lower is the system PFFO estimate from direct substitution.

Thus, an attempt to solve the problem via direct substitution of units' confidence limits is ineffective. The problem attracted the attention of American and Russian researchers in the past two decades and some proper approaches were discovered. Some of these approaches are described in the following section.

7.2.3 General Method

The method of constructing the confidence limits for a function of several arguments $R = R(\boldsymbol{\theta}) = R(\theta_1, \theta_2, \ldots, \theta_m)$ is a natural generalization of the method developed for a function with one argument (see Section 1.4).

Belyaev Method The following simplified method was considered in the literature (Belyaev, 1966b, 1968). Let S be a system statistic or some function of test results: $S = S(\mathbf{x}) = S(x_1, \ldots, x_m)$. We can use such a statistic as a point estimate of R, that is, $S = \hat{R}$. Consider the plane (R, S) represented in Figure 7.1. As we did in the one-dimensional case (see Figure 1.2), let us find (for each fixed value R) a corresponding γ-zone of H_R on the S-axis. The probability that statistic S will occur in H_R will be not less than γ.

Unlike a one-dimensional case, a multidimensional case may be degenerate in the sense that different values of a vector of parameters might correspond to the given fixed value $R = R(\boldsymbol{\theta})$. Let us denote

$$A_R = \{\boldsymbol{\theta}: R(\boldsymbol{\theta}) = R\},$$

the set of values of the vector of parameters $\boldsymbol{\theta}$, for which the value of the

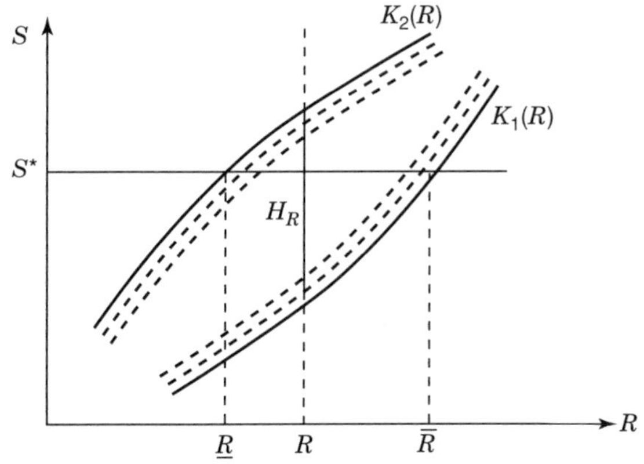

Figure 7.1 Illustration for the method of confidence limits.

reliability function $R(\boldsymbol{\theta})$ is equal to R. The distribution function of r.v. S for the given value of a vector of parameters $\boldsymbol{\theta}$ is denoted by

$$F(t, \boldsymbol{\theta}) = P_\theta(S \leq t). \tag{7.7}$$

For the sake of simplicity, let us assume that function (7.7) is continuous and strictly monotonic in t.

For a given fixed $\boldsymbol{\theta}$, let us choose $t_1(\boldsymbol{\theta})$ and $t_2(\boldsymbol{\theta})$ in the same manner as in the one-dimensional case (Section 1.4):

$$F(t_1, \boldsymbol{\theta}) = \alpha, \qquad F(t_2, \boldsymbol{\theta}) = 1 - \beta; \tag{7.8}$$

that is, $t_1(\boldsymbol{\theta})$ and $t_2(\boldsymbol{\theta})$ are respective quantiles of levels α and β of distribution (7.7). By construction, the probability that statistic S will be covered by the interval $[t_1(\boldsymbol{\theta}), t_2(\boldsymbol{\theta})]$ is equal to $\gamma = 1 - \alpha - \beta$. For a fixed value of the reliability index R, we construct the interval $[K_1(R), K_2(R)]$ by joining all intervals $[t_1(\boldsymbol{\theta}), t_2(\boldsymbol{\theta})]$ for different $\boldsymbol{\theta}$ from A_R. This interval is considered as a γ-zone H_R of the reliability index R. The lower and the upper limits of the γ-zone have the form (see Figure 7.1)

$$K_1(R) = \min_{\boldsymbol{\theta} \in A_R} t_1(\boldsymbol{\theta}), \qquad K_2(R) = \min_{\boldsymbol{\theta} \in A_R} t_2(\boldsymbol{\theta}). \tag{7.9}$$

Thus, the probability that the γ-zone H_R covers statistic S satisfies the inequality

$$P_\theta\{S \in H_r\} = P_\theta\{K_1(R) \leq S \leq K_2(R)\} \geq \gamma \tag{7.10}$$

for each $\boldsymbol{\theta} \in A_R$ and for any values of R.

Let $\mathbf{x}^* = (x_1^*, \ldots, x_m^*)$ denote the value of a random vector $\mathbf{x} = (x_1,$

. . . x_m) obtained as a test result and $S^* = S(\mathbf{x}^*)$ denote the corresponding statistic $S = S(\mathbf{x})$. We call the values $\mathbf{x}^*$ and S^* the *observed value of vector* $\mathbf{x}$ and the *observed value of statistic* S, respectively. We assume further that the limits of the γ-zone, $K_1(R)$ and $K_2(R)$, are continuous and montonically increasing in R. The lower and upper limits for $R = R(\boldsymbol{\theta})$ are defined from the conditions

$$K_2(\underline{R}) = S^*, \qquad K_1(\overline{R}) = S^*, \tag{7.11a}$$

where S^* is the observed value of statistic S (see Figure 7.1). Then, by construction and taking into account (7.10), we have

$$P_\theta\{\underline{R} \leq R(\boldsymbol{\theta}) \leq \overline{R}\} = P_\theta\{S^* \in H_R\} = P_\theta\{K_1(R) \leq S^* \leq K_2(R)\} \geq \gamma$$

for each $\boldsymbol{\theta} \in A_R$ and for any allowed value of R. From here it follows that

$$P_\theta\{\underline{R} \leq R(\boldsymbol{\theta}) \leq \overline{R}\} \geq \gamma$$

for possible values of $\boldsymbol{\theta}$. It means that the interval $[\underline{R}, \overline{R}]$ is the confidence interval for $R = R(\boldsymbol{\theta})$ with confidence coefficient not less than $\gamma - 1 - \alpha - \beta$.

If the limits of the γ-zone monotonically decrease in R, we can write, repeating the above arguments, that the lower and upper limits can be found from the equations

$$K_1(\underline{R}) = S^*, \qquad K_2(\overline{R}) = S^*. \tag{7.11b}$$

Bol'shev–Loginov Method A closely related approach was suggested elsewhere (Bol'shev, 1965; Bol'shev and Loginov, 1966). In the Belyaev approach, the main idea was to find the maximum limits of the γ-zone for each fixed value of R. It is also possible to find these limits in another way, namely, for each given test result S. Roughly speaking, we are finding the limits for a given fixed R in the vertical direction (i.e., along the S-axis in Figure 7.1) in the first case. In the second case, for the given fixed S^*, the limits are found along the R-axis. Let us introduce the functions

$$\Phi_1(S^*, R) = \min_{\boldsymbol{\theta} \in A_R} F(S^*, \boldsymbol{\theta}), \qquad \Phi_2(S^*, R) = \max_{\boldsymbol{\theta} \in A_R} F(S^*, \boldsymbol{\theta}). \tag{7.12}$$

Assume that these functions are continuous and strictly monotonically decreasing in R. Then (7.13) gives the lower ($\underline{R}$) and upper ($\overline{R}$) limits of the confidence interval for $R(\boldsymbol{\theta})$,

$$\Phi_1(\underline{R}, S^*) = 1 - \beta, \qquad \Phi_2(\overline{R}, S^*) = \alpha, \tag{7.13}$$

obtained from the test results S^*.

If these functions are monotonically increasing in R, the lower and upper limits of the confidence interval can be found from the following equations

$$\Phi_1(\underline{R}, S^*) = \alpha, \qquad \Phi_2(\overline{R}, S^*) = \beta.$$

Neyman Method A general approach to constructing confidence sets, which includes Belyaev and Bol'shev–Loginov methods as particular cases, originates from well-known works (Neyman, 1935, 1937). Let $S = S(\mathbf{x})$ again be some statistic. The set C_θ of values of statistic S defined as

$$C_\theta = \{S: t_1(\boldsymbol{\theta}) \leq S \leq t_2(\boldsymbol{\theta})\} \tag{7.14}$$

corresponds to some possible values of parameter $\boldsymbol{\theta} = (\theta_1, \ldots, \theta_m)$. As done previously, $t_1(\boldsymbol{\theta})$ and $t_2(\boldsymbol{\theta})$ are chosen under condition (7.8). The following equality can be established for each $\boldsymbol{\theta}$:

$$P_\theta\{S \in C_\theta\} = \gamma. \tag{7.15}$$

(Here, as above, $\gamma = 1 - \alpha - \beta$.) Let S^* be an observed value of statistic S from the test. Each S has its reflection on set $H(S^*)$ of parameter $\boldsymbol{\theta}$ such that value S^* belongs to the set $C_{\boldsymbol{\theta}}$:

$$H(S^*) = \{\boldsymbol{\theta}: S^* \in C_\theta\}.$$

Owing to (7.14), the set $H(S^*)$ is given by the conditions

$$t_2(\boldsymbol{\theta}) \geq S^*, \qquad t_1(\boldsymbol{\theta}) \leq S^*. \tag{7.16}$$

Taking into account the definition of $t_1(\boldsymbol{\theta})$ and $t_2(\boldsymbol{\theta})$, the inequalities (7.16) can also be written directly via the distribution function of statistic S in the form

$$F(S^*, \boldsymbol{\theta}) \geq \alpha, \qquad 1 - F(S^*, \boldsymbol{\theta}) \geq \beta. \tag{7.17}$$

For each fixed $\boldsymbol{\theta}$, events $\{\boldsymbol{\theta} \in H(S^*)\}$ and $\{S^* \in C_\theta\}$ are equivalent by construction. It follow from (7.15) that, for any $\boldsymbol{\theta}$, the following equality holds:

$$P_\theta\{\boldsymbol{\theta} \in H(S^*)\} = \gamma \tag{7.18}$$

A collection of sets $H(S^*)$ that satisfies (7.18) for all possible $\boldsymbol{\theta}$ is called a collection of γ-confidence sets for $\boldsymbol{\theta}$.

Let us now determine the lower and upper limits of $R = R(\boldsymbol{\theta})$:

$$\overline{R} = \max_{\boldsymbol{\theta} \in H(S^*)} R(\boldsymbol{\theta}), \qquad \underline{R} = \min_{\boldsymbol{\theta} \in H(S^*)} R(\boldsymbol{\theta}). \tag{7.19}$$

The minimum and maximum are taken in the confidence set $H(S^*)$. [We have assumed that the minimum and maximum are attained on the set $H(S^*)$. If it is not the case, one must use infimum and supremum instead of min and max, respectively.] For each fixed $\boldsymbol{\theta}$ the following relationship is valid:

$$\{\boldsymbol{\theta} \in H(S^*)\} \subset \{\underline{R} \leq R(\boldsymbol{\theta}) \leq \overline{R}\}.$$

Taking (7.17) into account, we have

$$P_\theta\{\underline{R} \leq R(\boldsymbol{\theta}) \leq \overline{R}\} \geq \gamma.$$

It means that the interval $(\underline{R}, \overline{R})$ thus constructed is the confidence interval for $R = R(\boldsymbol{\theta})$ with confidence coefficient not less than $\gamma = 1 - \alpha - \beta$.

Remark 7.1 Under some additional monotonicity and continuity assumptions on $R(\boldsymbol{\theta})$, which almost always hold in reliability problems, all three approaches considered above can be shown to be factually equivalent. In other words, expressions (7.11a), (7.13), and (7.19) give the same confidence interval $(\underline{R}, \overline{R})$ if one uses the same α, β, and statistic S (see Pavlov, 1982a, pp. 82–83) under these restrictions. From a computational viewpoint, these different methods are dual descriptions of mathematical programming problems (conditional optimization problems). Thus, the function $F(S^*, \boldsymbol{\theta})$, for which the extremum in (7.12) is searched, represents restrictions in (7.17) for the last approach. The function $R(\boldsymbol{\theta})$, for which the extremum in (7.19) is searched, represents restrictions of the type $R = R(\boldsymbol{\theta})$ in the Belyaev and Bol'shev–Loginov approaches (for more details, see Pavlov, 1982a). □

7.2.4 Arbitrary Distribution of Statistic *S*

In the previous section we assumed that statistic S has continuous distribution. Now we consider a more general case where the distribution of test results $\mathbf{x} = (x_1, x_2, \ldots, x_m)$ and the statistic $S = S(\mathbf{x})$ are arbitrary (possibly, discrete). In this case denote

$$\begin{aligned} t_1(\boldsymbol{\theta}) &= \sup\{t\colon P_\theta(S < t) \leq \alpha\}, \\ t_2(\boldsymbol{\theta}) &= \inf\{t\colon P_\theta(S > t) \leq \beta\}, \end{aligned} \tag{7.20}$$

where $\alpha + \beta < 1$. Let us introduce the set

$$C_\theta = \{S\colon t_1(\boldsymbol{\theta}) \leq S \leq t_2(\boldsymbol{\theta})\}. \tag{7.21}$$

Continuity of the function $P_\theta(S > t)$ from the right and of the function $P_\theta(S < t)$ from the left, together with (7.20), implies that the inequality

$$P_\theta\{S \in C_\theta\} = P_\theta\{t_1(\boldsymbol{\theta}) \le S \le t_2(\boldsymbol{\theta})\} \ge \gamma \tag{7.22}$$

is valid for each fixed $\boldsymbol{\theta}$ where $\gamma = 1 - \alpha - \beta$.

Let S^* be an observed value of statistic S. Let the set

$$H(S^*) = \{\boldsymbol{\theta}: S^* \in C_\theta\} \tag{7.23}$$

correspond to each value of S^*. In accordance with (7.21), this set is given by the inequalities

$$t_1(\boldsymbol{\theta}) \le S^*, \qquad t_2(\boldsymbol{\theta}) \ge S^*. \tag{7.24}$$

Due to (7.22), the inequality

$$P_\theta\{\boldsymbol{\theta} \in H(S^*)\} = P_\theta\{S^* \in C_\theta\} \ge \gamma \tag{7.25}$$

holds for each fixed $\boldsymbol{\theta}$. In other words, sets $H(S^*)$ form a collection of confidence sets for $\boldsymbol{\theta}$ with confidence coefficient not less than γ.

Let us again determine the lower and upper limits $\underline{R}$ and $\overline{R}$ as the minimum and maximum values of the function $R(\boldsymbol{\theta})$, respectively, for all values of parameter $\boldsymbol{\theta}$ that belong to the confidence set $H(S^*)$. It can be shown (see Section 7.9.1) that the final expressions for the confidence limits have the form

$$\underline{R} = \min R(\boldsymbol{\theta}), \qquad \overline{R} = \max R(\boldsymbol{\theta}), \tag{7.26}$$

where the minimum and maximum are taken for all values of the parameters $\boldsymbol{\theta} = (\theta_1, \ldots, \theta_m)$ that satisfy the inequalities

$$P_\theta(S \le S^*) \ge \alpha, \qquad P_\theta(S \ge S^*) \ge \beta. \tag{7.27}$$

The interval $(\underline{R}, \overline{R})$ is the confidence interval for $R = R(\boldsymbol{\theta})$ with confidence coefficient not less than $\gamma = 1 - \alpha - \beta$. Formulas (7.26) and (7.27) include previously considered confidence limits as particular cases.

From formulas (7.26) and (7.27) we can easily obtain one-sided confidence limits for $R = R(\theta)$. For instance, let the reliability function $R(\boldsymbol{\theta}) = R(\theta_1, \ldots, \theta_m)$ be monotonically increasing in each argument θ_i and the function $P_\theta(S \le S^*)$ $[P_\theta(S \ge S^*)]$ monotonically decreasing (increasing) in each θ_i. Setting $\gamma = 1 - \beta$ and $\alpha = 0$ and from (7.26) and (7.27), we obtain the lower γ-confidence limit for R:

$$\underline{R} = \min R(\boldsymbol{\theta}), \tag{7.28}$$

where a minimum is taken for all parameters $\boldsymbol{\theta}$ satisfying the inequality

$$P_\theta(S \geq S^*) \geq 1 - \gamma. \tag{7.29}$$

In correspondence with (7.12) and (7.13) the value $\underline{R}$ is determined from the following equation in respect to R:

$$\max_{\boldsymbol{\theta} \in A_R} P_\theta(S \geq S^*) = 1 - \gamma. \tag{7.30}$$

where the maximum is taken in the set A_R of parameters for which $R(\boldsymbol{\theta}) = R$.

In an analogous way, setting $\gamma = 1 - \alpha$ and $\beta = 0$, we obtain, from (7.26) and (7.27), the upper γ-confidence limit for R:

$$\overline{R} = \max R(\boldsymbol{\theta}), \tag{7.31}$$

where the maximum is taken for all parameters $\boldsymbol{\theta}$ satisfying the inequality

$$P_\theta(S \leq S^*) \geq 1 - \gamma. \tag{7.32}$$

The value $\overline{R}$ is determined from the equation

$$\max_{\boldsymbol{\theta} \in A_R} P_\theta(S \leq S^*) = 1 - \gamma. \tag{7.33}$$

For the chosen statistic $S = S(\mathbf{x})$, the lower and upper confidence limits of (7.33) cannot be improved for more general conditions (see, Pavlov, 1977a,b). Different statistics $S' = S'(\mathbf{x})$ generate different confidence limits. The problem of choosing the best initial statistic is still open.

Example 7.3 (Binomial Test) Consider a standard binomial test scheme. The system consists of m types of units, and the reliability index of the ith type unit equals p_i. The system PFFO R is the function of the vector of unit reliability parameters $\mathbf{p} = (p_1, \ldots, p_m)$:

$$R = R(\mathbf{p}) = R(p_1, \ldots, p_m). \tag{7.34}$$

We assume that this function is increasing in each parameter p_i; that is, the system PFFO increases if the unit reliability increases. The Bernoulli test is invoked for determining each parameter p_i. During testing N_i units were tested and d_i failures were observed. All tests are supposed to be independent.

In this case the vector of unknown parameters $\boldsymbol{\theta}$ is the vector of binomial parameters $\mathbf{p}$, and the vector of test results $\mathbf{x}$ is the vector of the number of failures for units of different types, $\mathbf{d} = (d_1, \ldots, d_m)$. We need to construct the lower confidence limit with confidence coefficient not less than the given γ on the basis of the test results $\mathbf{d}$. □

Let us take the point estimate of the system PFFO for the initial statistic $S = S(\mathbf{x})$, that is,

$$S = \hat{R} = R(\hat{p}_1, \hat{p}_2, \ldots, \hat{p}_m), \tag{7.35}$$

where $\hat{p}_i = 1 - d_i/N_i$ is the standard point estimate of the parameter p_i, $\overline{1, m}$. Let $\mathbf{d}^* = (d_1^*, \ldots, d_m^*)$ be the observed values of the vector of failures and $S^* = S(\mathbf{d}^*)$ be the corresponding observed statistic S. From (7.28) and (7.29) we obtain the following formula for the lower confidence limit (for R with confidence coefficient not less than γ):

$$\underline{R} = \max R(p_1, p_2, \ldots, p_m), \tag{7.36}$$

where the minimum is taken over all parameters $\mathbf{p} = (p_1, \ldots, p_m)$ that satisfy the conditions

$$\sum_{S(\mathbf{d}) \geq S(\mathbf{d}^*)} \prod_{1 \leq i \leq m} \binom{N_i}{d_i} p_i^{N_i - d_i}(1 - p_i)^{d_i} \geq 1 - \gamma, \tag{7.37}$$

$$0 \leq p_i \leq 1, \qquad i = \overline{1, m}. \tag{7.38}$$

We notice that inequality (3.37) corresponds to (7.29).

In accordance with (7.30), we want to find maximum of (7.37); that is, the value of $\underline{R}$ can be found from the following equation in respect to R:

$$\max_{\mathbf{p} \in A_R} \sum_{S(\mathbf{d}) \leq S(\mathbf{d}^*)} \prod_{1 \leq i \leq m} \binom{N_i}{d_i} p_i^{N_i - d_i}(1 - p_i)^{d_i} = 1 - \gamma, \tag{7.39}$$

where the maximum is taken over the set A_R of parameters satisfying restrictions

$$R(p_1, p_2, \ldots, p_m) = R, \qquad 0 \leq p_i \leq 1, \qquad i = \overline{1, m}. \tag{7.40}$$

Sometimes it is more convenient to rewrite (7.36) in another form. Remember that the statistic $S(\mathbf{d}) = S(d_1, \ldots, d_m)$ is any function of test results that is monotonically nondecreasing for each d_i. For instance, $S(\mathbf{d})$ can be a point estimate of R. Order all possible values of the vector of test results $\mathbf{d} = (d_1, \ldots, d_m)$ with decreasing values of the statistic $S(\mathbf{d})$. In other words, each value of the m-dimensional vector $\mathbf{d} = (d_1, \ldots, d_m)$ with positive integer coordinates corresponds to a number $n = n(\mathbf{d})$ determined in such a way that, for any vector $\mathbf{K} = (K_1, \ldots, K_m)$ with $S(\mathbf{K}) > S(\mathbf{d})$, we have $n(\mathbf{d}) \geq n(\mathbf{K})$. Then the confidence limit (7.36) is given as

$$\underline{R} = \min R(p_1, \ldots, p_m), \tag{7.41}$$

where the minimum is taken under the constraints

$$\sum_{n(\mathbf{d}) \leq n(\mathbf{d}^*)} \prod_{1 \geq i \leq m} \binom{N_i}{d_i} p_i^{n_i - d_i}(1 - p_i)^{d_i} \geq 1 - \gamma, \tag{7.42}$$

$$0 \leq p_i \leq 1, \qquad i = \overline{1, m}. \tag{7.43}$$

The difference between (7.42) and (7.37) is in the summation limit. Here $n(\mathbf{d}) = n(d_i, \ldots, d_m)$ is such an ordering of test result vectors **d** that $n(\mathbf{d})$ montonically increases (nondecreases) in each d_i. The first number is as∃ signed to the null vector $\mathbf{d} = (0, 0, \ldots, 0)$; the next to vectors of the type $(0, \ldots, 0, 1, 0, \ldots, 0)$; and so on. This approach, connected with the ordering of test results, was used by Buehler (1957) for a parallel system consisting of two units (see Example 7.5 and Section 7.4).

Example 7.4 (Test with No Failures) Under the conditions of the previous example, consider a particular case where no failures were observed: $d_1^* = d_2^* = \cdots = d_m^* = 0$. In this case the formula for the lower γ-confidence limit for the system PFFO follows from (7.36)–(7.38) or from (7.41)–(7.43):

$$\underline{R} = \min R(p_1, p_2, \ldots, p_m), \tag{7.44}$$

where the minimum is taken under the restrictions

$$\prod_{1 \leq i \leq m} p_i^{N_i} \geq 1 - \gamma, \qquad 0 \leq p_i \leq 1, i = \overline{1, m}. \tag{7.45}$$

This is the best lower γ-confidence limit for the system PFFO in the case of a no-failure test (see Chapter 6). □

Example 7.5 (Buehler Problem) Find the PFFO for a parallel system consisting of two units. The system is tested by a binomial scheme. The index of interest is

$$R(\mathbf{p}) = R)p_1, p_2) = 1 - (1 - p_1)(1 - p_2). \tag{7.46}$$

Each vector of test results $\mathbf{d} = (d_1, d_2)$ corresponds to some number $n(\mathbf{d}) = n(d_1, n_2)$ such that $n(\mathbf{d})$ is monotonically decreasing in each d_i, $i = 1, 2$. The minimum number, naturally, is assigned to the vector (0, 0); the next to the vectors (0, 1) or (1, 0); and so on. From (7.41)–(7.43) we get the following lower γ-confidence limit for the system PFFO:

$$\underline{R} = \min_{\mathbf{p}}\{1 - (1 - p_1)(1 - p_2)\}, \tag{7.47}$$

where the minimum is taken over all $\mathbf{p} = (p_1, p_2)$ that satisfy (7.42). The solution to this problem was given by Buehler (1957). □

Example 7.6 (Series System, Binomial Test) For the conditions of Example 7.3, consider a particular case where a series system consists of m different units. In this case the system PFFO is $R = \Pi_{1\le i\le m} p_i$. Let the point estimate of the system PFFO be taken as an initial statistic $S = S(\mathbf{d})$, that is,

$$S(\mathbf{d}) = \hat{R}(\mathbf{d}) = \prod_{1\le i\le m} \left(1 - \frac{d_i}{N_i}\right).$$

Denote the observed value of the vector $\mathbf{d} = (d_1, \ldots, d_m)$ by $\mathbf{d}^* = (d_1^*, \ldots, d_m^*)$. Then from (77.36)–(7.38) we obtain the lower γ-confidence limit for the system PFFO,

$$\underline{R} = \min \prod_{1\le i\le m} p_i,$$

where the minimum is taken over all parameters $(p_1, \ldots, p_m)$ that satisfy the restrictions

$$\begin{aligned} &\sum_{\hat{R}(\mathbf{d})\ge\hat{R}(\mathbf{d}^*)} \prod_{1\le i\le 2} \binom{N_i}{d_i} p_i^{N_i-d_i}(1 - p_i)^{d_i} \ge 1 - \gamma, \\ &0 \le p_i \le 1, \qquad i = \overline{1, m}. \end{aligned} \tag{7.48}$$

The difference between (7.48) and (7.37) is in the limit of the summation. □

If probabilities of failure are small, that is, $q_i = 1 - p_i << 1$, and the number of tested units N_i is large, we can use the Poisson approximation for solving this problem. In this case we can consider that the number of failures d_i approximately has the Poisson distribution with parameter $\Lambda_i = N_i q_i$, $i = \overline{1, m}$.

For the conditions above, we can write an approximation for R:

$$R = \prod_{1\le i\le m} (1 - q_i) \approx \exp\left(\sum_{1\le i\le m} q_i\right) = \exp\left(-\sum_{1\le i\le m} \frac{\Lambda_i}{N_i}\right).$$

Thus, the problem of finding the lower confidence limit of R is reduced to the problem of finding the upper confidence limit $\overline{f}$ for the function of the Poisson parameters $\mathbf{\Lambda} = (\Lambda_1, \ldots, \Lambda_m)$:

$$f(\mathbf{\Lambda}) = \sum_{1\le i\le m} \frac{\Lambda_i}{N_i}.$$

It follows that the Poisson approximation leads us to an approximate lower γ-confidence limit for the system PFFO in the form $\underline{R} = e^{-\overline{f}}$, where $\overline{f} = \max \Sigma_{1 \le i \le m} (\Lambda_i / N_i)$. The maximum here is taken over all parameters $\mathbf{\Lambda} = (\Lambda_1, \ldots, \lambda_m)$ satisfying the conditions

$$\sum_{\hat{R}(\mathbf{d}) \ge \hat{R}(\mathbf{d}^*)} \prod_{1 \le i \le m} e^{-\Lambda_i} \left(\frac{\Lambda_i^{d_i}}{d_i!} \right) \ge 1 - \gamma, \qquad \lambda_i \ge 0, \qquad i = \overline{1, m}.$$

The problem of constructing an approximate confidence limit for the PFFO of a series system was considered by Bol'shev and Loginov (1966) for equal N_i's and by Pavlov (1973) and Sudakov (1974) for arbitrary N_i.

7.3 SERIES STRUCTURES

7.3.1 Binomial Model

Consider a series system consisting of units of m different types. The number of units of the ith type equals r_i, $i = \overline{1, m}$. All of the system's units are independent. The system PFFO for some fixed operation time t_0 is

$$R = \prod_{1 \le i \le m} p_i^{r_i}, \tag{7.49}$$

where p_i is the PFFO of a unit of the ith type. Assume that N_i units of type i have been tested during time t_0 and d_i failures were observed. The test results $d_1, \ldots, d_m$ are supposed to be independent. Let us construct the confidence limits of the system PFFO (7.49) on the basis of the vector of test results $\mathbf{d} = (d_1, \ldots, d_m)$. Notice that for practical purposes the lower confidence limit is most important.

Consider the case where the unit TTFs are exponentially distributed. The PFFO for the unit of the ith type during time t is given by the formula $p_i(t) = \exp(-\lambda_i t)$, where λ_i is the unit's failure rate. Then the system PFFO defined in (7.49) has the form

$$R = \prod_{1 \le i \le m} p_i^{r_i}(t_0) = e^{-t_0} \sum_{1 \le i \le m} r_i \lambda_i. \tag{7.50}$$

Assume that units of the ith type were tested by plan $[N_i\ R\ T_i]$. This notation means that initially there were N_i units, failed units were replaced by new ones, and the test duration was T_i. The number of observed failures equals d_i. We need to construct the lower γ-confidence limit for the system PFFO defined by (7.50) on the basis of test results $\mathbf{d} = (d_1, \ldots, d_m)$. We would like to emphasize, in constrast to the binomial case, that values T_i for different types of units may be different and do not necessarily coincide with the operational time t_0.

Random variable has the Poisson distribution with the parameter $\Lambda_i = N_i T_i \lambda_i$, $i = \overline{1, m}$. The problem is now reduced to finding the upper γ-confidence limit $\overline{f} = \overline{f}(\mathbf{d})$ of the following function of the Poisson parameters $\mathbf{\Lambda} = (\Lambda_1, \ldots, \Lambda_m)$:

$$f(\mathbf{\Lambda}) = \sum_{1 \le i \le m} \left(\frac{r_i}{N_i T_i} \right) \Lambda_i.$$

Finally the lower γ-confidence limit for the system PFFO can be found as

$$\underline{R} = \exp(-\overline{f} t_0).$$

Notice that we meet the same problem if we use the Poisson approximation for a binomial model. Indeed, in accordance with well-known limit theorems of probability theory, the binomial distribution of the r.v. d_i can be approximated by a Poisson distribution with the parameter $\Lambda_i = N_i q_i$ if $q_i \to 0$ and $N_i \to \infty$, so that $N_i q_i$ remains fixed. (Here $q_i = 1 - p_i$ is the probability of failure of a unit of the ith type.) It follows that for highly reliable units, that is, $q_i << 1$, $i = \overline{1, m}$, and a large number of tested units N_i the problem of an approximate confidence estimate of the system PFFO (7.49) can be reduced to the confidence estimation of the value

$$R = \prod_{1 \le i \le m} (1 - q_1)^{r_i} \approx \exp\left(\sum_{1 \le i \le m} r_i q_i \right) = \exp\left(- \sum_{1 \le i \le m} \frac{r_i}{N_i} \Lambda_i \right).$$

It means that the problem again is reduced to finding the upper γ-confidence limit $\overline{f} = \overline{f}(\mathbf{d})$ of the following function of Poisson parameters:

$$f(\mathbf{\Lambda}) = \sum_{1 \le i \le m} \frac{r_i}{N_i} \Lambda_i.$$

Again the lower γ-confidence limit for the system PFFO can be found as $\underline{R} = e^{-\overline{f}}$.

7.3.2 Lidstrem–Madden Method

The Lidstrem–Madden method is discussed in a well-known book by Lloyd and Lipov (1962). At that time, the method was considered heuristic and approximate because there was no proof that this is indeed the lower confidence limit $\underline{R}$ for the system PFFO (7.49). The correct confirmation of this method was obtained by Pavlov (1973), Sudakov (1974), and others (see Sections 7.6.1 and 7.6.6).

The main idea of this method is grounded on the hypothetical construction of possible outcomes of a system test based on the results of test of individual units. For the sake of simplicity, consider a simple case where $m = 2$ and

$r_1 = r_2 = 1$; that is, the system consists of two different units. Without any loss in generality we can assume that

$$\min_{1 \le i \le m} N_i = N_1; \qquad (7.51)$$

that is, units are ordered by the increasing number of tested units. The unit test results can be given by the two sets

$$(x_{11}, x_{12}, \ldots, x_{1N_1}), \qquad (7.52)$$

$$(x_{21}, x_{22}, \ldots, x_{2N_2}), \qquad (7.53)$$

where x_{ij} is an indicator of failure of the ith unit at the jth test,

$$x_{ij} = \begin{cases} 1 & \text{if the unit has failed,} \\ 0 & \text{if the unit has not failed.} \end{cases}$$

Using these notations, let us try to enumerate all possible outcomes of the system tests. For this purpose, for each test result of the first set, x_{1j}, there corresponds a randomly chosen test result of the second set, x_{2j}. Thus we obtain N_1 pairs

$$(x_{1j}, x_{2k}) \qquad (7.54)$$

each of which can be interpreted as the result of testing a system. The number of "tested" systems is equal to N_1 and the random number of system failures, ξ, is equal to the number of pairs for which at least one unit has failed (at least one of x_{1j} or x_{2j} equals 1). For a given set of test results (7.52) and (7.53), the number of such pairs, ξ, is random since we chose the pairs randomly. Therefore, we take as the number of system failures the value D_1 (later called the "equivalent number of failures"), which is the mathematical expectation of ξ. It is clear that

$$D_1 = E\xi = N_1(1 - \hat{R}),$$

where $\hat{R}$ is the point estimate of the system PFFO. In turn, this value of $\hat{R}$ is calculated as

$$\hat{R} = \hat{p}_1\hat{p}_2 = \left(1 - \frac{d_1}{N_1}\right)\left(1 - \frac{d_2}{N_2}\right).$$

Following such a heuristic procedure, we obtain N_1 system tests (remember that N_1 is the minimum number of tested units) and equivalent number of failures, D_1. The lower γ-confidence limit $\underline{R}$ for the system PFFO is defined as

$$\underline{R} = \underline{P}_\gamma(N_1, D_1), \tag{7.55}$$

where $\underline{P}_\gamma(N, D)$ is the standard Clopper–Pearson lower γ-confidence limit (Clopper and Pearson, 1934) for the binomial parameter p. Confidence limit (7.55) corresponds to the Bernoulli test with N units tested and d failures observed.

In the case of $m \geq 2$ and $r_1 = r_2 = \cdots = r_m = 1$ the lower confidence limit $\underline{R}$ for the system PFFO is found by formula (7.55) in an analogous way. Again N_1 is the minimum number of tested units defined in (7.51) and D_1 is the equivalent number of failures. The latter value is calculated using the formula $D_1 = N_1(1 - \hat{R})$. The point estimate $\hat{R}$ is calculated as

$$\hat{R} = \prod_{1 \leq i \leq m} \left(1 - \frac{d_i}{N_i}\right).$$

Obviously, the equivalent number of failures D_1 might be noninteger. In this case $\underline{P}_\gamma(N_1, D_1)$ can be calculated by interpolation of corresponding values in tables of binomial distribution or with the use of the beta function. In this case the lower γ-confidence limit $\underline{P}_\gamma(N_1, D_1)$ can be found by solving the following equation for p:

$$B_p(N_1 - D_1, D_1 + 1) = 1 - \gamma,$$

where $B_p(a, b)$ is the beta function defined as

$$B_p(a, b) = \frac{\int_0^p x^{a-1}(1 - x)^{b-1}\, dx}{\int_0^1 x^{a-1}(1 - x)^{b-1}\, dx}.$$

(The reader can find some details in Section 7.9.2.) Corresponding numerical tables of $\underline{P}_\gamma(N, d)$ can be found elsewhere (Pearson, 1934; National Bureau of Standards, 1950; U.S. Army, 1952; Romig, 1953; Bol'shev and Smirnov, 1965; Sudakov, 1975; Ushakov, 1994; and others).

Sometimes it is more convenient to write formula (7.55) for the lower γ-confidence limit of the system PFFO as

$$\underline{R} = \min_{1 \leq i \leq m} \underline{P}_\gamma(N_i, D_i), \tag{7.56}$$

where $D_i = N_i(1 - \hat{R})$. This expression allows us to give one more interpretation of this method. For each unit i let us calculate the equivalent number of failures D_i from the condition

$$\frac{D_i}{N_i} = 1 - \hat{R}, \qquad \hat{R} = \prod_{1 \le i \le m} \left(1 - \frac{d_i}{N_i}\right). \tag{7.57}$$

In other words, D_i is defined in such a way that the point estimate of the ith unit coincides with the point estimate of the system as a whole. For each unit we construct the lower γ-confidence limit for the PFFO, $\underline{P}_\gamma(N_i, D_i)$, based on the equivalent number of failures. The minimum of these confidence limits, obtained in this way, is taken as the lower confidence limit $\underline{R}$ for the entire system. This interpretation allows us to expand the method for series systems for $r_i >> 1$, $i = \overline{1, m}$, and for series–parallel systems (see Section 7.6.1). For a series system with $r_i >> 1$, $i = \overline{1, m}$, the lower γ-confidence limit of the PFFO (7.49) is calculated by the formula

$$\underline{R} = \min_{1 \le i \le m} [\underline{P}_\gamma(N_i, D_i)]^{r_i}, \tag{7.58}$$

where the equivalent number of failures D_i is found from the conditions

$$\left(1 - \frac{D_i}{N_i}\right)^{r_i} = \hat{R}, \qquad \hat{R} = \prod_{1 \le i \le m} \left(1 - \frac{d_i}{N_i}\right)^{r_i}. \tag{7.59}$$

It means that the value of D_i is chosen in such a way that the point estimate of the PFFO of the subsystem consisting of units of the ith type will coincide with the point estimate of the system as a whole. Formulas (7.56) and (7.58) represent a particular case of the general method of "equivalent tests" for series–parallel systems, which will be considered in Section 7.6.1.

Example 7.7 Consider a no-failure test where $d_1 = d_2 = \cdots = d_m = 0$. In this case the point estimate of the system PFFO is

$$\hat{R} = \prod_{1 \le i \le m} \left(1 - \frac{d_i}{N_i}\right)^{r_i} = 1$$

and, correspondingly, all equivalent numbers of failure $D_i = 0$, $i = \overline{1, m}$. The Clopper–Pearson lower γ-confidence limit for this case is

$$\underline{P}_\gamma(N_i, D_i) = \underline{P}_\gamma(N_i, 0) = (1 - \gamma)^{1/N_i}.$$

To be concrete, assume that

$$\min_{1\le i\le m}\left(\frac{N_i}{r_i}\right)=\frac{N_1}{r_1}.$$

From (7.58), the Lidstrem–Madden method (method of equivalent tests) delivers (in the no-failure case) the following lower γ-confidence limit of the system PFFO:

$$\underline{P}=\min_{1\le i\le m}(1-\gamma)^{r_i/N_i}=(1-\gamma)^{r_1/N_1}.$$

This coincides with the best lower confidence limit obtained by Mirnyi and Solovyev for the same case (see Chapter 6.) □

Example 7.8 Consider a system consisting of units of three different types ($m = 3$). The number of units of each type, r_i and test results, N_i and d_i, are presented in Table 7.1.

We need to construct the lower confidence limit with the confidence probability $\gamma = 0.9$ for the system PFFO, $R = p_1 \quad p_2 p_3$. The point estimate of the system PFFO in this case equals

$$\hat{R}=\hat{p}_1\hat{p}_2\hat{p}_3=\left(1-\frac{d_1}{N_1}\right)\left(1-\frac{d_2}{N_2}\right)\left(1-\frac{d_3}{N_3}\right)$$

$$=\left(1-\frac{1}{10}\right)\left(1-\frac{2}{40}\right)\left(1-\frac{1}{60}\right)=0.84.$$

The minimum number of tested units equals 10 ($N_1 = 10$). Applying (7.56) and (7.57), we obtain that the equivalent number of failures is

$$D_1=N_1(1-\hat{R})=10(1-0.84)=1.6.$$

The lower γ-confidence limit $\underline{R}$ for the system PFFO, obtained from the equation $B_p(N_1 - D_1, D_1 + 1) = 1 - \gamma$, is equal to $\underline{R} = \underline{P}_\gamma(N_1, D_1) = \underline{P}_{0.9}(10, 1.6) = 0.594$. □

Example 7.9 Let $m = 2$, $r_1 = r_2 = 1$. The test results are $N_1 = 100$, $d_1 = 1$ and $N_2 = 200$, $d_2 = 8$. We need to construct the lower confidence limit

Table 7.1 Input Data for Example 7.8

i	1	2	3
l_i	1	1	1
N_i	10	40	60
d_i	1	2	1

with the confidence probability $\gamma = 0.9$ for the system PFFO, $R = p_1 p_2$. The point estimate of the system PFFO in this case equals

$$\hat{R} = \hat{p}_1 \hat{p}_2 = \left(1 - \frac{d_1}{N_1}\right)\left(1 - \frac{d_2}{N_2}\right) = \left(1 - \frac{1}{100}\right)\left(1 - \frac{8}{200}\right) = 0.95.$$

The minimum number of tested units equals 100 ($N_1 = 100$). Applying (7.56) and (7.57), we obtain that the equivalent number of failures is

$$D_1 = N_1(1 - \hat{R}) = 100(1 - 0.95) = 5.$$

The lower γ-confidence limit $\underline{R}$ for the system PFFO equals

$$\underline{R} = \underline{P}_\gamma(N_1, D_1) = \underline{P}_{0.9}(100, 5) = 0.909. \ \square$$

Example 7.10 Consider a system consisting of three different types of units ($m = 3$). The number of units of each type, r_i, and test results, N_i and d_i, are presented in Table 7.2. The structure of the system is depicted in Figure 7.2.

We need to construct the lower confidence limit with the confidence probability $\gamma = 0.95$ for the system PFFO $R = p_1^2 p_2 p_3$. The point estimate of the system PFFO in this case equals

$$\hat{R} = \hat{p}_1^2 \hat{p}_2 \hat{p}_3 = \left(1 - \frac{d_1}{N_1}\right)^2 \left(1 - \frac{d_2}{N_2}\right)\left(1 - \frac{d_3}{N_3}\right)$$

$$= \left(1 - \frac{4}{100}\right)\left(1 - \frac{3}{150}\right) = 0.941.$$

The "minimum relative number of tested units" (i.e., the number of tested units N_i divided by the number of units r_i) corresponds to a unit of the first type:

$$\min_{1 \le i \le m} \left(\frac{N_i}{r_i}\right) = \frac{N_1}{r_1} = \frac{40}{2} = 20.$$

From (7.58) and (7.59) we have

Table 7.2 Input Data for Example 7.10

i	1	2	3
l_i	2	1	1
N_i	40	100	150
d_i	0	4	3

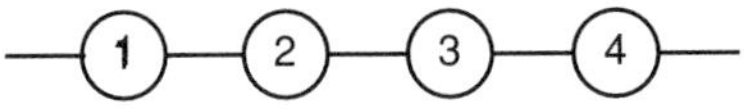

Figure 7.2 Reliability block diagram for Example 7.10.

$$\left(1 - \frac{D_1}{N_1}\right)^2 = \hat{R}$$

and from here the equivalent number of failures for the units of the first type is

$$D_1 = N_1(1 - \sqrt{\hat{R}}) = 40(1 - \sqrt{0.941}) = 1.2.$$

The lower γ-confidence limit for the system PFFO is

$$\underline{R} = [\underline{P}_\gamma(N_1, D_1)]^2 = [\underline{P}_{0.95}(40, 1.2)]^2 = (0.879)^2 = 0.771. \ \square$$

The above method can be applied to the exponential model in an analogous way. Denote the volume of the ith unit test (the total time of testing) by $S_i = N_i T_i$. Then the lower γ-confidence limit for the system PFFO (7.5) can be calculated as

$$\underline{R} = \min_{1 \le i \le m} \exp\{- t_0 r_i \bar{\lambda}_\gamma(S_i D_i)\}. \tag{7.60}$$

Here, $\bar{\lambda}_\gamma(S, a) = \Delta_{1-\gamma}(a)/S$ and $\Delta_{1-\gamma}(a)$ is the standard upper γ-confidence limit for the parameter of the Poisson distribution built by the test result a. Therefore, $\bar{\lambda}_\gamma(S, d)$ is the standard upper γ-cofidence limit for the parameter of the exponential distribution built on the basis of test result (S, d), where S is the total testing time and d is the number of failures (see Sections 1.4.8 and 2.2).

The equivalent number of failures D_i for units of the ith type is found from the condition

$$r_i \left(\frac{D_i}{S_i}\right) = \hat{f}, \qquad \hat{f} = \sum_{1 \le j \le m} r_i \left(\frac{d_j}{S_j}\right). \tag{7.61}$$

From (7.61) it follows that

$$D_i = \sum_{1 \le j \le m} \left(\frac{S_i l_j}{r_i S_j}\right) d_j. \tag{7.62}$$

Example 7.11 Consider a no-failure test where $d_1 = d_2 = \cdots = d_m = 0$.

To be concrete, assume that the first type of unit has the minimum relative volume of testing:

$$\min_{1\le i\le m}\left(\frac{S_i}{r_i}\right) = \frac{S_1}{r_1}.$$

In this case, the equivalent number of failures $D_i = 0$, $i = \overline{1, m}$. Taking into account that

$$\Delta_{1-\gamma}(0) = \ln\left(\frac{1}{1-\gamma}\right),$$

from (7.60) we obtain the following lower γ-confidence limit for the system PFFO:

$$\underline{R} = \min_{1\le i\le m} \exp\left[-t_0 \ln\left(\frac{1}{1-\gamma}\right)\frac{r_i}{S_i}\right] = \min_{1\le i\le m} (1-\gamma)^{t_0(r_i/S_i)} = (1-\gamma)^{t_0(r_1/S_1)}.$$

This limit coincides with the limit obtained in Example 7.3 for the binomial model if the volume of testing equals $N_i = S_i/t_0$, $i = \overline{1, m}$.

The Lidstrem–Madden method (or the method of equivalent testing) considered above is often used in practice for series systems and delivers rather effective confidence limits for the system PFFO. At the same time, its deficiency lies in its insensitivity to the test volume of units of any type except that of the minimum, N_1. As a simple example, consider two systems consisting of two units. Assume the following results for the first and second cases, respectively: $N_1 = 10$, $d_1 = 0$, $N_2 = 20$, $d_2 = 2$ and $N_1 = 10$, $d_1 = 0$, $N_2 = 100$, $d_2 = 10$. For both cases the point estimate of the system PFFO is the same and is equal to

$$\hat{R} = \left(1 - \frac{d_1}{N_1}\right)\left(1 - \frac{d_2}{N_2}\right) = 0.9.$$

The lower γ-confidence limit (e.g., for $\gamma = 0.95$) is the same for both cases:

$$\underline{R} = \underline{P}_\gamma(N_1, D_1),$$

where the equivalent number of failures $D_1 = N_1(1 - \hat{R}) = 10(1 - 0.9) = 1$; therefore $\underline{R} = \underline{P}_\gamma(10, 1) = \underline{P}_{0.95}(10, 1) = 0.606$.

Intuitively the confidence limit in the second situation should be better than in the first one, because the number of tested units of the second type is essentially higher (though the point estimate of the PFFO of the second type unit is the same: $\hat{p}_2 = 1 - d_2/N_2 = 0.9$). □

For the normal approximation, improvement of the Lidstrem–Madden method was obtained by Pavlov (1992a; see also Ushakov, 1994, pp. 346–350).

7.4 PARALLEL STRUCTURES

7.4.1 System of Different Units

Consider a parallel system consisting of m different units. The system PFFO is

$$R = R(\mathbf{p}) = 1 - \prod_{1 \le i \le m} (1 - p_i), \tag{7.63}$$

where p_i is the PFFO of the ith unit. Each unit was tested by a binomial plan: N_i units of the ith type were tested and d_i failures were observed, $i = \overline{1, m}$. The test results $\mathbf{d} = (d_1, \ldots, d_m)$ are assumed to be independent.

Confidence limits (7.63) were obtained by Buehler (1957) for $m = 2$. The Buehler lower γ-confidence limit for the system PFFO has the form

$$\underline{R} = \min \left[1 - \prod_{1 \le i \le m} (1 - p_i) \right], \tag{7.64}$$

where the minimum is taken over all parameters of units $\mathbf{p} = (p_1, \ldots, p_m)$ that satisfy the inequalities

$$\sum_{n(\mathbf{d}) \le n(\mathbf{d}^*)} \prod_{1 \le i \le m} \binom{N_i}{d_i} p_i^{N_i - d_i} (1 - p_i)^{d_i} \ge 1 - \gamma,$$
$$0 \le p_i \le 1, \qquad i = \overline{1, m}. \tag{7.65}$$

Here $\mathbf{d}^* = (d_1^*, \ldots, d_m^*)$ is the observed value of the vector $\mathbf{d} = (d_1, \ldots, d_m)$ obtained in the result of testing and $n(\mathbf{d}) = n(d_1, \ldots, d_m)$ is monotonically decreasing in each d_i and gives some ordering on the set of all possible results of testing.

As we mentioned above, for such an ordering of $n(\mathbf{d})$, the lower γ-confidence limit $\underline{R}$ defined in (7.64) and (7.65) cannot be improved (see Section 7.2). For different types of ordering of $n(\mathbf{d})$, we obtain different confidence limits $\underline{R}$. The question of how to choose the best (optimal) ordering is still open. We usually choose the principle of ordering based on qualitative arguments or convenience of calculation.

The numerical solution of the minimization problem (7.64) under restrictions (7.65) is comparatively not difficult for small m and d_i^*. For $m = 2$, numerical values of the lower confidence limit $\underline{R}$ for some test results (N_i, d_1^*) and (N_2, d_2^*) (mainly, for $N_1 = N_2$) are given elsewhere (Owen, 1962; Steck, 1957; Lipow, 1958, 1959; Schick, 1959). For m, N_i, and d_i^* arbitrary, an approach based on the idea that one can utilize results for series systems for constructing the corresponding limits for parallel system consisting of the same units have been suggested in the literature (Pavlov, 1982a; Sudakov, 1986).

Consider this approach in more detail. Let us take a supplementary (imaginary) series system consisting of the same m units. For this system the PFFO has the form

$$R' = R'(\mathbf{p}) = \prod_{1 \le i \le m} p_i. \tag{7.66}$$

Let $\mathbf{d}^* = (d_1^*, \ldots, \mathbf{d}_m^*)$ be the observed value of the vector of failures. Construct the lower γ-confidence limit $\underline{R}' = \underline{R}'(\mathbf{d}^*)$ for the PFFO of the series system described by (7.66), for instance, on the basis of the Lidstrem–Madden method:

$$\underline{R}' = \underline{P}_\gamma(N_1, D_1) \tag{7.67}$$

where $\underline{P}_\gamma(N, d)$ is the standard Clopper–Pearson lower γ-confidence limit, N_1 is the minimum number of tested units, and D_1 is the equivalent number of failures determined by the formula (see Section 7.3 for details)

$$D_1 = N_1 \left[1 - \prod_{1 \le i \le m} \left(1 - \frac{d_i^*}{N_i} \right) \right]. \tag{7.68}$$

After this, the lower γ-confidence limit for the PFFO of the initial parallel system, described by (7.63), can be found as

$$\underline{R} = \min \left[1 - \prod_{1 \le i \le m} (1 - p_i) \right]. \tag{7.69}$$

Here the minimum is taken over all parameters $\mathbf{p} = (p_1, \ldots, p_m)$ that satisfy the restrictions

$$\prod_{1 \le i \le m} p_i \ge \underline{R}', \qquad 0 \le p_i \le 1, \qquad i = \overline{1, m}. \tag{7.70}$$

It is easy to find that the obtained value $\underline{R}$ is the lower γ-confidence limit for the PFFO of the initial parallel system. Indeed, by construction of $\underline{R}$, the relation

$$\{\underline{R}' \le R(\mathbf{p})\} \subset \{\underline{R} \le R(\mathbf{p})\}$$

is valid for each fixed $\mathbf{p}$. It follows that

$$P_p\{\underline{R}(\mathbf{p})\} \geq P_p\{\underline{R}' \leq R(\mathbf{p})\} \geq \gamma;$$

that is, $\underline{R}$ is the lower γ-confidence limit of the PFFO of a parallel structure $R(\mathbf{p})$.

The minimum of (7.69) under restrictions (7.70) can be easily found analytically. (Notice that from the formal viewpoint, this problem coincides with the problem of finding the lower confidence limit for the no-failure test considered in Chapter 6.) From the solution obtained in Section 6.4, it follows that the minimum (7.69) is attained in the point $p_1 = p_2 = \cdots = p_m = \sqrt[m]{\underline{R}'}$ that corresponds to the equally reliable units.

So, the lower γ-confidence limit (7.69) of the parallel structure PFFO is

$$\underline{R} = 1 - (1 - \sqrt[m]{\underline{R}'})^m. \tag{7.71}$$

Example 7.12 Consider a no-failure test ($d_1^* = d_2^* = \cdots d_m^* = 0$). The number of tested units are $N_1 = N_2 = \cdots = N_m = N$. From (7.67) and (7.68) we obtain that $D_1 = 0$, and the lower γ-confidence limit for the PFFO of a supplementary series structure equals

$$\underline{R}' = \underline{P}_\gamma(N_1, D_1) = \underline{P}_\gamma(N, 0) = (1 - \gamma)^{1/N}.$$

From (7.71) we obtain the following lower γ-confidence limit of the parallel system PFFO:

$$\underline{R} = 1 - [1 - (1 - \gamma)^{1/mN}]^m.$$

This limit coincides with the best lower confidence limit (6.27) obtained in Section 6.4 for the same case. □

Example 7.13 Consider a parallel system consisting of two units. Test results are $N_1 = 20$, $d_1^* = 1$ and $N_2 = 40$, $d_2^* = 4$. We need to construct the lower confidence limit of the PFFO with confidence coefficient not less than $\gamma = 0.9$. Let us first construct the lower γ-confidence limit for the PFFO of a series system with the same units, that is, $R' = p_1 p_2$. In this case the minimum number of tested units is $N_1 = 20$. Applying formulas (7.67) and (7.68) for a series system, we have the equivalent number of failures

$$D_1 = N_1 \left[1 - \left(1 - \frac{d_1^*}{N_1}\right)\left(1 - \frac{d_2^*}{N_2}\right)\right]$$
$$= 20 \left[1 - \left(1 - \frac{1}{20}\right)\left(1 - \frac{4}{40}\right)\right] = 2.9.$$

The lower 0.9 confidence limit for the series system PFFO equals

$$\underline{R}' = \underline{P}_\gamma(N_1, D_1) = \underline{P}_{0.9}(20;\, 2.9) = 0.701.$$

Then applying formula (7.71), we obtain the lower 0.9 confidence limit for the parallel system PFFO,

$$\underline{R} = 1 - (1 - \sqrt[m]{\underline{R}'})^m = 1 - (1 - \sqrt{0.701})^2 = 0.974. \ \square$$

7.4.2 System with Replicated Units

Consider a parallel system consisting of m different types of units. There are n_i units of the ith type in the system. In this case the system PFFO has the form

$$R = R(\mathbf{p}) = 1 - \prod_{1 \le i \le m} (1 - p_i)^{n_i}, \tag{7.72}$$

where $\mathbf{p} = (p_1, \ldots, p_m)$ and p_i is the unit PFFO, $i = \overline{1, m}$.

To construct the lower confidence limit for the PFFO of this system, again consider first a corresponding supplementary series system consisting of the same units. For this series system, the PFFO is

$$R' = R'(\mathbf{p}) = \prod_{1 \le i \le m} p_i^{n_i}. \tag{7.73}$$

Let $\mathbf{d} = (d_1^*, \ldots, d_m^*)$ be an observed value of the vector of unit failures. Applying again the results above to series systems, let us construct the lower confidence limit for the PFFO of a series system, $\underline{R}' = \underline{R}'(\mathbf{d}^*)$, described by (7.73). By formulas (7.58) and (7.59) we have

$$\underline{R}' = \min_{1 \le i \le m} [\underline{P}_\gamma(N_i, D_i)]^{n_i}, \tag{7.74}$$

where the equivalent number of failures D_i is determined from the condition

$$\left(1 - \frac{D_i}{N_i}\right)^{n_i} = \prod_{1 \le j \le m} \left(1 - \frac{d_j^*}{N_j^*}\right)^{n_j}. \tag{7.75}$$

Then we find the lower γ-confidence limit for the PFFO of the initial parallel system described by (7.72):

$$\underline{R} = \min \left[1 - \prod_{1 \le i \le m} (1 - p_i)^{n_i}\right], \tag{7.76}$$

where the minimum is taken over all $\mathbf{p} = (p_1, \ldots, p_m)$ that satisfy the restrictions

$$\prod_{1 \le i \le m} p_i^{n_i} \ge \underline{R}',$$

$$0 \le p_i \le 1, \qquad i = \overline{1, m}. \tag{7.77}$$

The proof that this limit has a confidence coefficient not less than γ completely coincides with that given in Section 7.4.1.

For finding the minimum (7.76), we introduce new arguments $\mathbf{z} = (z_1, \ldots, z_m)$, where $z_1 = -\ln p_i$, $i = \overline{1, m}$. Then problem (7.76)–(7.77)) can be rewritten as follows: Find

$$\underline{R}' = 1 - e^{\bar{f}},$$

where

$$\bar{f} = \max f(\mathbf{z}) \tag{7.78}$$

and

$$f(\mathbf{z}) = \sum_{1 \le i \le m} n_i \ln(1 - e^{-z_i}).$$

The maximum in (7.78) is taken under the restrictions

$$\sum_{1 \le i \le m} n_i z_i \le -\ln \underline{R}',$$

$$z_i \ge 0, \qquad i = \overline{1, m}. \tag{7.79}$$

We can check by direct differentiation that function $f(z)$ is monotone decreasing in each z_i and strictly concave ("convex up") in $\mathbf{z} = (z_1, \ldots, z_m)$. It follows that the maximum (7.78) is attained in the unique point that satisfies the following Lagrange equation system:

$$\frac{\partial f}{\partial z_i} = \frac{n_i}{e^{z_i} - 1} = \lambda n_i, \qquad i = \overline{1, m}, \tag{7.80}$$

$$\sum_{1 \le i \le m} n_i z_i = -\ln \underline{R}', \tag{7.81}$$

where λ is the Lagrange multiplier. [Notice that in this case we do not need to analyze the boundary points of area (7.81) because $f(\mathbf{z}) = -\infty$ for any $z_i = 0$.] This system of equations has an obvious solution

$$z_1 = z_2 = \cdots = z_m = \frac{-\ln \underline{R}'}{\sum_{1 \le i \le m} n_i}.$$

In the force of strong concavity of function $f(\mathbf{z})$, this solution is unique. From here it follows that minimum (7.76) under restrictions (7.77) is attained at a "symmetrical point" that corresponds to equally reliable units, that is,

$$p_1 = p_2 = \cdots = p_m = \sqrt[n]{\underline{R}'},$$

where $n = n_1 + n_2 + \cdots + n_m$ is the total number of units within the system. Thus, the lower γ-confidence limit for the PFFO of a parallel system described by (7.72) has the form

$$\underline{R} = 1 - (1 - \sqrt[n]{\underline{R}'})^n. \tag{7.82}$$

Example 7.14 Consider a parallel system consisting of units of two types. The numbers of units of each type are $n_1 = 2$ and $n_2 = 1$ (see Figure 7.3). Test results are $N_1 = 10$, $d_1^* = 1$ and $N_2 = 20$, $d_2^* = 1$, respectively. The system PFFO is written as

$$R = 1 - (1 - p_1)^2(1 - p_2). \tag{7.83}$$

We need to construct the lower confidence limit for the PFFO with confidence coefficient not less than 0.9.

Let us introduce a supplementary series system consisting of the same units (see Figure 7.4), the PFFO of which is $R' = p_1^2 p_2$.

First we construct the confidence limit of the series system PFFO, R'. In this case, the point estimate for R' is

$$\hat{R}' = \left(1 - \frac{d_1^*}{N_1}\right)^2 \left(1 - \frac{d_2^*}{N_2}\right) = \left(1 - \frac{1}{10}\right)\left(1 - \frac{1}{20}\right) = 0.769.$$

Equivalent numbers of failures D_i can be found from (7.75):

$$D_1 = N_1(1 - \sqrt{\hat{R}'}) = 10 \times (1 - \sqrt{0.769}) = 1.25,$$

$$D_2 = N_2(1 - \hat{R}') = 20 \times (1 - 0.769) = 4.62.$$

In correspondence with (7.74), the lower γ-confidence limit for R' equals

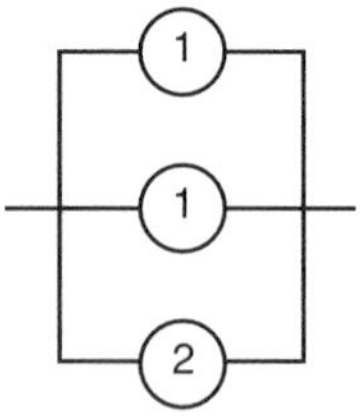

Figure 7.3 Reliability block diagram for Example 7.14.

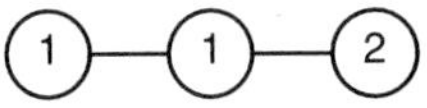

Figure 7.4 Reliability block diagram of a supplementary series system.

$$\underline{R}' = \min\{\underline{P}_{\gamma}^2(N_1, D_1), \underline{P}_{\gamma}(N_2, D_2)\}$$
$$= \min\{\underline{P}_{0.9}^2(10, 1.25), \underline{P}_{0.9}(20, 4.62)\} = \min\{(0.634)^2, 0.605\} = 0.401.$$

Applying formula (7.82), we find the lower 0.9 confidence limit for the PFFO of the considered parallel system (7.83), taking into account that the total number of units within the system is $n = n_1 + n_2 = 2 + 1 = 3$, can be written as

$$\underline{R} = 1 - (1 - \sqrt[n]{\underline{R}'})^n = 1 - (1 - \sqrt[3]{0.401})^3 = 0.982. \ \square$$

7.5 SERIES–PARALLEL SYSTEM

Consider a system that consists of m subsystems (redundant groups). In turn, each subsystem consists of n_i identical redundant units in parallel; that is, redundant units are working in the loading regime. The PFFO of this system can be written as

$$R = \prod_{1 \le i \le m} [1 - (1 - p_i)^{n_i}], \tag{7.84}$$

where p_i is the PFFO of a unit of the ith subsystem, $i = \overline{1, m}$.

Assume that each unit was tested in accordance with a binomial plan: N_i units were tested and d_i failures were registered. We need to construct the lower γ-confidence limit for the PFFO (7.84) on the basis of these test results. A simple solution of the problem exists only for a no-failure test: $d_1^* = d_2^* = \cdots = d_m^* = 0$ (see Section 6.3). In other cases the solution of the problem is rather difficult.

7.5.1 Superreliable System

Assume that all numbers of observed failures d_i^* satisfy the inequalities

$$d_i^* \le n_i - 1 \qquad \text{for all } i = \overline{1, m}. \tag{7.85}$$

It means that the number of failures does not exceed the number of redundant units for each subsystem. Such a system will be conditionally called "superreliable."

We apply the general method described above in Section 7.1 (see also Example 7.3). Take an unbiased point estimate $\hat{R}$ of the system PFFO as an initial statistic S. This unbiased estimate for the binomial scheme of testing is given as

$$\frac{d_i(d_i - 1) \times \cdots \times (d_i - n_i + 1)}{N_i(N_i - 1) \times \cdots \times (N_i - n_i + 1)}.$$

We further assume that inequality $N_i \geq n_i$ is valid for each type of unit; that is, the number of tested units of each type is not less than the number of these units within the ith subsystem. If unit tests are independent, then the unbiased estimate for the system PFFO can be written as

$$\hat{R} = \hat{R}(\mathbf{d}) = \prod_{1 \leq i \leq m} \left[1 - \frac{d_i(d_i - 1) \times \cdots \times (d_i - n_i + 1)}{N_i(N_i - 1) \times \cdots \times (N_i - n_i + 1)}\right]. \tag{7.86}$$

Assuming $S(\mathbf{d}) = \hat{R}(\mathbf{d})$ in (7.36)–(7.38), we obtain the following lower confidence limit for the system PFFO:

$$\underline{R} = \min \prod_{1 \leq i \leq m} [1 - (1 - p_i)^{n_i}], \tag{7.87}$$

where a minimum is taken over all values of unit parameters $\mathbf{p} = (p_1, \ldots, p_m)$ that satisfy the inequalities

$$\sum_{\hat{R}(\mathbf{d}) \geq \hat{R}(\mathbf{d}^*)} \prod_{1 \leq i \leq m} \binom{N_i}{d_I} p_i^{N_i - d_i}(1 - p_i)^{d_i} \geq 1 - \gamma, \tag{7.88}$$

$$0 \leq p_i \leq 1, \qquad i = \overline{1, m}. \tag{7.89}$$

From (7.85) and (7.86) it follows that $\hat{R}(\mathbf{d}^*) = 1$. So, the sum in (7.88) is taken over all $\mathbf{d} = (d_1, \ldots, d_m)$ that satisfy the inequality

$$\hat{R}(\mathbf{d}) \geq 1. \tag{7.90}$$

From the definition of estimate $\hat{R}(\mathbf{d})$, it follows that inequality (7.90) is equivalent to the inequalities

$$0 \leq d_i \leq n_i - 1 \quad \text{for all } i - \overline{1, m}. \tag{7.91}$$

Thus the sum in (7.88) is taken over the set of all vectors $\mathbf{d} = (d_1, \ldots, d_m)$ that satisfy inequalities (7.91). This set represents an m-dimensional hypercube in the space of test results $\mathbf{d}$. In this situation, inequality (7.88) takes the form

$$\prod_{1 \le i \le m} f_i(p_i) \ge 1 - \gamma, \tag{7.92}$$

where

$$f_i(p_i) = \sum_{0 \le d_i \le n_i - 1} \binom{N_i}{d_i} p_i^{N_i - d_i} (1 - p_i)^{d_i}.$$

We introduce new arguments,

$$z_i = -\ln f_i(p_i), \qquad i = \overline{1, m}, \tag{7.93}$$

for a more convenient calculation of $\underline{R}$.

After some simple transformations, the problem of calculating a minimum of (7.87) can be represented in the form

$$\underline{R} = \exp\left(-\max \sum_{1 \le i \le m} \varphi_i(z_i)\right), \tag{7.94}$$

where

$$\varphi_i(z_i) = -\ln[1 - [1 - p_i(z_i)]^{n_i}]$$

and $p_i(z_i)$ is an inverse function corresponding to (7.93). The maximum in (7.94) is taken under the restrictions

$$\sum_{1 \le i \le m} z_i \le -\ln(1 - \gamma),$$

$$z_i \ge 0, \qquad i = \overline{1, m}. \tag{7.95}$$

It is easy to show by direct differentiation that $\varphi_i'(z_i) \ge 0$, $\varphi_i''(z_i) \ge 0$, $i = \overline{1, m}$. So, the function

$$\varphi(\mathbf{z}) = \sum_{1 \le i \le m} \varphi_i(z_i)$$

is monotone increasing in each z_i and convex ("convex down") in $\mathbf{z} = (z_1, \ldots, z_m)$. In accordance with well-known results in the theory of convex programming (see Section 6.7.3), it follows that the maximum in (7.94) is attained at one of m "corner" points of area (7.95). This point has the form

$$(0_1, \ldots, 0_{i-1}, -\ln(1 - \gamma), 0_{i+1}, \ldots, 0_m),$$

where 0_i means the zero of the ith position.

After some simple transformations, we can obtain the lower confidence limit of the system PFFO in the form

$$\underline{R} = \min_{1 \le i \le m} \{1 - [1 - \underline{P}_\gamma(N_i, r_i)]^{n_i}\}, \tag{7.96}$$

where $\underline{P}_\gamma(N, d)$ is the standard Clopper–Pearson lower γ-confidence limit of a binomial parameter p. Here N is the number of units tested, d is the number of failures, and $r_i = n_i - 1$ is the number of redundant units within the ith subsystem. (For more details see Section 7.9.2.) If $n_1 = n_2 = \cdots = n_m = 1$, formula (7.96) is included for the Mirnyi–Solovyev result for a no-failure test as a particular case (see Sections 6.1 and 6.3).

7.5.2 Method of Hyperplane

Consider a solution of the problem for the case where the Poisson approximation is valid. We have N_i Bernoulli trials (the number of tested units) and the probability that an event has occurred equals q_i. As is well known in probability theory, the binomial distribution coverages to the Poisson distribution with parameter $\Lambda_i = N_i q_i$, if $q_i \to 0$ and $N_i \to \infty$ such a way that $N_i q_i = \text{const}$. (In our case, a binomial r.v. is represented by the number of failures d_i.) This statement allows us to use the Poisson approximation for a distribution of r.v. d_i if units, consisting a system, are highly reliable, $q_i << 1$, and the numbers of tested units N_i are large. In this case, we can use the following approximate formula for the system PFFO:

$$R = \prod_{1 \le i \le m} (1 - q^{n_i}) \approx \exp\left(-\sum_{1 \le i \le m} q_i^{n_i}\right) = \exp\left(-\sum_{1 \le i \le m} c_i \Lambda_i^{n_i}\right), \tag{7.97}$$

where the coefficients are $c_i = (1/N_i)^{n_i}$. So, the problem is reduced to constructing the upper confidence limit of the following function of Poisson parameters $\mathbf{\Lambda} = (\Lambda_1, \ldots, \Lambda_m)$,

$$f(\mathbf{\Lambda}) = \sum_{1 \le i \le m} c_i \Lambda_i^{n_i}, \tag{7.98}$$

on the basis of test results $\mathbf{d} = (d_1, \ldots, d_m)$. Here d_i is a r.v. with a Poisson distribution with unknown parameter Λ_i, $i = \overline{1, m}$. All these r.v.'s are mutually independent.

Notice that the construction of confidence limits for the system PFFO (7.84) can be reduced to the analogous problem if system units have exponential distribution of TTF, $p_i(t) = e^{-\lambda_i t}$, where λ_i is the failure rate, $i = \overline{1, m}$. Assume that the ith type units were tested in accordance with plan $[N_i\ R\ T_i]$ (see Chapter 2) and d_i failures have registered. Then r.v. d_i approximately has a Poisson distribution with parameter $\Lambda_i = n_i T_i \lambda_i$. The system PFFO (7.84) can be presented in the form

$$R = \prod_{1 \le i \le m} \{1 - [1 - p_i(t)]^{n_i}\} = \prod_{1 \le i \le m} \{1 - [1 - e^{-\lambda_i t}]^{n_i}\}$$

$$= \exp\left(\sum_{1 \le i \le m} f_i(\Lambda_i)\right),$$

where

$$f_i(\Lambda_i) = -\ln\{1 - [1 - e^{-t\Lambda_i/N_i T_i}]^{n_i}\}.$$

The problem is again reduced to construction of the upper confidence limit for the function

$$f(\mathbf{\Lambda}) = \sum_{1 \le i \le m} f_i(\lambda_i) \tag{7.99}$$

on the basis of the independence of Poisson r.v.'s $\mathbf{d} = (d_1, \ldots, d_m)$. In the case of highly reliable units ($\lambda_i t << 1$) function (7.99) approximately has a form analogous to (7.98):

$$f(\mathbf{\Lambda}) \approx \sum_{1 \le i \le m} c_i \Lambda_i^{n_i},$$

where the coefficients are $c_i = (t/N_i T_i)^{n_i}$, $i = \overline{1, m}$.

Further, let us take the total number of failures (for all types of units)

$$S = S(\mathbf{d}) = d_1 + d_2 + \cdots + d_m$$

as an initial statistic for constructing the confidence limit. Let $d_1^*, d_2^*, \ldots, d_m^*$ be the registered numbers of failures and $S = d_1^* + d_2^* + \cdots + d_m^*$ the corresponding value of observed statistic S. Applying (7.31) and (7.32), we find that the upper γ-confidence limit for $f(\mathbf{\Lambda})$ has the form

$$\overline{f} = \max f(\mathbf{\Lambda}). \tag{7.100}$$

Here the maximum is taken over all parameters $\mathbf{\Lambda} = (\Lambda_1, \ldots, \Lambda_m)$ in such a way that inequality

$$P_\Lambda(S \le S^*) \ge 1 - \gamma \tag{7.101}$$

holds.

Statistic S is the sum of independent Poisson r.v.'s. Thus, S itself has a Poisson distribution with parameters $\Lambda_1 + \Lambda_2 + \cdots + \Lambda_m$. After simple transformations, inequality (7.101) can be rewritten as

$$\sum_{1 \le i \le m} \Lambda_i \le \Delta_{1-\gamma}(d_1^* + d_2^* + \cdots + d_m^*), \tag{7.102}$$

where $\Delta_{1-\gamma}(\mathbf{d})$ is the standard upper γ-confidence limit of the parameters of the Poisson distribution based on test results $\mathbf{d}$. Besides, parameters Λ_i must satisfy the obvious restrictions

$$\Lambda_i \geq 0, \qquad i = \overline{1, m}. \tag{7.103}$$

Thus, the maximum in (7.100) is taken under restrictions (7.102) and (7.103). Function $f(\boldsymbol{\Lambda})$ is monotone increasing in each variable and convex (convex down) in vector $\boldsymbol{\Lambda}$. From here it follows (see Section 6.7.3) that the maximum mentioned above is attained at one of the corner points of the area described by (7.102) and (7.103). These points have the form

$$\left((\underbrace{0, \ldots, 0}_{i-1}, \Delta_{1-\gamma}\left(\sum_{1\leq i\leq m} d_i^* \right), 0_{i+1}, \ldots, 0_m \right), \qquad \overline{1, m},$$

where 0_i means the 0 at the ith position. From here we conclude that the upper γ-confidence limit of $f(\boldsymbol{\Lambda})$ has the form

$$\overline{f} = \max_{1\leq i\leq m} f\left[0_1, \ldots, 0_{i-1}, \Delta_{1-\gamma}\left(\sum_{1\leq i\leq m} d_i^* \right), 0_{i+1}, \ldots, 0_m \right]. \tag{7.104}$$

After this the lower γ-confidence limit for the system PFFO is calculated as $\underline{R} = e^{-\overline{f}}$.

This solution has the following meaning. First, we calculate a lower γ-confidence limit for each individual ith subsystem under the assumption that there were observed failures equal to $d_1^* + d_2^* + \cdots + d_m^*$. Then the minimum of such confidence limits is considered as the lower confidence limit of the system PFFO. Notice that for a series system and a no-failure test, $n_1 = n_2 = \cdots = n_m = 1$ and $d_1^* = d_2^* = \cdots = d_m^* = 0$, and the obtained solution coincides with the result obtained by the Lidstrem–Madden method (see Section 7.3) and by Mirnyi and Solovyev (Sections 6.1 and 6.3).

The area given by (7.102) and (7.103) in the space of parameters $\Lambda_1, \ldots, \Lambda_m$ is a simplex restricted by an m-dimensional plane:

$$\Lambda_1 + \Lambda_2 + \cdots + \Lambda_m = \Delta_{1-\gamma}\left(\sum_{1\leq i\leq m} d_i^* \right).$$

This is the reason why this method is often called the "method of hyperplane." [This method was considered elsewhere (Lipow, 1958, 1959; Mirnyi and Solovyev, 1964; Belyaev, 1966a,b; Bol'shev and Loginov, 1966; Belyaev et al., 1967; and others).]

This method effectively works in the following cases:

1. Series systems with equal (or very close) volumes of unit test, that is, for $n_1 = n_2 = \cdots = n_m = 1$ and $N_1 = N_2 = \cdots = N_m$ (or for the exponential model $N_1T_1 = N_2T_2 = \cdots = N_mT_m$). In this case the method of hyperplane produces confidence limits that coincide with the ones obtained by the Lidstrem–Madden method (see Section 7.3).
2. Series–parallel systems with equal numbers of units within different subsystems, $n_1 = n_2 = \cdots = n_m$, and equal volumes of tests, $N_1 = N_2 = \cdots = N_m$, in the case where all (or almost all) failures have occurred within a single subsystem (remaining subsystems are highly reliable).
3. Series–parallel systems with equal volumes of tests, $N_1 = N_2 = \cdots = N_m$, in the case where the system PFFO coincides with (or is close to) the PFFO of a subsystem with the minimum number of redundant units,

$$n_j = \min(n_1, \ldots, n_m)$$

(other subsystems are highly reliable).
4. Series–parallel systems with equal numbers of units within different subsystems, $n_1 = n_2 = \cdots = n_m$, in the case where the system PFFO coincides (or is close to) the PFFO of a subsystem with the minimum number of redundant units,

$$N_j = \min(N_1, \ldots, N_m)$$

(other subsystems are highly reliable).

In other cases, the efficiency of the method of hyperplane is significantly worse.

7.5.3 Method of Hypercube

Let $d_1^* = d_2^* = \cdots = d_m^*$ be observed number of failures of tested units. We calculate the upper γ-confidence limit of each parameter Λ_i of the ith unit by the standard formula $\overline{\Lambda}_i = \Delta_{1-\gamma}(d_i^*)$. Consider the m-dimensional cube

$$H(\mathbf{d}^*) = \{\mathbf{\Lambda}: 0 \le \Lambda_i \le \Delta_{1-\gamma}(d_i^*),\ i = \overline{1, m}\} \tag{7.105}$$

in the space of parameters $\mathbf{\Lambda} = (\Lambda_1, \ldots, \Lambda_m)$. Then, taking into account the test results for different types of units, we can write the relations

$$P\{\Lambda \in H(\mathbf{d}^*)\} = P\left\{\bigcap_{1\le i\le m} [\Lambda_i \le \Delta_{1-\gamma}(\mathbf{d}^*)]\right\}$$
$$= \prod_{1\le i\le m} P[\Lambda_i \le \Delta_{1-\gamma}(d_i^*)]\} \ge \gamma^m.$$

Thus, sets $H(\mathbf{d}^*)$ form a collection of confidence sets for $\mathbf{\Lambda}$ with confidence coefficient not less than γ^m. It follows that the value of

$$\overline{f} = \max_{\mathbf{\Lambda} \in H(\mathbf{d}^*)} f(\mathbf{\Lambda}) \tag{7.106}$$

produces the upper γ-confidence limit for $f(\mathbf{\Lambda})$ with confidence coefficient not less than γ^m.

Since function $f(\mathbf{\Lambda}) = f(\Lambda_1, \ldots, \Lambda_m)$ defined by (7.98) and (7.99) is monotone increasing in each parameter Λ_i, the maximum in (7.106) is calculated as

$$\overline{f} = f(\overline{\Lambda}_1, \ldots, \overline{\Lambda}_m).$$

The corresponding lower confidence limit for the system PFFO $\underline{R} = e^{-\overline{f}}$ is calculated by simple substitution of the γ-upper confidence limit for individual parameters Λ_i into a function that gives the dependence of system PFFO on unit parameters. An obvious deficiency of the method is in the fast decreasing of confidence coefficient γ^m with increasing m (number of different unit types). It normally leads to a very conservative confidence limit of the system PFFO.

7.5.4 Method of Truncated Hypercube

This method, suggested elsewhere (Belyaev, 1966a,b; Belyaev et al., 1967), represents a combination of the two methods: hyperplane and hypercube. Let $\gamma = \gamma_0$ in (7.102) and $\gamma = \gamma_1$ in (7.105). Form a confidence set in the space of unit parameters $\mathbf{\Lambda} = (\Lambda_1, \ldots, \Lambda_m)$ by the intersection of corresponding confidence sets obtained by the methods of hyperplane and hypercube. The resulting set is given by the restrictions

$$\sum_{1 \le i \le m} \Lambda_i \le \Delta_{1-\gamma_0}\left(\sum_{1 \le i \le m} d_i^*\right), \tag{7.107}$$

$$0 \le \Lambda_i \le \Delta_{1 \le \gamma_1}(d_i^*), \qquad i = \overline{1, m}. \tag{7.108}$$

The confidence coefficient γ for sets given by (7.107) and (7.108) satisfies the inequality

$$\gamma \ge \gamma_0 + \gamma_1^m - 1.$$

This inequality follows from the well-known formula for the intersection of two events

$$P(AB) = P(A) + P(B) - P(A \cup B) \ge P(A) + P(B) - 1.$$

Thus, the lower γ-confidence limit for the system PFFO with confidence coefficient not less than the value of $\gamma_0 + \gamma_i^m - 1$ is calculated as

$$\underline{R} = e^{-\bar{f}},$$

where

$$\bar{f} = \max f(\Lambda) \tag{7.109}$$

and the maximum is taken over the area given by constraints (7.107) and (7.108). This area in the m-dimensional space of parameters Λ is a hypercube truncated by a hyperplane. For series–parallel systems function $f(\Lambda)$ defined in (7.98) or (7.99) is monotone increasing in each Λ_i and convex (convex down). Thus, the maximum in (109) is attained at one of the corner points. This allows one to locate the maximum easily with the help of a computer. Using examples given in the literature (Belyaev et al., 1967), we show below that this method gives better results in comparison with the methods of hyperplane or hypercube.

Example 7.15 A system consists of 10 redundant groups (subsystems) connected in series ($m = 10$). The ith subsystem consists of n_i parallel identical units. Each unit was tested by the binomial plan: n_i is the number of tested units and d_i is the number of registered failures. Corresponding values are given in Table 7.3.

Values of the lower confidence limit with confidence coefficient $\gamma = 0.9$ for the system PFFO for this example, calculated with the help of different methods, are given in Table 7.4. □

Example 7.16 A system consists of 10 redundant groups (subsystems) connected in series ($m = 10$). Each unit was tested by the binomial plan analogously to Example 7.14. Corresponding values are given in Table 7.5. Values of the lower confidence limit with confidence coefficient $\gamma = 0.98$ for the system PFFO, calculated with the help of different methods for this example, are given in Table 7.4. □

Example 7.17 A system consists of 20 redundant groups (subsystems) connected in series ($m = 20$). Each group consists of two parallel units, $n_i = 2$.

Table 7.3 Input Data for Example 7.14

i	1	2	3	4	5	6	7	8	9	10
n_i	4	4	4	2	5	5	3	3	3	1
N_i	100	100	150	500	500	400	200	400	500	500
d_i	2	1	1	3	4	8	3	5	1	1

Table 7.4 Lower Confidence Limits for System PFFO Calculated by Different Methods

Method	Example 7.15	Example 7.16	Example 7.17
Hypercube	0.966	0.861	0.907
Hyperplane	0.967	0.896	0.955
Truncated hypercube	0.999	0.896	0.984
Equivalent testing	0.9998	0.956	0.994

An equal number of each unit were tested, $N_1 = N_2 = \cdots N_{20} = 100$. Corresponding numbers of failures are $d_1 = d_2 = \cdots = d_5 = 0$ and $d_6 = d_6 = \cdots = d_{20} = 1$.

Values of the lower confidence limit with confidence coefficient $\gamma = 0.9$ for the system PFFO, calculated with the help of different methods for this example, are given in Table 7.4. □

Notice that the method of truncated hypercube, nevertheless, still gives a to conservative confidence limit of the system PFFO.

7.5.5 Modified Hyperplane Method

The method of hyperlane, as it was shown by Pavlov (1972) can be improved for series–parallel systems if $n_i \geq 2$ for all i, $i = \overline{1, m}$; that is, if each subsystem has at least one redundant unit. Then for the confidence limit defined in (7.104) we can use the value

$$\tilde{S} = \max\left[\max_{1 \leq i \leq m} d_i^*, \sum_{1 \leq i \leq m} d_i^* - r\right] \tag{7.110}$$

instead of the total number of failures, $\Sigma_{1 \leq i \leq m} d_i^*$. Above we used the notation $r = \min_{1 \leq i \leq m} n_i - 1$, which denotes the minimum number of redundant units among all subsystems.

Table 7.5 Input Data for Example 7.16

i	1	2	3	4	5	6	7	8	9	10
n_i	2	2	2	2	2	1	1	1	1	2
N_i	200	200	200	200	200	300	300	300	300	300
d_i	0	1	2	4	2	4	1	0	1	5

Example 7.18 Consider a series system consisting of three subsystems (see Figure 7.5). All data related to the example are given in Table 7.6. We need to construct the lower confidence limit for the system PFFO with confidence coefficient $\gamma = 0.9$.

In this case $r = 2$ and $\tilde{S} = \max(2; 4 - 2) = 2$. The method of hyperplane gives the following lower γ-confidence limit for the system PFFO:

$$\underline{R} = \exp\left\{-\max_{1\le i\le m}\left[\frac{1}{N_i}\Delta_{1-\gamma}\left(\sum_{1\le i\le m} d_i^*\right)\right]^{n_i}\right\} = \exp\left[\frac{\Delta_{0.1}(4)}{25}\right]^3 = 0.965.$$

The modified method of hyperplane produces the lower confidence limit (with the same confidence coefficient $\gamma = 0.9$) equal to

$$\underline{R} = \exp\left\{-\max_{1\le i\le m}\left[\frac{\Delta_{1-\gamma}(\tilde{S})}{N_i}\right]^{n_i}\right\} = \exp\left[\frac{\Delta_{0.1}(2)}{25}\right]^3 = 0.991.$$

The latter estimate is significantly better than the previous one. □

Example 7.19 Consider a system consisting of 10 subsystems, $m = 10$. All data concerning unit tests result are presented in Table 7.7. In this case the total number of failures is $d_1^* + \cdots + d_{10}^* = 6$. The minimum number of redundant units is

$$r = \min(n_1, \ldots, n_m) - 1 = 2.$$

The lower γ-confidence limit with confidence coefficient $\gamma = 0.9$ calculated with the help of the hyperplane method is given as

$$\underline{R} = \exp\left[-\frac{\Delta_{0.1}(6)}{40}\right]^3 = 0.982.$$

Now from (7.110) we find that

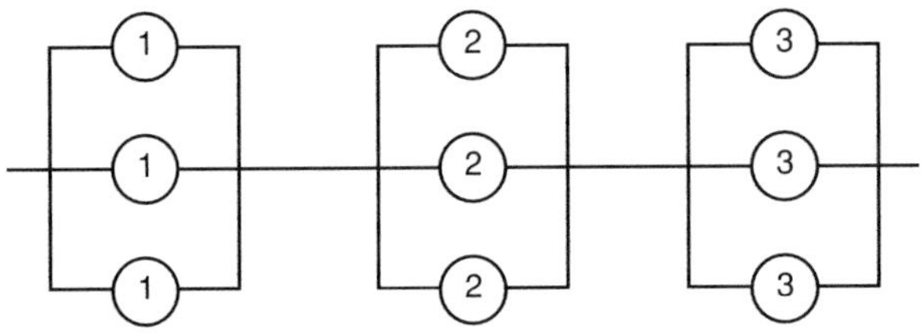

Figure 7.5 Reliability block diagram for Example 7.18.

Table 7.6 Input Data for Example 7.18

i	1	2	3
l_i	3	3	3
N_i	25	25	25
d_i^*	1	1	2

$$\tilde{S} = \max(1;\ 6 - 2) = 4.$$

Thus the modified method of hyperplane gives

$$\underline{R} = \exp\left[-\frac{\Delta_{0.1}(4)}{40}\right]^3 = 0.992. \ \square$$

Example 7.20 Consider again Example 7.17. For that example, the minimum number of redundant units among all subsystems is $r = \min(n_1, \ldots, n_m) - 1 = 1$, the total number of failures is $d_1^* + \cdots + d_m^* = 15$, and the value of $\tilde{S}$ is $\max(1, 15 - 1) = 14$.

For the confidence coefficient $\gamma = 0.9$ the method of hyperplane (see Table 7.4) gives the γ-confidence limit of the system PFFO as $\underline{R} = 0.955$. The modified method of hyperplane produces

$$\underline{R} = \exp\left[-\frac{\Delta_{0.1}(14)}{100}\right]^3 = 0.960. \ \square$$

All methods considered in this section are correct and give the guaranteed confidence coefficient for series–parallel systems. Nevertheless, in many cases they produce a to conservative estimate of system reliability to be of practical interest. It leads one to create heuristic or approximate methods that give better confidence limits. We consider these approaches in the next section. Besides, in Section 7.7 we suggest a more effective method though one that needs more sophisticated computations.

Table 7.7 Input Data for Example 7.19

i	1	2	3	4	5	6	7	8	9	10
n_i	3	3	3	3	3	3	3	3	4	60
N_i	40	60	60	100	100	60	150	150	100	60
d_i^*	0	1	0	1	1	0	1	1	1	0

7.6 APPROXIMATE METHODS

All algorithms considered previously deliver guaranteed value of the confidence coefficient for series–parallel systems, that is, they satisfy the condition

$$P_p\{\underline{R} \le R_p\} \ge \gamma \tag{7.111}$$

for all possible values of unit parameters $\mathbf{p} = (p_1, \ldots, p_m)$. At the same time, these strong methods give conservative estimates $\underline{R}$ of the system PFFO $R = R(\mathbf{p})$. Heuristic and approximate methods that can be described by the condition

$$P_p\{\underline{R} \le R_p\} \approx \gamma$$

might lead to an opposite effect. As discussed, overestimating the reliability indices is often unacceptable in practice. Thus, these methods must be used with care.

7.6.1 Method of Equivalent Tests

This method is a natural extension of the Lidstrem–Madden method (7.56) for series–parallel systems. Taking into account the vector of test results $\mathbf{d}^* = (d_1^*, d_2^*, \ldots, d_m^*)$, let us find point estimates of unit parameters $\hat{p}_i = 1 - d_i^*/N_i$ and a point estimate of the system PFFO (7.84):

$$\hat{R} = R(\hat{p}_1, \ldots, \hat{p}_m) = \prod_{1 \le i \le m} \left[1 - \left(\frac{d_i^*}{N_i}\right)^{n_i}\right].$$

Determine the equivalent number of failures D_i for the ith subsystem units from the condition

$$R\Bigl(\underbrace{1, \ldots, 1}_{i-1}, 1 - \frac{D_i}{N_i}, 1, \ldots, 1\Bigr) = \hat{R}, \tag{7.112}$$

where 1_i means that the ith component of the vector equals 1.

From (7.112), we obtain

$$1 - \left(\frac{D_i}{N_i}\right)^{n_i} = \prod_{1 \le i \le m} \left[1 - \left(\frac{d_i^*}{N_i}\right)^{n_i}\right]. \tag{7.113}$$

Now we determine the lower γ-confidence limit for the system PFFO (7.84) as

$$\underline{R} = \min_{1 \le i \le m} R\Big(\underbrace{1, \ldots, 1}_{i-1}, \underline{P}_\gamma(N_i, D_i), 1, \ldots, 1\Big)$$

$$= \min_{1 \le i \le m} \{1 - [1 - \underline{P}_\gamma(N_i, D_i)]^{n_i}\}, \tag{7.114}$$

where $\underline{P}_\gamma(N, d)$ is the Clopper–Pearson lower γ-confidence limit calculated on the basis of the binomial test (N units were tested and d failures were observed). If the value of D_i is not integer, the value of $\mathbf{P}_\gamma(N_i, D_i)$ is found from solution of the equation

$$B_p(N_i - D_i, D_i + 1) = 1 - \gamma,$$

where $B_p(a, b)$ is the beta function (see Section 7.9.2).

Thus, we first find the equivalent number of failures D_i for each ith subsystem in such a way that the point estimate of each subsystem, calculated on the basis of such a value, coincides with the point estimate $\hat{R}$ of the system PFFO. Then on the basis of this equivalent number of failures, we construct the lower γ-confidence limit of the PFFO of a unit of the ith type and for the ith subsystem itself. The minimum confidence limit among the lower γ-confidence limit for the subsystems is taken as the lower γ-confidence limit for the entire system. This method includes the Lidstrem–Madden method (7.56) for series systems as a particular case. In an analogous way, it is applied to the exponential model.

Explanations of this method can be obtained for the normal approximation in the case of a large test volume (see Section 7.6.5; also see Krol', 1974, 1975; Pavlov, 1982a).

This method relates to the general strong method considered in Section 7.2. Let us take the point estimate of the system PFFO

$$S(\mathbf{d}) = \hat{R} = \prod_{1 \le i \le m} \left[1 - \left(\frac{d_i}{N_i}\right)^{n_i}\right]$$

as an initial statistic $S = S(\mathbf{d})$ for constructing the confidence limit. Then in accordance with (7.36)–(7.38) the lower γ-confidence limit for the system PFFO has the form

$$\underline{R} = \min \prod_{1 \le i \le m} [1 - (1 - p_i)^{n_i}] \tag{7.115}$$

where a minimum is taken over all values of parameter $\mathbf{p} = (p_1, \ldots, p_m)$ that satisfy the inequalities

$$\sum_{S(\mathbf{d})\geq S(\mathbf{d}^*)} \prod_{1\leq i\leq m} \binom{N_i}{d_i} p_i^{N_i-d_i}(1-p_i)^{d_i} \geq 1-\gamma, \tag{7.116}$$

$$0 \leq p_i \leq 1, \qquad i=\overline{1,m}. \tag{7.117}$$

Now consider m corner points of the area (7.116)–(7.117), which has the form

$$\mathbf{p}^{(i)} = \Big(\underbrace{1,\ldots,1}_{i-1}, \pi_i, 1, \ldots, 1\Big), \qquad i=\overline{1,m}.$$

Substituting the ith corner point $p^{(i)}$ into inequality (7.116), the value π_i can be found from the equation

$$\sum_{0\leq d_i\leq L_i} \binom{N_i}{d_i} \pi_i^{N_i-d_i}(1-\pi_i)^{d_i} = 1-\gamma, \tag{7.118}$$

where L_i is the maximum integer quantity among r ones satisfying the inequality

$$S\Big(\underbrace{0,\ldots,0}_{i-1}, r, 0, \ldots, 0\Big) \geq S(\mathbf{d}^*),$$

from which we have

$$L_i = \max\left\{r\colon \left(1-\frac{1}{N_i}\right)^{n_i} \geq \prod_{1\leq i\leq m}\left(1-\frac{d_i^*}{N_i}\right)\right\}.$$

It shows that the equivalent number of failures D_i is connected with the quantity L_i by the relation

$$L_i = [D_i], \qquad i=\overline{1,m}, \tag{7.119}$$

where $[D_i]$ is the integer part of D_i.

From (7.118) it follows that the value of π_i coincides with the standard Clopper–Pearson lower γ-confidence limit calculated on the basis of L_i failures in N_i tests:

$$\pi_i = \underline{P}_\gamma(N_i, L_i).$$

All corner points belong to the area (7.115) in which a minimum is searched, so from here the following inequality for the strong lower confidence limit $\underline{R}$ follows:

$$\underline{R} = \min_{1 \le i \le m} R(p^{(i)}) = \min_{1 \le i \le m} R\left((\underbrace{1, \ldots, 1}_{i-1}, \pi_i, 1, \ldots, 1\right)$$

$$= \min_{1 \le i \le m} [1 - (1 - \pi_I)^{n_i}] = \min_{1 \le i \le m} \{1 - [1 - \underline{P}_\gamma(N_i, L_i)]^{n_i}\}.$$

The right part of this inequality coincides with the lower confidence limit calculated by the method of equivalent tests if the equivalent numbers of failures D_i are integer. [If D_i are noninteger, the difference is not significant since $D_i - L_i = D_i - [D_i]$ from (7.119).]

Thus, the lower confidence limit calculated by the method of equivalent tests is always as good as the strong lower confidence limit (7.115). The above argument shows that from a computational viewpoint this heuristic procedure factually reduces finding a minimum in (7.115) among the corner points of the area determined by (7.116) and (7.117) instead of searching within the entire area. Correspondingly, the proof of correctness of the method is reduced to the question: How much does the absolute minimum value within the entire area differ from the minimum value found in m corner points (see Section 7.6.6).

Example 7.21 Consider Example 7.18. Let us construct the lower confidence limit for the system PFFO by the method of equivalent tests. The point estimate of the system PFFO in this case is

$$\hat{R} = \prod_{1 \le i \le m} [1 - (1 - \hat{p}_i)^{n_i}] = \prod_{1 \le i \le m} \left[1 - \left(\frac{d_i^*}{N_i}\right)^{n_i}\right]$$

$$= \left[1 - \left(\frac{1}{25}\right)^3\right]\left[1 - \left(\frac{1}{25}\right)^3\right]\left[1 - \left(\frac{2}{25}\right)^3\right] = 0.99936.$$

From condition (7.113) we find that the equivalent number of failures D_1 for the first subsystem is

$$1 - \left(\frac{D_1}{N_1}\right)^{n_i} = \hat{R},$$

from which we get

$$D_1 = N_1(1 - \hat{R})^{1/n_1} = 25\sqrt[3]{64 \times 10^{-5}} = 2.16.$$

For the second and third subsystems, the equivalent number of failures D_2 and D_3 coincide with D_1. Further, from (7.114) we find the lower γ-confidence limit (with the confidence coefficient $\gamma = 0.9$) for the system PFFO

$$\underline{R} = 1 - [1 - \underline{P}_\gamma(N_1, D_1)]^3$$
$$= 1 - [1 - \underline{P}_\gamma(25; 2.160)^3 = 1 - (1 - 0.795)^3 = 0.991.$$

In this case the confidence limit coincides with that found in Example 7.18 by the modified method of hyperplane. □

Example 7.22 Consider again the same system as in the previous example but with different test results (see Table 7.8). In this case, the point estimate of the system PFFO is

$$\hat{R} = \left[1 - \left(\frac{1}{25}\right)^3\right]\left[1 - \left(\frac{1}{50}\right)^3\right]\left[1 - \left(\frac{2}{100}\right)^3\right] = 0.99992.$$

The minimum number of tested units is $n_1 = 25$. From (7.113) we find the equivalent number of failures D_1 for the first subsystem:

$$D_1 = N_1(1 - \hat{R})^{1/n_1} = 25 \times \sqrt[3]{0.00008} = 1.08.$$

Thus, the lower γ-confidence limit with confidence coefficient $\gamma = 0.9$ for the system PFFO calculated by the method of equivalent tests is

$$\underline{R} = 1 - [1 - \underline{P}_\gamma(N_1, D_i)]^3 = 1 - [1 - \underline{P}_{0.9}(25, 1.08)]^3$$
$$= 1 - (1 - 0.848)^3 = 0.96.$$

Notice that, in this case, the minimum number of tested units $N_1 = 25$ and the observed numbers of failures d_i^* are the same as in Example 7.18. Thus the methods of hyperplane and modified hyperplane method give in this case the previous quantities equal to 0.966 and 0.9991, respectively. It shows that they are not sensitive to increasing N_2 and N_3 if the minimum number N_1 is fixed. The method of equivalent tests delivers higher values of lower γ-confidence limit of the system PFFO. □

Table 7.8 Input Data for Example 7.22

i	1	2	3
n_i	3	3	3
N_i	25	50	100
d_i^*	1	1	2

Example 7.23 Apply the method of equivalent tests to the system considered in Example 7.17.

In this case the point estimate of the system PFFO is

$$\hat{R} = \prod_{1 \le i \le m} [1 - (1 - \hat{p}_i)^{n_i}] = \left[1 - \left(\frac{1}{100}\right)^2\right]^{15} = 0.9985.$$

From condition (7.113) we find the equivalent number of failures for the first subsystem:

$$D_1 = N_1(1 - \hat{R})^{1/n_1} = 100 \times \sqrt{0.0015} = 3.88.$$

In this case all values n_i and N_i are equal; therefore all equivalent number of failures D_i for different subsystems coincides with D_1. The lower γ-confidence limit of the system PFFO calculated with the help of the method of equivalent tests is

$$\underline{R} = 1 - [1 - \underline{P}_\gamma)N_1, D_i)]^2 = 1 - [1 - \underline{P}_{0.9}(100, 3.88)]^2$$
$$= 1 - (1 - 0.923)^2 = 0.994.$$

In this case the method of equivalent tests produces a γ-confidence limit much higher than other methods considered above (see Table 7.4 and Examples 7.17 and 7.20). □

7.6.2 Method of Reduction

This method was considered elsewhere (Martz and Duran, 1985; Tyoskin and Kursky, 1986). Let us begin with a series–parallel system described by reliability index (7.84). The idea of the method is in the following. First we construct the point estimate $\hat{R}_i$ and lower γ-confidence limit $\underline{R}_i$ for each subsystem PFFO $R_i = 1 - (1 - p_i)^{n_i}$. These values are constructed in a standard way as

$$\hat{R}_i = 1 - (1 - \hat{p}_i)^{n_i} = 1 - \left(\frac{d_i^*}{N_i}\right)^{n_i}, \qquad \underline{R}_i = 1 - [1 - \underline{P}_\gamma(N_i, d_i^*)]^{n_i},$$

where $\underline{P}_\gamma(N_i, d_i^*)$ is the standard Clopper–Pearson lower γ-confidence limit. Then we replace the ith subsystem (redundant group) by an "equivalent unit" with the equivalent number of tested units M_i and equivalent number of failures r_i, which are chosen from the conditions

$$1 - \frac{r_i}{M_i} = \hat{R}_i, \qquad \underline{P}_\gamma(M_i, r_i) = \underline{R}_i. \tag{7.120}$$

In other words, values of M_i and r_i are chosen in such a way that the point estimate and lower γ-confidence limit for an equivalent unit coincides with the corresponding quantities of the ith subsystem.

Thus, an initial series–parallel system is replaced by some supplementary (imaginary) series system consisting of m equivalent units for each of which we have test results M_i and r_i, $i = \overline{1, m}$. After this, the lower γ-confidence limit can be constructed with the help of any known method for a series system (e.g., by the Lidstrem–Madden method).

Notice that in contrast to the previous method of equivalent tests (see Section 7.6.1), where the number of tests N_i was kept constant, in this case both r_i and M_i, determined from (7.120), are varied. Both quantities might be noninteger. In this case the lower γ-confidence limit $\underline{P}_\gamma(M_i, r_i)$ is found as the solution of the equation

$$B_p(M_i - r_i, r_i + 1) = 1 - \gamma$$

where $B_p(a, b)$ is the beta function (see Section 7.9.2).

7.6.3 Method of Reduction for Complex Systems

In contrast to the previous method, the method of equivalent tests can be easily extended to systems with more complex structure than series–parallel.

Series–Parallel Systems with Different Units Consider a system consisting of m redundant groups (subsystems). The ith subsystem might consist of n_i different units in parallel. In this case, the system PFFO has the form

$$R = \prod_{1 \le i \le m} \left[1 - \prod_{1 \le j \le n_i} (1 - p_{ij}) \right],$$

where p_{ij} is the PFFO of the jth unit of the ith subsystem. For each parameter p_{ij}, we have results of independent binomial tests with the following results: N_{ij} units were tested and d_{ij}^* of them failed. The PFFO of the ith subsystem is denoted by

$$R_i = 1 - \prod_{1 \le j \le n_i} (1 - p_{ij}).$$

As before, we construct a corresponding point estimate $\hat{R}$ for this index,

$$\hat{R}_i = 1 - \prod_{1 \le j \le n_i} (1 - \hat{p}_{ij}) = 1 - \prod_{1 \le j \le n_i} \left(\frac{d_{ij}^*}{N_{ij}} \right),$$

and the lower γ-confidence limit $\underline{R}_i$, which can be constructed with the help of any known method for a parallel system, for instance, that considered in Section 7.4.1.

After this, the entire ith subsystem is replaced by the supplementary equivalent unit with equivalent number of tests M_i and equivalent number of failures r_i. The numbers M_i and r_i are again found from (7.120). Thus, we replace an initial series–parallel system by a series system consisting of m equivalent units. After this, the lower γ-confidence limit can be constructed by the Lidstrem–Madden method.

Systems with Reducible Structures The method of reduction can be extended on the so-called reducible structures. Remember that a reducible structure is such a structure that can be obtained from an initial simple series (or parallel) system by replacing units of this structure by series and parallel substructures. Such a procedure of replacement can be recurrently continued. It is obvious that a reducible structure can be "converted" up to a single unit by the inverse procedure (replacing simple series and parallel fragments by a single equivalent unit). In such a manner an entire initial reducible system can be transformed into a single equivalent unit. An example of such a structure is depicted in Figure 8.1.

The method of reduction is applied for statistical problems as follows. Each kth parallel structure (fragment of a system) is replaced by an equivalent unit with equivalent number of tests N_k and equivalent number of failures r_k. (These values are found as above.)

In an analogous manner, we replace each ith series structure (fragment of a system) by an equivalent unit with equivalent number of tests N_k and equivalent number of failures D_k, found as

$$D_k = N_k(1 - \hat{R}_k).$$

Here $\hat{R}_k$ is the point estimate of the PFFO of the fragment replaced. In this case the point estimate $1 - D_k/N_k$ and the lower γ-confidence limit $\underline{P}_\gamma(N_k, D_k)$ of the PFFO of the equivalent unit coincide with the corresponding characteristics of the replaced fragment.

As mentioned above, the procedure of recursive reduction allows one to represent an initial reducible system as a single equivalent unit.

Example 7.24 Consider a series–parallel system analogous to that in Examples 7.18 and 7.21 (each redundant group consists of identical units). The structure of the system is depicted in Figure 7.5 and input data are presented in Table 7.6

The reduction method gives for this system the lower γ-confidence limit of the system PFFO as 0.987 with confidence coefficient $\gamma = 0.9$. The method of hyperplane, the modified method of hyperplane, and the method of equivalent tests give for this case the quantities 0.966, 0.991, and 0.991, respectively. □

7.6.4 Method of Fiducial Probabilities

The idea of the method was suggested by Fisher (1935) and developed by many authors (e.g., see Rao, 1965; Fraser, 1961). Consider this approach on a simple example. Let θ_i be a reliability parameter of unit i and random variable x_i be the result of testing this unit. Denote the distribution function of random variable x_i for a fixed value of parameter θ_i by

$$F_i(t, \theta_i) = P_{\theta_i}(x_i \leq t) \tag{7.121}$$

For the sake of simplicity, assume that function (7.121) is continuous in t. Assume also that this function is continuous and monotone increasing in parameter θ_i.

Let x_i^* be a value of random variable x_i observed in the result of test. For fixed x_i^*, we consider parameter θ_i as a random variable with the distribution $F_i(x_i^*, \theta_i)$. The distribution of parameter θ_i defined in such a way is called a fiducial distribution. The upper and lower γ-fiducial limits $\underline{\theta}_i$ and $\overline{\theta}_i$ for parameter θ_i for given fixed test result x_i^* are determined from the conditions

$$F_i(x_i^*, \underline{\theta}_i) = 1 - \gamma, \qquad F_i(x_i^*, \overline{\theta}_i) = \gamma. \tag{7.122}$$

It means that they are corresponding quantiles of fiducial distribution $F_i(x_i^*, \theta_i)$. The γ-fiducial limit coincides with the corresponding lower γ-confidence limit for parameter θ_i, as can be seen from Equations (1.31) for the confidence limit.

Notice that in this approach parameter θ_i is not a random variable, but some unknown constant, and function $F_i(x_i^*, \theta_i)$ is a distribution function of the result of observations x_i for a given fixed value of parameter θ_i. So, we interpret $F_i(x_i^*, \theta_i)$ as a distribution function of parameter θ_i for a given fixed x_i. The correctness of the procedure is not obvious and, even more, doubtful. The probabilistic ("physical") sense of such a transform also is unclear. (For details see, e.g., Rao, 1965, Section 5b; Zacks, 1971, Section 10.6; and Stein, 1959. For application to reliability tests see Pavlov, 1982a, Sections 2.4, 4.5, and 4.6.) Nevertheless, independently of the interpretation and validation of the method, we might consider it a convenient formal approach that gives good practical results in many cases.

Let R be a reliability index of a system consisting of m units of different types. This index is a function $R = R(\boldsymbol{\theta}) = R(\theta_1, \ldots, \theta_m)$ of parameters $\boldsymbol{\theta} = (\theta_1, \ldots, \theta_m)$ of the units. Let $\mathbf{x}^* = (x_1^*, \ldots, x_m^*)$, be observed values of the vector of tet results $\mathbf{x}^* = (x_1^*, \ldots, x_m)$, where x_i is the test result for the ith type units. Test results for different units are assumed independent. For each parameter θ_i let us construct fiducial distribution $F_i(x_i^*, \theta_i)$ in a manner proposed above. Now the reliability index $R = R(\theta_1, \ldots, \theta_m)$ can be considered as a function of fiducial random variables θ_i with corresponding distributions. The fiducial distribution function for R for a given vector of test results x^* is determined by the formula

$$\Phi(\mathbf{x}^*, R) = \int_{\mathbf{R}(\theta_1,\ldots,\theta_m)\le R} \cdots \int \prod_{1\le i\le m} f_i(x_i^*, \theta_i)\, d\theta_i \tag{7.123}$$

where $f_i(x_i^*, \theta_i) = \partial F_i(x_i^*, \theta_i)/\partial\theta_i$ is the density function of the fiducial distribution of parameter θ_i, $i = \overline{1, m}$.

The lower and upper γ-fiducial limits $\underline{R}$ and $\overline{R}$ for the system PFFO R are determined from the conditions

$$\Phi(\mathbf{x}^*, \underline{R}) = 1 - \gamma, \qquad \Phi(\mathbf{x}^*, \overline{R}) = \gamma. \tag{7.124}$$

Analytical calculation of distribution (7.123) and limits $\underline{R}$ and $\overline{R}$ in (7.124) is usually too complicated. Nevertheless, in many practical cases these values can be found by Monte Carlo simulation.

The fiducial approach discussed above was used for reliability problems (mostly for binomial and exponential models) in Springer and Thompson (1964), Kredentser, et al. (1967), Farkhad-Zadeh (1979), Groisberg (1980a,b), and Pavlov (1980, 1981a,b). More available sources are Gnedenko (1983) and Ushakov (1994).

Example 7.25 (Binomial Model) Let $\theta_i = p_i$ be the unit PFFO, $x_i = d_i$ be the number of failures during the test, and N_i be the number of tested units of type i. In this case, distribution function (7.121) has the form

$$F_i(d_i, p_i) = \sum_{0\le j\le d_i} \binom{N_i}{j} p_i^{N-j}(1 - p_i)^j. \tag{7.125}$$

Function (7.125) is a distribution function of the test result d_i for a given fixed value of parameter p_i and, at the same time, is a fiducial distribution function of parameter p_i for a given fixed test result d_i. Further, let the system consist of m different types of units with parameters $\mathbf{p} = (p_1, \ldots, p_m)$ and function $R = R(\mathbf{p}) = R(p_1, \ldots, p_m)$ reflect the dependence of system PFFO on unit parameters. For instance, for the series–parallel system considered above,

$$R = \prod_{1\le i\le m} [1 - (1 - p_i)^{n_i}].$$

The value of R is considered below as a fiducial random variable with a distribution determined by distributions of parameters (7.125). Then, the γ-fiducial limit for R is constructed as mentioned above. □

Example 7.26 (Exponential Model) A system consists of m different units each of which has the exponential distribution of TTF. The failure rate of the ith unit is λ_i, $i = \overline{1, m}$. In this case $\theta_i = \lambda_i$ and $x_i = S_i$, and S_i is the total time of testing of all units of the ith type untl occurrence of r_i failures. In this case

$$F_i(S_i, \lambda_i) = 1 - e^{-\lambda_i S_i} \sum_{0 \le j \le r_i - 1} \frac{(\lambda_i S_i)^j}{j!} \tag{7.126}$$

is a function of test results S_i for a given fixed value of parameter λ_i and simultaneously the same function is a fiducial distribution of parameter λ_i for a given fixed value of test result S_i. Let $R = R(\lambda) = R(\lambda_1, \ldots, \lambda_m)$ be a function that gives the dependence of the system PFFO on unit parameters. Then λ-fiducial unit for R can be constructed on the basis of fiducial distribution (7.126) for parameters λ_i, $i = \overline{1, m}$, as described above. □

Sometimes the fiducial approach produces obviously ineffective limits (which is illustrated by the example below).

Example 7.27 (Binomial Model, No-Failure Test) Take the series system considered in Example 7.25 for the case $n_1 = n_2 = \cdots = n_m = 1$. All units are tested in equal numbers, $N_1 = N_2 = \cdots = N_m = N$, and no failures have been observed, $d_1 = d_2 = \cdots = d_m = 0$.

In this case the lower γ-fiducial limit for the system PFFO can be easily found analytically. Indeed, from (7.125) it follows that parameter p_i is a fiducial random variable with distribution function

$$F_i(d_i, p_i) = F_i(0, p_i) = p_i^{N_i} = p_i^N, \qquad i = \overline{1, m}. \tag{7.127}$$

The system PFFO in this case is

$$R = p_1 p_2 \times \cdots \times p_m. \tag{7.128}$$

Thus, the lower γ-fiducial limit $\underline{R}$ for R equals a quantile of the level of $1 - \gamma$ for random variable (7.128) under the condition that each parameter p_i has distribution (7.127). This random variable can be found, for instance, by the following way. Introduce random variables $\xi_i = p_i^N$, $i = \overline{1, m}$. □

Each random variable ξ_i has the uniform distribution on interval [0, 1]. Introduce also random variables $\eta_i = -\ln \xi_i$, $i = \overline{1, m}$. Since ξ_i has the uniform distribution on interval [0, 1], random variable η_i has the exponential distribution $1 - e^{-x}$. Thus, R can be written in the form

$$R = p_1 \times p_2 \times \cdots \times p_m = \prod_{1 \le i \le m} \xi_i^{\frac{1}{N}} = \exp\left(-\frac{1}{N} \sum_{1 \le i \le m} \eta_i\right).$$

Then use the well-known fact that random variable $2(\eta_1 + \eta_2 + \cdots + \eta_m)$

has the standard χ^2 distribution wth $2m$ degrees of freedom (see Section 1.2). Inequality $R \geq \underline{R}$ is equivalent to inequality

$$2 \sum_{1 \leq i \leq m} \eta_i \leq -2N \ln \underline{R}.$$

After simple transformations we obtain that the lower γ-fiducial limit $\underline{R}$ for the system PFFO is

$$\underline{R} = e^{-\chi^2_\gamma(2m)/2N}, \tag{7.129}$$

where χ^2_γ $(2m)$ is the quantile of the level γ for the χ^2 distribution with $2m$ degrees of freedom. The lower limit (7.129) rapidly decreases with increasing number of system units, m. On the first glance, this fact seems natural, since the system has a series structure. However, the best lower γ-confidence limit obtained by Mirnyi and Solovyev (1964) for the same situation (see Sections 6.1–6.3) is

$$\underline{R} = (1 - \gamma)^{1/N} = e^{-\chi^2_\gamma(2)/2N}, \tag{7.130}$$

and it does not depend on m and for $m > 1$ is always better than (7.129). The more m, the worse the fiducial limit (7129). (Notice incidentally that a Bayesian approach possesses the same deficiency.)

Nevertheless, if we do not deal with no-failure tests, the fiducial method (as well as the Bayesian one) can deliver good results (see Chapter 8). Besides, this method is rather universal and can be applied for systems with various types of structures. It makes this method very popular in engineering applications.

Notice also that the fiducial method is approximate in the sense that for the γ-fiducial limit (e.g., lower) it does not guarantee the inequality for the confidence coefficient.

$$P_\theta\{\underline{R} \leq R(\boldsymbol{\theta})\} \geq \gamma \tag{7.131}$$

for all possible values of parameters $\boldsymbol{\eta} = (\eta_1, \ldots, \eta_m)$. Although for $m = 1$, that is for one unknown parameter, the γ-fiducial limit simultaneously represents the γ-confidence limit, for $m > 1$ justification of (7.131) does not follow from anywhere (for all $\boldsymbol{\eta}$). In chapter 8 it will be shown that there are examples where the fiducial method does not work. However, at the same time one can find many examples where this method is valid. For instance, for a wide class of exponential models for systems with complex structures, the fiducial approach delivers strong lower γ-confidence limit; that is, its application is correct and effective.

7.6.5 Bootstrap Method

The idea of this method (Efron, 1979, 1982) will be illustrated using an example of the binomial model. Let $R = R(\mathbf{p}) = R(p_1, \ldots, p_m)$ be a function expressing the dependence of the system PFFO on binomial parameters of its units $\mathbf{p} = (p_1, \ldots, p_m)$. The point estimate $\hat{R}$ of the reliability index R is written as

$$\hat{R} = R(\hat{\mathbf{p}}) = R(\hat{p}_1, \ldots, \hat{p}_m),$$

where $\hat{p}_i = 1 - d_i/N_i$ is the point estimate of parameter p_i.

Denote the distribution function of the point estimate $\hat{R}$ for a given vector of parameters $\mathbf{p} = (p_1, \ldots, p_m)$ by $\Phi(t, \mathbf{p}) = P_p(\hat{R} \le t)$. We need to construct the lower γ-confidence limit $\underline{R}$ for the system PFFO $R = R(\mathbf{p})$. Let us define $\underline{R}$ as the quantile of the level $1 - \gamma$ of the distribution function of estimate $\hat{R}$, that is, from the condition

$$\Phi(\underline{R}, \mathbf{p}) = 1 - \gamma. \tag{7.132}$$

The distribution of estimate $\hat{R}$ depends on parameters $\mathbf{p} = (p_1, \ldots, p_m)$, which are unknown by the formulation of the problem. Therefore a direct determination of the value of $\underline{R}$ from (7.132) is impossible. To bypass this obstacle, set $\mathbf{p} = \mathbf{p}^*$, where $\mathbf{p}^* = (p_1^*, \ldots, p_m^*)$ and $p_i^* = 1 - d_i^*/N_i$, $i = \overline{1, m}$, where d_i^* is the number of observed failures. Function $\Phi(t, \mathbf{p}^*)$ is called a bootstrap distribution of estimate $\hat{R}$. So, value $\underline{R}$ is determined from the condition

$$\Phi(\underline{R}, \mathbf{p}^*) = 1 - \gamma. \tag{7.133}$$

In other words, to find $\underline{R}$, we use an estimate of the distribution $\Phi(t, \mathbf{p}^*)$ instead of the distribution of the estimate, $\Phi(t, \mathbf{p})$.

Finding an exact analytical expression for a distribution function in the left side of (7.133) is usually a difficult task. Therefore most often Monte Carlo simulation is used for estimating $\Phi(t, \mathbf{p}^*)$. In this case, for simulation we use estimates $p_i^* = 1 - d_i^*/N_i$, which are taken from real tests (see Section 2.6) instead of unknown values of parameters p_i.

The main benefit of this method is its universality, simplicity, and clearness. However, this method sometimes produces (for lower limits) estimates that are to optimistic (see, e.g., Martz and Duran, 1985).

7.6.6 Normal Approximation

Consider a binomial testing plan for a series system. For the case of highly reliable units the system PFFO can be approximately written in the form

$$R = \prod_{1 \le i \le m} (1 - q_i) \approx \exp\left(\sum_{1 \le i \le m} q_i\right),$$

where $q_i = 1 - p_i$ is the probability of the ith unit PFFO, $q_i << 1/m$, $i = 1, \ldots, m$. The construction of the lower confidence level for the system PFFO is reduced to the construction of the UCL for the value

$$f = f(q) = \sum_{1 \le i \le m} q_i.$$

As an initial statistic S that is used for the construction of the confidence limit, take the unbiased point estimate $\hat{f}$ for f, that is,

$$S = \hat{f} = \sum_{1 \le i \le m} \hat{q}_i, \tag{7.134}$$

where $\hat{q} = d_i/N_i$ is the standard point estimate for parameter q_i, $i = 1, \ldots, m$. Denote the mathematical expectation and variance of statistics S for a given vector of parameters $\mathbf{q} = (q_1, \ldots, q_m)$ by $E(\mathbf{q})$ and $V(\mathbf{q})$:

$$\begin{aligned} E(\mathbf{q}) &= ES = E\left(\sum_{1 \le i \le m} \hat{q}_i\right) = \sum_{1 \le i \le m} q_i, \\ V(\mathbf{q}) &= \mathrm{Var}\{S\} = \sum_{1 \le i \le m} \mathrm{Var}\{\hat{q}_i\} = \sum_{1 \le i \le m} \frac{q_i(1 - q_i)}{N_i} \approx \sum_{1 \le i \le m} \frac{q_i}{N_i}. \end{aligned} \tag{7.135}$$

The distribution function of statistic S for given $\mathbf{q}$ is denoted by $F(t, \mathbf{q}) = P_q(S \le t)$. Random variable S represents the sum of i independent random variables with finite moments. Therefore for large enough m and N_i the distribution of statistic S can be approximated by the normal distribution with the mean $E(\mathbf{q})$ and variance $V(\mathbf{q})$

$$F(t, q) \approx \Phi\left(\frac{t - E(\mathbf{q})}{\sqrt{V(\mathbf{q})}}\right), \tag{7.136}$$

where $\Phi(u)$ is the standard normal distribution.

Applying general functions (7.9) and (7.11), we obtain that the upper γ-confidence limit $\hat{f}$ for $f = f(\mathbf{q})$ is defined from the equation

$$K_1(\bar{f}) = S^*, \tag{7.137}$$

where $S^* = \Sigma_{1 \le i \le m} (d_i^*/N_i)$, d_i^* is the number of observed failures, and function $K_1(f)$ is defined as

$$K_1(f) = \min_{\mathbf{q} \in A_f} t_1(\mathbf{q}),$$

where the minimum is taken over the set $A_f = \{\mathbf{q}: f(\mathbf{q}) = f\}$, and the value $f_1(\mathbf{q})$ is the quantile of level $1 - \gamma$ of the distribution function of statistic S. In accordance with (7.136), $t_1(\mathbf{q})$ is defined from the equation

$$\Phi\left(\frac{t_1 - E(\mathbf{q})}{\sqrt{V(\mathbf{q})}}\right) = 1 - \gamma,$$

from where

$$t_1(\mathbf{q}) = E(\mathbf{q}) - u_\gamma \sqrt{V(\mathbf{q})},$$

where U_γ is the quantile of the level γ of the standard normal distribution. (Here the difference is understood as operations over vector components.)

Thus, for the normal approximation, the upper γ-confidence limit $\bar{f}$ is defined from (7.137), where

$$K_1(f) = \min_{\mathbf{q} \in A_f} E(\mathbf{q}) - u_\gamma \sqrt{V(\mathbf{q})} \tag{7.138}$$

is the minimum taken over all $\mathbf{q} = (q_1, \ldots, q_m)$ that satisfy the constraints

$$\sum_{1 \le i \le m} q_i = f, \; 0 \le \gamma_i \le 1, \qquad i = 1, \ldots, m. \tag{7.139}$$

The value of $U_\gamma > 0$ if $\gamma > \frac{1}{2}$. Taking into account inequality (7.135), the minimum (7.138) is written in the form

$$K(f) = E(f) - u_\gamma \sqrt{V^+(f)},$$

where

$$V^+(f) = \max_{\mathbf{q} \in A_f} V(\mathbf{q}) = \max_{\mathbf{q} \in A_f} \sum_{1 \le i \le m} \frac{q_i}{N_i} \tag{7.140}$$

is the maximum value of the variance $V(\mathbf{q})$ within the area of parameter $\mathbf{q}$ satisfying constraints (7.139). The maximum (7.140) belongs to one of the corner points of the form $(\underbrace{0, \ldots, 0}_{i-1}, f, 0, \ldots, 0)$. It produces $V^+(f) =$

f/N_1, where $N_1 = \min_{1 \le i \le m} N_i$ is the minimum volume of tests among all units.

Thus, the upper γ-confidence limit $\bar{f}$ for the normal approximation can be defined from the equation

$$\bar{f} - u_\gamma \sqrt{\frac{\bar{f}}{N_1}} = \sum_{1 \le i \le m} \frac{d_i^*}{N_1}. \tag{7.141}$$

Let us introduce an equivalent number of failures

$$\tilde{D}_1 = N_1 S^* = \sum_{1 \le i \le m} \left(\frac{N_1}{N_i}\right) d_i^*.$$

For small numbers of failures $d_i << N_i$, i 1, . . . , m, the value $\tilde{D}_1$ coincides with the equivalent number of failures D_1 given by formula (7.57), differing only in the infinitesimally small values of second and higher orders:

$$\sum_{1 \le i \le m} \left(\frac{N_1}{N_i}\right) d_i^* \approx N_1 \left[1 - \prod_{1 \le i \le m} \left(1 - \frac{d_i^*}{N_i}\right)\right].$$

Equation (7.141) for the confidence limit can also be written in the form

$$\bar{f} - u_\gamma \sqrt{\frac{\bar{f}}{N_1}} = \frac{\tilde{D}_1}{N_1}. \tag{7.142}$$

Finally, the lower confidence level for the system PFFO can be computed as $\underline{R} = \exp(-\bar{f})$.

From (7.142) it follows that the lower confidence level for the system PFFO is computed in the same way as the lower confidence level for a single unit of type 1 with the minimum test volume N_1 and the equivalent number of failures $\tilde{D}_1$. This is the idea of equivalent tests for the series system under present consideration. Thus, to construction the lower confidence level of the PFFO of a series system, if we take the point statistic (7.134) as the initial statistic S, then the normal approximation of this confidence limit corresponds to that obtained by the method of equivalent tests (for $\gamma > \frac{1}{2}$). [See Krol (1974, 1975) and Pavlov, (1982a)] for construction of analogous series–parallel systems.]

7.7 METHOD OF THE USE OF BASIC STRUCTURES

Accurate methods of confidence estimation considered in Section 7.2 and 7.5 [i.e., the methods for which a given value of γ is guaranteed by inequality

(7.111)] in many cases deliver very conservative estimates for system reliability. This is the reason for the use of different heuristic or approximate methods. based in some sense on asymptotic results (see Section 7.6). Notice that the major existing methods of confidence estimation of complex system reliability do not take into account optimization approaches due to added complications. In addition, asymptotic methods for large samples, for instance, the maximum likelihood method [see Wilks, 1938; Madansky, 1965; and others), do not appear to be a good approximation for small samples.

Below suggest an approach based on the construction of an algorithm of system confidence limit estimation based on methods for systems with simpler structures. This approach reflects the development of the simple idea (already used in Sections 4.5 and 7.4) of constructing system confidence limits on the basis of a simpler system consisting of the same units.

The construction of the algorithm is realized in two stages. At first, an ensemble of simple auxiliary structures is introduced, each of which consists of the same units as the investigated system. For each of these structures, the confidence limits are constructed with the help of "trivial algorithms" used for developing the algorithm for the complex structure confidence estimation. Then some operations are introduced (see Theorem 7.1) that allow us to combine these trivial algorithms. Then the best algorithm is searched among the class of these combined algorithms. It is shown that for some general conditions there is an optimal algorithm (Theorem 7.2) that gives the confidence limits that are better for arbitrary test results than any trivial algorithm for the basic ensemble (Theorem 7.3).

7.7.1 Basic Ensemble of Trivial Algorithms

Let a system consist of m units of different types and $\mathbf{z} = (z_1, z_2, \ldots, z_m)$ be the vector or reliability parameters of units, where z_i is the parameter of the ith type unit, $i = 1, \ldots, m$. Vector $\mathbf{z}$ takes its values from the area $Z = \{\mathbf{z}: z_i \geq 0, i = 1, \ldots, m\}$ of m-dimensional Euclidean space. Let $\mathbf{x} = (x_1, \ldots, x_m) \in X$ be the vector of test results and $P_{\mathbf{z}}\{\cdot\}$ be the distribution of test results $\mathbf{x} \in X$ for given values of vector $\mathbf{z} \in Z$. Let $f = f(\mathbf{z})$ denote the function expressing the dependence between the reliability index and unit reliability parameters $\mathbf{z}$. This function is called the system reliability function. We are interested in constructing the confidence limit (say, upper) with confidence coefficient not less than given γ for $f = f(\mathbf{z})$; that is, we are searching for a function of the test results $\overline{f} = \overline{f}(\mathbf{x})$ such that the inequality

$$P_{\mathbf{z}}\{\overline{f} \geq f(\mathbf{z})\} \geq \gamma \tag{7.143}$$

holds for all $\mathbf{z} \in Z$.

In correspondence with the general method of confidence sets (see Sections 1.4 and 7.2), the limit $\overline{f} = \overline{f}(\mathbf{x})$ is found as the maximum value of the reliability function $f(\mathbf{z})$ over the entire γ-confidence site $H_x \subset Z$. A collection of sets in the space of parameters

$$H_x \subset Z, \qquad \mathbf{x} \in X \tag{7.144}$$

is called a collection of γ-confidence sets for $\mathbf{z} \in Z$ if inequality

$$P_z\{\mathbf{z} \in H_x\} \geq g \tag{7.145}$$

holds for all $\mathbf{z} \in Z$. The upper γ-confidence limit $\overline{f} = \overline{f}(\mathbf{x})$ for $f = f(\mathbf{z})$ is computed as

$$\overline{f} = \overline{f}(\mathbf{x}) = \max_{\mathbf{z} \in H_x} f(z) \tag{7.146}$$

It follows directly from (7.145) that $\overline{f}$ constructed in such a way satisfies inequality (7.143) for all $\mathbf{z} \in Z$, that is, $\overline{f}$ is the γ-UCL for $f = \underline{f}(\mathbf{z})$.

Thus, an algorithm for computation of the confidence limit $\overline{f} = \overline{f}(\mathbf{x})$ is defined by a given collection of confidence sets (7.144) and by the operation of taking the optimum in (7.146). Different collections of γ-confidence sets correspond to different algorithms of computing the confidence set.

Further assume that there are several different collections of γ-confidence sets

$$\{H_x^{\mu} \subset Z, x \in X\}, \qquad \boldsymbol{\mu} \in M, \tag{7.147}$$

such that

$$P_z\{\mathbf{z} \in H_x^m\} \geq \gamma$$

for all $\mathbf{z} \in Z$ and all for $\boldsymbol{\mu} \in M$. The value of m is the "mark" of a collection of confidence sets in (7.147). Each collection of confidence sets H_x^m, $\mathbf{x} \in X$, corresponds to its own algorithm of computation of the following kind of the confidence limit:

$$\overline{f}_{\mu} = \overline{f}_{\mu}(x) = \max_{z \in H_x^{\mu}} f(z), \qquad \boldsymbol{\mu} \in M. \tag{7.148}$$

Thus, the value of $\boldsymbol{\mu}$ is the mark of the algorithm of computation of the confidence limit (7.148), and M defines a collection of such algorithms. An aggregation of all algorithms (7.139) with $\boldsymbol{\mu} \in M$ is called a basis collection of algorithms for computation of the confidence limit for the system reliability function $f(\mathbf{z})$.

Examples of such basis collections are collections of algorithms for systems with the simplest structures as series and parallel. Let us consider, for instance, a binomial model. Let $z_i = -\ln p_i$, where p_i is the binomial parameter equal to the PFFO of a unit of the ith type, and $x_i = d_i$ be the number of failures observed during a test of N_i units of the ith type, $i = 1, \ldots, m$, $\mathbf{x} = (d_1, \ldots, d_m)$. Assume that we use these units of an initial complex system for constructing some series system in which a unit of the ith type is used μ_i, times, $i = 1, \ldots, m$. The PFFO of this series system is

$$R_\mu = \prod_{1 \le i \le m} p_i^{\mu_i} = \exp\left(-\sum_{1 \le i \le m} \mu_i z_i\right) \tag{7.149}$$

where $\boldsymbol{\mu} = (\mu_1, \ldots, \mu_m)$ is the vector defining the number of different units in the system. Let us construct the lower confidence level on the basis of a known method, for instance, the Lidstrem–Madden method (7.58) and (7.59) for the system reliability (7.140):

$$\underline{R}_\mu = \underline{R}_\mu(x = \min_{1 \le i \le m} [\underline{P}_\gamma(N_i, D_i)]^{\mu_i}, \tag{7.150}$$

where values D_i are found from the conditions

$$\left(1 - \frac{D_i}{N_i}\right)^{\mu_i} = \hat{R}, \qquad \hat{R} = \prod_{1 \le j \le m} \left(1 - \frac{d_j}{N_j}\right)^{\mu_j}.$$

By definition of the lower confidence level $\underline{R}_\mu$, we have

$$P_z\,\{\underline{R}_\mu \le R_\mu\} \ge \gamma.$$

The above inequality can be written for all $\mathbf{z} \in Z$ in the equivalent form

$$P_z\left\{\sum_{1 \le i \le m} \mu_i z_i \le \bar{g}_\mu(x)\right\} \ge \gamma, \tag{7.151}$$

where $\bar{g}_\mu(x) = -\ln \underline{R}_\mu(x)$.

Consider a collection of sets in the space Z of the unit parameters $\mathbf{z} = (z_1, \ldots, z_m)$:

$$H_x^\mu = \left\{\mathbf{z}\colon \sum_{1 \le i \le m} \mu_i z_i \le \bar{g}_\mu(x)\right\}, \qquad x \in X, \tag{7.152}$$

where $\mathbf{x} = (d_1, \ldots, d_m)$ is the vector of the numbers of failures observed at the different unit tests, X is the set of all vectors $\mathbf{x}$ with positive integer coordinates. By (7.151) the collection of sets of the type (7.152) forms the collection of γ-confidence sets for $\mathbf{z} \in Z$. It means that the value

$$\bar{f}_\mu = \bar{f}_\mu(x) = \max_{z \in H_x^\mu} f(z) \tag{7.153}$$

is the upper γ-confidence limit for the reliability function $f = f(\mathbf{z})$.

Further, consider an ensemble of all such series structures with various numbers of units of different types, $\boldsymbol{\mu} = (\mu_1, \ldots, \mu_m)$. For any fixed $\boldsymbol{\mu}$, each structure of such type is characterized by its own collection of confidence sets (7.152) and its own algorithm for confidence limit computation (7.153) for the reliability function of the initial system. We thus obtain a basis set of computational algorithms for the confidence limit of the form

$$\bar{f}_\mu(x) = \max_{z \in H_x^\mu} f(\mathbf{z}), \qquad \boldsymbol{\mu} \in M, \tag{7.154}$$

where $\boldsymbol{\mu} = (\mu_1, \ldots, \mu_m)$ is the mark of an algorithm (and simultaneously it is the vector of numbers of units in the corresponding auxiliary series system). The set M represents the set of all m-dimensional vectors $\boldsymbol{\mu}$ with positive integer coordinates:

$$M = \{\boldsymbol{\mu}: \mu_i = 0, 1, \ldots; i = 1, \ldots, m\}.$$

Notice that the confidence limit $\underline{R}_m$ is not defined just for an integer but for any positive value μ_i. Therefore below we will use M defined as

$$M = \{\boldsymbol{\mu}: \mu_i \geq 0; i = 1, \ldots, m\}.$$

We will call a collection of algorithms of the kind (7.154) basis collection of "trivial" algorithms. Further the basis collection of confidence sets (7.152) and corresponding collection of trivial algorithms (7.154) will be significantly extended, and within this new extended class of algorithms, we will find the optimal algorithm (Theorem 7.1), which delivers for all test results $\mathbf{x}$ the best confidence limit for $f(\mathbf{z})$ in comparison with any trivial algorithm of the basis set (7.154).

7.7.2 Constructing New Algorithms from the Basis Collection of Trivial Algorithms

Denote the boundary hyperplane for the confidence set in (7.152) by

$$\Gamma_x^\mu = \left\{ z \in Z: \sum_{1 \leq i \leq m} \mu_i z_i = g_\mu(x) \right\}. \tag{7.155}$$

Assume that the reliability function of a system $f(\mathbf{z}) = f(z_1, \ldots, z_m)$ is monotone increasing in each variable, has continuous partial derivatives of the first order, and is convex in $\mathbf{z} \in Z$. Assume also that the following conditions hold:

A. Function $g_m(\mathbf{x})$ is continuous in $\boldsymbol{\mu} \in M$.
B. For any $\boldsymbol{\mu}$ and $\boldsymbol{\nu}$ in M, sets Γ_x^{μ} and Γ_x^{ν} have nonempty intersection for any $\mathbf{x} \in X$.
C. $H_x^{\mu} = H_x^{c\mu}$ for any $c > 0$, $\mathbf{x} \in X$, $\boldsymbol{\mu} \in M$.

The justness of these conditions can be proved directly on the basis of (7.150), (7.152), and (7.155). Notice that the latter condition shows that the set

$$M = \{\boldsymbol{\mu}: \|\boldsymbol{\mu}\| = 1; \mu_i \geq 0; i = 1, \ldots, m\}$$

can be considered as the set M. In other words, a set of all "directing" unitary vectors $\mathbf{m}$ with positive coordinates where

$$\|\boldsymbol{\mu}\| = \sqrt{\sum_{1 \leq i \leq m} \mu_i^2}$$

is the standard norm in m-dimensional Euclidean space.

Now let s be a curve in Z given in parametric form:

$$s = \{\mathbf{z}: z_i = j_i(t), i = 1, \ldots, m; t \geq 0\}, \tag{7.156}$$

where $j_i(t)$ are continuous, monotone functions increasing in t.

Let any ensemble S of nonintersected curves of type (7.156) and covering the entire space Z,

$$S = \{s_v: \mathbf{v} \in V\}, \tag{7.157}$$

be called the S-division of the parameter space Z. Arbitrary curves s^v and s_u above do not intersect if $v \neq u$ [may be with the exception of a unique point of the origin, $\mathbf{z} = (0, 0, \ldots, 0)$] and

$$\bigcup_{v \in V} s_v = Z.$$

An example of such a division is $S = \{s_v: \mathbf{v} \in V\}$ of the space Z by various rays (semilines) initiating at the origin:

$$s_v = \{\mathbf{z}: z_i = v_i t; 1 \leq i \leq m; t \geq 0\}$$

where $\mathbf{v} = (v_1, \ldots, v_m)$ is a "directing" unitary vector $\|\mathbf{v}\| = 1$ that determines the direction of ray s_v and V is the set of all directing unitary vectors with positive coordinates

$$V = \{\mathbf{v}: \|\mathbf{v}\| = 1;\ v_i \geq 0;\ i = 1, \ldots, m\}.$$

Let some S-division (7.157) of the parameter space Z exist and $a = a(\mathbf{v})$ be a one-to-one reflection

$$a: V \Rightarrow M.$$

Then the following theorem holds.

Theorem 7.1 The collection of sets

$$H_x(S, \alpha) = \bigcap_{v \in V} \{s_v \cap H_x^{\alpha(v)}\}, \qquad \mathbf{x} \in X, \tag{7.158}$$

represents a collection of γ-confidence sets for $\mathbf{z} \in Z$. □

Proof Let $\mathbf{z}$ be a fixed point in Z. Then by the definition of the S-division there is a unique $\mathbf{v} \in V$ such that $\mathbf{z} \in s_v$. For these fixed $\mathbf{z}$ and $\mathbf{v}$, we have

$$P_z\{\mathbf{z} \in H_x(S, a)\} = P_z\{\mathbf{z} \in (s_v \cap H_x^{a(v)})\} = P_z\{\mathbf{z} \in H_x^{a(v)}\} \geq \gamma.$$

The proof of the theorem follows from this statement. □

Thus each pair S and a

$$S = \{s_v, \mathbf{v} \in V\}, \qquad \text{a: } V \Rightarrow M$$

produces a new collection of γ-confidence sets formed in accordance with operation (7.149) on the basis of initial confidence limits H_x, $\mathbf{x} \in X$, $\boldsymbol{\mu} \in M$. For instance, a trivial (identically equal to a constant) reflection

$$a(\mathbf{v}) \equiv \mathbf{v} \in M$$

in accordance with (7.154) gives the collection of confidence sets $H_x(S, a)$, $\mathbf{x} \in X$, belonging to the basis collection of sets (7.147). However, an arbitrary reflection $a = a(\mathbf{v})$, not identically equal to a constant, produces a new limit for the system reliability function, $f = f(\mathbf{z})$.

Pair $A = (S, a)$ is called an algorithm of constructing the upper confidence limits (7.149) and corresponding confidence limit $\overline{f}$ for $f = f(\mathbf{z})$. This upper confidence limits $\overline{f} = \overline{f}_A(\mathbf{x})$, computed by the algorithm A, has the form

$$\overline{f}_A(\mathbf{x}) = \sup_{z \in H_x(A)} f(\mathbf{z}) = \sup_{v \in V} \sup_{z \in l_v \cap H_x^{\alpha(v)}} f(\mathbf{z}) \tag{7.159}$$

where $H_x(A) = H_x(S, a)$ is the confidence limit formed in accordance with (7.158).

The construction given above for computing a new confidence limit can be explained in the following way. Assume that we know a priori that parameter $\mathbf{z}$ belongs to some subset $s_v \subset Z$. Then the confidence limit (7.154) for $f = f(\mathbf{z})$ can be improved by the formula.

$$\overline{f}'_{\mu} = \sup_{z \in s_v \cap H_x^{\mu}} f(\mathbf{z}) \tag{7.160}$$

Notice that a collection of confidence limits H_x^m with an arbitrary mark $\boldsymbol{\mu}$, $\boldsymbol{\mu} \in M$, can be used in the formula (7.160). If there is no prior information concerning parameter $\mathbf{z}$, then to compute the UCL, we need to find the maximum of the value (7.151) over various subsets s_v that cover the entire parametric space Z. The mark m, generally speaking, can change dependent on v as some function $\boldsymbol{\mu} = a(\mathbf{v})$. In other words, we use a "best" collection of confidence limits from the basis ensemble (7.147) for estimation of the function $f(\mathbf{z})$ on each subset s_v. This leads us to constructions (7.158) and (7.159) where the choice of the function $a(\mathbf{v})$ dividing the space Z onto subsets s_v can be arbitrary. Thus the problem is in finding the function $a(\mathbf{v})$ and optimal division of L, that is, in choosing the optimal algorithm $A = (S, a)$ among various pairs (S, a).

7.7.3 Constructing the Optimal Algorithm

Let S^* be a class of all S-divisions of the parameter space Z and W be a class of all reflections $a = a(\mathbf{v}) \in M$ such that

$$\{\boldsymbol{\mu}: \boldsymbol{\mu} = a(\mathbf{v}), v \in V\} = M. \tag{7.161}$$

This condition means that all collections of confidence limits from the basis ensemble (7.147) are used for constructing the confidence limit (7.150). Let $A = S^* \times W$ be a class of corresponding algorithms $A = (S, a)$ of computation of the confidence limit (7.150) for the system reliability function $f = f(\mathbf{z})$.

Denote the gradient of function $f(\mathbf{z})$ by

$$\nabla f(\mathbf{z}) = \left(\frac{\partial f}{\partial z_1}, \ldots, \frac{\partial f}{\partial z_m} \right)$$

and introduce set V_f, $V_f \subset M$, of all normed gradients of function $f(\mathbf{z})$,

$$V_f = \left\{v: v = \frac{\nabla f(\mathbf{z})}{\|\nabla f(\mathbf{z})\|}, z \in Z\right\}.$$

Let $A^* = (S^*, \alpha^*)$, where division $S^* = \{s_v^*, v \in V_f\}$ of space Z and reflection $\alpha^* = \alpha^*(\mathbf{v}) \in M$ are defined

$$s_v^* = \{\mathbf{z}: \nabla f(z) = tv, t \geq 0\}, v \in V_f \tag{7.162}$$

$$\alpha^*(\mathbf{v}) = \mathbf{v}, \qquad \mathbf{v} \in V_f. \tag{7.163}$$

The following theorem shows that the confidence limit $\overline{f}_{A*}(x)$ that is calculated with respect to the algorithm $A^* = (S^*, \alpha^*)$ from (7.159) is the best one for $\mathbf{x} \in X$ in comparison with the limit $\overline{f}_A(\mathbf{x})$, that is, calculated by any other algorithm $A \in A^*$.

Theorem 7.2 For any algorithm $A \in A^*$ the inequality

$$\overline{f}_A(x) \geq \overline{f}_{A*}(x) \tag{7.164}$$

holds for all $\mathbf{x} \in X$. □

Proof Let $\Psi = \Psi(\boldsymbol{\mu})$, $\boldsymbol{\mu} \in M$, be the reflection inverse to $\alpha = \alpha(\mathbf{v})$, $\mathbf{v} \in V$, that is,

$$\Psi(\boldsymbol{\mu}) = \{\mathbf{v}: \alpha(\mathbf{v}) = \boldsymbol{\mu}\} \subset V.$$

If the reflection $\alpha: V \Rightarrow M$ is not unique, then subsets $\Psi(\boldsymbol{\mu}) \subset V$ might include more than one point. The formula (7.150) can be rewritten as

$$\overline{f}_A(x) = \sup_{v \in V} \max_{z \in l_v \cap H_x^{\alpha(v)}} f(z) = \sup_{\mu \in M} \sup_{v \in \Psi(\mu)} \max_{z \in l_v \cap H_x^{\mu}} f(z). \tag{7.165}$$

However, for the limit corresponding to algorithm A^*, we have

$$\overline{f}_{A*}(x) = \sup_{v \in V_f} \max_{z \in l_v^* \cap H_x^{\alpha^*(v)}} f(z) = \sup_{v \in V_f} \max_{z \in l_v^* \cap H_x^{\mu}} f(z).$$

Let $\mathbf{z}^*$ be the crossing point of subset s_v^* with the border Γ_v^* of the set H_v^*. At this point, by the definition of division S^*, the condition

$$\nabla f(z^*) = v$$

holds due to the convexity of function $f(\mathbf{z})$, and at this point the minimum

$$\min_{z \in \Gamma_x^v} f(z) = f(z^*) \tag{7.166}$$

is attained. Taking into account the monotone increase of the function $f(\mathbf{z})$ in each variable z_i, it follows that

$$\max_{z \in l_v^* \cap H_x^v} f(z) = \max_{z \in l_v^* \cap \Gamma_x^v} f(z) = f(z^*) = \min_{z \in \Gamma_x^v} f(z).$$

From here, with condition $V_f \in M$, we have

$$\overline{f}_{A^*}(x) = \sup_{v \in V_f} \min_{z \in \Gamma_x^v} f(z) \leq \sup_{\mu \in M} \min_{z \in \Gamma_x^v} f(z). \tag{7.167}$$

For any $\mathbf{v} \in V$ and $\boldsymbol{\mu} \in M$ the inequality

$$\max_{z \in l_v \cap H_x^v} f(z) = \max_{z \in s_v \cap \Gamma_x^v} f(z) \geq \min_{z \in \Gamma_x^v} f(z)$$

and, moreover, the inequality

$$\sup_{v \in \psi(\mu)} \max_{z \in s_v \cap H_x^v} f(z) \geq \min_{z \in \Gamma_x^v} f(z) \tag{7.168}$$

hold. The proof now follows from (7.165), (7.167), and (7.168). □

Notice that, by condition (7.161), class A^* does not contain algorithms $A = (S, a)$ with trivial reflections $a = a(\mathbf{v})$ of the form $a(\mathbf{v}) \equiv \mu$. Therefore from Theorem 7.2 it does not follow directly that the limit $\overline{f}_{A^*}(x)$ is optimal in comparison with any limits $\overline{f}_\mu(x)$ of the form (7.154), which corresponds to the initial collection of trivial algorithms. This fact follows from the next theorem.

Theorem 7.3 For any $\boldsymbol{\mu} \in M$ the limit $\overline{f}_{S^*}(x)$ satisfied the inequality

$$\overline{f}_{A^*}(x) \leq \overline{f}_\mu(x)$$

for all $\mathbf{x} \in X$. □

Proof In correspondence with (7.167), the value $\overline{f}_{A^*}(x)$ can be written in the form

$$\overline{f}_{A^*}(x) = \sup_{\mu \in V_f} \min_{z \in \Gamma_x^v} f(z). \tag{7.169}$$

Therefore for any $e > 0$ there exists a $\mu(\varepsilon) \in V_f \subset M$ such that

$$\min_{z\in\Gamma_x^{v(\varepsilon)}} f(z) > \sup_{\mu\in V_f} \min_{z\in\Gamma_x^v} f(z) - \varepsilon = \overline{f}_{A*}(x) - \varepsilon.$$

For any $\boldsymbol{\mu} \in M$, the set Γ_x^m due to the condition B has at least one common point $\tilde{z}$ with the set $\Gamma_x^{m(e)}$. Thus,

$$\max_{z\in\Gamma_x^{v(\varepsilon)}} f(z) \le f(\tilde{z}) \le \max_{z\in\Gamma_x^v} f(z) = \max_{z\in H_x^\mu} f(z) = \overline{f}_\mu(x), \qquad \boldsymbol{\mu} \in M,$$

and taking into account the previous inequality, we have

$$\overline{f}_{A*}(x) < \overline{f}_\mu(x) + \varepsilon, \qquad \boldsymbol{\mu} \in M.$$

Due to the arbitrariness of ε, the needed inequality is proved. □

As an example, consider an application of the optimal algorithm A^* for computation of the lower confidence level of the PFFO of a series–parallel system for a binomial model. The PFFO of this system (7.84) can be written via parameters $z_i = -\ln p_i$ in the form

$$R = \prod_{1\le i\le m} [1 - (1 - p_i)^{n_i}] = \exp[-f(\mathbf{z})],$$

where

$$f(\mathbf{z}) = \sum_{1\le i\le m} f_i(z_i),$$

$$f_i(z_i) = -\ln[1 - (1 - e^{z_i})^{n_i}].$$

Confidence estimation of the system reliability index R from below is reduced to constructing the UCL for $f(\mathbf{z})$. Function $f(\mathbf{z})$ is monotonically increasing in each z_i and convex in $\mathbf{z} \in Z$. The S^*-division of the space of parameters Z in correspondence with (7.162) has the form

$$s_v^* = \{\mathbf{z}\colon f_i'(z_i) = tv_i,\ i = 1, \ldots, m;\ t \equiv 0\},$$

where $\mathbf{v} = (v_1, \ldots, v_m) \in V_f \subset M$. The upper confidence limit (7.159) for $f(\mathbf{z})$ corresponding to the optimal algorithm A^* has the form

$$\overline{f}_{A*}(x) = \max_{\mu\in V_f} \max_{z\in l_\mu^*\cap H_x^\mu} f(z), \tag{7.170}$$

or, in correspondence to (7.167),

$$\overline{f}_{A*}(x) = \max_{\mu\in V_f} \min_{z\in\Gamma_x^\mu} f(z). \tag{7.171}$$

Notice that if the number of units of different subsystems $n_i > 1$, $i = 1, \ldots, m$, then the set $V_f = M$, that is, the external maximum in (7.171) or (7.170), is taken over the set M.

Example 7.28 Consider the series–parallel system from Example 7.15. The system consists of $m = 10$ parallel subsystems connected in series. The number of units n_i in the subsystems and the test results N_i, d_i are given in Table 7.2. In this case the lower γ-confidence level (for $\gamma = 0.9$) for system reliability computed with the help of algorithm A^* is $\underline{R} = 0.9998$. This value coincides with the limit found for this example by the heuristic method of equivalent tests (see Table 7.4). □

7.8 BAYES METHOD

Let us demonstrate this method applied to a binomial testing plan. Consider a system consisting of units of m different types. The system reliability index $R = R(\mathbf{p}) = R(p_1, \ldots, p_m)$ is a function of vector $\mathbf{p} = (p_1, \ldots, p_m)$ representing the results of tests of individual units (here p_i is the PFFO of a unit of the ith type). For each unit of each type we have the results of tests: number of failures d_i and N_i tests. Unit test results are assumed to be independent.

Assume also that for each parameter p_i a prior density function $h_i(p_i)$ is given and parameters p_i are also assumed independent. Thus the prior density of the distribution $h(\mathbf{p})$ has the form

$$h(\mathbf{p}) = \prod_{1 \le i \le m} h_i(p_i).$$

In correspondence with the standard Bayes procedure [see Equation (4.1)] the posterior density of the distribution of vector $\mathbf{p} = (p_1, \ldots, p_m)$ for given test results $\mathbf{d} = (d_1, \ldots, d_m)$ has the form

$$h(\mathbf{p}|\mathbf{d}) = \frac{h(\mathbf{p})L(\mathbf{d}|\mathbf{p})}{\varphi(\mathbf{d})}, \tag{7.172}$$

where $L(\mathbf{d}|\mathbf{p})$ is the likelihood function, the probability of the test result $\mathbf{d}$ for given vector of parameters $\mathbf{p}$,

$$L(\mathbf{d}|\mathbf{p}) = \prod_{1 \le i \le m} \binom{d_i}{N_i} (1 - p_i)^{d_i} p_i^{N_i - d_i},$$

and

$$\varphi(\mathbf{d}) = \int_0^1 \cdots \int_0^1 h(\mathbf{p})L(\mathbf{d}|\mathbf{p})\, dp_1 \cdots dp_m.$$

It follows that the posterior density of distribution (7.172) is expressed by the formula

$$h(\mathbf{p}|\mathbf{d}) = \prod_{1 \le i \le m} h_i(p_i|d_i), \tag{7.173}$$

where $h_i(p_i|d_i)$ is the posterior density of the distribution of parameter p_i:

$$h_i(p_i|d_i) = \frac{h_i(p_i)(1 - p_i)^{d_i} p_i^{N_i - d_i}}{\int_0^1 h_i(u)(1 - u)^{d_i} u^{N_i - d_i}\, du}. \tag{7.174}$$

In the binomial scheme of testing considered here, one usually assumes that the prior distribution $h_i(p_i)$ is a standard beta distribution with parameters (a_i, b_i), that is,

$$h_i(p_i) = \frac{(1 - p_i)^{a_i - 1} p_i^{b_i - 1}}{B(a_i, b_i)}, \tag{7.175}$$

where $B(a_i, b_i)$ is the beta function. In this case the posterior distribution (7.174) is also beta but with different parameters (see Section 4.3):

$$h_i(p_i|d_i) = \frac{(1 - p_i)^{a_i + d_i - 1} p_i^{b_i + N_i - d_i - 1}}{B(a_i + d_i, b_i + N_i - d_i)}.$$

The Bayesian γ-confidence limit (e.g., lower) $\underline{R} = \underline{R}(\mathbf{d})$ of the system PFFO $R = R(\mathbf{p})$ can be found from the equation

$$\iiint_{R(p_1, \ldots, p_m) \ge \underline{R}} \prod_{1 \le i \le m} h_i(p_i|d_I)\, dp_i = \gamma, \tag{7.176}$$

that is, $\underline{R}$ is the quantile of level $1 - \gamma$ of the prior distribution of $R(\mathbf{p}) = R(p_1, \ldots, p_m)$. Analytical evaluation of limit $\underline{R}$ on the basis of (7.176) might be too complicated, although it can be easily found with Monte Carlo simulation. For a Monte Carlo simulation, independent random values of parameters $p_i^{(j)}$, $i = 1, \ldots, m$, are generated on the basis of the posterior distributions (7.174). After this the value of system reliability $R^{(j)} = R(p_1^{(j)}, \ldots, p_m^{(j)})$ is calculated, and the computational process goes to the next $(j + 1)$th step. On the basis of n realizations one can construct a corresponding empirical distribution function, and afterward the lower Bayesian γ-confidence limit $\underline{R}$ is taken equal to the quantile of the level $1 - \gamma$ of this empirical distribution.

The Bayesian approach might deliver a conservative confidence limit $\underline{R}$ for the reliability of series systems consisting of a large number of subsystems and units if there were few failures; that is, the case is close to the nonfailure tests for almost all types of units (see Example 7.27). Nevertheless, for "medium" numbers of failures, this approach gives admissible results. Besided, the merit of the Bayesian approach for this problem is that it is universal and visual. The Bayesian approach was used for analysis of the reliability of complex systems in many works (e.g., Barlow, 1985; Cole, 1975; Dostal and Iannuzzelli, 1977; Mann et al., 1974; Martz and Waller, 1982, 1990; Martz et al. 1988; Mastran, 1976; Mastran and Singpurwalla, 1978; Natvig and Eide, 1987; Savchuk, 1989; Springer and Thompson, 1964, 1966, 1967a-c; Smith and Springer, 1976; and others).

An additional merit of the Bayesian approach lies in a possibility to use mixed information to test subsystems and units of the same system. Consider, for instance, a situation where, in addition to the unit test results $d_1, \ldots, d_m$, we also know the results of test of K different subsystems of the system. These subsystems could be of series–parallel type. During the tests only subsystem failures were registered, not its units. Let $R_j = R_j(\mathbf{p})$ be the PFFO of the lth subsystem, $1 \geq j \geq K$.

If the jth subsystem has a series structure, then

$$R_j(\mathbf{p}) = \prod_{1 \leq i \leq m} p_i^{\nu_{ij}}$$

ν_{ij} is the number of units of the ith type comprising the jth subsystem. If the jth subsystem has a parallel structure, then

$$R_j(\mathbf{p}) = 1 - \prod_{1 \leq i \leq m} (1 - p_i)^{\nu_{ij}}.$$

In addition, suppose we have information about the test of M_j subsystems of the jth type and it is known that there were D_j failures. The vector of the test results has the form

$$\mathbf{z} = (\mathbf{d}, \mathbf{D}) = (d_1, \ldots, d_m; D_1, \ldots, D_K),$$

where $\mathbf{d} = (d_1, \ldots, d_m)$ is the vector of test results for units and $\mathbf{D} = (D_1, \ldots, D_K)$ is the vector of the test results for the subsystems. As usual, all test results are assumed to be independent. We need to construct, for instance, the lower confidence limit $\underline{R}$ for the system PFFO $R = R(\mathbf{p})$.

In the frame of the Bayesian method such problems are solved in the same manner as for more simple problems where only unit test results were known.

However, pure calculational difficulties can arise. Indeed, the posterior density in this case is again determined by the formula equivalent to (7.172):

$$h(\mathbf{p}|\mathbf{d}, \mathbf{D}) = \frac{h(\mathbf{p})L(\mathbf{d}, \mathbf{D}|\mathbf{p})}{\varphi(\mathbf{d}, \mathbf{D})}, \tag{7.177}$$

where $L(\mathbf{d}, \mathbf{D}|\mathbf{p})$ is the probability of test results $(\mathbf{d}, \mathbf{D})$ for the given vector of parameters $\mathbf{p}$:

$$L(\mathbf{d}, \mathbf{D}|\mathbf{p}) = \prod_{1 \le i \le m} \binom{d_i}{N_i} (1 - p_i)^{d_i} p_i^{N_i - d_i} \prod_{1 \le l \le K} \binom{D_l}{M_l}$$

$$[1 - R_l(\mathbf{p})]^{D_l} [R_l(\mathbf{p})]^{-M_l - D_l}, \tag{7.178}$$

$$\varphi(\mathbf{d}, \mathbf{D}) = \int_0^1 \cdots \int_0^1 h(\mathbf{p})L(\mathbf{d}, \mathbf{D}|\mathbf{p})\, dp_1 \cdots dp_m.$$

Now the lower Bayesian γ-confidence limit $\underline{R}$ for the system PFFO $R(\mathbf{p}) = R(p_1, \ldots, p_m)$ is determined similarly to (7.176):

$$\int \cdots \int_{R(p1,\ldots,p_m) \ge \underline{R}} h(\mathbf{p})L(\mathbf{d}, \mathbf{D}|\mathbf{p})\, dp_1 \cdots dp_m = \gamma.$$

7.9 APPENDIX

7.9.1 Derivation of Formulas (7.26) and (7.27) for Confidence Limits in the Arbitrary Distribution of Test Results x and Statistic S = S(x)

Let

$$\underline{R} = \inf_{\boldsymbol{\theta} \in H(S^*)} R(\boldsymbol{\theta}), \qquad \overline{R} = \sup_{\boldsymbol{\theta} \in H(S^*)} R(\boldsymbol{\theta}),$$

where set $H(S^*)$ is defined by inequalities

$$t_1(\boldsymbol{\theta}) \le S^*, \qquad t_2(\boldsymbol{\theta}) \ge S^*.$$

In correspondence with (7.20) and (7.25), values $\underline{R}$ and $\overline{R}$ give the confidence interval for $R = R(\boldsymbol{\theta})$ with confidence coefficient not less than $\gamma = 1 - \alpha - \beta$.

Further, assume that the following conditions hold:

Condition A: Function $R(\boldsymbol{\theta})$ is continuous in $\boldsymbol{\theta}$.

Condition B: Functions of the form $P_\theta(S \le S^*)$ and $P_\theta(S \ge S^*)$ are continuous in $\boldsymbol{\theta}$.

Directly from the definition of $t_1(\boldsymbol{\theta})$ and $t_2(\boldsymbol{\theta})$ in (7.20) the following applications follow:

$$
\begin{aligned}
t_1(\boldsymbol{\theta}) \le S^* &\Rightarrow P_\theta(S \le S^*) \ge \alpha, \\
t_2(\boldsymbol{\theta}) \ge S^* &\Rightarrow P_\theta(S \ge S^*) \ge \beta,
\end{aligned}
\tag{7.179}
$$

and

$$
\begin{aligned}
P_\theta(S \le S^*) > \alpha &\Rightarrow t_1(\boldsymbol{\theta}) \le S^*, \\
P_\theta(S \ge S^*) > \beta &\Rightarrow t_2(\boldsymbol{\theta}) \ge S^*.
\end{aligned}
\tag{7.180}
$$

Denote the set of parameters $\boldsymbol{\theta}$ that satisfy the conditions $P_\theta(S \le S^*) \ge \alpha$ and $P_\theta(S \ge S^*) \ge \beta$ by $G(S^*)$. Introduce also the set $L(S^*)$ of parameters $\boldsymbol{\theta}$ that satisfy the conditions $P_\theta(S \le S^*) > \alpha$ and $P_\theta(S \ge S^*) > \beta$. Due to (7.170) and (7.171) the following relations are valid:

$$
L(S^*) \subset H(S^*) \subset G(S^*). \tag{7.181}
$$

By continuity of the function $R(\boldsymbol{\theta})$, the set $L(S^*)$ is open and the set $G(S^*)$ is closed.

Now assume that together with conditions A and B above, the following conditions also hold:

Condition C: The minimum and maximum of function $R(\boldsymbol{\theta})$ are attained on the set $G(S^*)$. [This condition surely holds if, for instance, the set $G(S^*)$ is restricted.]

Condition D: Closure $L^*(S^*)$ of set $L(S^*)$ coincides with set $G(S^*)$.

All conditions from A through D given above are not very restrictive in practice for common reliability problems. From (7.172) due to condition C the following inequalities follow:

$$
\min_{\boldsymbol{\theta} \in G(S^*)} R(\boldsymbol{\theta}) \le \underline{R} \le \inf_{\boldsymbol{\theta} \in L(S^*)} R(\boldsymbol{\theta}),
$$

$$
\sup_{\boldsymbol{\theta} \in L(S^*)} R(\boldsymbol{\theta}) \le \overline{R} \le \max_{\boldsymbol{\theta} \in G(S^*)} R(\boldsymbol{\theta}).
$$

It gives the following lower and upper confidence limits due to the continuity of $R(\boldsymbol{\theta})$ and condition D:

$$\underline{R} = \min R(\boldsymbol{\theta}), \qquad \overline{R} = \max R(\boldsymbol{\theta}),$$

where the minimum and maximum are taken over set $G(S^*)$. This delivers the proof of the statement.

Notice in conclusion that formulas (7.26) and (7.26) for the confidence limits $\underline{R}$ and $\overline{R}$ are valid for the more general case where statistic S depends on the test results and a parameter $\boldsymbol{\theta}$, that is, $S = S(\mathbf{x}, \boldsymbol{\theta})$. In this case inequalities (7.27) can be written in the form

$$P_{\theta}\{S(\mathbf{x}, \boldsymbol{\theta}) \leq S(\mathbf{x}^*, \boldsymbol{\theta})\} \geq \alpha,$$

$$P_{\theta}\{S(\mathbf{x}, \boldsymbol{\theta}) \geq S(\mathbf{x}^*, \boldsymbol{\theta})\} \geq b\beta,$$

where $\mathbf{x}^*$ is the observed value of random vector $\mathbf{x}$.

7.9.2 Computation of Confidence Limits for Binomial Testing Plan

The lower γ-confidence Clopper–Pearson limit for the binomial parameter p (the PFFO) is defined from the equation

$$\sum_{0 \leq j \leq d} \binom{N}{j} (1 - p)^j p^{N-j} = 1 - \gamma,$$

where N is the number of tests and d is the observed number of failures. For integer N and d the left part of this equation can be written also in the form

$$\sum_{0 \leq j \leq d} \binom{N}{j} (1 - p)^j p^{N-j} = B_p(N - d, d + 1),$$

where

$$B_p(a, b) = \frac{\int_0^p x^{a-1}(1 - x)^{b-1}\, dx}{\int_0^1 x^{a-1}(1 - x)^{b-1}\, dx}$$

is the beta function. Thus, the equation for funding the lower confidence level can be written in the form $B_p(N - d, d + 1) = 1 - \gamma$. The left part of this equation is defined for all positive values $d < N$ (not necessarily integer). The solution of this equation relative to p is denoted by $\underline{P}_g(N, d)$. Thus the value of $\underline{P}_g(N, d)$ is the lower γ-confidence Clopper-Pearson limit for parameter p obtained on the basis of N tests with d failures. This limit is valid for all positive $d < N$, which allows one to apply it for construction of confidence limits for the system reliability using various methods of equivalent tests (see Sections 7.6.1 and 7.6.2).

Table 7.9 Input Data for Example 7.2

i	1	2	3
l_i	2	1	1
N_i	200	150	250
d_i	2	0	4

PROBLEMS

7.1. A series system consists of two different units. Test of $N_1 = 100$ units of the first type has no failures, $d_1 = 0$. Test of $N_1 = 200$ units of the second type found $d_1 = 4$. Construct the lower γ-confidence level with confidence coefficient not less than $\gamma = 0.9$ for the system PFFO $R = p_1 p_2$.

7.2. Consider a series system consisting of units of different type, $m = 3$. The number of units of each type, r_i, and results of tests, N_i and d_i, are given in Table 7.9. Construct the lower γ-confidence level with confidence coefficient $\gamma = 0.95$ for the system PFFO $R = p_1^2 p_2 p_3$.

7.3. Consider a parallel system consisting of two units, $m = 2$. Test results are $N_1 = 10$, $d_1 = 0$ and $N_2 = 20$, $d_2 = 1$, respectively. Construct the lower γ-confidence level with confidence coefficient $\gamma = 0.9$ for the system PFFO $R = 1 - (1 - p_1)(1 - p_2)$.

7.4. Consider a parallel system consisting of units of two types, $m = 2$. There are $n_1 = 2$ units of the first type and a single unit of the second type, $n_2 = 1$. Test results are $N_1 = 12$, $d_1 = 0$ and $N_2 = 6$, $d_2 = 1$, respectively. Construct the lower γ-confidence level with confidence coefficient $\gamma = 0.95$ for the system PFFO $R = 1 - (1 - p_1)^2(1 - p_2)$.

7.5. Consider the series–parallel system consisting of $m = 10$ subsystems from Example 7.14. Find the lower γ-confidence level for the system PFFO by the method of equivalent tests (for $\gamma = 0.9$).

7.6. Find the lower γ-confidence level for the PFFO of the system considered in Example 7.16 (for $\gamma = 0.9$) by the method of equivalent tests.

CHAPTER 8

CONFIDENCE LIMITS FOR SYSTEMS CONSISTING OF UNITS WITH EXPONENTIAL DISTRIBUTION OF TIME TO FAILURE

8.1 INTRODUCTION

The methods considered in previous chapters are valid for constructing confidence limits of simple series or series–parallel structures. For systems with more complex structures one uses usually different heuristic or approximate methods. However, the correctness of application of these methods remains unclear; in other words, it is not known if this method produces confidence limits with guaranteed confidence coefficients for a complex system. The answer can be obtained by Monte Carlo simulation of the test process and constructing the confidence limit of a particular system for a particular set of parameters. But a specific solution for this particular case does not deliver any information about results for other parameters. Moreover, there is no information for other types of system structures. Enumerating all the specter of structures of interest is practically impossible. Thus, Monte Carlo simulation is not the best way to validate one method or another, though it is frequently used for engineering purposes in reliability analysis.

In this chapter we suggest methods of constructing confidence limits which for all realizations of parameters are accurate in the sense that they guarantee the confidence level not less than given value γ. Besides, these methods work for a wide class of complex systems, particularly for systems with repair if an additional suggestion is made; namely, units have exponential distribution of TTF.

8.2 METHOD OF REPLACEMENT

8.2.1 Description of Method

Let a system consist of m different types and the distribution of TTF of the ith unit be exponential: $F_i(t) = 1 - \exp(-\lambda_i t)$ with unknown parameter, Failure Rate (FR) λ_i, $1 \leq i \leq m$. Assume that a test of unit i was performed by the standard plan of type $[N_{i'}, U\ r_i]$, that is, without "replacement" of failed units, or by the plan $[N_{i'}, R\ r_i]$, that is, with replacement of failed units. (See details in Section 2.1.) The test results in the form of summarized unit testing time are S_i, $1 \leq i \leq m$. The test results are assumed independent.

Let R be some system reliability index and

$$R(\boldsymbol{\lambda}) = R(\lambda_1, \ldots, \lambda_m) \tag{8.1}$$

be a function expressing the dependence of this reliability index on element parameters $\boldsymbol{\lambda} = (\lambda_1, \ldots, \lambda_m)$. It is necessary to construct the confidence limit (e.g. the lower) for the system reliability $R = R(\lambda)$ on the basis of unit test results.

Let us introduce the notation

$$\overline{\lambda}_i = \frac{\chi^2_\gamma(2r_i)}{2S_i}, \tag{8.2}$$

where $\chi^2_\gamma(2r_i)$ is a quantile of level γ for the χ^2 distribution with $2r_i$ degrees of freedom and r_i is the number of failures of the units of the ith type, $1 \leq i \leq m$. Let us also introduce the vector $\overline{\boldsymbol{\lambda}} = (\overline{\lambda}_1, \ldots, \overline{\lambda}_m)$ of upper γ-confidence limits for the separate unit parameters. The function (8.1), as a rule, is monotone decreasing for each of its parameter λ_i. That is, it satisfies the following natural condition: System reliability decreases with decreasing unit reliability.

Take $\underline{R}$ as a lower confidence limit of the reliability index R: a value computed by direct substitution of the upper γ-confidence limit into the function (8.1), that is,

$$\underline{R} = R(\overline{\boldsymbol{\lambda}}) = R(\overline{\lambda}_1, \ldots, \overline{\lambda}_m). \tag{8.3}$$

Since $R(\boldsymbol{\lambda})$ is monotonically decreasing in each parameter, the following relation is valid:

$$\bigcap_{1\leq i\leq m} \overline{\lambda}_i \geq \lambda_i) \subset \{R(\overline{\lambda}_1, \ldots, \overline{\lambda}_m) \leq R(\lambda_1, \ldots, \lambda_m)\}.$$

Taking into account the unit test independence, we have

$$P\{R(\bar{\boldsymbol{\lambda}}) \leq R(\boldsymbol{\lambda})\} \geq P\{\bigcap_{1\leq i\leq m} (\bar{\lambda}_i \leq \lambda_i)\} = \prod_{1\leq i\leq m} P(\bar{\lambda}_i \geq \lambda_i) = \gamma^m. \quad (8.4)$$

Therefore the lower confidence limit (8.3) has confidence coefficient not less than the value of γ^m. Such a procedure of confidence limit construction, obviously, corresponds to the previously considered rectangular method (Section 7.5.3).

This simple procedure is universal enough and can be applied if the function $R(\lambda)$ is monotone. Notice that this condition is almost always valid. Nevertheless, the confidence coefficient for this procedure decreases very fast with growing number of system unit types m. As a result, this simple approach gives too conservative a confidence estimate of system reliability.

Pavlov (1980, 1982a) and others have shown that for many of the models of complex systems considered below (including systems with repairable units), the lower confidence limit of system reliability (8.3) can be used with preservation of the initial confidence coefficient for $\gamma \geq 1 - e^{3/2} \cong 0.778$). In other words, the lower confidence limit of system reliability with the given confidence coefficient can be done by simple substitution of confidence limits $\bar{\lambda}_i$ for unit parameters (with the same confidence coefficient) into the function (8.1).

This procedure is called the method of substitution. In many practical cases this method allows us to obtain a simple and effective solution immediately. Although, as we will show below, sometimes this method still produces very conservative estimates of system reliability (see Sections 8.5 and 8.6).

8.2.2 Conditions of Method Application

We will further assume that function $R(\boldsymbol{\lambda}) = R(\lambda_1, \ldots, \lambda_m)$ is monotone decreasing in each of its parameter λ_i, $1 \leq i \leq m$. To apply the replacement method, we also need some conditions of convexity of the function $R(\boldsymbol{\lambda})$.

Function $R(\boldsymbol{\lambda}) = R(\lambda_1, \ldots, \lambda_m)$ is called quasi-concave (convex) if the region of parameters $\boldsymbol{\lambda}$ of the form

$$\{\boldsymbol{\lambda}: R(\boldsymbol{\lambda}) \geq (\leq)\ C\}$$

is concave (convex) for any constant C. It is easy to see that a convex (concave) function is simultaneously quasi-convex (quasi-concave) although the inverse is not true. (One can find details in Section 8.7.1.)

In Sections 8.2–8.4 we will assume that the confidence coefficient γ satisfies the inequality $\gamma \geq 1 - \exp(-\frac{3}{2}) \approx 0.778$. We also assume that function $R(\boldsymbol{\lambda}) = R(\lambda_1, \ldots, \lambda_m)$ is monotonically increasing in each parameter λ_i and quasi-concave. Then the lower γ-confidence limit $\underline{R}$ for $R(\boldsymbol{\lambda})$ can be calculated by the method of replacement, that is, by formula (8.3). (See Theorem 8.6 in Section 8.7.)

8.2.3 Systems with Series–Parallel Structure

Series Structure Let a system consist of m different units connected in series. The PFFO of such a system with independent units is defined as

$$R = \prod_{1 \le i \le m} p_i(t),$$

where $p_i(t) = \exp(-\lambda_i t)$ is the PFFO of the ith unit, $1 \le i \le m$. This expression can be rewritten via parameters $\boldsymbol{\lambda} = (\lambda_1, \ldots, \lambda_m)$ as

$$R(\boldsymbol{\lambda}) = \exp\left(-t \sum_{1 \le i \le m} \lambda_i\right). \tag{8.5}$$

Function (8.5) monotonically decreases in each λ_i and is quasi-concave in $\boldsymbol{\lambda}$, since the region of parameters

$$\{\boldsymbol{\lambda}\colon R(\boldsymbol{\lambda}) \ge C\} = \left\{\boldsymbol{\lambda}\colon \sum_{1 \le i \le m} \lambda_i \le -\ln \frac{C}{t}\right\}$$

is convex. So, by (8.4), the lower γ-confidence limit for R can be calculated as

$$\underline{R} = \exp\left(-t \sum_{1 \le i \le m} \overline{\lambda}_i\right).$$

Series Structure with Replicated Units In an analogous manner we can consider a series structure where units of some type are replicated several times. Lt r_i be the number of units of the ith type. Then the system PFFO equals

$$R = \prod_{1 \le i \le m} p_i^{r_i}(t) = \exp\left(-t \sum_{1 \le i \le m} r_i \lambda_i\right).$$

The lower γ-confidence limit for the PFFO can again be computed with the help of the substitution method as

$$\underline{R} = \exp\left(-t \sum_{1 \le i \le m} r_i \overline{\lambda}_i\right).$$

Series–Parallel Structure (Loaded Redundancy) Consider a system of m parallel subsystems in series. Each subsystem consists of n_i parallel identical and independent units. Each unit has an exponential d.f. of TTF with parameter λ_i, $1 \leq i \leq m$. In this case the system PFFO can be written as

$$R = \prod_{1 \leq i \leq m} \{1 - [1 - p_i(t)]^{n_i}\},$$

where $p_i(t) = \exp(-\lambda_i t)$. Using vector parameters $\boldsymbol{\lambda} = (\lambda_1, \ldots, \lambda_m)$, we can write

$$R(\boldsymbol{\lambda}) = \exp\{-f(\boldsymbol{\lambda})\}$$

where $f(\boldsymbol{\lambda}) = \Sigma_{1 \leq i \leq m} f_i(\lambda_i)$, $f_i(\lambda_i) = -\ln\{1 - [1 - \exp(-\lambda_i t)]^{n_i}\}$. It is easy to show, for instance by direct differentiation, that functions $f_i(\lambda_i)$ are monotone increasing and convex. It follows that the area of the form

$$\{\boldsymbol{\lambda}\colon R(\boldsymbol{\lambda}) \geq C\} = \{\boldsymbol{\lambda}\colon f(\boldsymbol{\lambda}) \leq -\ln C\}$$

is convex. Thus, in accordance with (8.4), the lower γ-confidence limit for $R(\boldsymbol{\lambda})$ can be calculated by the method of substitution as

$$\underline{R} = \exp\left\{\sum_{1 \leq i \leq m} f_i(\overline{\lambda}_i)\right\}.$$

***Series Connection of Systems of Type* k *Out of* n** Consider a system consisting of m parallel subsystems in series. Each subsystem consists of n_i parallel identical and independent units. The ith subsystem failure occurs if k_i or more units of this fail, $1 \leq k_i \leq n_i$. Each unit again has an exponential d.f. of TTF with parameter λ_i, $1 \leq i \leq m$. The system described above is a particular case of this general model. In this case the system PFFO can be written as

$$R = \prod_{1 \leq i \leq m} R_i(p_i),$$

where $p_i = p_i(t) = \exp(-\lambda_i t)$ and

$$R_i(p_i) = \sum_{0 \leq d \leq k_i - 1} \binom{d}{n_i} (1 - p_i)^d \, p_i^{n_i - d}.$$

The system PFFO expressed via parameters λ_i can be written as

$$R(\boldsymbol{\lambda}) = \exp\left\{-\sum_{1\le i\le m} f_i(\lambda_i t)\right\}, \tag{8.6}$$

where

$$f_i(\lambda_i t) = -\ln R_i[\exp(-\lambda_i t)]. \tag{8.7}$$

It is easy to show by direct differentiation that each function in (8.7) is monotone increasing and convex in λ_i. Notice that the convexity of these functions follows from the known fact that a system of type *k out of n*, consisting of identical units with an exponential d.f. of TTF, has an IFR distribution of TTF. Moreover, this fact is also correct even for systems consisting of identical units with an IFR d.f. of TTF (see Barlow and Proschan, 1965). From here, by definition of the IFR distribution, it follows that each function in (8.7) is convex in t and, consequently, in λ_i for fixed t.

It follows that (8.6) is monotone decreasing in each λ_i and quasi-concave in $\boldsymbol{\lambda}$. Thus, the lower γ-confidence limit for $R(\boldsymbol{\lambda})$ can be calculated with the help of the method of substitution as

$$\underline{R} = \exp\left\{-\sum_{1\le i\le m} f_i(\overline{\lambda}_i t)\right\}.$$

Series–Parallel Structure (Unloaded Redundancy) Consider a system consisting of m parallel subsystems in series. Each subsystem consists of n_i identical and independent units, one of them operating and other spare (unloaded redundancy). The TTF of the ith subsystem represents the sum of i.i.d. exponential r.v.'s with parameter λ_i, $1 \le i \le m$:

$$\tau_i = \sum_{1\le j\le n_i} \xi_j,$$

where ξ_j is the TTF of unit j of subsystem i. The PFFO for the ith subsystem has the form

$$R_i(\lambda_i t) = P(\tau_i > t) = e^{-\lambda_i t} \sum_{0\le d\le n_i-1} \frac{(\lambda_i t)^d}{d!}.$$

The system PFFO is determined as

$$R(\boldsymbol{\lambda}) = \prod_{1\le i\le m} R_i(\lambda_i t) = \exp\left\{-\sum_{1\le i\le m} \varphi_i(\lambda_i t)\right\},$$

where

$$\varphi_i(\lambda_i t) = -\ln R_i(\lambda_i t). \tag{8.8}$$

It is easy again to verify that each function in (8.8) is monotone increasing and convex in λ_i. Notice that the convexity follows from the fact that the sum of i.i.d IFR distributed r.v.'s has IFR distribution (see Barlow and Proschan, 1965). Thus we again obtain that the lower confidence limit for $R(\boldsymbol{\lambda})$ can be calculated with the help of the method of substitution as

$$\underline{R} = \exp\left\{-\sum_{1\le i\le m} \varphi_i(\overline{\lambda}_i t)\right\}.$$

8.3 SYSTEMS WITH COMPLEX STRUCTURE

8.3.1 "Recurrent" Structures

For all previously considered systems, we assumed that the units within a redundant group are identical to the main ones. Now consider a more general case where the number of redundancy levels is arbitrary and redundant units might differ from main units.

Consider, at first, a separate redundant group consisting of n units. If $\lambda_1, \ldots, \lambda_n$ are the unit failure rates, then the PFFO for loaded redundancy for this redundant group is

$$R = 1 - \prod_{1\le i\le n} (1 - e^{-\lambda_i t}). \tag{8.9}$$

Further, instead of (8.9), we will use an approximate formula for highly reliable groups, that is, for $t^n \Pi_{1\le i\le n} \lambda_i << 1$, $1 \le i \le n$,

$$R \approx \exp\left(-t^n \prod_{1\le i\le n} \lambda_i\right). \tag{8.10}$$

This formula gives the lower estimate for index R for all $\lambda_1, \ldots, \lambda_n$ (Gnedenko et al., 1969).

In an analogous way, one can obtain an expression for unloaded redundancy. Instead of the clumsy and practically useless formula

$$R = \int_{y_1+\cdots+y_n\ge t} \cdots \int \prod_{1\le i\le n} \lambda_i \exp(-\lambda_i y)\, dy_1 \cdots dy_n,$$

we will use an approximation,

$$R \approx \exp\left(-\frac{t^n}{n!}\prod_{1\le i\le n} \lambda_i\right), \tag{8.11}$$

which gives the lower estimate for index R for all $\lambda_1, \ldots, \lambda_n$ if

$$\frac{t^n}{n!}\prod_{1\le i\le n} \lambda_i << 1$$

(see Gnedenko et al., 1969) and delivers the lower limit for PFFO.

Let a system consist of n units. The system PFFO for some given time t has the form

$$R = H(p_1, \ldots, p_n),$$

where $p_i = \exp(-\lambda_i t)$, λ_i being the failure rate of the ith unit. The system PFFO, expressed via parameter $\boldsymbol{\lambda} = (\lambda_1, \ldots, \lambda_n)$ can be written in the form

$$R = R(\boldsymbol{\lambda}) = R(\lambda_1, \ldots, \lambda_n), \tag{8.12}$$

where $R(\lambda_1, \ldots, \lambda_n) = H[\exp(-\lambda_i t) \cdots \exp(-\lambda_n t)]$.

Introduce operation $\mathbf{S}(i, n)$ of substituting the subsystem of n units with failure rates $\alpha_1, \ldots, \alpha_n$ instead of the ith unit. After this substitution the PFFO can be written in the form

$$R = R[\lambda_1, \ldots, \lambda_{i-1}, \Lambda(\boldsymbol{\alpha}), \lambda_{i+1}, \ldots, \lambda_m] \tag{8.13}$$

where $\Lambda(\boldsymbol{\alpha}) = -(1/t) \ln \pi(\boldsymbol{\alpha})$, $\pi(\boldsymbol{\alpha})$ being the PFFO of the new subsystem, $\boldsymbol{\alpha} = (\alpha_1, \ldots, \alpha_n)$.

Let $\mathbf{S}_1 = \mathbf{S}_1\ (i, n)$ denote the operation where the subsystem is series. For this operation

$$\pi(\boldsymbol{\alpha}) = \prod_{1\le j\le n} e^{-\alpha_j t} = \exp\left(-t \sum_{1\le j\le n} \alpha_j\right)$$

and transformation (8.13) has the form

$$\Lambda(\boldsymbol{\alpha}) = \sum_{1\le j\le n} \alpha_j. \tag{8.14}$$

Now let $\mathbf{S}_2 = \mathbf{S}_2(i, n)$ denote the operation where the subsystem is parallel. For this case, if redundancy is loaded, due to (8.10), we have

$$\pi(\boldsymbol{\alpha}) \approx \exp\left(-t^n \prod_{1\le j\le n} \alpha_j\right).$$

If the redundancy is unloaded, then, by (8.11),

$$\pi(\boldsymbol{\alpha}) \approx \exp\left(-\frac{t^n}{n!} \prod_{1\le j\le n} \alpha_j\right).$$

Thus, transformation (8.13) for operation $\mathbf{S}_2$ has the form

$$\Lambda(\boldsymbol{\alpha}) \approx t^{n-1} \prod_{1\le j\le n} \alpha_j, \qquad \Lambda(\boldsymbol{\alpha}) \approx \frac{t^{n-1}}{n!} \prod_{1\le j\le n} \alpha_j \tag{8.15}$$

for loaded and unloaded redundancy, respectively.

Say that a system has recurrent or reducible structure if it can be obtained from some series structure by sequential application of procedure $\mathbf{S}_1$ and $\mathbf{S}_2$ in a finite number of iterations. Such structures were considered in Chapter 7. Obviously the class recurrent structures includes series, parallel, series–parallel, and parallel-series structures as particular cases. A more sophisticated recurrent structure is depicted in Figure 8.1, where numbers denote unit types. There are no restrictions on the number of levels of "recurrence" or on the identity of main and redundant units. Let us introduce a class of functions of the type

$$R(\boldsymbol{\lambda}) = \exp\left(-\sum_{1\le i\le M} A_i \prod_{j\in G_i} \lambda_j^{n_{ij}}\right), \tag{8.16}$$

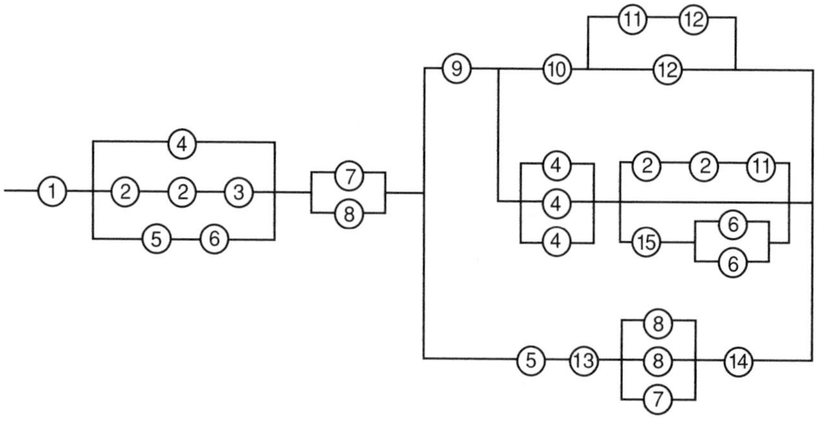

Figure 8.1 Example of recurrent structure.

where $A_i \geq 0$, $i = 1, \ldots, M$; n_{ij} are arbitrary integer positive numbers; and G_i, $i = 1, \ldots, M$, are arbitrary (possibly intersecting) subsets of subscripts from the set $\{1, 2, \ldots, m\}$.

Assume that some recurrent structure has reliability function of type (8.16). Then application of operation $\mathbf{S}_1$ or $\mathbf{S}_2$ to each of its units corresponds to the substitution of some parameters λ_j in the initial expression by sums of type (8.14) or products of type (8.15) that again produce a formula of type (8.16) although from a larger number of variables. Notice that the reliability function of a series system

$$R(\boldsymbol{\lambda}) = \exp\left(-\sum_{1 \leq i \leq M} \lambda_i t\right)$$

obviously is function of type (8.16). It follows that for any recurrent structure the reliability function has the form (8.16).

Thus the lower confidence limit for the PFFO of the system with recurrent structure of type (8.16) is reduced to the construction of the upper confidence limit for the function of type

$$f(\boldsymbol{\lambda}) = \sum_{1 \leq i \leq M} A_i \prod_{j \in G_i} \lambda_j^{n_{ij}}, \tag{8.17}$$

where A_i are positive coefficients and G_i are arbitrary subsets of the unit's subscripts. Notice that if we have prior information about identity of some units within the system, function (8.17) again transforms to another function of the form of (8.17) but with a smaller number of parameters. Function (8.17) of the general type is not quasi-convex or quasi-concave in $\boldsymbol{\lambda} = (\lambda_1, \ldots, \lambda_m)$. Nevertheless, as Pavlov (1982) showed (see Section 8.7.2 and Theorem 8.7), the upper γ-confidence limit for a function of type (8.17) can be constructed by the method of substitution, that is, using the formula

$$\overline{f} = f(\overline{\boldsymbol{\lambda}}) = \sum_{1 \leq i \leq M} A_i \prod_{j \in G_i} \overline{\lambda}_j^{n_{ij}}, \tag{8.18}$$

where $\overline{\boldsymbol{\lambda}} = (\overline{\lambda}_1, \ldots, \overline{\lambda}_m)$ is the vector of standard upper γ-confidence limits for the unit parameter [here $\gamma \geq 1 - \exp(-\frac{3}{2})$]. It means that the lower γ-confidence limit for the PFFO of a recurrent structure, that is, the index of type (8.16), can also be constructed with the method of substitution as

$$\underline{R} = R(\overline{\boldsymbol{\lambda}}) = R(\overline{\lambda}, \ldots, \overline{\lambda}_m).$$

Example 8.1 Consider a system with the recurrent structure depicted in Figure 8.1. The number of different unit types is equal, $m = 16$. All redundant units are in a loaded regime. Test results r_i, S_i, $1 \leq i \leq m$, for different unit types are given in Table 8.1. The table also contains upper 0.9 confidence limit $\overline{\lambda}_i$ for parameters λ_i. Application of the method of substitution in this case gives the lower confidence limit with confidence coefficient not less than 0.9 for system PFFO (for $t = 1$) equal to $\underline{R} = 0.953$. □

Example 8.2 Consider a system with the same input data as above but with all redundant units in an unloaded regime. In this case the method of substitution gives $\underline{R} = 0.963$. □

8.3.2 Monotone Structures

A wide class of real systems can be described with a model of monotone structures (see Section 4.2 and Barlow & Proschan, 1965). Let a system consist of m units each of which is characterized by the failure probability during time t: $q_i = 1 - p_i = 1 - \exp(-\lambda_i t)$, $1 \le i \le m$. Let us introduce the following notation:

$H_i \subset \{1, 2, \ldots, m\}$ = a set of the unit's subscripts belonging to a system's minimum cut i;

C_i = an event that the ith unit has failed;

$B_i = \bigcap_{j \in H_i} C_j$ = an event that all units belonging to a system's minimum cut i have failed; and

B = an event that the system has failed.

Table 8.1 Input Data and Obtained Upper 0.9 Confidence Limit for Unit Failure Rates

Unit Type, i	Number of Units, r_i	Total Test Time, S_i	Unit $\overline{\lambda}_i$
1	2	420	0.009
2	5	125	0.064
3	2	32	0.122
4	8	50	0.235
5	1	220	0.010
6	9	65	0.199
7	4	54	0.124
8	6	54	0.171
9	2	80	0.049
10	3	27	0.197
11	5	56	0.143
12	3	32	0.166
13	3	54	0.098
14	3	43	0.124
15	1	230	0.010
16	3	20	0.089

In this notation the probability of system failure, Q, can be written in the form

$$Q = P(B) = P\left(\bigcup_{1\le i\le M} B_i\right), \tag{8.19}$$

where M is the total number of the system's minimum cuts.

For a highly reliable system ($\lambda_i t << 1$) the following approximation can be used: $q_i \approx \lambda_i t$. Notice that $\lambda_i t$ gives an overestimate of q_i. From (8.19) in the case of independent units, we have

$$Q \approx \sum_{1\le i\le M} P(B_i) = \sum_{1\le i\le M} P\left(\bigcap_{j\in H_i} C_i\right) = \sum_{1\le i\le M}\prod_{j\in H_i} q_i \approx \sum_{1\le i\le M}\prod_{j\in H_i} (\lambda_i t). \tag{8.20}$$

This approximation is valid under the assumption that the probability of two or more failures of cuts is negligibly small. Notice that (8.20) gives an overestimate for Q:

$$Q \le \sum_{1\le i\le M}\prod_{j\in H_i} q_i \le \sum_{1\le i\le M}\prod_{j\in H_i} (\lambda_i t).$$

Thus, for highly reliable systems the confidence estimation of Q from above is reduced to the construction of the upper confidence limit for

$$Q \le \sum_{1\le i\le M}\prod_{j\in H_i} (\lambda_i t), \tag{8.21}$$

where H_i are some (in the general case intersecting) subsets of the unit's subscripts. For a series–parallel system each parameter λ_i belongs to only one of the products in (8.21). In other words, each unit belongs to only one system's cut. For complex systems, each unit might belong to different intersecting cuts.

Sometimes we can assume in advance that some units within a system are identical, that is, they have same reliability parameters. In this case, for the probability of system failure, (8.21) has the form

$$Q \approx \sum_{1\le i\le M}\prod_{j\in H_i} (\lambda_i t)^{n_{ij}}, \tag{8.22}$$

where n_{ij} is the number of identical units with equal parameter λ_i in cut H_i. Inequality (8.21) is a particular case of (8.22).

Thus, for the case of highly reliable systems the problem (with the approximation made above) is reduced to the construction of the upper confidence limit of a function of type (8.22). This function has the same form as (8.17) considered in the previous section. For the latter we showed that (8.18) can be used to calculate the upper confidence limit. So, for arbitrary complex monotone structures the upper γ-confidence limit for the probability of system failure can be calculated (if $\gamma \geq 1 - \exp(-\frac{3}{2})$ by the method of substitution with the formula

$$\overline{Q} = Q(\overline{\boldsymbol{\lambda}}) = \sum_{1\leq i\leq M} \prod_{j\in H_i} (\overline{\lambda}_j t)^{n_{ij}}.$$

8.4 SYSTEMS WITH REPAIRABLE UNITS

For repairable systems often the system MTTF is much larger than the mean time of renewal. Usually it is easier to estimate renewal (repair) time than MTTF, because it can be done by special and simple experiments. We will assume that the mean renewal time is known with accuracy.

8.4.1 System with Series Structure

Let a system consist of n units in series. The d.f. of the ith unit's TTF is assumed exponential with unknown parameter λ_i, that is, the MTTF of this unit is equal to $T_i = 1/\lambda_i$. The mean renewal time a_i is assumed known. All units are assumed independent in the sense of failures and repair.

Consider the construction of the upper confidence limit for the main system reliability indices for this case. The most widely used reliability index in this case is the stationary availability coefficient, the probability that the system is in the up state at arbitrary stationary time t ($t \to \infty$). The availability coefficient of each unit i is determined by the well-known formula

$$K_i(T_i) = \frac{T_i}{T_i + a_i}.$$

For independent units the system's availability coefficient has the form

$$K(\mathbf{T}) = \prod_{1\leq i\leq m} K_i(T_i) = \prod_{1\leq i\leq m} \left(\frac{T_i}{T_i + a_i}\right), \tag{8.23}$$

where $\mathbf{T} = (T_1, \ldots, T_m)$.

Since repair parameters are assumed known, the problem is reduced to the confidence estimation of (8.23) depending on unknown parameters $\mathbf{T} = (T_1, \ldots, T_m)$. Introduce now a vector of lower confidence limits for unit parameters: $\underline{\mathbf{T}} = (\underline{T}_1, \ldots, \underline{T}_m)$, where

$$\underline{T}_i = \frac{1}{\overline{\lambda}_i} = \frac{\lambda S_i}{\chi^2_\gamma(2r_i)}$$

is the lower γ-confidence limit for parameter T_i, $1 \le i \le m$. As shown below (see Section 7.8.1 and Theorem 8.4), the lower γ-confidence limit for reliability index $R = R(\mathbf{T})$ can be calculated by the method of substitution (for $\gamma \ge 1 \exp[-\frac{3}{2}]$), that is, by the formula

$$\underline{R} = R(\underline{\mathbf{T}}) = R(\underline{T}_1, \ldots, \underline{T}_m)$$

if the function $R(\underline{T}_1, \ldots, \underline{T}_m)$ is monotone increasing in each T_i and quasi-concave in vector $\mathbf{T}$. It is easy to see that the reliability index (8.23) satisfies all these conditions. Indeed, the monotone increase of $K(\mathbf{K})$ in each T_i is obvious. Besides, by direct differentiation we can prove that the function

$$\ln K_i(T_i) = \ln \frac{T_i}{T_i + a_i}$$

is concave in T_i. Thus, the area of parameters

$$\{\mathbf{T}: K(\mathbf{T}) \ge C\} = \left\{\mathbf{T}: \sum_{1 \le i \le m} \ln K_i(T_i) \ge \ln C\right\}$$

is convex, and, consequently, it follows that function $K(\mathbf{T})$ is quasi-concave. So, the lower γ-confidence limit for the system availability coefficient (8.23) can be calculated by the method of substitution as follows:

$$\underline{K} = K(\underline{\mathbf{T}}) = \prod_{1 \le i \le m} \frac{\underline{T}_i}{\underline{T}_i + a_i}.$$

Analogously, we can prove that this method is valid for confidence estimation from above for other main reliability indices. Consider the coefficient of operative availability probability of failure-free operation on the time interval $(t, t + \tau)$ in the stationary regime for $t \to \infty$. This reliability index has the form

$$K_\tau = K_\tau(\mathbf{T}) = K(\mathbf{T}) \exp\left(-\tau \sum_{1 \le i \le m} \frac{1}{T_i}\right).$$

Another standard reliability index, MTBF, has the form

$$L = L(\mathbf{T}) = \left(\sum_{1 \le i \le m} \frac{1}{T_i}\right)^{-1}.$$

It is easy to see that functions $K_\tau(\mathbf{T})$ and $L(\mathbf{T})$ are monotone increasing in each T_i and quasi-concave in $\mathbf{T}$. So, the lower γ-confidence limit for each of these reliability indices can be calculated by the method of substitution as follows:

$$\underline{K}_\tau = K_\tau(\underline{\mathbf{T}}) = \prod_{1\le i\le m} \left(\frac{\underline{T}_i}{\underline{T}_i + a_i}\right) \exp\left(-\tau \sum_{1\le i\le m} \frac{1}{\underline{T}_i}\right),$$

$$\underline{L} = L(\underline{\mathbf{T}}) = \left(\sum_{1\le i\le m} \frac{1}{\underline{T}_i}\right)^{-1}.$$

8.4.2 Series–Parallel System with Loaded Redundancy

Consider a system of m subsystems (redundant groups) in series. Each group consists of $n_i + 1$ identical units in parallel. All failures and renewals are independent. The TTF of unit i has exponential d.f. with unknown mean T_i. The mean repair time for each group is known and equals a_i. Random repair time has an arbitrary d.f. The repair process for all units is independent, that is, repair (replacement) on all units begins immediately after failure.

In the frame of the assumption made above, the availability coefficient of each unit of type i equals $T_i/T_i + a_i$, $1 \le i \le m$. For the ith group we can write

$$K_i(T_i) = 1 - \left[1 - \frac{T_i}{T_i + a_i}\right]^{n_i+1}$$

and for the entire system

$$K(\mathbf{T}) = \prod_{1\le i\le m} K_i(T_i) = \prod_{1\le i\le m} \left[1 - \left(\frac{a_i}{a_i + T_i}\right)^{n_i+1}\right]. \tag{8.24}$$

In this case (see Gnedenko et al., 1969, Sections 2.4 and 6.2) the MTBF of each redundant group can be found as

$$l_i(T_i) = \frac{a_i}{n_i + 1}\left[\left(1 + \frac{T_i}{a_i}\right)^{n_i+1} - 1\right]. \tag{8.25}$$

It follows that (Gnedenko et al., 1965, Section 2.4)

$$L(\mathbf{T}) = \left(\sum_{1\le i\le m} \frac{1}{l_i(T_i)}\right)^{-1} \left(\sum_{1\le i\le m} \frac{n_i + 1}{a_i[(1 + T_i/a_i)^{n_i+1} - 1]}\right)^{-1}.$$

Consider the PFFO on a time interval of length τ. Denote the random TTF (in the stationary regime) of the ith redundant group by ξ_i. The d.f. of this r.v. in general differs from the exponential if $n_i + 1 \geq 2$. Nevertheless (Gnedenko, Belyaev and Solovyev, 1969, Section 6.2), if the group MTBF is much larger than the mean repair time, that is, $l_i(T_i) >> a_i$, then the d.f. of r.v. ξ_i is approximately exponential:

$$P(\xi_i > \tau) \approx \exp\left[-\frac{\tau}{l_i(T_i)}\right];$$

therefore, the stationary PFFO for the system as a whole can be calculated by the formula

$$P_\tau(\mathbf{T}) \approx \exp\left[-\sum_{1\leq i\leq m} \frac{\tau}{l_i(T_i)}\right],$$

where $l_i(T_i)$ is calculated by (8.25).

The coefficient of operative availability (unconditional stationary probability of failure-free operation on an interval of length τ) can be found from

$$K_\tau(\mathbf{T}) = K(\mathbf{T}),$$

$$P_\tau(\mathbf{T}) \approx \prod_{1\leq i\leq m}\left[1 - \left(\frac{a_i}{a_i + T_i}\right)^{n_i+1}\right] \exp\left[-\sum_{1\leq i\leq m}\frac{\tau}{l_i(T_i)}\right].$$

The main reliability indices, such as $K(\mathbf{T})$, $L(\mathbf{T})$, $K_\tau(\mathbf{T})$, and $P_\tau(\mathbf{T})$, are monotone increasing in T_i and can be seen directly from the preceding equations. In Section 7.8.3, we show that all these functions are quasi-concave in the vector $\mathbf{T} = (T_1, \ldots, T_m)$. Therefore the lower confidence limits for them (for $\gamma \geq \exp\{-\frac{3}{2}\}$) can be calculated by the method of substitution as $\underline{K} = K(\underline{\mathbf{T}})$, $\underline{L} = L(\underline{\mathbf{T}})$, $\underline{K}_\tau = K_\tau(\underline{\mathbf{T}})$, and $\underline{P}_\tau = P_\tau(\underline{\mathbf{T}})$.

8.4.3 General Model of Series Connection of Renewal Subsystems

The results obtained above can be expanded on a wider class of systems. Consider a system of m subsystems in series. All failures and renewals are independent. Let subsystem i consist of $A_i + B_i + C_i$ identical units with the same unknown parameter—MTBF T_i. Among these units, A_i of them are main, B_i are loaded redundant units, and C_i are unloaded redundant units. There are D_i repair facilities each of which can repair a single failed unit at a time. The mean time of repair of unit i equals a_i, $1 \leq i \leq m$. For this model we assume that the d.f. of random TTF as well as that for repair time are exponential.

Subsystem i is in the operational state if at least A_i of its units are in the up state. Stochastic process $N_i(t)$, describing the change of the number of nonfailed units within the ith subsystem, is a birth-and-death process. The ith subsystem failure corresponds to the moment when process $N_i(t)$ crosses the threshold $n_i = B_i + C_i$ from below. Processes $N_i(t)$, $1 \leq i \leq m$, are independent. The system has failed if at least one subsystem has failed.

Consider the process for $N_i(t)$. The transitive intensity from state k to state $k + 1$ equals α_{ki}/T_i and from state k to state $k - 1$ equals β_{ki}/T_i, where coefficients α_{ki} and β_{ki} are defined by A_i, B_i, C_i, and D_i and also by the regime of repair and replacement within the ith subsystem. Assume, for example, that subsystem i operates in the following manner. If a main unit has failed, it is replaced by a unit from an unloaded redundant group. If a loaded redundant unit has failed or is directed for replacement of a failed main unit, an unloaded unit takes its place. Repair of failed units begins immediately. Coefficients α_{ki} and β_{ki} for such a model are defined as

$$\alpha_{ki} = \begin{cases} A_i + B_i, & k \leq C_i, \\ A_i + B_i + C_i, & k > C_i; \end{cases}$$

$$\beta_{ki} = \begin{cases} k_i, & k \leq D_i, \\ D_i, & k > D_i. \end{cases}$$

It is easy to write down transitive intensities for other possible regimes.

Let us evaluate the main reliability indices using known results from theory of stochastic processes. Let π_{ki} be the stationary probability that $N_i(t) = k$. This value is determined (see, e.g., Gnedenko et al., 1969, Section 6.3) as

$$\pi_{ki} = \frac{c_{ki}(a_i/T_i)^k}{\sum_{0 \leq l \leq z_i} c_{li}(a_i/T_i)^{l}},$$

where $z_i = A_i + B_i + C_i$ is the total number of units in the ith subsystem and coefficients c_{ki} are defined via α_{ki} and β_{ki} as

$$c_{ki} = \frac{\alpha_{0i} \times \alpha_{1i} \times \cdots \alpha_{k-1,i}}{\beta_{1i} \times \beta_{2i} \times \cdots \beta_{ki}}, \qquad 1 \leq k \leq Z.$$

The availability coefficient of subsystem i is the stationary probability that the number of failed units has not exceeded the system failure level n_i:

$$K_i = \lim_{t \to \infty} P\{N_i(t) \leq n_i\} = \sum_{0 \leq k \leq n_i} \pi_{ki}.$$

Thus, the availability coefficient of subsystem i represents a ratio of polynomials with positive coefficients,

$$K_i(T_i) = \frac{\sum_{0\le k\le n_i} c_{ki}(a_i/T_i)^k}{\sum_{0\le k\le z_i} c_{ki}(a_i/T_i)^k}, \tag{8.26}$$

where $c_{ki} \ge 0$. The system availability coefficient is defined as

$$K(\mathbf{T}) = \prod_{1\le i\le m} K_i(T_i). \tag{8.27}$$

Formulas (8.23) and (8.24) are particular cases of (8.26) and (8.27).

The MTBF of subsystem i, that is, process $N_i(t)$ transition time from state n_i to state $n_i + 1$, is given by (see Gnedenko et al., 1969, Section 6.3)

$$l_i(T_i) = \frac{T_i}{\alpha_{n_i,i} c_{n_i,i}} \left(\frac{T_i}{a_i}\right)^{n_i} \sum_{0\le k\le n_i} c_{ki} \left(\frac{a_i}{T_i}\right)^k. \tag{8.28}$$

For the system as a whole, the MTBF (for the stationary regime) is determined as

$$L(\mathbf{T}) = \left(\sum_{1\le i\le m} \frac{1}{l_i(T_i)}\right)^{-1}.$$

Distribution of TTF of the system or its individual subsystems is not, in general, exponential, though the exponent d.f. gives a good approximation. For instance, the system's availability coefficient for time interval τ has the form of the following approximation (see Gnedenko and Ushakov, 1995):

$$K_\tau(\mathbf{T}) \approx \prod_{1\le i\le m} K_i(T_i) \exp\left(-\sum_{1\le i\le m} \frac{\tau}{l_i(T_i)}\right),$$

where characteristics of individual subsystems are defined by (8.26) and (8.28).

The stationary system PFFO on time interval τ is given as

$$P_\tau \approx \exp\left[-\sum \frac{\tau}{l_i(T_i)}\right].$$

Consider the moment $t = 0$ at which all system units are operable. The mean time of the transition of process $N_i(t)$ from state zero to failure state $n_i + 1$ is determined by

$$l_i^0(T_i) = \sum_{0\le k\le n_i} \frac{T_i}{\alpha_{ki} c_{ki}} \left(\frac{T_i}{a_i}\right)^k \sum_{0\le r\le k} c_{ri} \left(\frac{a_i}{T_i}\right)^r$$

(see Gnedenko et al., 1969, Ushakov, 1994), which after changing the order of summation gives

$$l_i^0(T_i) = T_i \sum_{0 \le r \le n_i} E_{ri} \left(\frac{T_i}{a_i}\right)^r, \tag{8.29}$$

where coefficients E_{ri} are defined as

$$E_{ri} = \sum_{r \le k \le n_i} \frac{c_{k-r,i}}{\alpha_{ki} c_{ki}}.$$

The system PFFO on the time interval $(0, \tau)$ can be calculated by the approximate formula

$$P_\tau^0(\mathbf{T}) \approx \exp\left[-\sum_{1 \le i \le m} \frac{\tau}{l_i^0(T_i)}\right],$$

where characteristics of individual subsystems $l_i^0(T_i)$ are found from (8.29).

All main reliability indices $K(\mathbf{T})$, $K_\tau(\mathbf{T})$, $P_\tau(\mathbf{T})$, $L(\mathbf{T})$, and $P_{0\tau}(\mathbf{T})$ are monotone increasing in each T_i, as can be seen directly from corresponding formulas. In addition (see Section 7.8.3) these functions are concave in vector $\mathbf{T} = (T_1, \ldots, T_m)$. Thus, the lower γ-confidence limit [for $\gamma > 1 - \exp(-\frac{3}{2})$] for those reliability indices can be calculated by the substitution method as $K = K(\mathbf{T})$, $K_\tau = K_\tau(\mathbf{T})$, $P_\tau = P_\tau(\mathbf{T})$, $L = L(\mathbf{T})$, and $P_{0\tau} = P_{0\tau}(\mathbf{T})$. It allows one to calculate the lower confidence limits of main reliability indices for sufficient general models of renewable systems.

8.5 METHOD OF FIDUCIAL PROBABILITIES

8.5.1 Introduction and Method Description

As before in this chapter, we assume that system units have exponential d.f. of TTF and the test of units of type i continues up until r_i failures. Test is assumed to be performed in accordance with plans $[N_i\ U\ r_i]$ or $[N_i\ R\ r_i]$, $1 \le i \le m$. Let S_i be the total test time of units of type i. Random variable S_i has gamma d.f. of the form

$$F_i(S_i, \lambda_i) = 1 - e^{-\lambda_i S_i} \sum_{0 \le j \le r_i - 1} \frac{(\lambda_i S_i)^j}{j!}, \tag{8.30}$$

where λ_i is an unknown parameter of the failure rate of a unit of type i.

For each fixed S_i, Equation (8.30), considered a function of parameter λ_i, possesses all formal properties of a d.f. with respect to parameter λ_i. Let S_i^* be the observed value of r.v. S_i obtained in the test. For given fixed valuue S_i^* we consider parameter λ_i as a r.v. with d.f.

$$F_i(S_i^*, \lambda_i) = 1 - e^{-\lambda_i S_i^*} \sum_{1 \le j \le r_i - 1} \frac{(\lambda_i S_i^*)^j}{j!}. \tag{8.31}$$

Defined in such a way d.f. (8.31) of parameter λ_i is called fiducial (see also Section 7.6.4). The corresponding density of the fiducial distribution of parameter λ_i has the form

$$\varphi_i(S_i^*, \lambda_i) = \frac{\partial}{\partial \lambda_i} F_i(S_i^*, \lambda_i) = e^{-\lambda_i S_\text{i}^*} \frac{(S_i^*)^{r_i}}{(r_i - 1)!} \lambda_i^{r_i - 1},$$

that is, it is a density of the standard gamma distribution with parameters (S^*, r_i).

Let, further, R be a reliability index that is a function

$$R = R(\boldsymbol{\lambda}) = R(\lambda_1, \ldots, \lambda_m) \tag{8.32}$$

of unit parameters $\boldsymbol{\lambda} = (\lambda_1, \ldots, \lambda_m)$, where m is the number of different system unit types. Let also $\mathbf{S}^* = (S_1^*, \ldots, S_m^*)$ be a vector of test results for different types of units. As above, test results for different unit types are assumed independent.

For given fixed $\mathbf{S}^*$ let us consider reliability indices (8.32) as a function of independent r.v.'s λ_i with fiducial d.f.'s (8.31) mentioned above. The corresponding fiducial d.f. of R for given $\mathbf{S}^*$ has the form

$$\Phi(\mathbf{S}^*, R) = \int \cdots \int_{R(\lambda) \le R} \prod_{1 \le i \le m} \varphi_i(S_i^*, \lambda_i)\, d\lambda_i,$$

where $\varphi_i(S^*, \lambda_i)$ mentioned above are the fiducial densities of parameters λ_i. The lower and upper (one-sided) fiducial limits with confidence coefficient γ, $\underline{R}$ and $\overline{R}$ for system reliability index $R = R(\lambda)$, are defined from the conditions

$$\Phi(\mathbf{S}^*, \underline{R}) = 1 - \gamma, \qquad \Phi(\mathbf{S}^*, \overline{R}) = \gamma,$$

that is, as corresponding quantiles of fiducial d.f.'s of r.v. R. The d.f. $\Phi(\mathbf{S}^*, R)$ and limits $\underline{R}$ and $\overline{R}$ can be easily found with Monte Carlo simulation.

As mentioned in Section 7.6.4, in the multidimensional case, where $m > 1$, the fiducial method is approximate in the sense that the fiducial limit constructed by this method (e.g., upper) $\overline{R}$ may not be with confidence coefficient equal to γ. Its real confidence coefficient might be essentially smaller than γ. The following simple example illustrates this fact.

Example 8.3 Let the reliability index have the form $R = R(\boldsymbol{\lambda})$,

$$R = \min(\lambda_1, \ldots, \lambda_m), \tag{8.33}$$

and we need to construct the UCL for R. This task can be interpreted as follows. Reliability index (8.33) has the sense of the best (minimum) parameter of failure rate among unit failure rates of different types. Thus, on the basis of results of independent unit tests of m different types, we need to construct the UCL of the failure rate for the best (most reliable) unit type. (We do not know in advance which unit type is the best.)

To solve this problem, let us apply the fiducial method considered above. In accordance with this method, parameters λ_i are assumed independent r.v.'s with d.f.'s (8.31). The corresponding fiducial d.f. reliability index (8.33) has the form

$$\begin{aligned}\Phi(\mathbf{S}^*, R) &= P\left\{\min_i \lambda_i \le R\right\} = 1 - P\left\{\min_i \lambda_i > R\right\} \\ &= 1 - \prod_{1 \le i \le m} P\{\lambda_i > R\} \\ &= 1 - \prod_{1 \le i \le m} [1 - F_i(\mathbf{S}^*, R)].\end{aligned}$$

So, the upper fiducial limit $\overline{R}$ with confidence coefficient γ for reliability index R is determined from

$$1 - \prod_{1 \le i \le m} [1 - F_i(\mathbf{S}^*, \overline{R})] = \gamma. \tag{8.34}$$

Notice that the standard γ-UCL $\hat{\lambda}_i$ (coinciding with the upper fiducial limit) for each individual parameter λ_i is determined from

$$F_i(\mathbf{S}^*, \overline{\lambda}_i) = \gamma. \tag{8.35}$$

The left sides of (8.34) and (8.35) satisfy the inequality

$$1 - \prod_{1 \le i \le m} [1 - F_i(\mathbf{S}^*, t)] > F_i(\mathbf{S}^*, t)$$

for all $t > 0$. From here the inequality

$$\overline{R} < \overline{\lambda}_i$$

for all $1 \le i \le m$. So,

$$\overline{R} < \min_i \overline{\lambda}_i,$$

and the following inequality for the confidence probability holds:

$$P_\lambda\{\overline{R} \ge R(\boldsymbol{\lambda})\} = P_\lambda \left\{\overline{R} \ge \min_i \lambda_i\right\} \le P_\lambda \left\{\min_i \overline{\lambda}_i \ge \min_i \lambda_i\right\}.$$

Taking in this inequality $\lambda_1 = \cdots = \lambda_m$, we obtain that in this case the confidence probability satisfies the inequality

$$P_\lambda\{\overline{R} \ge R(\boldsymbol{\lambda})\} \le P_\lambda \left\{\min_i \overline{\lambda}_i \ge \lambda_i\right\} = P_\lambda \left\{\bigcap_{1 \le i \le m} (\overline{\lambda}_i \ge \lambda_i)\right\}$$

$$= \prod_{1 \le i \le m} P_\lambda(\overline{\lambda}_i \ge \lambda_i) = \gamma^m.$$

From here it follows that the upper fiducial limit has confidence coefficient not larger than γ^m. For instance, for $m = 10$ and $\gamma = 0.9$ the value of the confidence coefficient is not larger than $\gamma^m \approx 0.35$. □

Thus, in this example, using the fiducial method for confidence estimation of the reliability index leads to an increase in reliability, and the larger the increase, the larger the problem dimension is, that is, the larger is the number of types of units.

Remark 8.1 In the example considered above, $\min_i \overline{\lambda}_i$ is equal to the UCL for R calculated with the method of "substitution." Due to this, using an analogous argument, we can show that, for this example, the method of substitution is not recommended. □

Thus, in general, the fiducial method for confidence estimation of complex system might lead to errors. Therefore application of this method to one or another concrete situation needs to be verified. Further we will show, using results from the literature (Pavlov, 1980, 1981,a,b), that application of the fiducial method is correct for a sufficiently wide class of complex systems for confidence estimation of system reliability indices from below (which is most important in practice).

8.5.2 Conditions of Method Correctness

Let

$$R = R(\boldsymbol{\lambda}) = R(\lambda_1, \ldots, \lambda_m) \tag{8.36}$$

be a function expressing the dependence of system reliability index R on parameters of units $\boldsymbol{\lambda} = (\lambda_1, \ldots, \lambda_m)$. Consider whether, for each reliability index (8.36), it is possible to state that application of the fiducial method for confidence estimation is correct (or, in other words, that the fiducial limit for R is simultaneously a γ-confidence limit for this reliability index).

It is convenient to write (8.36) in the form

$$R = \exp\{-f(\boldsymbol{\lambda})\}. \tag{8.37}$$

Estimating R from below is reduced to estimation of the function $f(\boldsymbol{\lambda}) = f(\lambda_1, \ldots, \lambda_m)$ from above. Let us introduce new parameters $\mathbf{z} = (z_1, \ldots, z_m)$, where $z_i = \ln \lambda_i$, $1 \le i \le m$.

Assume that the following condition holds.

Condition A: Function $f(\boldsymbol{\lambda})$ expressed via parameters $\mathbf{z} = f(z_1, \ldots, z_m)$, that is, the function

$$\tilde{f}(\mathbf{z}) = f(e^{z_1}, \ldots, e^{z_m}) \tag{8.38}$$

is convex in $\mathbf{z}$.

Then the upper fiducial limit $\overline{f}$ for $f(\boldsymbol{\lambda})$ is simultaneously γ-UCL, that is, it has confidence coefficient not smaller ;γ (see the proof of Theorem 8.8 in Section 7.8.4). In other words, if condition A holds, the use of method of fiducial propabilities is correct for confidence estimation of $f(\boldsymbol{\lambda})$ from above. [It corresponds to estimation of the reliability index (8.37) from below.]

Let us now show that condition A [function (8.38) convexity] is valid for a satisfactory wide class of complex systems, including systems with recurrent structures and with renewal.

8.5.3 Systems with Series–Parallel Structure

Consider a system consisting of m subsystems in series. Each subsystem i consists of n_i identical units in parallel, each unit characterized by failure rate

λ_i, $1 \le i \le m$. The redundant units regime might be any: loaded or unloaded. Assume that within subsystems with subscripts $1 \le i \le l$ redundant units are unloaded and within subsystems with subscripts $l + 1 \le i \le m$ redundant units are load. The system PFFO during time t_0 can be written in the form

$$R = R(\boldsymbol{\lambda}) = \prod_{1 \le i \le m} h_i(\lambda_i), \tag{8.39}$$

where $h_i(\lambda_i)$ is the PFFO of the ith subsystem that depends on the redundant unit regime. For a subsystem with unloaded redundant units, the PFFO can be written as

$$h_i(\lambda_i) = e^{-\lambda_i t_0} \sum_{0 \le d \le n_i - 1} \frac{(\lambda_i t_0)^d}{d!}, \qquad 1 \le i \le l,$$

and for loaded redundant units as

$$h_i(\lambda_i) = 1 - (1 - e^{-\lambda_i t_0})^{n_i}, \qquad l + 1 \le i \le m.$$

Obviously, (8.39) includes as particular cases systems with all loaded or all unloaded redundant units. It is easy to check by direct differentiation that for both types of redundancy functions

$$f_i(\lambda_i) = -\ln h_i(\lambda_i)$$

is monotone increasing and convex in λ_i, $1 \le i \le m$. Consequently, the system reliability index (8.39) can be written as

$$R(\boldsymbol{\lambda}) = \exp\{-f(\boldsymbol{\lambda})\},$$

where

$$f(\boldsymbol{\lambda}) = \sum_{1 \le i \le m} f_i(\lambda_i)$$

is the sum of monotone increasing convex functions. It follows that the function

$$f(\mathbf{z}) = f(e^{z_1}, \ldots, e^{z_m}) = \sum_{1 \le i \le m} f_i(e^{z_i})$$

is convex in $\mathbf{z}$, that is, condition A holds. It means (see Section 8.5.2) that the fiducial method can be used for confidence estimation of $f(\boldsymbol{\lambda})$ from above or for confidence estimation of the reliability index (8.39) from below.

Example 8.4 Consider a series–parallel system consisting of $m = 10$ subsystem in series. Within subsystems 1–4, redundant units are unloaded, and within subsystems 5–10, redundant units are loaded. The numbers of units in subsystems, n_i, and the test results (the number of failures, r_i, and the total test time, S_i) are given in Table 8.2.

We need to construct the LCL with confidence coefficient $\gamma = 0.95$ for the system reliability index (8.39) for $t_0 = 1$. The use of the method of fiducial probabilities in this case gives $\underline{R}_{0.95} = 0.856$. Notice that the method of substitution, considered above, gives for the same case an essentially worse lower limit: $\underline{R}_{0.95} = 0.786$. □

Example 8.5 Consider the same system as in the previous example with the difference that all subsystems consist of unloaded redundant units. In this case the method of fiducial probabilities gives $\underline{R}_{0.95} = 0.918$, and the method of substitution $\underline{R}_{0.95} = 0.876$. □

8.5.4 Systems with Complex Structure

As mentioned in Section 8.3, the confidence estimate of a system with complex structure can be reduced to the construction of the UCL of the function

$$f(\boldsymbol{\lambda}) = \sum_{1 \le i \le M} A_i \prod_{j \in G_i} \lambda_j^{n_{ij}}, \tag{8.40}$$

where A_i are positive coefficients, G_i are some (in the general case, intersecting) subsets of subscripts, $G_i \subset (1, 2, \ldots, m)$, where m is the number of different types of units within the system, and λ_j is the failure rate of unit j [see formulas (8.17) and (8.22)].

Function (8.40), written via parameters $z_j = \ln \lambda_j$, $1 \le j \le m$, has the form

$$\tilde{f}(\mathbf{z}) = f(e^{z_1}, \ldots, e^{z_m}) = \sum_{1 \le i \le M} A_i \exp\left(\sum_{j \in G_i} n_{ij} z_j \right).$$

Table 8.2 Data for Example 8.4

Type of Redundancy	Unloaded Standby				Loaded Standby				
i	1	2	3	4	5	6	7	8	9
n_i	2	2	2	3	3	3	2	2	2
S_i	25	40	40	75	15	15	75	75	20
r_i	3	2	1	2	1	3	2	1	3

This function is convex in $\mathbf{z} = (z_1, \ldots, z_m)$ because it represents the sum of convex functions, that is, condition A above holds. Thus, the UCL for reliability index of type (8.40) can be constructed by the method of fiducial probabilities. It means that this method works for the confidence estimation of complex system PFFO from below.

8.5.5 Systems with Renewal

For a system with independent renewal, consider that a failed unit begins to be repaired immediately after its failure. Let the system consist of m subsystems in series. Each subsystem i consists of n_i identical redundant units in parallel. Processes of failure and renewal in all subsystems are independent.

The unit's TTF and the unit's intensity of repair have exponential d.f. with parameters λ_i and μ_i, respectively, $1 \leq i < m$.

Parameters $\boldsymbol{\lambda} = (\lambda_1, \ldots, \lambda_m)$ and $\boldsymbol{\mu} = (\mu_1, \ldots, \mu_m)$ are unknown but there are test results: the total testing time for unit i, S_i, up to the occurrence of r_i failures, and the total time of repair V_i of l_i units of this type. In other words, each S_i and V_i represents the sum of corresponding r.v.'s. Notice that these test results might be obtained in the result of unit tests, the test of a system as a whole, or as a combination of data of both types of testing.

The system's reliability index is a function of mentioned parameters

$$R = R(\boldsymbol{\lambda}, \boldsymbol{\mu}) = R(\lambda_1, \ldots, \lambda_m, \mu_1, \ldots, \mu_m). \tag{8.41}$$

We need to construct the confidence limit (lower or upper) for reliability index (8.41) based on the test results. Systems with renewal are quite different from systems without renewal. They are characterized by different reliability indices. Nevertheless, from a formal viewpoint, parameters μ_i are the same "exponential" parameters as λ_i. Therefore all results obtained above for systems without renewal can be extended to systems with renewal.

We again write reliability index (8.41) in the form

$$R = \exp\{-f(\boldsymbol{\lambda}, \boldsymbol{\mu})\}.$$

Estimation of R from below is reduced to estimate of the function $f(\boldsymbol{\lambda}, \boldsymbol{\mu}) = f(\lambda_1, \ldots, \lambda_m, \mu_1, \ldots, \mu_m)$ from above.

Introduce $z_i = \ln \lambda_i$ and y_i and $y_i = \ln \mu_i$, $1 \leq i \leq m$. As above, denote $\mathbf{z} = (z_1, \ldots, z_m)$ and $\mathbf{y} = (y_1, \ldots, y_m)$. Condition A given in Section 8.5.2 in this case transforms into the following.

Condition A′: Function $f(\boldsymbol{\lambda}, \boldsymbol{\mu})$ expressed via parameters $(\mathbf{z}, \mathbf{y})$, that is

$$\tilde{f}(\mathbf{z}, \mathbf{y}) = f(e^{z_1}, \ldots, e^{z_m}, e^{y_1}, \ldots, e^{y_m})$$

is convex in $(\mathbf{z}, \mathbf{y})$.

The condition of applicability of the method of fiducial probabilities for confidence estimation of renewal system reliability indices is formulated in a way similar to above. If condition A' holds, then the upper fiducial limit for $f(\boldsymbol{\lambda}, \boldsymbol{\mu})$ is simultaneously γ-UCL (see the proof of Theorem 8.8 in Section 8.7.4). In other words, if condition A$'$ holds, the use of the method of fiducial probabilities is correct for confidence estimation of $f(\boldsymbol{\lambda}, \boldsymbol{\mu})$ from above. [It corresponds to estimation of reliability index (8.41) from below.]

Condition A$'$ is valid for many practical cases. For instance, one of the most frequently used reliability indices for renewal systems is the availability coefficient. For the case of independent unit repair and failures,

$$K = \prod_{1 \le i \le m} K_i(\lambda_i, \mu_i), \tag{8.42}$$

where $K_i(\lambda_i, \mu_i)$ is the availability coefficient of subsystem i, defined as (see Section 8.4.2)

$$K_i(\lambda_i, \mu_i) = 1 - \left(\frac{\lambda_i}{\lambda_i + \mu_i}\right)^{n_i}.$$

The reliability index (8.42) can be presented in the form

$$K = e^{-f(\lambda, \mu)},$$

where

$$f(\boldsymbol{\lambda}, \boldsymbol{\mu}) = \sum_{1 \le i \le m} f_i(\lambda_i, \mu_i), \tag{8.43}$$

and in turn

$$f_i(\lambda_i, \mu_i) = -\ln\left[1 - \left(\frac{\lambda_i}{\lambda_i + \mu_i}\right)^{n_i}\right].$$

Confidence estimation of the availability coefficient from below is reduced to the construction of the UCL for function (8.43). This function can be written via parameters $(\mathbf{z}, \mathbf{y})$ in the form

$$\tilde{f}(\mathbf{z}, \mathbf{y}) = f(e^{z_1}, \ldots, e^{z_m}, e^{y_m}) = \sum_{1 \le i \le m} \varphi_i(z_i - y_i), \tag{8.44}$$

where

$$\varphi_i(u) = -\ln\left[1 - \left(\frac{e^u}{1 + e^u}\right)^{n_i}\right].$$

By direct differentiation we can show that functions $\varphi_i(u)$ are convex in u, $1 \leq i \leq m$, and the convexity of function (8.44) follows. Thus, condition A′ holds and, consequently, γ-UCL $\overline{f}$ for $f(\boldsymbol{\lambda}, \boldsymbol{\mu})$ can be calculated by the fiducial method. After this, the γ-LCL for the system availability coefficients are calculated as

$$\underline{K} = e^{-\overline{f}}.$$

Example 8.6 Let a system consist of $m = 10$ subsystems in series. The number of units within the subsystems and the test results (i.e., values of S_i, r_i, V_i, l_i) are given in Table 8.3.

We need to construct the LCL for the system availability coefficient (8.42) with confidence coefficient $\gamma = 0.9$. In this case, the use of the fiducial method gives the LCL $\underline{K}_{0.9} = 0.678$. □

Consider now construction of the LCL for another standard system reliability index with renewal, the MTBF (in the stationary regime). For the model with independent renewal, this reliability index is defined by the formula (see Section 8.4.2; see also Gnedenko et al., 1969, Ushakov, 1994)

$$L = \left(\sum_{1 \leq i \leq m} \frac{n_i \mu_i}{(1 + u_i/\lambda_i)^{n_i} - 1} \right)^{-1}.$$

Confidence estimation of L from below is reduced to construction of the UCL for

$$f(\boldsymbol{\lambda}, \boldsymbol{\mu}) = \sum_{1 \leq i \leq m} \frac{n_i \mu_i}{(1 + \mu_i/\lambda_i)^{n_i} - 1}. \tag{8.45}$$

In a most interesting case of highly reliable systems, where $\lambda_i << \mu_i$ for all i, from (8.45) the following approximation holds:

Table 8.3 Data for Example 8.6

i	1	2	3	4	5	6	7	8	9
n_i	2	2	2	2	3	3	2	2	2
S_i	415	450	240	155	170	225	340	215	478
r_i	1	1	4	2	1	2	3	1	2
V_i	19	11	7	4	5	12	8	3	15
l_i	1	1	4	2	1	2	3	1	2

$$f(\boldsymbol{\lambda}, \boldsymbol{\mu}) \approx \sum_{1 \le i \le m} n_i \left(\frac{\lambda_i^{n_i}}{\mu_i^{n_i - 1}} \right).$$

Expressing this formula via parameters $(\mathbf{z}, \mathbf{y})$, we have

$$\tilde{f}(\mathbf{z}, \mathbf{y}) \approx \sum_{1 \le i \le m} n_i \exp[n_i z_i - (n_i - 1) y_i].$$

This function is convex in $(\mathbf{z}, \mathbf{y})$. Thus, condition A′ holds and the UCL $\overline{f}$ for $f(\boldsymbol{\lambda}, \boldsymbol{\mu})$ can be calculated by the fiducial method. After this the LCL for the system MTBF, L, can be found as $\underline{L} = 1/\overline{f}$.

In an analogous way, we can show the correctness of the fiducial method for a lower confidence estimate for another standard reliability index, the operative availability coefficient K_τ. Remember that this reliability index represents the probability of successful operation during interval τ in the stationary regime. For highly reliable systems ($\lambda_i << \mu_i$) the following approximation is valid:

$$K_\tau \approx K e^{-\tau/L},$$

where K and L are the availability coefficient and the MTBF of the system. Using previous formulas, we can rewrite the latter formula as

$$K_\tau \approx e^{-f(\boldsymbol{\lambda}, \boldsymbol{\mu})}$$

where $f(\boldsymbol{\lambda}, \boldsymbol{\mu})$ expressed via parameters $(\mathbf{z}, \mathbf{y})$ takes the form

$$\tilde{f}(\mathbf{z}, \mathbf{y}) = \sum_{1 \le i \le m} \varphi_i(z_i - y_i) + \tau \sum_{1 \le i \le m} e^{n_i z_i - (n_i - 1) y_i}.$$

This function is convex in $(\mathbf{z}, \mathbf{y})$, so it follows that the LCL for K_τ can be constructed by the fiducial method.

In an analogous way, on the basis of condition A′, we can show the correctness of the fiducial method for confidence estimation from below standard reliability indices, such as availability coefficient K, operative availability coefficient K_τ, MTBF, and L, for a general model of series–parallel systems with renewal (see Section 8.4.3).

8.6 METHOD OF TANGENT FUNCTIONS

Consider an approach that is more complex in sense of computations but allows, in particular, to improve the fiducial method. In addition, this approach is sufficiently general and can be applied not only to the exponential case considered in this chapter, that is, to a system consisting of units with exponential d.f. of TTF. (See Remark 8.2.)

Let $f(\boldsymbol{\lambda}) = f(\lambda_1, \ldots, \lambda_m)$ be system reliability indices where λ_i is the failure rate of a unit i. Again introduce new parameters $z_i = \ln \lambda_i$, $1 \le i \le m$, and assume that the function

$$\tilde{f}(\mathbf{z}) = f(e^{z_1}, \ldots, e^{z_m}) \tag{8.46}$$

is convex in $\mathbf{z} = (z_1, \ldots, z_m)$. As we show above, this condition holds for many standard reliability indices and models of complex systems.

Consider a class of all linear functions of the type

$$h_b(\mathbf{z}) = b_0 + \sum_{1\le i\le m} b_i z_i, \qquad \mathbf{b} \in B, \tag{8.47}$$

where $\mathbf{b} = (b_0, b_1, \ldots, b_m)$ is a vector of coefficients and $B = \{\mathbf{b}\colon -\infty < b_i < \infty, 1 \le i \le m\}$. Since function (8.46) is convex, it can be presented as a maximum in some subset of linear functions of the type (8.47), that is,

$$\tilde{f}(\mathbf{z}) = \max_{\mathbf{b}\in B'} h_b(\mathbf{z}), \tag{8.48}$$

where $B' \subset B$. Obviously, the function $h_b(\mathbf{z})$ has the sense of tangent "planes" for function $\tilde{f}(\mathbf{z})$. We call functions $h_b(\mathbf{z})$ "tangent functions" for function $\tilde{f}(\mathbf{z})$.

Denote by $\mathbf{b}(\mathbf{z}) \in B'$ a vector of coefficients for which (for given fixed $\mathbf{z}$) the maximum in (8.48) attains, that is,

$$\tilde{f}(\mathbf{z}) = \max_{\mathbf{b}\in B'} h_b(\mathbf{z}) \tag{8.49}$$

Let $\overline{h}_b = \overline{h}_b(S)$ be the upper fiducial limit with confidence coefficient γ for function $h_b(\mathbf{z})$ for a given vector of test results $\mathbf{S} = (S_i, \ldots, S_m)$. Value $\overline{h}_b$ simultaneously represents the γ-UCL for $h_b(\mathbf{z})$. Denote also by $\overline{f} = \overline{f}(S)$ the upper fiducial limit for f. From (8.48) the inequality

$$\tilde{f}(\mathbf{z}) \ge h_b(\mathbf{z}), \qquad \mathbf{b} \in B',$$

follows for all $\mathbf{z}$. For fiducial limits the inequality.

$$\overline{f}(\mathbf{S}) \ge \overline{h}_b(\mathbf{S}),\ \mathbf{b} \in B',$$

follows for all $\mathbf{S}$. So, inequality

$$\overline{f}(\mathbf{S}) \ge \sup_{\mathbf{b}\in B'} \overline{h}_b(\mathbf{S}) \tag{8.50}$$

holds for all $\mathbf{S}$. Introduce

$$H(\mathbf{S}) = \sup_{\mathbf{b}\in B'} \overline{h}_b(\mathbf{S}), \tag{8.51}$$

which we call a "confidence majorant." Taking into account (8.49), we have, for arbitrary fixed $\mathbf{z}$,

$$P\{H(\mathbf{S}) \geq \tilde{f}(\mathbf{z})\} = P\left\{\sup_{\mathbf{b}\in B'} \overline{h}_b(\mathbf{S}) \geq h_{\mathbf{b(z)}}(\mathbf{z})\right\} \geq 'P\{\overline{h}_{\mathbf{b(z)}}(\mathbf{S}) \geq h_{\mathbf{b(z)}}(\mathbf{z})\} \geq \gamma;$$

that is, $H(\mathbf{S})$ is γ-UCL for function f. Due to (8.50), the inequality

$$\overline{f}(\mathbf{S}) \geq H(\mathbf{S}) \tag{8.52}$$

holds for all $\mathbf{S}$.

The latter inequality shows, first, that the upper fiducial limit $\overline{f}(\mathbf{S})$ presents the γ-ULC for f because $H(\mathbf{S})$ is such a limit. It proves the correctness of using the fiducial method for confidence estimation from above [if the condition of convexity of function (8.46) holds]. Second, from (8.52), it follows that the γ-UCL $H(\mathbf{S})$ is better (for all test results $\mathbf{S}$) in comparison with the upper fiducial limit $\overline{f}(\mathbf{S})$ with confidence coefficient γ.

The precise calculation of a "confidence majorant" $H(\mathbf{S})$ is sufficiently complicated if the maximum in (8.48) and (8.51) is taken over the infinite set B'. Nevertheless, convex function $\tilde{f}(\mathbf{z})$ can be always presented approximately with an accuracy (given in advance) in the form of a maximum over a finite set of tangent linear functions:

$$\tilde{f}(\mathbf{z}) = \max\{h_1(\mathbf{z}), \ldots, h_N(\mathbf{z})\}.$$

Then the confidence majorant is calculated by the formula

$$H(\mathbf{S}) = \max[\overline{h}_1(\mathbf{S}), \ldots, \overline{h}_N(\mathbf{S})\}.$$

where each of confidence limits $\overline{\mathrm{h}}_j(\mathbf{S})$ for individual tangent linear functions $h_j(\mathbf{z})$, $1 \leq j \leq N$, is calculated (if it is impossible to do analytically), for example, by a common fiducial method using Monte Carlo simulation.

This procedure of confidence limit construction for $H(\mathbf{S})$ is called the method of "tangent functions," or the method of "confidence majorant." Thus, if the condition of function (8.46) convexity holds, then the method of tangent functions allows improved confidence estimation of function f from above in comparison with the fiducial method. (See examples 8.7 and 8.8.) Remember that this corresponds to the confidence estimation of the system PFFO from below (see Section 8.5).

Example 8.7 Consider the system from Example 8.4. The fiducial method gives $\underline{R}_{0.95} = 0.856$. The method of tangent functions gives $\underline{R}_{0.95} = 0.914$. □

Example 8.8 Consider a series–parallel system with renewal, similar to that in Example 8.5. Let the system consist of $m = 4$ subsystems in series. Each subsystem consists of n_i parallel identical units. A failed unit is repaired immediately after its failure independent of the state of other units. All needed data for this example are in Table 8.4.

We need to find the LCL with confidence coefficient not less than $\gamma = 0.95$ for the system availability coefficient

$$K = \prod_{1 \leq i \leq m} \left[1 - \left(\frac{\lambda_i}{\lambda_i + \mu_i} \right)^{n_i} \right].$$

The fiducial method gives $\underline{K}_{0.95} = 0.533$, and the method of tangent functions gives $\underline{K}_{0.95} = 0.686$. □

Remark 8.2 It is easy to see that the method of tangent functions suggested by Pavlov (1981a,b; 1982a, pp. 157–159) can be applied not just for the exponential case investigated in this chapter. Consider the general function

$$f(\boldsymbol{\theta}) = f(\theta_1, \ldots, \theta_m) \tag{8.53}$$

reflecting the dependence of reliability index f from the vector of parameters $\boldsymbol{\theta} = (\theta_1, \ldots, \theta_m)$, where θ_i is the parameter of unit reliability. We need to construct γ-UCL for f on the basis of test results $\mathbf{x}$.

Assume that there is a basis set of the "tangent functions"

$$h_b(\boldsymbol{\theta}) = h_b(\theta_1, \ldots, \theta_m), \qquad b \in B, \tag{8.54}$$

such that function (8.53) can be represented in the form of a maximum over some subset of tangent functions. In other words, for arbitrary $\boldsymbol{\theta}$ inequality

$$f(\boldsymbol{\theta}) = \max_{\mathbf{b} \in B'} h_b(\theta) \tag{8.55}$$

Table 8.4 Data for Example 8.8

i	1	2	3	4
n_i	2	2	2	2
S_i	159	115	177	191
r_i	2	3	4	1
V_i	12	14	8	15
l_i	1	1	4	1

where $B' \subset B$, and maximum (8.55) is attained on some $b(\boldsymbol{\theta}) \in B'$ for arbitrary b. Assume also that for each tangent function, $h_b(\theta)$ γ-UCL $\overline{h}_b = \overline{h}_b(x)$, $b \in B$, can be constructed. Then the value

$$H(\mathbf{x}) = \sup_{\mathbf{b} \in B'} \overline{h}_{\mathbf{b}}(\mathbf{x})$$

gives γ-UCL for function (8.53).

The proof of this fact is completely similar to that for the exponential case. The problem is in the "optimal" (more precisely, rational) choice of a sufficiently constructive class of tangent functions (8.54) for a concrete situation. □

In conclusion note that the three methods (substitution, fiducial, and tangent functions) represents methods each of which improves on the previous one but consumes more calculating time. The method of substitution calculates the function $f(\boldsymbol{\lambda})$ only in a single point $\boldsymbol{\lambda} = \overline{\boldsymbol{\lambda}}$ in the space of parameters $\boldsymbol{\lambda}$. The fiducial method requires multiple calculations of function $f(\boldsymbol{\lambda})$ at different points of the parametric space. These points may be calculated using Monte Carlo simulation. Finally, the method of tangent functions needs multiple repetitions of the fiducial method for different tangent functions.

8.7 APPENDIX

8.7.1 Confidence Limits for Quasi-Convex and Quasi-Concave Functions

Let Ω be a convex area in m-dimensional Euclidean space. The function of m variables

$$g(\mathbf{x}) = g(x_1, \ldots, x_m)$$

determined in area Ω is called quasi-concave if, for any $\alpha > 0$ and $\beta > 0$, $\alpha + \beta = 1$, $\mathbf{x} \in \Omega$ and $\mathbf{y} \in \Omega$, the inequality

$$g(\alpha\mathbf{x} + \beta\mathbf{y}) \geq \min[g(\mathbf{x}), g(\mathbf{y})]$$

holds, or, equivalently, if area

$$\{\mathbf{x} \in \Omega: g(\mathbf{x}) \geq C\}$$

is convex for arbitrary constant C. Function $g(\mathbf{x})$ is called quasi-convex if function $-g(\mathbf{x})$ is concave. It is easy to see that a concave (convex) function is simultaneously quasi-concave (quasi-convex), although, generally, an inverse statement is not true. Notice also that if function $g(\mathbf{x})$ is convex (concave) and the function of a single variable $h(u)$ is monotone decreasing in u, then function $h[g(\mathbf{x})]$ is quasi-concave (convex).

Let $F_\xi(t) = P(\xi < t)$ be the d.f. off r.v. ξ. Denote by Φ a class of d.f.'s $F(t)$ such that $-\ln[1 - F(t)$ is convex in t such that $F(t) < 1$ and $\tilde{\Phi}$ a class of d.f.'s $F(t)$ such that $-\ln F(t)$ is convex in t such that $F(t) > 0$. If r.v. ξ has d.f. $F_\xi(t)$ belonging to Φ ($\tilde{\Phi}$), we will denote it by $\xi \in \Phi$ (or by $\xi \in \tilde{\Phi}$).

From the definition of classes above, it follows that, if $\xi \in \Phi$, then $-\xi \in \tilde{\Phi}$, and, on the contrary, if $\xi \in \tilde{\Phi}$, then $-\xi \in \Phi$. If $\xi \in \Phi$ ($\xi \in \tilde{\Phi}$) and $C > 0$, then $C\,\xi \in \Phi$ ($C\,\xi \in \tilde{\Phi}$). Classes Φ and $\tilde{\Phi}$ include such d.f.'s as normal, exponential, Weibull–Gnedenko (with shape parameter $\alpha \geq 1$), and some others. Notice that under the additional condition $F(0) = 0$, class Φ coincides with the class of IFR distributions.

Lemma 8.1 Let $\xi_1, \ldots, \xi_m$ be independent r.v.'s, and $u_1, \ldots, u_m$ be such constants that $P\{\xi_k \leq (\geq)\, u_k\} \geq \gamma$, $1 \leq k \leq m$. It follows that, if $\xi_k \in \Phi$ ($\xi_k \in \tilde{\Phi}$) and $\gamma \geq 1 - \exp(-\frac{3}{2})$, then

$$P\left\{\sum_{1\leq k\leq m} \xi_k \leq (\geq) \sum_{1\leq k\leq m} u_k\right\} \geq \gamma. \;\square \tag{8.56}$$

Proof It is enough to consider the case where $\xi_k \in \Phi$, $1 \leq k \leq m$, since the second half of the proof follows from a simple transition to r.v.'s $-\xi_k$. Since class Φ is closed with respect to the convolution of independent r.v.'s, it is enough to prove (8.56) for $m = 2$. Let $F_k(t) = P(\xi_k < t)$ and $F_k(u_k) = \gamma$, $1 \leq k \leq 2$. Due to convexity of $\Lambda_k'(t) = -\ln[1 - F_k(t)]$, it lies above the tangent at point $[u_k, \Lambda_k'(u_k)]$, and it follows that, for all t,

$$\Lambda_l(t) \geq \max[0, \Lambda_k(u_k) + \Lambda_k'(u_k)(t - u_k)],$$

or, taking into account that $\Lambda_k(u_k) = -\ln(1 - \gamma)$, we obtain

$$\Lambda_k(t) \geq \lambda_k^*(t) = \max[0, \beta_k(t - u_k)],$$

where $\tau_k = u_k + \ln(1 - \gamma)[\Lambda_k'(u_k)]^{-1}$, $\beta_k = \Lambda_k'(u_k)$, $1 \leq k \leq 2$. Denoting $F_k^*(t) = 1 - \exp[-\Lambda_k^*(t)]$, we obtain $F_k^*(t) \leq F_k(t)$ for all t, which gives

$$\int_{t_1+t_2\leq u_1+u_2} \cdots \int dF_1(t_1)\, dF_2(t_2) \geq J^* = \int_{it_1+t_2\leq u_1+u_2} \cdots \int dF_1^*(t_1)\, dF_2^*(t_2)$$

$$= \int_{t_1+t_2\leq C} \cdots \int \beta_1 e^{-\beta_1 t_1} \beta_2 e^{-\beta_2 t_2}\, dt_1\, dt_2,$$

where

$$c = -\ln(1 - \gamma)\,\frac{\beta_1 + \beta_2}{\beta_1\beta_2}.$$

Inequality $J^* \geq \gamma$ after integration can be written in the form

$$\frac{\beta_1 e^{-\beta_1 C} - \beta_2 e^{-\beta_2 C}}{\beta_1 - \beta_2} \leq 1 - \gamma.$$

Setting $\beta_1 \geq \beta_2$ and denoting $\mu = \beta_2/\beta_1$, we can write

$$h(\mu) = \mu + e^{-B\mu} - \mu e^{-B/\mu} \leq 1,$$

where $B = -\ln(1 - \gamma)$. It is easy to show that, for $B \geq \frac{3}{2}$ and $0 < \mu \leq 1$,

$$h''(\mu) = B^2\left(e^{-B\mu} - \frac{1}{\mu^3}\,e^{-B/\mu}\right) \geq 0$$

holds. From here it follows that, if $B \geq \frac{3}{2}$, function $h(\mu)$ is convex on interval (0, 1). Since $h(+0) = h(1) = 1$, it follows that, for $B \geq \frac{3}{2}$ and $0 < \mu \leq 1$, inequality $h(\mu) \leq 1$ holds, and inequality $J^* \geq \gamma$ and (8.56) follow. □

We now have m independent test results, where $X_k = \{x_k\}$ is the result of the kth test and P_{a_k} is a family of d.f.'s on X_k, where $\underline{a}_k \in A_k$, $1 \leq k \leq m$. Let $\theta_k(a_k)$ be given on A_k real functions and $\underline{\theta}_k(x_k)$ and $\overline{\theta}_k(x_k)$ be r.v.'s defined on X_k such that

$$\inf_{a_k \in A_k} P_{a_k}\{\underline{\theta}_k(x_k) \leq \theta_k(a_k)\} \geq \gamma,$$

$$\inf_{a_k \in A_k} P_{a_k}\{\overline{\theta}_k(x_k) \geq \theta_k(a_k)\} \geq \gamma,$$

$1 \leq k \leq m$, $0 < \gamma < 1$. Further we use an abbreviated notation: $\theta_k = \theta_k(a_k)$, $\underline{\theta}_k = \underline{\theta}_k(x_k)$, $\overline{\theta}_k = \overline{\theta}_k(x_k)$, $\boldsymbol{\theta} = (\theta_1, \ldots, \theta_m)$, $\underline{\boldsymbol{\theta}} = (\underline{\boldsymbol{\theta}}_1, \ldots, \underline{\boldsymbol{\theta}}_m)$, $\overline{\boldsymbol{\theta}} = (\overline{\theta}_1, \ldots, \overline{\theta}_m)$.

Theorem 8.1 Let $\underline{\theta}_k \in \Phi$, $1 \leq k \leq m$, and function $R(\boldsymbol{\theta}) = R(\theta_1, \ldots, \theta_m)$ be strictly increasing (decreasing) monotone in each variable and quasi-concave (quasi-convex). If $\gamma \geq 1 - \exp(-\frac{3}{2})$, then

$$\inf_{a \in A} P_a\{R(\underline{\boldsymbol{\theta}}) \leq (\geq)\, R(\boldsymbol{\theta})\} \geq \gamma. \;\square \tag{8.57}$$

Theorem 8.2 Let $\overline{\theta}_k \in \tilde{\Phi}$, $1 \leq k \leq m$, and function $R(\boldsymbol{\theta}) = R(\theta_1, \ldots, \theta_m)$

be strictly increasing (decreasing) monotone in each variable and quasi-convex (quasi-concave). If $\gamma \geq 1 - \exp(-\frac{3}{2})$, then

$$\inf_{a \in A} P_a\{R(\bar{\boldsymbol{\theta}}) \geq (\leq)\, R(\boldsymbol{\theta})\} \geq \gamma. \ \square$$

Proof Consider at the beginning the proof of Theorem 8.1 if $R(\boldsymbol{\theta})$ is increasing in each variable and quasi-concave function. Let $a \in A$ and $\boldsymbol{\theta}$ and $= (a)$. Since set $H = \{\underline{\boldsymbol{\theta}}: R(\underline{\boldsymbol{\theta}}) \geq R(\boldsymbol{\theta})\}$ is convex, then there exists a plane in m-dimensional space of the form

$$L = \left\{\underline{\boldsymbol{\theta}}: \sum_{1 \leq k \leq m} c_k(\underline{\boldsymbol{\theta}}_k - \theta_k) = 0\right\},$$

which comes via a border point $\underline{\boldsymbol{\theta}} = \boldsymbol{\theta} \in H$ and the area H locates on one side of this plane, that is,

$$H \subset \left\{\underline{\boldsymbol{\theta}}: \sum_{1 \leq k \leq m} c_k \underline{\theta}_k \geq \sum_{1 \leq k \leq m} c_k \theta_k\right\}.$$

From here it follows that

$$\left\{\underline{\boldsymbol{\theta}}: \sum_{1 \leq k \leq m} C_k \underline{\theta}_k < \sum_{1 \leq k \leq m} C_k \theta_k\right\} \subset \{\underline{\boldsymbol{\theta}}: R(\underline{\boldsymbol{\theta}}) < R(\underline{\boldsymbol{\theta}})\} \subset \{\underline{\boldsymbol{\theta}}: R(\underline{\boldsymbol{\theta}}) \leq R(\underline{\boldsymbol{\theta}})\}$$

due to the monotonicity of $R(\cdot)$ coefficients $c_k \geq 0$, $1 \leq k \leq m$. If function $R(\cdot)$ has continuous partial derivatives, then coefficients c_k are defined in a unique way:

$$c_k = \alpha \frac{\partial R(\boldsymbol{\theta})}{\partial \theta_k}, \qquad 1 \leq k \leq m,$$

with the accuracy of a positive multiplier α. Using Lemma 8.1, we obtain

$$P_a\{R(\underline{\boldsymbol{\theta}}) \leq R(\boldsymbol{\theta})\} \geq P_a\left\{\sum_{1 \leq k \leq m} C_k \underline{\theta}_k < \sum_{1 \leq k \leq m} C_k \theta_k\right\}$$

$$= P_a\left\{\sum_{1 \leq k \leq m} C_k \underline{\theta}_k \leq \sum_{1 \leq k \leq m} C_k \theta_k\right\} \geq \gamma,$$

which proves (8.57) for the case where $R(\boldsymbol{\theta})$ is increasing in each variable and quasi-concave function. The proof of the statement in braces follows from transition from function $R(\boldsymbol{\theta})$ to $R(-\boldsymbol{\theta})$ in Theorem 8.1. Notice also that function $R(-\boldsymbol{\theta})$ is monotone decreasing (increasing) in each variable and quasi-convex (quasi-concave) if function $R(\boldsymbol{\theta})$ is monotone increasing (decreasing) in each variable and quasi-convex (quasi-concave). Since, additionally, r.v.'s $-\xi_k \in \Phi$ and $\xi_k \in \tilde{\Phi}$, the proof of Theorem 8.2 follows from transition to variables $-\theta_1, -\theta_2, \ldots, -\theta_m$ in Theorem 8.1. □

The generalization of previous statements for the case where function $R(\boldsymbol{\theta})$ is monotone increasing in some of the variables and decreasing in others is given by the following theorem.

Theorem 8.3 Let function $R(\boldsymbol{\theta})$ be strictly monotone increasing in $\theta_1, \ldots, \theta_n$ and decreasing in $\theta_{n+1}, \ldots, \theta_m$ and let $\gamma \geq 1 - e^{-3/2}$. Then:

1. If $(\underline{\theta}_1, \ldots, \underline{\theta}_n) \in \Phi$ and $(\overline{\theta}_{n+1}, \ldots, \overline{\theta}_m) \in \tilde{\Phi}$ and function $R(\boldsymbol{\theta})$ is quasi-concave, then

$$\inf_{a \in A} P_a\{R(\underline{\theta}_1, \ldots, \underline{\theta}_n, \underline{\theta}_{n+1}, \ldots, \overline{\theta}_m) \leq R(\boldsymbol{\theta})\} \geq \gamma.$$

2. If $(\overline{\theta}_1, \ldots, \overline{\theta}_n) \in \tilde{\Phi}$ and $(\underline{\theta}_{n+1}, \ldots, \underline{Q}_m) \in \Phi$ and function $R(\boldsymbol{\theta})$ is quasi-convex, then

$$\inf_{a \in A} P_a\{R(\overline{\theta}_1, \ldots, \overline{\theta}_n, \underline{\theta}_{n+1}, \ldots, \underline{\theta}_m) \geq R(\boldsymbol{\theta})\} \geq \gamma.$$

The proof is analogous to the one given above.

Consider now an application of the results above for the exponential case. Let λ_k be a parameter (failure rate) for a unit of type k and

$$\overline{\lambda}_k = \frac{\chi^2_\gamma(2r_k)}{2S_k}$$

be the standard γ-UCL for λ_k. (See Section 8.2.1.) Denote also by $\boldsymbol{\lambda} = (\lambda_1, \ldots, \lambda_m)$ a vector of unit parameters and by $\overline{\boldsymbol{\lambda}} = (\overline{\lambda}_1, \ldots, \overline{\lambda}_m)$ a vector of corresponding UCLs. In some cases, it is convenient to use parameters $\mathbf{T} = (T_i, \ldots, T_m)$, where $T_k = 1/\lambda_k$ is MTBF of a unit of type k, $1 \leq k \leq m$. Vectors of corresponding confidence limits for individual parameters are denoted by $\underline{\mathbf{T}} = (\underline{T}_1, \ldots, \underline{T}_m)$ and $\overline{\mathbf{T}} = (\overline{T}_1, \ldots, \overline{T}_m)$, where

$$\underline{T}_k = \frac{2S_k}{\chi^2_\gamma(2r_k)}, \qquad \overline{T}_k = \frac{2S_k}{\chi^2_{1-\gamma}(2r_k)}$$

are the standard lower and upper γ-confidence limits for T_k.

The r.v. S_k (the total test time of units of type k) has gamma d.f. with density

$$f_k(t) = \lambda_k \frac{(\lambda_k t)^{r_k - 1}}{(r_k - 1)!} e^{-\lambda_k t}. \square$$

On the basis of this it is easy to show that $\underline{S}_k \in \Phi$ and $\overline{S}_k \in \tilde{\Phi}$. From here, it follows that confidence limits $\underline{T}_k \in \Phi$ and $\underline{T}_k \in \tilde{\Phi}$ and $\overline{T}_k \in \Phi$ and $\overline{T}_k \in$, $\tilde{\Phi}$, $1 \le k \le m$. From previous results the following theorem follows.

Theorem 8.4 If function $R(\mathbf{T})$ is monotone increasing (decreasing) in each variable and quasi-concave (quasi-convex), then, for $\gamma \ge 1 - e^{-3/2}$,

$$\inf_{\mathbf{T}} P_T\{R(\underline{\mathbf{T}}) \le (\ge) R(\mathbf{T})\} \ge \gamma.$$

In addition assume one of the following conditions holds:

(a) For some C, et $\varphi_c = \{\mathbf{T}: R(\mathbf{T}) \le (\ge) C\}$ is restricted.
(b) For some C set

$$\Psi_C = \left\{\boldsymbol{\lambda}: R\left(\frac{1}{\lambda_1}, \ldots, \frac{1}{\lambda_m}\right) \le (\ge) C\right\} \tag{8.58}$$

is restricted.

Then

$$\inf_{\mathbf{T}} P_T\{R(\underline{\mathbf{T}} \le (\ge) R(\underline{\mathbf{T}})\} = \gamma. \tag{8.59}$$

Proof Since $\underline{T}_k \in \Phi$, $1 \le k \le m$, and function $R(\mathbf{T})$ is monotone increasing (decreasing) in each variable and quasi-concave (quasi-convex), the proof inequality (8.58) follows from Theorem 8.1. Let us show (8.59). For instance, let condition (a) hold. Then there exists

$$0 < \sup_{\mathbf{T} \in \varphi_C} T_k = \eta_k < \infty, \qquad 1 \le k \le m.$$

From here the relation

$$\{\underline{\mathbf{T}}: R(\underline{\mathbf{T}}) \le (\ge) C\} \subset \{\underline{\mathbf{T}}: \underline{T}_k \le \eta_k\}$$

follows, due to which, for any $T \in \varphi_C$

$$P_T\{R(\underline{\mathbf{T}}) \le (\ge) R(\underline{\mathbf{T}})\} \le P_T\{R(\underline{\mathbf{T}}) \le (\ge) C\} \le P_T\{\overline{T}_k \le \eta_k\}$$
$$= P_{T_k}\{\underline{T}_k \le \eta_k\}.$$

For any $\varepsilon \geq 0$ such $T_\varepsilon = (T_{\varepsilon_1}, \ldots, T_{\varepsilon_m}) \in \varphi_c$ it exists that $\eta_k - \varepsilon < T_{k_e} < \eta_k$. Since $P_{T_k}\{\underline{T}_k \leq \eta_k\}$ is continuous in T_k, then

$$P_{T_k}\{R(\underline{\mathbf{T}}) \leq (\geq)\, R(T_\varepsilon)\} \leq P_{T_k}\{\underline{T}_k \leq \eta_k\} + \delta_\varepsilon = \gamma + \delta_\varepsilon,$$

where $\delta_\eta \geq 0$ and $\lim_{\varepsilon \to 0} \delta_\varepsilon = 0$. From here, (8.59) follows. The proof for condition (b) is similar. □

Theorem 8.5 If function $R(\mathbf{T})$ is monotone increasing (decreasing) in each variable and quasi-convex (quasi-concave), then, for $\gamma \geq 1 - e^{-3/2}$,

$$\inf_{\mathbf{T}} P_T\{R(\overline{\mathbf{T}}) \geq (\leq)\, R(\mathbf{T})\} = \gamma.$$

Proof Since $\overline{T}_k \in \tilde{\Phi}$, $1 \leq k \leq m$, in accordance with Theorem 8.2,

$$\inf_{\mathbf{T}} P_T\{R(\overline{\mathbf{T}}) \geq (\leq)\, R(\mathbf{T})\} \geq \gamma.$$

Let further $\mathbf{u} = (u_1, \ldots, u_m)$ be some fixed point $u_k > 0$, $1 \leq k \leq m$, and $c = R(\mathbf{u})$. Due to the quasi-convexity of $R(\mathbf{T})$ there exists a plane in m-dimensonal space of type

$$\sum_{1 \leq k \leq m} b_k(T_k - u_k) = 0, \qquad b_k > 0, \qquad T_k > 0, \qquad 1 \leq k \leq m,$$

which passes via point $\mathbf{T} = \mathbf{u}$ and such that

$$\varphi_c = \{\mathbf{T}\colon R(\mathbf{T}) \leq (\geq)\, c\} \subset \left\{\mathbf{T}\colon \sum_{1 \leq k \leq m} b_k(T_k - u_k) \leq 0,\ T_k > 0\right\} = G.$$

Since area G is restricted, area φ_c is also restricted. After this the proof is analogous to that for Theorem 8.4. □

Let $\overline{\boldsymbol{\lambda}} = (\overline{\lambda}_1, \ldots, \overline{\lambda}_m)$ be a vector of γ-UCLs. The following theorem can be formulated.

Theorem 8.6 If function $f(\boldsymbol{\lambda})$ is monotone increasing (decreasing) in each variable and quasi-concave (quasi-convex), then, for $\gamma \geq 1 - r^{-3/2}$,

$$\inf_{\boldsymbol{\lambda}} P_\lambda\{f(\overline{\boldsymbol{\lambda}}) \leq (\geq)\, f(\boldsymbol{\lambda})\} = \gamma.$$

Proof It is enough to consider the case where $f(\boldsymbol{\lambda})$ is quasi-concave and monotone decreasing in each variable function. Introduce function $g_f(\mathbf{T}) = f(1/T_1, \ldots, 1/T_m)$. For any $\mathbf{T}$, $\mathbf{u}$ and $\alpha > 0$, $\beta > 0$, $\alpha + \beta = 1$, we have

$$g_f(\alpha\mathbf{T} + \beta\mathbf{u}) = f\left(\frac{1}{\alpha T_1 + \beta u_1}, \ldots, \frac{1}{\alpha T_m + \beta u_m}\right)$$

$$\geq f\left(\frac{\alpha}{T_1} + \frac{\beta}{u_1}, \ldots, \frac{\alpha}{T_m} + \frac{\beta}{u_m}\right) \geq \min[g_f(\mathbf{T}), g_f(\mathbf{u})].$$

From here it follows that $g_f(\mathbf{T})$ is quasi-concave and monotone increasing in each variable function. In addition,

$$\left\{\boldsymbol{\lambda}: g_f\left(\frac{1}{\lambda_1}, \ldots, \frac{1}{\lambda_m}\right) \geq g_f\left(\frac{1}{u_1}, \ldots, \frac{1}{u_m}\right)\right\} = \{\boldsymbol{\lambda}: f(\boldsymbol{\lambda}) \geq f(\mathbf{u})\}$$

is restricted for any $\mathbf{u}$ and it follows that the function satisfies condition (b) of Theorem 8.4, and the proof follows from this theorem. □

8.7.2 Construction of Confident Limits by Substitution Method for Function of Type (8.17)

We need additional knowledge about some distribution families.

The continuous distribution density $f(t)$ or r.v. ξ is called a Polya second-order density (abbreviated P2-density) if one of the following equivalent conditions holds:

1. Function $f((t - \Delta)/f(t)$ is monotone increasing in t for $\Delta > 0$ and t such that $f(t) > 0$.
2. Function $\varphi(t) = -\ln f(t)$ is convex in t such that $f(t) > 0$.

It is easy to see that d.f.'s such as normal, exponential, Gnedenko–Weibull (with shape parameter $a \geq 1$), and some other frequently used distributions have P2-density (for more details, see Barlow and Proschan, 1965). It is easy to show that if r.v. ξ has P2-density, then $\xi \in \Phi$ and $\xi \in \tilde{\Phi}$, where Φ and $\tilde{\Phi}$ are classes of logarithmically convex distributions. Indeed, let ξ have P2-density and $F(t) = \int_{-\infty}^{t} f(u)\, du$ be the corresponding d.f. Then the function

$$-\frac{d}{dt}\ln[1 - F(t)] = \frac{f(t)}{1 - F(t)} = \frac{f(t)}{\displaystyle\int_t^\infty f(x)\, dx} = \left(\int_t^\infty \frac{f(x)}{f(t)}\, dx\right)^{-1}$$

$$= \left(\int_0^\infty \frac{f(t + u)}{f(t)}\, du\right)^{-1}$$

is monotone increasing in t such that $F(t) < 1$, and it follows that $\xi \in \Phi$. (In particular, from here it follows and is well known that if r.v. ξ is nonnegative and has P2-density, then it has IFR distribution.) Besides, from the definition above it follows that if r.v. ξ has P2-density, then $-\xi$ also has P2-density. Consequently, $-\xi \in \Phi$ and if follows that $\xi = -(-\xi) \in \tilde{\Phi}$. Thus, if r.v. ξ has P2-density, then $\xi \in \Phi$ and $\xi \in \tilde{\Phi}$.

Consider, further, the function

$$f(\boldsymbol{\lambda}) = \sum_{1 \le i \le M} A_i \prod_{j \in G_i} \lambda_j^{n_{ij}}, \tag{8.60}$$

where A_i are positive coefficients and G_i are arbitrary subsets of a set if subscripts, $G_i \subset (1, 2, \ldots, m)$. Show that for an arbitrary function of type (8.60) the γ-UCL can be calculated [for $\gamma \ge 1 - \exp(-\frac{3}{2})$] by the method of substitution, that is, by the formula $\overline{f} = f(\overline{\boldsymbol{\lambda}})$. For the sake of convenience, introduce new parameters $z_i = \ln \lambda_i$, $1 \le i \le m$. For parameter z_i,

$$\overline{z}_i = \ln \overline{\lambda}_i = \ln \left[\frac{\chi_\gamma^2(2r_i)}{2S_i} \right]$$

is γ-UCL. It is not difficult to find that $\overline{z}_i$ has P2-density. For this purpose, we need to show that the r.v. $\eta_i = \ln S_i$ has P2-density. The r.v. S_i gamma d.f. (or Erlang d.f. of order $r_i - 1$) with density

$$f_i(t) = \frac{\lambda_i^{r_i} t^{r_i - 1}}{(r_i - 1)!} e^{-\lambda_i t}.$$

The r.v. η_i has d.f.

$$H_i(t) = P(\eta_i \le t) = P(\ln S_i \le t) = P(S_i \le e^t) = F_i(e^t),$$

where $F(\cdot)$ is the d.f. of S_i. After simple transforms, it gives the following density of r.v. η_i:

$$h_i(t) = H_i'(t) = f_i(e^t)e^t = \frac{\lambda_i^{r_i}}{(r_i - 1)!} e^{-\varphi(t)},$$

where $\varphi(t) = \lambda_i e^t - r_i t$ is a convex function. So, r.v. η_i (and, consequently, confident limit $\overline{z}_i$) has P2-density.

Introduce the vector of parameters $\mathbf{z} = (z_1, \ldots, z_m)$ and the vector of γ-UCLs $\overline{\mathbf{z}} = (\overline{z}_1, \ldots, \overline{z}_m)$. From the previous results the next theorem follows.

Theorem 8.7 Let function $g(\mathbf{z}) = g(z_1, \ldots, z_m)$ be monotone increasing in each z_i and quasi-convex; then, for $\gamma \geq 1 - e^{-3/2}$,

$$P\{g(\bar{\mathbf{z}}) \geq g(\mathbf{z})\} \geq \gamma.$$

Proof As we showed above, the UCL $\bar{z}_i$ has P2-density. Then $\bar{z}_i \in \tilde{\Phi}$, $1 \leq i \leq m$, and the needed proof follows from Theorem 8.2.

Express function (8.60) via parameters $\mathbf{z}$ as

$$g(\mathbf{z}) = f(e^{z_1}, \ldots, e^{z_m}) = \sum_{1 \leq i \leq M} A_i \exp\left(\sum_{j \in G_i} h_{ij} z_j\right). \tag{8.61}$$

This function is convex in $\mathbf{z}$ being the sum of convex functions. Thus, using Theorem 8.7, we show that for a function of type (8.61) for parameters $\mathbf{z}$, or for a function of type (8.60) of parameters of type $\boldsymbol{\lambda}$, γ-UCL can be calculated by the method of substitution. □

8.7.3 Proof of Quasi-Convexity of Reliabilty Index for Renewable Systems

Let $g(\mathbf{T})$ be a function of type (8.27), where $K_i(T_i)$ are given in (8.26). Notice that (8.23) and (8.24) are particular cases of (8.26) and (8.27). For the proof of quasi-concavity of functions (8.27), it is sufficient to show convexity in $\mathbf{z}$ of the function

$$\varphi_{nm}(\mathbf{z}) = -\ln\left(\frac{\sum_{0 \leq k \leq m} A_k / z_k}{\sum_{0 \leq k \leq n} A_k / z_k}\right)$$

where $m < n$, $A_k > 0$, $0 \leq k \leq n$. Denote

$$\varphi_l(\mathbf{z}) = \frac{\sum_{0 \leq k \leq l} k A_k / z_k}{\sum_{0 \leq k \leq l} A_k / z_k}.$$

After differentiation of $\varphi_{nm}(\mathbf{z})$, we obtain

$$\varphi'_{nm}(\mathbf{z}) = \frac{1}{z}[\Psi_m(\mathbf{z}) - \Psi_n(\mathbf{z})] = \frac{1}{z} \sum_{m+1 \leq l \leq n} [\Psi_{l-1}(\mathbf{z}) - \Psi_l(\mathbf{z})]$$

$$= \frac{1}{z} \sum_{m+1 \leq l \leq n} \varphi'_{l,l-1}(\mathbf{z}).$$

It follows that it is sufficient to show the convexity of function $\varphi_{l,l-1}(z)$, $m + 1 \leq l \leq n$. We can write

$$\varphi_{l,l-1}(z) = -\ln[1 - h(z)], \qquad \varphi'_{l,l-1}(z) = \frac{h'(z)}{1 - h(z)},$$

where

$$h(z) = \frac{A_l}{A_l + A_{l-1}z + \cdots + A_0 z^l}.$$

Because $A_k > 0$ for all $0 \le k \le m$, function $f(z)$ is convex for $z \ge 0$, which gives the convexity of functions $\varphi_{l,l-1}(z)$ and $\varphi_{nm}(z)$. Consequently, function (8.27) can be represented as a monotone decreasing function of the sum of monotone decreasing convex functions and is quasi-concave.

Let $g(\mathbf{T})$ be a function of the form

$$g(\mathbf{T}) = \kappa \left\{ \sum_{1 \le i \le m} \left(\frac{1}{\sum_{1 \le k \le n_i} A_{ki} T_i^k} \right) \right\}, \tag{8.62}$$

where $\kappa(u)$ is monotone decreasing in u and $A_{ki} > 0$. [Functions of type (8.62) are $L(T)$ and $P_\tau(T)$.] The function $(\Sigma_{1 \le k \le n_i} A_{ki} T_i^k)^{-1}$ is monotone decreasing and convex in T_i, $1 \le i \le m$, from where it follows that (8.62) is a quasi-concave function. Now let

$$g(\mathbf{T}) = g_1(T) \exp \left\{ - \sum_{1 \le i \le m} \frac{1}{\sum_{1 \le k \le n_i} A_{ki} T_i^k} \right\} \tag{8.63}$$

where $g_1(T)$ is a function of type (8.27). Function $K_\tau(T)$ belongs to type (8.63).

Formula (8.63) can also be represented in the form of a monotone decreasing function of the sum of monotone decreasing convex functions and, consequently, is quasi-concave.

8.7.4 Conditions of Correctness of Method of Fiducial Probabilities for Calculation of Confidence Limits for System PFFO

Let us investigate the following question: For what functions $f(\boldsymbol{\lambda}) = f(\lambda_1, \ldots, \lambda_m)$ does the fiducial limit with confidence coefficient γ coincide with the γ-confidence limit? It is not difficult to show that such functions are, for instance,

$$\varphi_b(\boldsymbol{\lambda}) = b_0 \prod_{1 \le i \le m} \lambda_i^{b_i}, \qquad \mathbf{b} \in B \tag{8.64}$$

where $\mathbf{b} = (b_0, \ldots, b_m)$ is the vector of constant coefficients and $B = \{\mathbf{b}: b_0 > 0, -\infty < b_i < \infty, 1 \le i \le m\}$. For an arbitrary function of type (8.64), the fiducial approach is correct for the construction of the UCL and LCL of $\varphi_b(\lambda)$. Indeed, for each fixed λ_i the r.v. $\xi_i = \lambda_i S_i$ has a distribution independent of λ_i. For given fixed test results S_i^*, the fiducial distribution of parameter $\underline{\lambda}_i$ coincides with the distribution of r.v. ξ_i / S_i^*. Thus, the upper fiducial limit φ_b with confidence coefficient γ for $\varphi_b(\boldsymbol{\lambda})$ is equal to quantile of level γ for the r.v.

$$\eta_b = b_0 \prod_{1 \le i \le m} \left(\frac{\xi_i}{S_i^*} \right)^{b_i},$$

and from here it follows that

$$\bar{\varphi}_b = b_0 \prod_{1 \le i \le m} \left(\frac{1}{S_i^*} \right)^{b_i} t_b(\gamma)$$

where $t_b(\gamma)$ is the quantile of level γ for the r.v.

$$\tau_b = \prod_{1 \le i \le m} \xi_i^{b_i}.$$

It follows that for arbitrary fixed $\boldsymbol{\lambda} = (\lambda_1, \ldots, \lambda_m)$, the confidence probability corresponding to upper limit $\bar{\varphi}_b$ is equal to

$$\begin{aligned} P_\lambda\{\bar{\varphi}_b \ge \varphi_b(\boldsymbol{\lambda})\} &= P_\lambda \left\{ b_0 \prod_{1 \le i \le m} \left(\frac{1}{S_i^*} \right)^{b_i} t_b(\gamma) \ge b_0 \prod_{1 \le i \le m} \lambda_i^{b_i} \right\} \\ &= P_\lambda\{\tau_b \le t_b(\gamma)\} = \gamma; \end{aligned}$$

that is, the upper fiducial limit with confidence coefficient γ for $\bar{\varphi}_b(\lambda)$ coincides with the γ-ICL. Analogously, we can prove a similar relation for the lower fiducial limit.

A set of all function (8.64) for different $\mathbf{b} \in B$ is called a basis set of functions. For any function from this set, as was shown above, the fiducial method is correct for constructing the UCL and LCL. Obviously, the basis set is somewhat narrow. For instance, it does not contain simplest linear functions of the type

$$f(\boldsymbol{\lambda}) = \sum_{1 \le i \le m} C_i \lambda_i,$$

which are related, for instance, to the problem of reliability estimation of series systems. Nevertheless, the basis set of functions (8.64) can be used to expand the class of functions for which the fiducial method can be applied for constructing upper confidence limits.

For the sake of convenience, introduce new parameters $\mathbf{z} = (z_1, \ldots, z_m)$, where $z_i = \ln \lambda_i$, $1 \leq i \leq m$. This vector of parameters, $\mathbf{z}$, takes its values from $Z = \{\mathbf{z}\text{:}\ -\infty < z_i < \infty,\ 1 \leq i \leq m\}$. In these new variables, a set of functions (8.64) transforms into a set of linear functions

$$h_b(\mathbf{z}) = b_0 + \sum_{1 \leq i \leq m} b_i z_i, \qquad \mathbf{b} \in B, \tag{8.65}$$

where $B = \{\mathbf{b}\text{:}\ -\infty < b_i < \infty,\ 1 \leq i \leq m\}$.

The upper fiducial limit for $h_b(\mathbf{z})$ is denoted by $\overline{h}_b = \overline{h}_b(\mathbf{S})$. Function $f(\boldsymbol{\lambda})$ can be written via parameters $\mathbf{z}$ as

$$\tilde{f}(\mathbf{z}) = f(e^{z_1}, \ldots, e^{z_m}). \tag{8.66}$$

Now let $\overline{f} = \overline{f}(\mathbf{S})$ be the upper fiducial limit with confidence coefficient γ for $f = f(\boldsymbol{\lambda})$. The next theorem shows that the upper fiducial limit coincides with the γ-UCL for f if function (8.66) is represented in the form of a maximum over a set of functions of type (8.65). Let

$$\tilde{f}(\mathbf{z}) = \sup_{\mathbf{b} \in B'} h_b(\mathbf{z}) \tag{8.67}$$

where $B' \subset B$. Then we have the following theorem.

Theorem 8.8 Let the supremum in (8.67) be attained for any $\mathbf{z} \in Z$ at some point $\mathbf{b}(\mathbf{z}) \in B'$, that is,

$$\sup_{\mathbf{b} \in B'} h_b(\mathbf{z}) = h_{\mathbf{b}(\mathbf{z})}(\mathbf{z}). \tag{8.68}$$

Then, for any $\mathbf{z} \in Z$, inequality

$$P\{\overline{f}(\mathbf{x}) \geq \tilde{f}(\mathbf{z})\} \geq \gamma. \ \square$$

Proof Due to (8.67), the inequality

$$\tilde{f}(\mathbf{z} \geq h_b(\mathbf{z}), \ \mathbf{b} \in B',$$

holds for all $\mathbf{z} \in Z$. From the construction of fiducial limits $\overline{f}$ and $\overline{h}_b$, the following analogous inequality for these limits follows:

$$\bar{f}(\mathbf{S}) \geq \bar{h}_b(S), \qquad \mathbf{b} \in B'$$

for all S. It follows that

$$\bar{f}(S) \leq \sup_{\mathbf{b} \in B'} \bar{h}_b(S) = H(S) \tag{8.69}$$

holds for all S. Thus, it is sufficient to show that

$$P\{H(S) \leq \tilde{f}(\mathbf{z})\} \geq \gamma, \qquad \mathbf{z} \in Z.$$

Further, for arbitrary fixed $\mathbf{z} \in Z$, we have

$$P\{H(S) \geq \tilde{f}(\mathbf{z})\} = P\left\{\sup_{\mathbf{b} \in B'} \bar{h}_b(S) \geq h_{\mathbf{b}(\mathbf{z})}(\mathbf{z})\right\} \geq P\{\bar{h}_b(S) \geq h_{\mathbf{b}(\mathbf{z})}(\mathbf{z})\} = \gamma.$$

This completes the proof. □

Thus, the fiducial method is correct for confidence estimation of f from above if function (8.66) can be represented in the form

$$\tilde{f}(\mathbf{z}) = \max_{\mathbf{b} \in B'} h_b(\mathbf{z}). \tag{8.70}$$

This condition holds if function (8.66) is convex in $\mathbf{z}$. Functions $f_b(\mathbf{z})$ in (8.70) represent tangent "planes." Thus, Theorem 8.8 establishes that the fiducial method is correct for confidence estimation of function $f(\lambda)$ from above if function (8.66) is convex in $\mathbf{z}$. This result can be, in a natural way, expanded to the case where functions are concave. For this purpose, one must consider function $-f$. From the proof of theorem 8.8 it follows that the fiducial method is not correct for strictly convex (concave) functions $\tilde{f}(\boldsymbol{\lambda})$ for confidence estimation from below (above). In this case we obtain confidence limits with confidence coefficient smaller than the γ used for this method. Moreover, this difference between a real confidence coefficient and a given γ cannot be evaluated.

As one can see from the proof of Theorem 8.8,

$$H(S) = \sup_{\mathbf{b} \in B'} \bar{h}_b(S) \tag{8.71}$$

is also γ-UCL for $\underline{f}$, and, moreover, better (lower) in comparison with the upper fiducial limit $\bar{f}(\mathbf{S})$ with the same γ for all test results S, as follows from (8.69). So, confidence limit (8.71) improves the fiducial method although needs more complicated calculations (see also Section 8.6 above).

CHAPTER 9

SEQUENTIAL CRITERIA OF HYPOTHESIS TESTING AND CONFIDENCE LIMITS FOR RELIABILITY INDICES

9.1 INTRODUCTION

Different objects (like units, subsystems or systems) are tested sequentially in many practical cases. In other words, statistical information about objects of interest is collected in sequential bits, not at once. The moment of test termination is not determined in advance. It is determined during the test, depending on the data obtained thus far.

The first result in this direction was obtained for two simple hypotheses by Wald (1947). The sequential criterion he formulated is known as the Wald criterion. This method gives an opportunity to decrease the necessary average test volume in comparison with the case where the test volume is fixed in advance. However, the classical Wald analysis does not "work" for some problems arising in reliability (hypothesis testing and construction of confidence limits for reliability indices of complex systems). It leads to the necessity of modifications and development of new methods.

9.1.1 Sequential Confidence Limits and Hypothesis Testing Criteria

The problems of sequential test of hypotheses for reliability indices are close to the problems of construction of sequential confidence limits. Assume that at each test step n, $n = 1, 2, \ldots$, we can construct the confidence limit (say, the lower)

$$\underline{R}_n = \underline{R}_n(x^{(n)}) = \underline{R}_n(x_1, \ldots, x_n) \tag{9.1}$$

for the reliability index of interest R by some known method. Here x_n is the test result at the nth step and $x^{(n)}$ is a set of all test results, $x^{(n)} = (x_1, \ldots, x_n)$, obtained at the first n steps, $n = 1, 2, \ldots$. Thus, for each fixed test step n, the inequality

$$P(\underline{R}_n \leq R) \geq \gamma \tag{9.2}$$

holds. Here γ is the confidence coefficient of the lower limit $\underline{R}_n$, $n = 1, 2, \ldots$.

Assume that the test is continued until some (generally speaking, random) step ν, and we observe the set of test results $x^{(\nu)} = (x_1, \ldots, x_\nu)$, and the test is stopped. The value ν is called the moment of test stop. Consider a situation where the stopping moment ν is not fixed in advance but determined during the test in accordance with some stopping rule. At first glance, it seems natural to use the already known confidence limit (9.1) for the estimation of the reliability index R in this case. In other words, if the test is terminated at $\nu = n$, then we take the value of $\underline{R}_n$ as the γ-LCL for R. Nevertheless, that simple solution is not correct in general. Indeed, the confidence probability corresponding to this procedure is equal to

$$P(\underline{R}_\nu \leq R) = \sum_{1 \leq n < \infty} P(\nu = n) P(\underline{R}_n \leq R | \nu = n)$$

where $\underline{R}_\nu = \underline{R}_\nu(x_1, \ldots, x_\nu)$ is the confidence limit (9.1) calculated at the random moment ν and $P(\underline{R}_n \leq R | \nu = n)$ is the conditional probability of the event $\{\underline{R}_n \leq R\}$ under the condition that $\nu = n$. Conditional and unconditional probabilities $P(\underline{R}_n \leq R | \nu = n)$ and $P(\underline{R}_n \leq R)$ do not coincide (except for the case where the random stopping moment ν does not depend on the process of the test). Therefore, analogous inequalities

$$P(\underline{R}_\nu \leq R) \geq \gamma \tag{9.3}$$

for $\underline{R}_\nu$ do not follow from the inequalities (9.2) in general. It means that the confidence coefficient of the confidence limit $\underline{R}_\nu$ can be smaller than γ on an unknown value. Therefore, the above procedure can lead to errors (except for the situation where the stopping moment is determined in advance or ν is an r.v. that does not depend on the text results $x_1, \ldots, x_n, \ldots$). Consider the following simple example for illustrative purposes. Assume that we need to confirm that the reliability index R exceeds some specified level b from the test results $x_1, \ldots, x_n, \ldots$. The standard method of solving such a problem consists in the proof that the γ-confidence limit $\underline{R}$ constructed from the test satisfies the inequality $\underline{R} \geq b$. This can be proved, for instance, by the following way. At the current kth step we construct by some known method the γ-LCL $\underline{R}_k = \underline{R}_k(x_k)$. Here x_n is the test result at the nth step. We continue the test until at some nth step the inequality

$$\underline{R}_n \geq b \tag{9.4}$$

holds. The moment of test stop is determined as

$$\nu = \min\{n: \underline{R}_n \geq b\}.$$

After this step the test is terminated, and due to inequality (9.4), we declare that the reliability requirement $R \geq b$ is confirmed with confidence level γ. However, it is clear that if, for instance, the test results $x_1, \ldots, x_n, \ldots$ at different steps are independent, then using such a way may lead to confusion: We can always confirm any reliability level although the real level of reliability is arbitrarily low! This follows from the fact that inequality (9.4) will be satisfied sooner or later for some random step ν due to randomness and independence of values $\underline{R}_n$, $n = 1, 2, \ldots$. Thus, a direct transition from usual confidence limits (9.4) found for the fixed test stop to the confidence limit $\underline{R}_\nu$ in the sequential test can lead to erroneous results. Therefore the confidence limits for sequential tests must be constructed by special procedure such that confidence coefficient γ will be preserved for any stopping rule.

Remark 9.1 The question of whether the confidence limit (9.1) found for random moment ν satisfies inequality (9.3), which is qualified as a γ-confidence limit, is a particular case of a more general problem. Assume we have a sequence of functions depending on test results

$$g_n = g_n(x_1, \ldots, x_n), \qquad n = 1, 2, \ldots \tag{9.5}$$

such that at each step n the inequality

$$E\{g_n\} \geq C, \qquad n = 1, 2, \ldots, \tag{9.6}$$

holds. Here C is a constant that does not depend on n. Let ν be some test stopping moment defined as the moment of the first entry into the "stopping area" D, that is,

$$\nu = \min|n: (n, x_n) \in D\}.$$

The question is: When can we say that the inequality

$$E\{g_\nu\} \geq C \tag{9.7}$$

follows from inequalities (9.6). Here $g_\nu = g_\nu(x_1, \ldots, x_\nu)$ is the function (9.5) depending on test results and is computed at random moment ν. This problem is discussed in detail, in particular, by Pavlov (1995). □

9.1.2 Sequential Independent and Identical Tests

Assume that the test result for some object is a sequence of independent and identically distributed random values (vectors)

$$x_1, x_2, \ldots, x_n, \ldots, \tag{9.8}$$

where $\mathbf{x}_n = (x_{n_1}, \ldots, x_{n_l})$ is a vector r.v. observed at the nth step of the test. The distribution of x_n is defined by the density function $f(x, \boldsymbol{\theta})$ depending on some unknown vector parameter $\boldsymbol{\theta} = (\theta_1, \ldots, \theta_m)$. We need to construct the confidence limits for some function of parameter $\boldsymbol{\theta}$ on the basis of data (9.8):

$$R = R(\boldsymbol{\theta}) = R(\theta_1, \ldots, \theta_m) \tag{9.9}$$

Function R is a reliability index as above. Sometimes we need to test hypotheses about $R(\boldsymbol{\theta})$ on the basis of (9.8):

$$H_0\colon R(\boldsymbol{\theta}) \le R_0, \qquad H_1\colon R(\boldsymbol{\theta}) \ge R_1, \tag{9.10}$$

where $R_0 < R_1$ are some specified critical levels of index R.

Example 9.1 Let a unit have TTF ζ with d.f. $F(x, \boldsymbol{\theta})$ and density $f(x, \boldsymbol{\theta})$ that depend on parameter $\boldsymbol{\theta}$ (in the general case a vector). During the test the unit is immediately replaced by a new identical one after a failure. As a result we have the following sequence of i.i.d. r.v.'s:

$$x_1, x_2, \ldots, x_n, \ldots, \tag{9.11}$$

where x_n is the TTF (random value of ζ) at the nth test step, $n = 1, 2, \ldots$. We need to construct the confidence limits on the basis of (9.11) and check hypotheses (9.10) for unit reliability index $R = R(\boldsymbol{\theta})$. The reliability index can be one of the following:

1. the PFFO for time interval t_0, that is,

$$R(\boldsymbol{\theta}) = 1 - F(t_0\boldsymbol{\theta}); \tag{9.12}$$

2. the unit lifetime guarantee $t_q(\boldsymbol{\theta})$ with the level q, which is defined by the equation

$$1 - F(t, \boldsymbol{\theta}) = q;$$

and

3. the unit MTTF

$$R(\boldsymbol{\theta}) = \int_0^\infty x f(x, \boldsymbol{\theta})\, dx. \square$$

Example 9.2 (Exponential Tests) In the condition of the previous example, consider a particular case where a unit has exponential d.f. of TTF with density $f(x, \lambda) = \lambda \exp(-\lambda x)$ with unknown parameter $\theta = \lambda$. In this case any reliability index can be expressed via parameter λ. It means that the problem is reduced to confidence estimation and hypotheses testing for parameter λ on the basis of (9.11) where i.i.d. r.v.'s have exponential d.f.

Notice that this case is the most investigated in the literature. □

Example 9.3 (Binomial Test) In the condition of Example 9.1 consider a particular case where the index of interest is (9.12). Denote the unit PFFO by q, $q = 1 - R = F(t_0, \boldsymbol{\theta})$. Assume that at each test step a unit is tested for time t_0.

Introduce an indicator function, δ_n, that equals 1 if a failure has been observed at the nth step and 0 otherwise: $\delta_n = I(x_n \leq t_0)$. The moment of failure occurrence x_n is not registered.

In this case we are observing a sequence of i.i.d. r.v.'s

$$\delta_1, \delta_2, \ldots, \delta_n, \ldots \tag{9.13}$$

where δ_ν takes value 0 or 1 with probabilities $P(\delta_n = 0) = 1 - q$ and $P(\delta_n = 1) = q$, $n = 1, 2, \ldots$. Thus, in this case the problem is reduced to confidence estimation and hypothesis testing for parameter q on the basis of (9.13). This "binomial" scheme represents the well-known classical scheme of sequential independent trials (Bernoulli trials). This is also one of the best investigated cases in probability theory. □

Example 9.4 (Continuous Time) Again refer to Example 9.1. Let

$$t_n = x_1 + \cdots + x_n \tag{9.14}$$

denote the moment of the nth failure and $d(u)$ the number of failures up to moment u. (In other words, n satisfies the condition $t_n \leq u$.) We consider a situation where the failure moments (9.14) are continuous in time. Actually, we observe a standard renewal process $d(u)$, $u \geq 0$. At each moment u we have the complete information about previous failure moments

$$t_1 < t_2 < \cdots < t_{d(u)} < u$$

(see Figure 9.1). Notice that if the d.f. of TTF is exponential, the renewal process represents a Poisson process. In that case, due to the memoryless

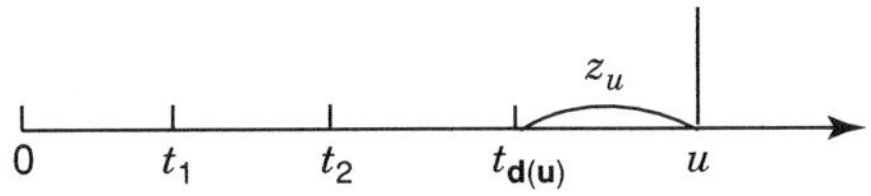

Figure 9.1 Sample of a time diagram for renewal process.

property, this model can be reduced to the Bernoulli scheme considered above by dividing the time axis into intervals of length h, $h \to 0$.

In the general case the process behavior after moment u depends on the prehistory, namely, on the residual time $z_u = u - t_{d(u)}$. It means that this scheme cannot be reduced to the Bernoulli scheme and must be studied in the frame of more general models. (This case will be discussed in later sections.) □

Example 9.5 (System with Loaded Redundant Units—Exponential Distribution) Consider a system consisting of m different types of units. There are N_i units of type i within the system, $1 \le i \le m$. The distribution function of TTF for each unit is exponential:

$$F_i(x, \lambda_i) = 1 - \exp(-\lambda_i x) \tag{9.15}$$

with unknown parameter λ_i, $1 \le i \le m$. All units are in the operational state, that is, under load. All unit failures are independent. Any failed unit is immediately replaced by a new identical one. In other words, we observe m independent tests of different units, and for each type i unit there are N_i independent renewal processes, that is, units of type i are tested by plan $[N_i\ R]$ (see Section 2.1). Due to the exponentiality of d.f. (9.15), each unit is characterized by the standard Poisson process of failures with failure rate λ_i, $1 \le i \le m$.

We need to construct confidence limits or/and test some hypotheses for the system reliability index $R = R(\boldsymbol{\lambda}) = R(\lambda_1, \ldots, \lambda_m)$ on the basis of this test. For instance, for a series–parallel system consisting of m parallel subsystems in series, the PFFO for time t_0 has the form

$$R = \prod_{1 \le i \le m} [1 - (1 - e^{-\lambda_i t_0})^{N_i}],$$

where N_i is the number of parallel units in subsystem i, $1 \le i \le m$.

Although we have the model with continuous time, this model can again be reduced to the discrete model (9.8) by dividing the time axis into intervals of length h. Random value x_n in (9.8) will be presented by the set of all test results on intervals $[(n - 1)h, nh]$, or in other words, by the random vector

$$\mathbf{x}_n = (d_{1n}, d_{2n}, \ldots, d_{mn}), \tag{9.16}$$

where d_{in} is the number of failures of units of type i on the interval of length h, $1 \leq i \leq m$, $n = 1, 2, \ldots$. Random value d_{in} has the Poisson d.f. with parameter $N_i\lambda_i h$, $1 \leq i \leq m$. The vector of parameters, $\boldsymbol{\theta}$, is represented by $\boldsymbol{\lambda} = (\lambda_1, \ldots, \lambda_m)$. Due to the memoryless property for Poisson processes in (9.18), random vectors $\mathbf{x}_1, \ldots, \mathbf{x}_n, \ldots$ are independent and identically distributed. Thus, we have reduced the initial continuous model to the discrete version (9.16). Results for the discrete model can be transferred back to the continuous-time case with $h \to 0$. □

Example 9.6 (System with Loaded Redundancy—General Case) Consider a test scheme analogous to the previous one but without the assumption of exponentiality of the d.f. of TTF (9.15). In this case we observe m independent processes of unit tests; each unit is tested by plan $[N_i, R]$. In other words, we have $N_1 + N_2 + \cdots + N_m$ independent renewal processes. Each group of N_i processes is characterized by the same d.f. $F_i(x, \boldsymbol{\theta})$.

We need to construct confidence limits or/and to test some hypotheses for system reliability index $R = R(\boldsymbol{\theta}) = R(\theta_1, \ldots, \theta_m)$. It is obvious that this model is a multidimensional analogue of Example 9.5. This model cannot be reduced to (9.8) and must be studied with the help of more general methods. □

Example 9.7 (Binomial Test) Consider the PFFO in conditions of the previous example. Let the test be presented by cycles of length t_0; in other words, units of type i are tested by plan $[N_i\ U\ t_0]$, $1 \leq i \leq m$. The result at the nth test step is presented by the random vector

$$x_n = (d_{1n}, d_{2n}, \ldots, d_{mn}) \tag{9.17}$$

where d_{in} is the number of failed units of type i within the nth cycle, $1 \leq i \leq m$, $n = 1, 2, \ldots$. The random value d_{in} has the binomial distribution with parameters (N_i, q_i), where $q_i = F_i(t_0, \theta_i)$ is the unit PFFO during time t_0. All failed units are replaced at the end of each cycle t_0. Thus we have a sequence of i.i.d. random vectors (9.17) that is used for constructing confidence limits or/and testing some hypotheses for reliability index $R = R(\mathbf{q}) = R(q_1, \ldots, q_m)$, where $\mathbf{q} = (q_1, \ldots, q_m)$ is vector of binomial parameters of units and $R = R(\mathbf{q})$ is the system PFFO for time t_0. For instance, for a system consisting of m parallel subsystem in series, the reliability index has the form

$$R = \prod_{1 \leq i \leq m} (1 - q_i^{N_i}),$$

where N_i, $1 \leq i \leq m$, is the number of parallel units in subsystem i.

This model is a multidimensional analogue of the model in Example 9.3. □

Example 9.8 (System with Spare Units) Consider a system of m subsystems each of which comprises r_i identical units, one main unit and $r_i - 1$ spares (unloaded redundancy). The distribution function of TTF of the ith unit is denoted by $F_i(x, \theta_i)$, where θ_i is an unknown parameter (in the general case a vector). All failures are independent.

Let d_i denote the number of units of type i having failed up to the moment $t = 0$. The process $d_i(t)$ is, obviously, the standard renewal process with d.f. $F_i(x, \theta_i)$. Thus we observe the m-dimensional stochastic process

$$d(t) = \{d_1(t), \ldots, d_m(t)\}. \tag{9.18}$$

Each cycle of testing is terminated at the moment of system failure, that is, at the moment τ:

$$\tau = \min\{t: d_i(t) \geq r_i \text{ at least for one } i, 1 \leq i \leq m\}.$$

The moment τ is the moment at which the process (9.18) first enters the "stopping area" D. Area D is defined in the space of m-dimensional vectors $\mathbf{d} = (d_1, \ldots, d_m)$ with integer coordinates

$$D = \bigcup_{1 \leq i \leq m} \{\mathbf{d}: d_i \geq r_i\} \tag{9.19}$$

(for $m = 2$; see also Figure 9.2). Thus we observe a sequence of i.i.d. test cycles of the form

$$x_n = \{d^{(n)}(t), 0 \leq t \leq \tau_n\}, \qquad n = 1, 2, \ldots, \tag{9.20}$$

where $d^{(n)}(t)$ is a realization of process $d(t)$ at the nth test cycle and τ_n is the test duration at this cycle:

$$\tau_n = \min\{t: d^{(n)}(t) \in D\}$$

[see Figure 9.2, where two different realizations of reaching the stopping area D by process $d(t)$ are depicted].

Using this information, we need to construct confidence limits or/and to test some hypotheses for system reliability index $R = R(\boldsymbol{\theta})$, where $\boldsymbol{\theta} = (\theta_1, \ldots, \theta_m)$ is the vector of unit parameters. For instance,

$$R(\boldsymbol{\theta}) = \prod_{1 \leq i \leq m} [1 - F_i^{(r_i)}(t_0, \theta_i)],$$

where $F_i^{(r_i)}(t, \theta_i)$ is the d.f. of the ith subsystem TTF, which is r_i times the convolution of d.f. $F_i(t, \theta_i)$. Tests of such systems were considered by Ushakov and Gordienko (1978), and others. □

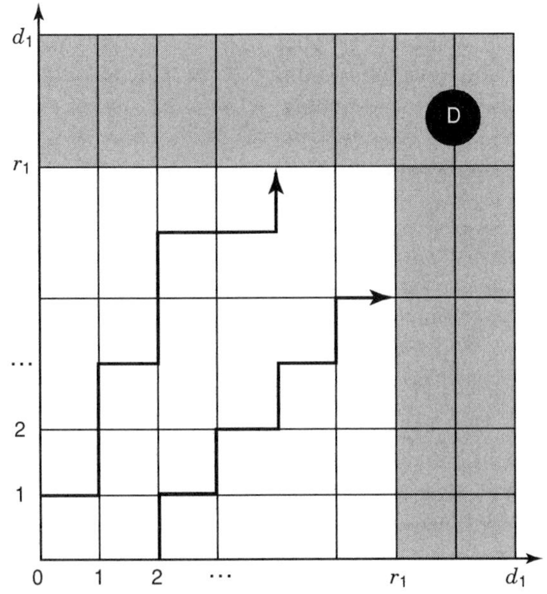

Figure 9.2 Two realizations of reaching stop area D by process $d(t)$.

9.2 SEQUENTIAL WALD CRITERION

9.2.1 Introduction

A frequently encountered problem in reliability practice is testing hypotheses such as

$$H_0: R(\boldsymbol{\theta}) \leq R_0, \qquad H_1: R(\boldsymbol{\theta}) \geq R_1 \tag{9.21}$$

on the basis of test results of type (9.8). Here $\boldsymbol{\theta}$ is an unknown parameter (in general, a vector), $R(\boldsymbol{\theta})$ is a reliability index, and R_0 and R_1 are some critical levels, where R_0 and R_1 play the roles of rejection and acceptance levels of the tested object by reliability index R. (Of course, if R is an index characterizing failure, for instance, failure rate (FR), then H_0 corresponds to acceptance and H_1 to rejection.) The probabilities of errors of type I (α) and of type II (β) correspond to manufacturer risk and consumer risk. Testing criteria are defined by the stopping rule and decision-making rule at the moment of the test stopping.

Hypotheses of type (9.18) are composite. First consider a more simple task, namely, a test of simple hypotheses of the type:

$$h_0: \boldsymbol{\theta} = \theta_0, \qquad h_1: \boldsymbol{\theta} = \theta_1,$$

where θ_0 and θ_1 are two points of the parametric space. The test stopping moment is denoted by ν. It is clear that the stopping moment determines the test volume. If this moment is defined in advance, that is, it is a constant $\nu = n$, the optimal decision rule is determined by the well-known Neyman–Pearson criterion (see Section 1.5 and Lehman, 1959). In correspondence with this criterion, hypothesis h_1 is accepted if the test results $x_1, \ldots, x_n$ satisfy the inequality

$$\varphi_n(x_1, \ldots, x_n) \leq C, \tag{9.22}$$

where the function of test results (statistics) $\varphi_n(x_1, \ldots, x_n) \geq C$ is determined as the likelihood ratio

$$\varphi_n = \frac{f(x_1, \theta_1) \times f(x_2, \theta_1) \times \cdots \times f(x_n, \theta_1)}{f(x_1, \theta_0) \times f(x_2, \theta_0) \times \cdots \times f(x_n, \theta_0)} = \alpha \tag{9.23}$$

and constant C is found so that the probability of error of type I equals the given value of α, that is, it satisfies the condition

$$P_0(\varphi_n \geq C) = \alpha. \tag{9.24}$$

(Probability P_0 corresponds to the value of parameter $\boldsymbol{\theta} = \theta_0$.)

For the sake of simplicity we consider the continuous case and assume that such a constant C can be found that satisfies equality (9.24) for any $0 < \alpha < 1$.

If inequality (9.22) does not hold, that is,

$$\varphi_n(x_1, \ldots, x_n) < C \tag{9.25}$$

then hypothesis H_0 is accepted.

The Neyman–Pearson criterion (9.22) and (9.25) possesses a known optimal property; namely, it has minimum probability of error of type II,

$$\beta = P_1(\varphi_n < C),$$

among all criteria with fixed constant test volume $\nu = n$ and fixed value of the probability of error of type I, α. (Here P_1 is the probability corresponding to the value of parameter $\boldsymbol{\theta} = \theta_1$.)

Let $n^* = n^*(\alpha, \beta)$ denote the test volume that is necessary for constructing the Neyman–Pearson criterion with given values of probability of errors of types I and II (α and β, $\alpha + \beta < 1$). The value of n^* is determined as the minimum integer value n for which the two following inequalities hold:

$$P_0(\varphi_n \geq C) \leq \alpha, \qquad P_1(\varphi_n < C) \leq \beta.$$

Notice that due to the optimal property of the Neyman–Pearson criterion, the necessary test volume $n^* = n^*\ (\alpha, \beta)$ cannot be decreased if one takes a criterion with deterministic test volume $\nu = n$.

The outstanding result obtained by Wald (1947) was that he constructed a sequential criterion significantly decreasing the *average volume* of the necessary test volume in comparison with the value above $n^*(\alpha, \beta)$ for the same values of probabilities of errors α and β. The Wald criterion is defined as follows. At current test step n, taking into account the results already obtained, $x_1, \ldots, x_n$, one calculates $\varphi_n(x_1, \ldots, x_n)$ considered in (9.23). This value is the likelihood ratio for hypotheses h_0 and h_1 on the mth test step. At each nth step, one checks the inequalities

$$B < \varphi_n(x_1, \ldots, x_n) < A, \tag{9.26}$$

where B and A are some given constants such that $0 < B < 1 < A$. If both inequalities (9.26) hold, then the test continues, that is, the next r.v. x_{n+1} is subject to observation. In other words, inequality (9.26) determines the "area of test continuation" for the Wald criterion.

The test terminates at the first violation of at least one of the inequalities in (9.26). If the left inequality is violated, hypothesis h_0 is accepted. The violation of the right inequality leads to acceptance of hypothesis h_1. Thus the moment of test stop ν for the Wald criterion is determined from the conditions

$$\varphi_n(x_1, \ldots, x_n) \in (A, B) \quad \text{for all } 1 \leq n \leq \nu - 1,$$

$$\varphi_\nu(x_1, \ldots, x_\nu) \notin (A, B),$$

or in more compact notation,

$$\nu = \min\{n\colon \varphi_n(x_1, \ldots, x_n) \notin (A, B)\}. \tag{9.27}$$

The vector of test results for any sequential criterion (including the Wald criterion) has the form $(\nu, x_1, \ldots, x_\nu)$, where ν is the stopping moment and $(x_1, \ldots, x_\nu)$ is the set of all test results obtained up to this step. Let

$$\varphi_\nu = \varphi_\nu(x_1, \ldots, x_\nu) = \frac{f(x_1, \theta_1) \times f(x_2, \theta_1) \times \cdots \times f(x_\nu, \theta_1)}{f(x_1, \theta_0) \times f(x_2, \theta_0) \times \cdots \times f(x_\nu, \theta_0)}$$

denote the value of function φ_n (likelihood ratio) at the test stopping moment ν. For the Wald criterion the decision-making rule at the stopping moment ν on the basis of test results $(\nu, x_1, \ldots, x_\nu)$ has the following form:

$$\begin{aligned} &\text{If } \varphi_\nu \leq B, \text{ hypothesis } h_0 \text{ is accepted.} \\ &\text{If } \varphi_\nu \geq A, \text{ hypothesis } h_1 \text{ is accepted.} \end{aligned} \tag{9.28}$$

The probabilities of error (risks) of types I and II for any criterion are $\alpha = P_0(h_1)$ and $\beta = P_1(h_0)$, respectively, where the probabilities $P_j(h_j)$ are taken for $\boldsymbol{\theta} = \theta_i$, $i = 0, 1$; $j = 0, 1$. For the Wald criterion, these values, in correspondence with (9.28), are equal to $\alpha = P_0(\varphi_\nu \geq A)$ and $\beta = P_1(\varphi_\nu \leq B)$.

The precise values of risks α and β for the Wald criterion can be approximately estimated with the help of the following known Wald formulas (Wald, 1947):

$$\frac{\beta}{1 - \alpha} = E_0(\varphi_\nu | h_0), \qquad \frac{1 - \beta}{\alpha} = E_0(\varphi_\nu | h_1), \tag{9.29}$$

where E_0 denotes the mathematical expectation for $\boldsymbol{\theta} = \theta_0$ and $E_0(\varphi_\nu / h_j)$ is the conditional mathematical expectation of r.v. φ_ν under the condition that at the stopping moment ν by the test results $(\nu, x_1, \ldots, x_n)$ hypothesis h_j is accepted, $j = 0, 1$. Notice that formulas (9.29) are valid for any criterion with the stopping moment ("not depending on the future") such that $P_i(\nu < \infty) = 1$, $i = 1, 2$.

It is easy to obtain corresponding inequalities and approximate estimates for α and β of the Wald criterion from (9.29). Indeed, from (9.29) for the Wald criterion, it follows by construction that the following conditions are valid:

$$\begin{aligned} &\varphi_\nu \leq B \text{ if hypothesis } h_0 \text{ is accepted.} \\ &\varphi_\nu \geq A \text{ if hypothesis } h_1 \text{ is accepted.} \end{aligned}$$

Thus, corresponding inequalities are valid for mathematical expectations:

$$E_0\left(\frac{\varphi_\nu}{h_0}\right) \leq B \quad \text{and} \quad E_0\left(\frac{\varphi_\nu}{h_1}\right) \geq A.$$

Taking into account (9.29), well-known inequalities for precise values of risks α and β for the Wald criterion follow:

$$\frac{\beta}{1 - \alpha} \leq B, \qquad \frac{1 - \beta}{\alpha} \geq A. \tag{9.30}$$

The area of (α, β) satisfying inequalities (9.30) are shown in Figure 9.3. From these inequalities it also follows that

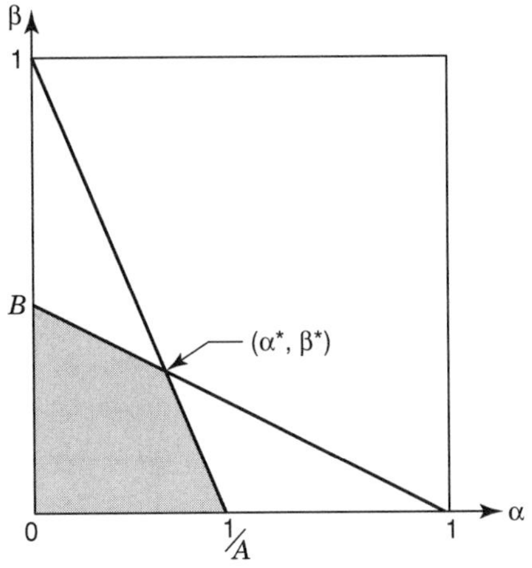

Figure 9.3 Area of (α, β) satisfying inequalities (9.30).

$$\beta \leq B, \qquad \alpha \leq \frac{1}{A}, \tag{9.31}$$

inequalities that are often used in practice for approximate risk estimation.

Notice that to construct the Wald criterion, the test is stopped at such a step $\nu = n$ when the process φ_n exits interval (A, B) the first time; that is when r.v. φ_n "jumps" over level A from below or under level B from above. Let us neglect the process of that jump and assume that at the stopping moment one of the following two approximate conditions are fulfilled:

$$\begin{aligned} \varphi_\nu &\approx B \text{ if hypothesis } h_0 \text{ is accepted.} \\ \varphi_\nu &\approx A \text{ if hypothesis } h_1 \text{ is accepted.} \end{aligned} \tag{9.32}$$

Then from (9.29) we obtain the known approximate Wald equalities

$$\frac{\beta}{1-\alpha} \approx B, \qquad \frac{1-\beta}{\alpha} \approx A. \tag{9.33}$$

These approximate solutions are often used in practice for risk estimation. Let us assume that precise values of risks α and β are close to the approximate values α^* and β^*, that is,

$$\alpha \approx \alpha^* \quad \text{and} \quad \beta \approx \beta^*,$$

where approximate risks are found from the equations

$$\frac{\beta^*}{1-\alpha^*} = B, \qquad \frac{1-\beta^*}{\alpha^*} = A, \tag{9.34}$$

from which

$$\alpha^* = \frac{1-B}{A-B}, \qquad \beta^* = \frac{B(A-1)}{A-B}.$$

[See Figure 9.3, where point (α^*, β^*) is found as the intersection of lines $\beta = B - B_\alpha$ and $\beta = 1 - A_\alpha$.] From Figure 9.3 one can see that the precise values of risks α and β are always within the shadowed area, that is,

$$\alpha + \beta \leq \alpha^* + \beta^*.$$

Besides, taking into account (9.34), inequalities (9.31) can be written in the form

$$\beta \leq \frac{\beta^*}{1-\alpha^*}, \qquad \alpha \leq \frac{\alpha^*}{1-\beta^*}.$$

The precise values of risks α, β can be found using computer programs. (For more details, see Section 9.6.1.)

Average Test Volume In accordance with (9.27), the stopping moment ν for Wald's criterion can be written as

$$\nu = \min\{n: Z_n \notin (-b, a)\},$$

where $a = \ln A > 0$, $b = -\ln B > 0$, and Z_n is the logarithm of the likelihood ratio at the nth step:

$$Z_n = \ln \varphi_n = \ln \frac{f(\theta_1, x_1) \times \cdots \times f(\theta_1, x_n)}{f(\theta_0, x_1) \times \cdots \times f(\theta_0, x_n)}.$$

Let us introduce the notation

$$\begin{aligned} z &= z(x) = \ln \frac{f(\theta_1, x)}{f(\theta_0, x)}, \\ z_n &= z(x_n) = \ln \frac{f(\theta_1, x_n)}{f(\theta_1, x_n)}, \qquad n = 1, 2, \ldots. \end{aligned} \tag{9.35}$$

Random values $z_1, \ldots, z_n, \ldots$ are i.i.d. and z_n is the nth observation of r.v. z. Here, Z_n is defined as

$$Z_n = z_1 + \cdots + z_n. \tag{9.36}$$

Let us find the mathematical expectation $E_0\nu$ for the case where hypothesis h_0 is true, that is, $\boldsymbol{\theta} = \theta_0$. In accordance with (9.36), the equality

$$E_0 Z_\nu = E_0(z_1 + \cdots + z_\nu), \tag{9.37}$$

where E_0 denotes the mathematical expectation under condition $\boldsymbol{\theta} = \theta_0$. Applying a well-known Wald's equivalence (Wald, 1944) to the right part of (9.37) for the mathematical expectation of the sum of the random number of i.i.d r.v.'s, we obtain

$$E_0(z_1 + \cdots + z_\nu) = E_0\nu E_0 z. \tag{9.38}$$

(For a proof of Wald's equivalence, see Section 9.6.2.) For the left side of the equality (9.37), we have

$$E_0 Z_\nu = P_0(h_0)E_0\left(\frac{Z_\nu}{h_0}\right) + P_0(h_1)E_0\left(\frac{Z_\nu}{h_1}\right), \tag{9.39}$$

from where

$$E_0 Z_\nu = (1 - \alpha)E_0\left(\frac{\ln \varphi_\nu}{h_0}\right) + \alpha E_0\left(\frac{\ln \varphi_\nu}{h_2}\right). \tag{9.40}$$

Neglecting again a jump mentioned above and using approximate equalities (9.32) and (9.33), we obtain from the latter equality

$$E_0 Z_\nu \approx (1 - \alpha)\ln\frac{\beta}{1 - \alpha} + \alpha \ln\frac{1 - \beta}{\alpha}. \tag{9.41}$$

After this, the known approximate equality (Wald, 1947) for the average test volume under condition $\boldsymbol{\theta} = \theta_0$ follows from (9.38) and (9.41):

$$E_0\nu \approx \frac{\omega(\alpha, \beta)}{E_0(-z)}, \tag{9.42}$$

where

$$\omega(\alpha, \beta) = (1 - \alpha)\ln\frac{1 - \alpha}{\beta} + \alpha \ln\frac{\alpha}{1 - \beta}.$$

In the same way we have an analogous approximate equality for the average test volume for hypothesis h_1, that is, for $\boldsymbol{\theta} = \theta_1$:

$$E_1\nu \approx \frac{\omega(\beta, \alpha)}{E_1(z)}. \tag{9.43}$$

Taking into account the definition of r.v. z in (9.35), formulas for the average test volume can also be represented in the form

$$E_0\nu \approx \frac{\omega(\alpha, \beta)}{\rho(\theta_0, \theta_1)}, \qquad E_1\nu \approx \frac{\omega(\beta, \alpha)}{\rho(\theta_1, \theta_0)}, \tag{9.44}$$

where

$$\rho(\theta_1, \theta_0) = E_1 z = E_1 \ln \frac{f(\theta_1, x)}{f(\theta_0, x)} \tag{9.45}$$

represents the "Kullback–Leiber information distance" between θ_1 and θ_0 [or, more exactly, between d.f.'s with densities $f(\theta_1, x)$ and $f(\theta_1, x)$]. Notice that in general (Kullback and Leiber, 1951; Kullback, 1959) $\rho(\theta_1, \theta_0) \neq \rho(\theta_0, \theta_1)$. Formulas (9.42)–(9.44) for the average test volumes are approximate (without accounting for jumps). See Section 9.6.2 for the precise calculation of test volumes using computer programs.

Lower Limit for Average Test Volume From (9.40) one can easily obtain the lower limit of the average test volume for any criterion with fixed risk values α and β. Indeed, this equality is valid for any criterion, not only for Wald's. Taking into account that function $\ln u$ is concave in u and applying Jensen's inequality for the mathematical expectation of a concave function, we obtain

$$\begin{aligned} E_0(\ln \varphi_\nu | h_0) &\leq \ln E_0(\varphi_\nu | h_0), \\ E_0(\ln \varphi_\nu | h_1) &\leq \ln E_0(\varphi_\nu | h_1). \end{aligned} \tag{9.46}$$

Taking into account (9.40) and (9.29), we obtain, from (9.46), the inequality

$$E_0 Z_\nu \leq (1 - \alpha) \ln \frac{\beta}{1 - \alpha} + \alpha \ln \frac{1 - \beta}{\alpha}.$$

Then taking into account (9.37) and (9.38), we obtain

$$E_0\nu \geq \frac{\omega(\alpha, \beta)}{E_0(-z)}. \tag{9.47}$$

In the same way, we can obtain an analogous inequality for the average test volume under condition $\boldsymbol{\theta} = \theta_1$:

$$E_1\nu \geq \frac{\omega(\beta, \alpha)}{E_1(z)}. \tag{9.48}$$

Inequalities (9.47) and (9.48) give us the lower limit for the average test volume under conditions $\boldsymbol{\theta} = \theta_0$ and $\boldsymbol{\theta} = \theta_1$ for any (not depending on the future) criterion for hypotheses h_0 and h_1 for given risks α and β. These inequalities were obtained by Wald (1947) and Hoeffding (1960).

As one can see from approximate equalities (9.42) and (9.43), the average test volume for Wald's criterion reaches the lower limit in (9.47) and (9.48) at least approximately (without taking into account the jump mentioned above). The accurate proof of Wald's criterion optimality was obtained elsewhere (Wald, 1947; Wald and Wolfowitz, 1948), where the following optimum property of this criterion was shown. Let us adopt some Wald's criterion for hypothesis h_0 and h_1 with stopping moment ν and with risks α and β. Then for any other criterion with stopping moment ν' and with risks α' and β' such that $\alpha' \leq \alpha$ and $\beta' \leq \beta$, the following inequalities are valid:

$$E_0\nu' \geq E_0\nu, \qquad E_1\nu' \geq E_1\nu.$$

The significant property of Wald's criterion is that it simultaneously minimizes two quality indices of the criteria $E_0\nu$ and $E_1\nu$ among all criteria with risks not larger than α and β. The generalization of Wald's results for the case of continuous time was obtained afterward by Dvoretsky et al. (1953), Epstein and Sobel (1955), and others.

9.2.2 Standard Sequential Wald's Plans of Reliability Index Control

Suppose we need to test two composite hypotheses about some one-dimensional parameter $\boldsymbol{\theta}$,

$$H_0\colon \boldsymbol{\theta} \leq \theta_0, \qquad H_1\colon \boldsymbol{\theta} \geq \theta_1, \tag{9.49}$$

on the basis of the test results (9.8). Here $\theta_0 \leq \theta_1$ are specified levels of parameter $\boldsymbol{\theta}$. Levels θ_0 and θ_1 mean, respectively, levels of "acceptance" and "rejection" of a tested object by parameter $\boldsymbol{\theta}$ if $\boldsymbol{\theta}$ is the index of type of failure probability, q, or FR, λ, (see examples below) or if $\boldsymbol{\theta}$ is the index of type of PFFO.

The standard approach for criterion construction (or in other terms, plans of control of parameter $\boldsymbol{\theta}$) for the case of composite hypotheses of type (9.49) consists in the following. Let us construct Wald's criterion for two simple hypotheses,

$$h_0\colon \boldsymbol{\theta} = \theta_0, \qquad h_1\colon \boldsymbol{\theta} = \theta_1, \tag{9.50}$$

in the same way as it was considered above. Apply this criterion to the test of initial composite hypotheses (9.49). In other words, let us continue the test up to the stopping moment ν determined above in (9.27). At the stopping moment we make the following decisions:

$$\begin{aligned} &\text{If } \varphi_\nu \leq B, \text{ hypothesis } H_0 \text{ is accepted.} \\ &\text{If } \varphi_\nu \geq A, \text{ hypothesis } H_1 \text{ is accepted.} \end{aligned} \tag{9.51}$$

Introduce the function

$$L(\boldsymbol{\theta}) = P_\theta(H_0) \tag{9.52}$$

equal to the probability of acceptance of hypothesis H_0 under the condition that the true value of the parameter is $\boldsymbol{\theta}$. Function (9.52) is called the *operative characteristic* of a criterion (plan of control). Since the rule of decision making for Wald's plan has the form of (9.51), the operative characteristic for this plan is determined by

$$L(\boldsymbol{\theta}) = P_\theta(\varphi_\nu \leq B), \tag{9.53}$$

or, in other words, the operative characteristic $L(\boldsymbol{\theta})$ for Wald's plan is equal to the probability that the function of the likelihood ratio φ_ν will exit the lower limit B (limit of the "acceptance" area) for a specified value of parameter $\boldsymbol{\theta}$. (See Figure 9.4.)

In many cases (see examples below) operative characteristic (9.53) is monotone decreasing in parameter $\boldsymbol{\theta}$. (See Figure 9.5.)

Since hypotheses (9.49) are composite, the probabilities of errors of types I and II are some functions of parameter $\boldsymbol{\theta}$ expressed via the operative characteristic as

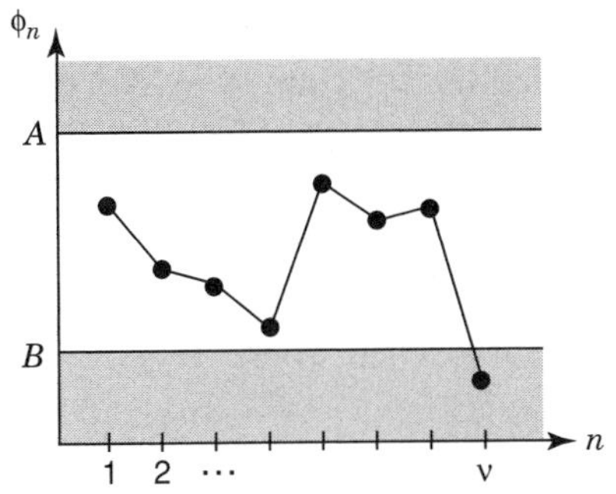

Figure 9.4 Sample of the process of decision making for Wald's plan.

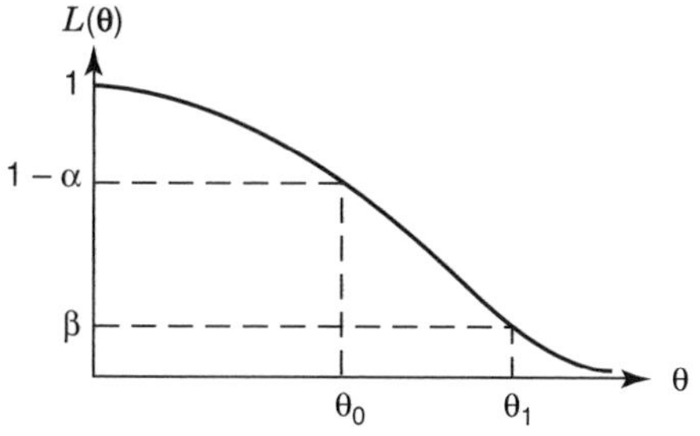

Figure 9.5 Operative characteristic $L(\boldsymbol{\theta})$.

$$\begin{aligned} \alpha(\boldsymbol{\theta}) &= 1 - L(\boldsymbol{\theta}) \quad \text{for } \boldsymbol{\theta} \leq \theta_0, \\ \beta(\boldsymbol{\theta}) &= L(\boldsymbol{\theta}) \qquad\quad \text{for } \boldsymbol{\theta} \geq \theta_1. \end{aligned} \tag{9.54}$$

The maximum values of the probability of error, that is, the values

$$\alpha = \max_{\theta \leq \theta_0} \alpha(\theta), \qquad \beta = \max_{\theta \geq \theta_1} \beta(\theta), \tag{9.55}$$

are called, respectively, risks of types I and II. If the operative characteristic $L(\boldsymbol{\theta})$ is monotone decreasing in $\boldsymbol{\theta}$, then

$$\alpha = 1 - L(\theta_0), \qquad \beta = L(\theta_1).$$

(See also Figure 9.5.) Notice that the values α and β coincide with risks for Wald's criterion for simple hypotheses (9.50). Due to (9.54) and (9.55), the following inequalities are valid:

$$\alpha(\boldsymbol{\theta}) \leq \alpha \quad \text{for } \boldsymbol{\theta} \leq \theta_0, \qquad \beta(\boldsymbol{\theta}) \leq \beta \quad \text{for } \boldsymbol{\theta} \geq \theta_1.$$

Thus, Wald's criterion applied for simple hypotheses h_0 and h_1 (corresponding to the boundary point of the acceptance and rejection levels θ_0 and θ_1) simultaneously gives the criterion for composite hypotheses of type (9.49) with errors of types I and II not larger than values α and β.

Average Test Volume Consider calculation of the average test volume for Wald's plan. Denote this plan by

$$N(\boldsymbol{\theta}) = E_\theta \nu, \tag{9.56}$$

where E_θ is the mathematical expectation under the given value of parameter $\boldsymbol{\theta}$. To estimate (9.56), use formulas of type (9.39):

$$E_\theta Z_\nu = P_\theta(H_0)E_\theta\left(\frac{Z_\nu}{H_0}\right) + P_\theta(H_1)E_\theta\left(\frac{Z_\nu}{H_1}\right)$$

$$= L(\theta)E_\theta\left(\frac{\ln \varphi_\nu}{H_0}\right) + [1 - L(\boldsymbol{\theta})]E_\theta\left(\frac{\ln \varphi_\nu}{H_1}\right).$$

Applying approximate formula (9.32), we obtain

$$E_\theta Z_\nu \approx [1 - L(\boldsymbol{\theta})]a - L(\boldsymbol{\theta})b, \tag{9.57}$$

where $a = \ln A$, $b = -\ln B$. On the other hand, applying Wald's equivalency to the left side of (9.57) (see Section 9.6.2), we have

$$E_\theta Z_\nu = E_\theta(z_1 + \cdots + z_\nu) = E_\theta \nu E_\theta z$$

The well-known Wald's formula for the average test volume follows from here:

$$N(\boldsymbol{\theta}) \approx \frac{[1 - L(\boldsymbol{\theta})]a - L(\boldsymbol{\theta})b}{E_\theta z} \tag{9.58}$$

for $E_\theta z \neq 0$. Taking into account

$$E_\theta z = E_\theta \ln \frac{f(\theta_1, x)}{f(\theta_0, x)}$$

$$= E_\theta \ln \frac{f(\boldsymbol{\theta}, x)}{f(\theta_0, x)} - E_\theta \ln \frac{f(\boldsymbol{\theta}, x)}{f(\theta_1, x)} = \rho(\boldsymbol{\theta}, \theta_0) - \rho(\boldsymbol{\theta}, \theta_1),$$

(9.58) can be rewritten in the form

$$N(\boldsymbol{\theta}) \approx \frac{[1 - L(\boldsymbol{\theta})]a - L(\boldsymbol{\theta})b}{\rho(\boldsymbol{\theta}, \theta_0) - \rho(\boldsymbol{\theta}, \theta_1)}. \tag{9.59}$$

Formulas (9.58) and (9.59) are true for such $\boldsymbol{\theta}$ that $E_\theta z \neq 0$, or in other words, for $\boldsymbol{\theta} \neq \boldsymbol{\theta}'$, where $\boldsymbol{\theta}'$ is determined from the condition $E_\theta, z = 0$, or from $\rho(\boldsymbol{\theta}, \theta_0) = \rho(\boldsymbol{\theta}', \theta_1)$.

Point $\boldsymbol{\theta}'$ is called "equally distant" (in sense of the Kullback–Leibler distance) from points θ_0 and θ_1. In most standard situations, point $\boldsymbol{\theta}'$ is located between points θ_0 and θ_1. For $\boldsymbol{\theta} = \boldsymbol{\theta}'$ for the average test volume the approximate formula found by Wald (1947),

$$N(\boldsymbol{\theta}') \approx \frac{ab}{E_{\theta'} z^2}, \tag{9.60}$$

is valid.

The precise values of the operative characteristic $L(\boldsymbol{\theta})$ and average test volume $N(\boldsymbol{\theta})$ for Wald's plan for given fixed boundaries A and B and for other sequential plans (with other stopping areas) can be calculated using special computer programs. Analogously, with a computer one can solve an inverse problem of finding the boundaries for stopping (A and B) in such a way that the precise values of risks coincide with the given values α and β. (For details see Section 9.6.1.)

A typical graph of dependence of the average test volume, $N(\boldsymbol{\theta})$, of parameter $\boldsymbol{\theta}$ for sequential Wald's plan is depicted in Figure 9.6. As one can see, a sequential Wald's plan is the most effective for $\boldsymbol{\theta} \leq \theta_0$ and $\boldsymbol{\theta} \geq \theta_1$. That is, Wald's plan for statistical control allows to quickly (on average) accept "good" objects (with parameter $\boldsymbol{\theta} \leq \theta_0$) and reject "bad" objects (with parameter $\boldsymbol{\theta} \geq \theta_1$). In areas of values $\boldsymbol{\theta} \leq \theta_0$ and $\boldsymbol{\theta} \geq \theta_1$ Wald's criterion gives an essential gain in the average test volume, in comparison with the best Neyman–Pearson criterion for the case of fixed test volume. However, Wald's criterion becomes less effective in the intermediate area of parameter values, that is, if $\theta_0 < \boldsymbol{\theta} < \theta_1$. This area is also called the "area of uncertainty" or the "area of indifference."

Example 9.9 (Binomial Test—Control of Failure Probability) Consider the binomial test considered in Example 9.1. A sequence of i.i.d. r.v.'s is observed,

$$\delta_1, \ldots, \delta_n, \ldots, \tag{9.61}$$

where δ_n is the indicator of unit failure at the nth step. As was mentioned

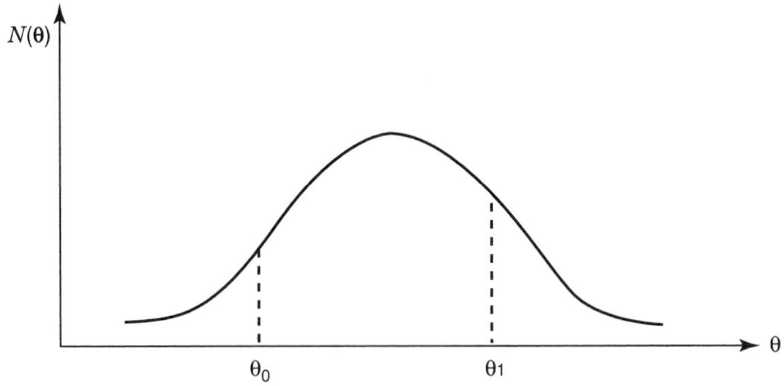

Figure 9.6 Dependence of average test volume on parameter $\boldsymbol{\theta}$.

above, this value takes 0 or 1 with probabilities $P(\delta_n = 0) = 1 - q$ and $P(\delta_n = 1) = q$, where q is the failure probability. On the basis of test results (9.61) we need to test hypotheses

$$H_0\colon q \le q_0, \qquad H_1\colon q \ge q_1,$$

where $q_0 \le q_1$ are the given critical levels of index q. These values are the acceptance and rejection levels of this index.

In this case the parameter is $\boldsymbol{\theta} = q$, the observed r.v. at the nth step is $x_n = \delta_n$, and the function $f(x, \boldsymbol{\theta}) = f(\delta, q) = q^{\delta}(1 - q)^{1-\delta}$, $\delta = 0, 1$. From here we obtain that the function of likelihood ratio (9.23) at the nth test step has the form

$$\varphi_n = \frac{f(\delta_1, q_1) \times f(\delta_2, q_1) \times \cdots \times f(\delta_n, q_1)}{f(\delta_1, q_0) \times f(\delta_2, q_0) \times \cdots \times f(\delta_n, q_0)} = \left(\frac{q_1}{q_0}\right)^{d_n} \left(\frac{1 - q_1}{1 - q_0}\right)^{n-d_n},$$

where $d_n = \delta_1 + \cdots + \delta_n$ is the total number of failures during n steps. The testing area continuation for Wald's plan (9.26) is given by the inequalities

$$B < \left(\frac{q_1}{q_0}\right)^{d_n} \left(\frac{1 - q_1}{1 - q_0}\right)^{n-d_n} < A,$$

or after simple transformations by

$$C_1 n - C_2 b < d_n < C_1 n + C_2 a, \tag{9.62}$$

where $a = \ln A > 0$, $b = -\ln B > 0$,

$$C_2 = \left[\ln\left(\frac{q_1(1 - q_0)}{q_0(1 - q_1)}\right)\right]^{-1'} \qquad C_1 = C_2 \ln\left(\frac{1 - q_0}{1 - q_1}\right).$$

The test continues if both inequalities (9.62) are true and stops at such a step $\nu = n$ at which at least one of these inequalities is violated. The violation of the left inequality means that hypothesis $H_0\colon q \le q_0$ holds (acceptance), and for the violation of the right inequality the decision is to accept hypothesis $H_1\colon q \ge q_1$ (rejection). Thus, the boundaries of the stopping area (see Figure 9.7) have the form of lines on the plane (n, d_n). The equations of these lines are

$d_n = C_1 n - C_2 b$ is the boundary of the acceptance area and
$d_n = C_1 n + C_2 b$ is the boundary of the rejection area.

In this test scheme the r.v. (9.35) has the form

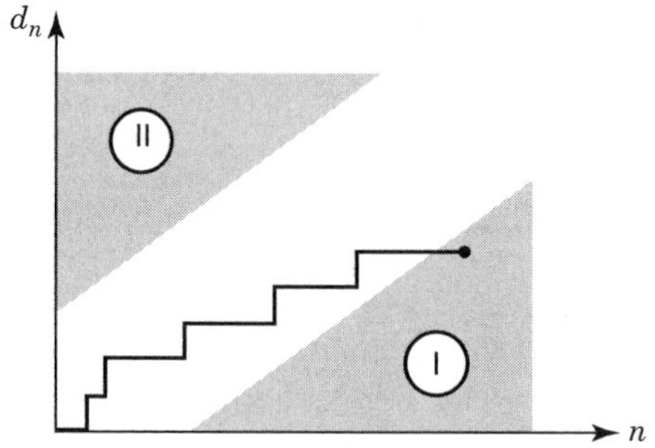

Figure 9.7 Boundaries of the stopping area (binomial distribution).

$$z = \ln \frac{f(\delta, q_1)}{f(\delta, q_0)} = \delta \ln \left(\frac{q_1}{q_0} \right) + (1 - \delta) \ln \left(\frac{1 - q_1}{1 - q_0} \right).$$

Taking into account that $E_0\delta = q_0$, $E_1\delta = q_1$, we obtain that (9.42)–(9.44) for the average test volume for $q = q_0$ and q_1 have the form

$$E_0\nu \approx \frac{\omega(\alpha, \beta)}{\rho(q_0, q_1)}, \qquad E_1\nu \approx \frac{\omega(\beta, \alpha)}{\rho(q_1, q_0)}, \tag{9.63}$$

where

$$\rho(q_0, q_1) = q_0 \ln \left(\frac{q_0}{q_1} \right) + (1 - q_0) \ln \left(\frac{1 - q_0}{1 - q_1} \right). \ \square$$

Example 9.10 (Exponential Test—Control of Failure Rate) Suppose we observe a sequence of i.i.d. r.v.'s

$$x_1, \ldots, x_n, \ldots \tag{9.64}$$

each of which has the exponential d.f. with density $f(x, \lambda) = \lambda \exp(-\lambda x)$. (See also Examples 9.1 and 9.2.) We need to test the following hypotheses on the basis of test results (9.64):

$$H_0: \lambda \leq \lambda_0, \qquad H_1: \lambda \geq \lambda_1, \tag{9.65}$$

where $\lambda_0 < \lambda_1$ are the given critical levels of acceptance and rejection.

In this case the likelihood ratio function (9.23) at the nth test step has the form

$$\varphi_n = \frac{f(x_1, \lambda_1) \times f(x_2, \lambda_1) \times \cdots \times f(x_n, \lambda_1)}{f(x_1, \lambda_0) \times f(x_2, \lambda_0) \times \cdots \times f(x_n, \lambda_0)} = \left(\frac{\lambda_1}{\lambda_0} \right)^n e^{-(\lambda_1 - \lambda_0)S_n},$$

where $S_n = x_1 + \cdots + x_n$ is the total test time during n steps. The area of test continuation for Wald's criterion in this case is given by the inequalities

$$B < \left(\frac{\lambda_1}{\lambda_0}\right)^n e^{-(\lambda_1-\lambda_0)S_n} < A$$

or the inequalities

$$C_1 n - C_2 a < S_n < C_1 n + C_2 b, \tag{9.66}$$

where $a = \ln A$, $b = -\ln B$,

$$C_2 = \frac{1}{\lambda_1 - \lambda_0}, \qquad C_1 = C_2 \ln\left(\frac{\lambda_1}{\lambda_0}\right).$$

The test continues if both inequalities (9.66) are true and test stops when at least one of these inequalities is violated the first time. If the right inequality in (9.66) does not hold, hypothesis H_0 is accepted. If the right inequality in (9.66) is violated, hypothesis H_1 is accepted (rejection). The boundaries of the stopping area (see Figure 9.8) in this case have the form of lines on the plane (n, S_n):

$S_n = C_1 n + C_2 b$ is the boundary of "area of acceptance" and
$S_n = C_1 n - C_2 a$ is the boundary of "area of rejection."

Let us also estimate the average test volume in this case. Random variable (9.35) in this case has the form

$$z = \ln \frac{f(x, \lambda_1)}{f(x, \lambda_0)} = \ln\left(\frac{\lambda_1}{\lambda_0}\right) - x(\lambda_1 - \lambda_0).$$

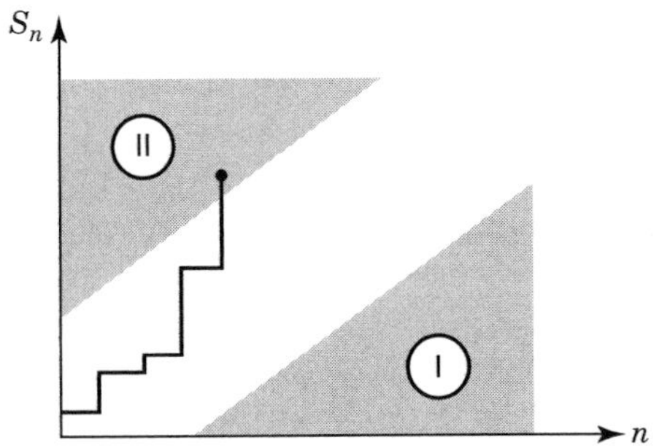

Figure 9.8 Boundaries of the stopping area (exponential distribution).

Taking into account that $E_0 x = 1/\lambda_0$ and $E_1 x = 1/\lambda_1$, we obtain that formulas (9.42)–(9.44) for $\boldsymbol{\lambda} = \lambda_0$ and $\boldsymbol{\lambda} = \lambda_1$ in this case have the form

$$E_0 \nu \approx \frac{\omega(\alpha, \beta)}{\rho(\lambda_0, \lambda_1)}, \qquad E_1 \nu \approx \frac{\omega(\beta, \alpha)}{\rho(\lambda_1, \lambda_0)}, \tag{9.67}$$

where

$$\rho(\lambda_0, \lambda_1) = \frac{\lambda_0 - \lambda_1}{\lambda_0} - \ln\left(\frac{\lambda_1}{\lambda_0}\right). \ \square$$

Example 9.11 (Poisson Process—Control of Failure Rate) Assume that we observe a Poisson process of failures d_t, $t > 0$, with unknown parameter λ, where d_t is the number of failures observed up to the moment t. We need to accept one of two hypotheses on the basis of the results of observation:

$$H_0 \colon \boldsymbol{\lambda} \le \lambda_0, \qquad H_1 \colon \boldsymbol{\lambda} \ge \lambda_1,$$

where $\lambda_0 < \lambda_1$ are the given critical levels of acceptance and rejection. The likelihood ratio φ_t at the moment t (an analogue of value φ_n in the discrete scheme considered above n) has the form

$$\varphi_t = \left(\frac{\lambda_1}{\lambda_0}\right)^{d_t} e^{-(\lambda_1 - \lambda_0)t}.$$

Thus the area of test continuation for Wald's criterion at the moment t is given by inequalities

$$B < \left(\frac{\lambda_1}{\lambda_2}\right)^{d_t} e^{-(\lambda_1 - \lambda_0)t} < A,$$

or, after simple transformations,

$$C_1 t - C_2 b < d_t < C_1 t + C_2 a, \tag{9.68}$$

where $a = \ln A$, $b = -\ln B$,

$$C_2 = \frac{1}{\ln(\lambda_1/\lambda_0)}, \qquad C_1 = C_2(\lambda_1 - \lambda_0). \tag{9.69}$$

The test continues until both inequalities (9.68) are true, and the test stops at moment $\tau = t$ when at least one of these inequalities is violated for the first time. If the right inequality in (9.69) does not hold, hypothesis H_0 is accepted.

If the right inequality in (9.66) does not hold, hypothesis H_1 is accepted (rejection). The boundaries of the stopping area (see Figure 9.9) in this case have the form of lines on the plane (t, d_t):

$d_t = C_1t - C_2b$ is the boundary of area of acceptance and
$d_t = C_1t + C_2a$ is the boundary of area of rejection.

The test stopping moment τ is the moment of exit of the process d_t from one of the boundaries mentioned above. From (9.42)–(9.44), in the scheme with discrete time, one can easily obtain analogous formulas for the average test time if $\boldsymbol{\lambda} = \lambda_0$ and $\boldsymbol{\lambda} = \lambda_1$ in this scheme with the continuous time:

$$E_0\tau \approx \frac{\omega(\alpha, \beta)}{\rho(\lambda_0, \lambda_1)}, \qquad E_1\tau \approx \frac{\omega(\beta, \alpha)}{\rho(\lambda_1, \lambda_0)},$$

where

$$\rho(\lambda_0, \lambda_1) = (\lambda_0 - \lambda_1) - \lambda_0 \ln\left(\frac{\lambda_1}{\lambda_0}\right).$$

Let us consider a numerical example that illustrates a gain in the average test volume obtained by the sequential Wald's criterion in comparison with the best Clopper–Pearson criterion with fixed test volume. □

Example 9.12 (Gain in Average Test Volume) In the scheme of Example 9.10 we need to test the hypothesis of type (9.65) where the critical levels of acceptance and rejection are $\lambda_0 = 0.1$ and $\lambda_1 = 0.2$, respectively. Given are risk values of types I and II of $\alpha = 0.1$ and $\beta = 0.1$.

If the test stopping moment is determined in advance and is a constant $\nu = n$, then the best Neyman–Pearson criterion has the form

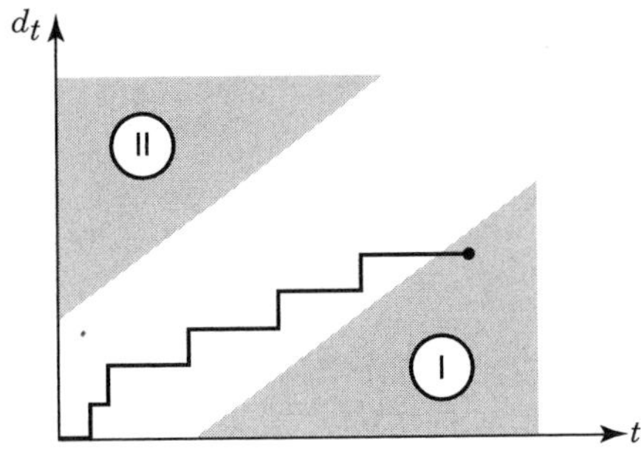

Figure 9.9 Boundaries of the stopping area (Poisson process).

$$\text{if } S_n < C, \text{ hypothesis } H_1 \text{ is accepted,}$$

$$\text{if } S_n \geq C, \text{ hypothesis } H_0 \text{ is accepted,}$$

where $S_n = x_1 + \cdots + x_n$ is the total time during n steps. Risks of types I and II for this criterion equal $P_0(S_n < C)$ and $P_1(S_n \geq C)$, where P_j denotes the probability that the parameter value is $\boldsymbol{\lambda} = \lambda_j$, $j = 0, 1$. Thus, the test volume $n^* = n^*(\alpha, \beta)$ needed for delivering given risk levels α and β is determined as a minimum integer number n that satisfies the following two inequalities:

$$P_0(S_n < C) \leq \alpha \quad \text{and} \quad P_1(S_n \geq C) \leq \beta.$$

These inequalities can be written in the form

$$P_0(2\lambda_1 S_n < 2\lambda_0 C) \leq \alpha, \qquad P_1(2\lambda_1 S_n \geq 2\lambda_1 C) \leq \beta,$$

or, taking into account that r.v. $2\lambda S_n$ has the standard χ^2 distribution with $2n$ degrees of freedom, we have

$$\chi^2(2\lambda_0 C, 2n) \leq \alpha, \qquad 1 - \chi^2(2\lambda_1 C, 2n) \leq \beta.$$

where $\chi^2(\cdot, 2n)$ is the function of the χ^2 distribution with $2n$ degrees of freedom. After simple transformations it gives the needed test volume n^*, equal to the minimum integer number n that satisfies the inequality

$$\chi^2_{1-\beta}(2n) \leq \left(\frac{\lambda_1}{\lambda_0}\right) \chi^2_\alpha(2n),$$

where $\lambda^2_\gamma(2n)$ is the quantile of level γ of the χ^2 distribution with $2n$ degrees of freedom. Using Table E.16 in the Appendix (χ^2 distribution), we find

$$n^* = \min\{n\colon \sigma^2_{0.9}(2n) \leq 2\chi^2_{0.1}(2n)\} = 15.$$

Applying (9.67), we find the average test volume for $\boldsymbol{\lambda} = \lambda_0$ and $\boldsymbol{\lambda} = \lambda_1$ for the sequential Wald's criterion (for the same values of λ_0, λ_1, α, and β):

$$E_0\nu \approx \frac{(1-\alpha)\ln[(1-\alpha)/\beta] + \alpha \ln[\alpha/(1-\beta)]}{(\lambda_1/\lambda_0) - 1 - \ln(\lambda_1/\lambda_0)}$$

$$= \frac{0.9 \ln 9 - 0.1 \ln 9}{1 - \ln 2} = 5.7,$$

$$E_1\nu \approx \frac{(1-\beta)\ln[(1-\beta)/\alpha] + \beta \ln[\beta/(1-\alpha)]}{(\lambda_0/\lambda_1) - 1 - \ln(\lambda_0/\lambda_1)}$$

$$= \frac{0.9 \ln 9 - 0.1 \ln 9}{\ln 2 - 0.5} = 9.2.$$

Thus, for $\boldsymbol{\lambda} = \lambda_0$ and $\boldsymbol{\lambda} = \lambda_1$ the average gains in test volumes obtained with Wald's criterion are equal, respectively, to

$$\frac{n^*}{E_0\nu} \approx \frac{15}{5.7} \approx 2.63, \qquad \frac{n^*}{E_1\nu} \approx \frac{15}{9.2} \approx 1.63.$$

9.2.3 Truncated Sequential Wald's Plans

The area of test continuation for Wald's plan is infinite. Therefore, in practice, one often uses truncated Wald's plans with a finite area of test continuation. Consider, for instance, the Poisson process scheme considered in Example 9.11. A typical area of test continuation for a truncated Wald's plan is depicted in Figure 9.10. The boundary of the acceptance area (if the process d_t reaches this area, then hypothesis H_0 is accepted) is given by the conditions

$$d_t = C_1 t - C_2 b \quad \text{for } t \leq T,$$
$$d_t \leq r \qquad\qquad \text{for } t = T.$$

The boundary of the rejection area (if the process d_t reaches this area, then hypothesis H_1 is accepted) is given by the conditions

$$d_t = C_1 t + C_2 a \quad \text{for } d_t \leq r,$$
$$t_t \leq T \qquad\qquad \text{for } d_t = r,$$

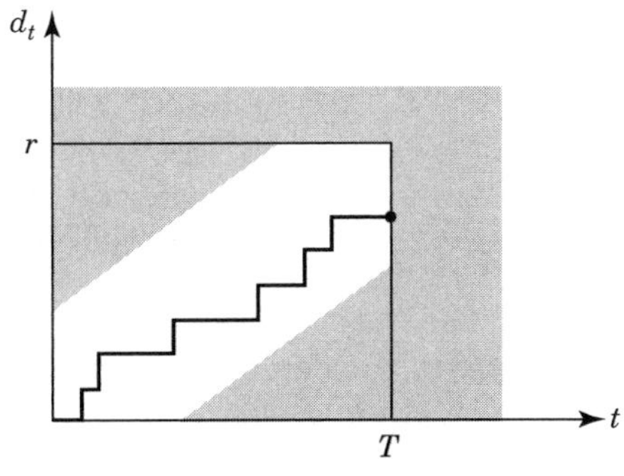

Figure 9.10 Stopping area boundaries for the truncated Wald's test plan.

where (T, r) is the truncating point (see Figure 9.10) and coefficients C_1 and C_2 are determined by (9.69).

Truncated Wald's plans lose, generally speaking, their property of a minimum average test volume. If the truncation is not "too severe," that is, if the coordinates of the truncating point (T, r) are large enough, then we believe that that property is preserved approximately. Analogous arguments are applied to the risks α and β.

The truncated Wald's plan is specified by four parameters A, B, T, and r. Precise values of risks α and β, operative characteristic $L(\boldsymbol{\theta})$, and average test volume $N(\boldsymbol{\theta})$ for the truncated Wald's plan can be calculated with a computer. (For details, see Section 9.6.1.)

9.3 SEQUENTIAL TEST OF COMPOSITE HYPOTHESES FOR MULTIDIMENSIONAL PARAMETERS

The sequential Wald's criterion discussed above has minimum average test volumes $E_0\nu$ and $E_1\nu$ for parameter values $\boldsymbol{\theta} = \theta_0$ and $\boldsymbol{\theta} = \theta_1$ (among all criteria with given risks α and β). However, the deficiency in the Wald's criterion is its low efficiency in the area of indifference, that is, where values of the parameter are $\theta_0 < \boldsymbol{\theta} < \theta_1$. This has led to a set of works in which sequential criteria with other properties were constructed (e.g., minimization of maximum average in the $\boldsymbol{\theta}$ test volume, $E_\theta\nu$): Kiefer and Weiss (1957), Chernoff (1959), Albert (1961), Weiss (1962), Schwarz (1962), Kiefer and Sacks (1963), Ayvazyan (1965), Shiryaev (1965, 1976), Lai (1973), Robbins and Siegmund (1973, 1974), Lorden (1976), Liptser and Shiryaev (1981), Huffman (1983), Pavlov (1985, 1987a, 1990), Draglin and Novikov (1987, 1996), and others.

An important application of the generalization of a sequential Wald's criterion (including reliability) is in testing composite hypotheses for multidimensional parameters, for instance, of the type

$$H_0\colon R(\boldsymbol{\theta}) \le R_0, \qquad H_1\colon R(\boldsymbol{\theta}) \ge R_1 \tag{9.70}$$

where $R(\boldsymbol{\theta}) = R(\theta_1, \ldots, \theta_m)$ is a function (reliability index) that depends on the multidimensional parameter $\boldsymbol{\theta} = (\theta_1, \ldots, \theta_m)$ and $R_0 < R_1$ are specified levels of index $R(\boldsymbol{\theta})$. Such problems are typical for complex system analysis (see Section 9.1 and Examples 9.5–9.8). At the same time such a problem arises for a single unit if its d.f. of TTF, $F(x, \boldsymbol{\theta})$ depends on vector parameter $\boldsymbol{\theta}$ (see Examples 9.1 and 9.4). The construction of approximately optimal sequential criteria for composite hypotheses of type (9.70) and the associated problem of constructing approximate optimal sequential confidence limits for $R(\boldsymbol{\theta})$ were studied in a general form in Pavlov (1983a, 1985, 1987a, 1990, 1993), as applications to reliability problems in Pavlov (1982a, 1983b,

1984a,b, 1986, 1987b), and, in particular, for renewal systems in Pavlov (1988) and Pavlov and Ushakov (1989).

Another direction of the generalization of Wald's criterion is in the study of more general test plans, differing from (9.8), including dependent tests. This direction is extremely important for reliability problems because the test scheme of identical and independent objects applies to a very restricted class of real tasks (see Examples 9.1–9.8). Problems of sequential testing of composite hypotheses of type (9.70) and construction of confidence limits for reliability indices for dependent tests (for Markov-type dependence) were considered by Pavlov (1982b, 1983b, 1985).

Let us first consider the sequential test of composite hypotheses of type (9.70) for the simple case of independent identical tests.

9.3.1 Sequential Rules of Composite Hypothesis Exclusion

Consider the sequential test of composite hypotheses for the multidimensional parameter $\boldsymbol{\theta} = (\theta_1, \ldots, \theta_m)$ for a simple scheme of independent and identical sequential tests (9.8). Let there be a composite hypothesis of the type

$$H_0: \boldsymbol{\theta} \in D, \tag{9.71}$$

where D is an area in the space parameter $\boldsymbol{\theta}$. Determine for any hypothesis of type (9.71) some random moment (step) ν that is called the moment (or the rule) of exclusion of this hypothesis. In other words, at moment ν one makes a decision that hypothesis H is not true. This moment ν is determined during the test process,

$$x_1, x_2, \ldots, x_n, \ldots, \tag{9.72}$$

in dependence on the current test results. It means that ν is a random Markovian moment, that is, a random moment independent of the future. (For more details see Section 3.8.1.)

For the moment ν, excluding hypothesis (9.71), we require that the condition

$$P_\theta(\nu < \infty) \leq \alpha \quad \text{for any } \boldsymbol{\theta} \in D \tag{9.73}$$

holds, where α is a given small constant (probability of error). Event $\{\nu < \infty\}$ means that hypothesis H is excluded at some step ν. So, condition (9.73) has the following meaning: If hypothesis H is true (the true value of parameter θ belongs to area D), then the probability of excluding this hypothesis by test results (9.72) at some future time does not exceed constant α.

Condition (9.73) does not guarantee usefulness of the moment ν. Indeed, a trivial exclusion moment $\nu = \infty$, for instance, satisfies this condition: Hypothesis H never excludes, that is, it is considered as true for any test results. Therefore, in addition to (9.73), we require that the condition

$$\text{for } \boldsymbol{\theta} \notin D, \quad E_{\theta}\nu \to \min \tag{9.74}$$

must be held. In other words, if this hypothesis is not true ($\boldsymbol{\theta} \notin D$), we expect the mathematical expectation $E_{\theta}\nu$ of the moment of exclusion for hypothesis to be small.

It can be shown (see Section 9.6.3) that from (9.73) it follows that the mathematical expectation $E_{\theta}\nu$ of the moment of hypothesis (9.71) must satisfy the inequality

$$E_{\theta}\nu \geq \frac{\ln(1/\alpha)}{\rho(\theta, D)} \quad \text{for all } \boldsymbol{\theta} \notin D, \tag{9.75}$$

where

$$\rho(\boldsymbol{\theta}, D) = \min_{\boldsymbol{\theta}' \notin D} \rho(\boldsymbol{\theta}, \boldsymbol{\theta}') \tag{9.76}$$

and $\rho(\boldsymbol{\theta}, \boldsymbol{\theta}')$ is the Kullback–Leibler "information distance" between $\boldsymbol{\theta}$ and $\boldsymbol{\theta}'$:

$$\rho(\boldsymbol{\theta}, \boldsymbol{\theta}') = E_{\theta} \ln \left[\frac{f(x, \boldsymbol{\theta})}{f(x, \boldsymbol{\theta}')} \right].$$

The value $\rho(\boldsymbol{\theta}, D)$ represents the Kullback–Leibler information distance between $\boldsymbol{\theta}$ and D. Notice that $\rho(\boldsymbol{\theta}, \boldsymbol{\theta}') = 0$ if $\boldsymbol{\theta} = \boldsymbol{\theta}'$ and $\rho(\boldsymbol{\theta}, \boldsymbol{\theta}') > 0$ if $\boldsymbol{\theta} \neq \boldsymbol{\theta}'$. Therefore, $\rho(\boldsymbol{\theta}, D) = 0$ if $\boldsymbol{\theta} \in D$ and $\rho(\boldsymbol{\theta}, D) \geq 0$ if $\boldsymbol{\theta} \notin D$. So, (9.75) delivers the lower limit for the mathematical expectation $E_{\theta}\nu$ for any (Markovian) moment ν that satisfies (9.74). Following Pavlov (1985, 1990), consider the rule of construction of ν where the lower limit on the right side of (9.75) is asymptotically reached for $\alpha \to \infty$. It allows one to obtain asymptotic solution of the problem of optimization (9.74).

The moment of hypothesis H: $\theta \in D$ exclusion is determined as

$$\nu(\alpha, D) = \min \left\{ n: \prod_{1 \leq r \leq n} f(x_r, \hat{\theta}_{r-1}) \geq \frac{1}{\alpha} \max_{\boldsymbol{\theta} \in D} \prod_{1 \leq r \leq n} f(x_r, \boldsymbol{\theta}) \right\}, \tag{9.77}$$

where $\hat{\boldsymbol{\theta}} = \hat{\boldsymbol{\theta}}(x_1, \ldots, x_n)$ is the point estimate of parameter $\boldsymbol{\theta}$ on the basis of test results $x_1, \ldots, x_n$ on the first n steps. The moment determined in such a way satisfies the inequality

$$P_{\theta}\{\nu(\alpha, D) < \infty\} \leq \alpha \tag{9.78}$$

for any H: $\boldsymbol{\theta} \in D$. (See the proof in Section 9.6.3, Lemmas 9.1 and 9.2.) In addition to the mathematical expectation of this moment, $E_\theta \nu(\alpha, D)$ for some general conditions reaches the lower limit in (9.75) asymptotically for $\alpha \to 0$, namely,

$$E_\theta \nu(\alpha, D) = \frac{\ln(1/\alpha)}{\rho(\boldsymbol{\theta}, D)}(1 + \varepsilon_\alpha)$$

for any $\boldsymbol{\theta} \notin D$, where $\varepsilon_\alpha \to 0$ if $\alpha \to 0$. (See the proof in Lemmas 9.1 and 9.2, Section 9.6.3.) Thus for a small probability of error α, the approximate formula for the average moment of exclusion for hypothesis H when it is not true (i.e., for $\boldsymbol{\theta} \in D$) is

$$E_\theta \nu(\alpha, D) \approx \frac{\ln(1/\alpha)}{\rho(\boldsymbol{\theta}, D)}. \tag{9.79}$$

The right side of this approximation coincides with the right side of inequality (9.75), so moment (9.77) is approximately (for $\alpha << 1$) optimal.

With construction of the moment of exclusion $\nu(\alpha, D)$ by rule (9.77), we can solve problems of tests for composite hypotheses.

9.3.2 Sequential Criterion for Composite Hypothesis Testing

Suppose we need to test two composite hypotheses of the type (9.70) for some function (reliability index) $R(\boldsymbol{\theta})$ of vector parameter $\boldsymbol{\theta} = (\theta_1, \ldots, \theta_m)$. These hypotheses can be written in the form

$$H_0: \boldsymbol{\theta} \in D_0, \qquad H_1: \boldsymbol{\theta} \in D_1, \tag{9.80}$$

where D_0 and D_1 are areas in the parameter space determined as $D_0 = \{\boldsymbol{\theta}: R(\boldsymbol{\theta}) \le R_0\}$ and $D_1 = \{\boldsymbol{\theta}: R(\boldsymbol{\theta}) \ge R_1\}$. The intermediate area $I = \{\boldsymbol{\theta}: R_0 < R(\boldsymbol{\theta}) < R_1\}$ is called the area of indifference.

Let us construct moments of exclusion for hypotheses H_0 and H_1, respectively, on the basis of (9.77),

$$\nu_0 = \nu(\alpha, D_0), \qquad \nu_1 = \nu(\beta, D_1), \tag{9.81}$$

where α and β are given risks of types I and II. The sequential criterion for tests of these hypotheses is determined as follows. Let us continue the test until such moment (step) when one of the hypotheses will be excluded, that is, until moment

$$\nu = \min(\nu_0, \nu_1). \tag{9.82}$$

At this moment the test is stopped and the remaining hypothesis is accepted. In other words, the rule of decision making at the stopping moment (9.82) has the form

$$\begin{aligned}&\text{If } \nu = \nu_1 < \nu_0, \text{ then hypothesis } H_0 \text{ is accepted.}\\&\text{If } \nu = \nu_1 > \nu_0, \text{ then hypothesis } H_1 \text{ is accepted.}\end{aligned} \tag{9.83}$$

(If $\nu_1 = \nu_0$, i.e., both hypotheses are excluded simultaneously, either of these hypotheses may be accepted.) In accordance with (9.77) the sequential criterion (9.82)–(9.83), determined as shown above, can be formulated as follows. The area of test continuation is given by the inequalities

$$U_n - V_{0n} < \ln\left(\frac{1}{\alpha}\right), \qquad U_n - V_{1n} < \ln\left(\frac{1}{\beta}\right), \tag{9.84}$$

where

$$U_n = \sum_{1\le r\le n} \ln f(x_r, \hat{\theta}_{r-1}) \tag{9.85}$$

$$V_{in} = \max_{\boldsymbol{\theta}\in D_i} \sum_{1\le r\le n} \ln f(x_r, \boldsymbol{\theta}), \qquad i = 0, 1. \tag{9.86}$$

The test continues until the first violation of at least one of the inequalities in (9.84). If the first one is violated, then hypothesis H_1 is accepted. If the second one is violated, then hypothesis H_0 is accepted.

It is easy to find that the sequential criterion constructed in such a way for test hypotheses (9.80) has risks of types I and II not larger than α and β, respectively. Indeed, let $\boldsymbol{\theta} \in D_0$, that is, hypothesis H_0 is true. Then, by construction of the criterion, the probability of erroneous decision (acceptance of hypothesis H_1) satisfies the inequality

$$P_\theta\{\text{to accept } H_1\} \le P_\theta(\nu_0 < \infty) \le \alpha.$$

Analogously, let $\boldsymbol{\theta} \in D_1$, that is, hypothesis H_1 is true. Then the probability of erroneous acceptance of hypothesis H_0 satisfies the inequality

$$P_\theta\{\text{to accept } H_0\} \le P_\theta(\nu_1 < \infty) \le \beta.$$

(For more details see Section 9.6.3.) In addition, this sequential criterion for $\alpha \to 0$ and $\beta \to 0$ is asymptotically optimal by average test volume $N(\boldsymbol{\theta}) = E_\theta \nu$ for all $\boldsymbol{\theta}$. The average test volume $N(\boldsymbol{\theta})$ for small values of α and β can be estimated by the following approximate formulas (see Section 9.6.3, Theorem 9.2):

$$N(\boldsymbol{\theta}) \approx \frac{\ln(1/\alpha)}{\rho(\boldsymbol{\theta}, D_1)} \quad \text{for } \boldsymbol{\theta} \in D_0, \tag{9.87}$$

$$N(\boldsymbol{\theta}) \approx \frac{\ln(1/\alpha)}{\rho(\boldsymbol{\theta}, D_0)} \quad \text{for } \boldsymbol{\theta} \in D_1, \tag{9.88}$$

$$N(\boldsymbol{\theta}) \approx \frac{\ln(1/\alpha)}{\Lambda(\boldsymbol{\theta})} \quad \text{for } \boldsymbol{\theta} \in I, \tag{9.89}$$

where

$$\Lambda(\boldsymbol{\theta}) = \max[\rho(\boldsymbol{\theta}, D_0), \rho(\boldsymbol{\theta}, D_1)].$$

9.4 SEQUENTIAL CONFIDENCE LIMITS

9.4.1 Construction of Sequential Confidence Limits

Let $R(\boldsymbol{\theta}) = R(\theta_1, \ldots, \theta_m)$ be some function (reliability index) depending on the m-dimensional vector parameter $\boldsymbol{\theta} = (\theta_1, \ldots, \theta_m)$ whose true value is unknown. Assume that, during the test, results are obtained at each step n. On the basis of test results $x_1, \ldots, x_n, \ldots$, we construct the confidence limit (e.g., lower) $\underline{R}_n = \underline{R}_n(x_1, \ldots, x_n)$ for index $R(\boldsymbol{\theta})$. The random sequence

$$\underline{R}_n = \underline{R}_n(x_1, \ldots, x_n), \qquad n = 1, 2, \ldots \tag{9.90}$$

is called the *sequential* γ-*LCL* for index R($\boldsymbol{\theta}$) if it satisfies the following conditions:

A. Sequence $\underline{R}_n$ is monotone nondecreasing in n,

$$\underline{R}_1 \leq \underline{R}_2 \leq \cdots \leq \underline{R}_n \leq \underline{R}_{n+1} \leq \ldots,$$

for any test results $(x_1, \ldots, x_n, \ldots)$.

B. For any Markovian stopping moment ν such that $P_\theta(\nu < \infty) = 1$, the inequality

$$P_\theta\{\underline{R}_\nu \leq R(\boldsymbol{\theta})\} \geq \gamma \tag{9.91}$$

holds for all $\boldsymbol{\theta}$, where $\underline{R}_\nu = \underline{R}_\nu(x_1, \ldots, x_\nu)$ is the LCL (9.90) calculated at the stopping moment ν. Notice that due to the monotonicity of sequence $\underline{R}_n$ condition (b) can be substituted by the weaker condition.

C. For any fixed step n the inequality

$$P_\theta\{\underline{R}_n \leq R(\boldsymbol{\theta})\} \geq \gamma, \qquad n = 1, 2, \ldots, \tag{9.92}$$

holds for all $\boldsymbol{\theta}$.

Indeed, from monotonicity of sequence $\underline{R}_n$ it follows that there exists (for any test results $x_1, \ldots, x_n, \ldots$) a limit for $n \to \infty$,

$$\underline{R} = \lim \underline{R}_n,$$

and due to (9.92) this limit satisfies the inequality

$$P_\theta\{\underline{R} \leq R(\boldsymbol{\theta})\} = \lim P_\theta\{\underline{R}_n \leq R(\boldsymbol{\theta})\} \geq \gamma,$$

and, taking into account the monotone decreasing sequence of $\underline{R}_n$, it follows that for arbitrary (not necessarily Markovian) stopping moment ν inequalities $\underline{R}_\nu \leq \underline{R}$ and

$$P_\theta\{\underline{R}_\nu \leq R(\boldsymbol{\theta})\} \geq P_\theta\{\underline{R} \leq R(\boldsymbol{\theta})\} \geq \gamma$$

hold. The latter inequality is true for any moment ν such that $P_\theta(\nu < \infty) < 1$ if $\underline{R}_\infty = \lim \underline{R}_n = \underline{R}$. Thus, if sequence (9.90) is monotone nondecreasing (in all $x_1, \ldots, x_n, \ldots$), then for demonstration, that (b) is true, it is enough to prove condition (c).

Let us now construct the sequential confidence limit for $R(\boldsymbol{\theta})$ based on the exclusion moment $\nu(\alpha, D)$ for the composite hypotheses given in (9.77). For this purpose, consider the composite hypothesis

$$H_c\colon R(\boldsymbol{\theta}) \leq C \tag{9.93}$$

where C is a constant. The exclusion moment for this hypothesis is denoted as

$$\nu_c = \nu(\alpha, D_c), \tag{9.94}$$

where D_c is the area in space of parameter $\boldsymbol{\theta}$ of the form $D_c = \{\boldsymbol{\theta}\colon R(\boldsymbol{\theta}) \leq C\}$. The LCL $\underline{R}_n = \underline{R}_\nu(x_1, \ldots, x_n)$ for $R(\boldsymbol{\theta})$ at step n is defined as

$$\underline{R}_n = \max\{C\colon \nu_c \leq n\}, \qquad n = 1, 2, \ldots, \tag{9.95}$$

that is, as the maximum value C for which a hypothesis of type (9.93) is found untrue at step n.

It is easy to find that r.v.'s $\underline{R}_n$, $n = 1, 2, \ldots$, defined in (9.95) form γ-LCL for $R(\boldsymbol{\theta})$, where $\gamma = 1 - \alpha$. Indeed, let $\boldsymbol{\theta}'$ be an arbitrary point in the space of parameters. Denote $b = R(\boldsymbol{\theta}')$. For each fixed n due to (9.95), the following relation is valid:

$$\{\nu_b > n\} \subset \{\underline{R}_n \leq b\}. \tag{9.96}$$

Taking into account that $\boldsymbol{\theta}'$ belongs to area $D_b = \{\boldsymbol{\theta}: R(\boldsymbol{\theta}) \leq b\}$ and due to (9.78), the following inequality holds:

$$\begin{aligned} P_{\theta'}(\nu_b > n) &= 1 - P_{\theta'}(\nu_b \leq n) \\ &\geq 1 - P_{\theta'}(\nu_b < \infty) = 1 - P_{\theta'}\{\nu(\alpha, D_b) < \infty\} \geq 1 - \alpha. \end{aligned}$$

From here, due to (9.96), it follows that

$$\begin{aligned} P_{\theta'}\{\underline{R}_n \leq R(\boldsymbol{\theta}')\} &= P_{\theta'}(\underline{R}_n \leq b) \\ &\geq P_{\theta'}(\nu_b > n) \geq 1 - \alpha \end{aligned}$$

for each fixed $n = 1, 2, \ldots$ (i.e., condition C holds). The monotone nondecreasing sequence of $\underline{R}_n$ (condition A) follows directly from its definition (9.95). Thus r.v.'s (9.95) form sequential $(1 - \alpha) -$ LCL for $R(\boldsymbol{\theta})$.

Analogously define the UCL. The sequence of r.v.'s

$$\overline{R}_n = \overline{R}_n(x_1, \ldots, x_n), \qquad n = 1, 2, \ldots.$$

is called the sequential γ-UCL for $R(\boldsymbol{\theta})$ if the following conditions hold:

A′. The sequence of $\overline{R}_n$ is monotone nonincreasing:

$$\overline{R}_1 \geq \overline{R}_2 \geq \cdots \geq \overline{R}_n \geq \overline{R}_{n+1} \geq \cdots$$

for any test results $x_1, \ldots, x_n, \ldots$.

B′. For any (Markovian) stop test ν such that $P_\theta(\boldsymbol{\theta} < \infty) = 1$, the inequality

$$P_\theta\{\overline{R}_\nu \geq R(\boldsymbol{\theta})\} \geq \gamma$$

holds for all $\boldsymbol{\theta}$ where

$$\overline{R}_\nu = \overline{R}_\nu(x_1, \ldots, x_\nu).$$

Analogously, we can show that if the sequence $\overline{R}_n$, $n = 1, 2, \ldots$, satisfies the first condition of monotonicity A', then the second condition of monotonicity B is weaker.

C′. For arbitrary fixed $n = 1, 2, \ldots$, the inequality

$$P_\theta\{\overline{R}_n \geq R(\boldsymbol{\theta})\} \geq \gamma$$

holds for all $\boldsymbol{\theta}$.

In an analogous way we define the two-sided sequence γ-confidence interval $(\underline{R}_n, \overline{R}_n)$ for $R(\boldsymbol{\theta})$. It is easy to show that if r.v.'s $\underline{R}_n$ and $\overline{R}_n$ form, respectively, $(1 - \alpha)$-*LCL* and $(1 - \beta)$-*UCL* for $R(\boldsymbol{\theta})$, then $(\underline{R}_n, \overline{R}_n)$ form a sequential $(1 - \alpha - \beta)$-confidence interval for $R(\boldsymbol{\theta})$.

To construct a sequential $(1 - \beta)$-UCl for $R(\boldsymbol{\theta})$, consider again composite hypotheses of the type

$$H_c'\colon R(\boldsymbol{\theta}) \geq C. \tag{9.97}$$

The exclusion moment for hypothesis (9.97) is denoted by

$$\nu_c' = \nu(\beta, D_c'), \tag{9.98}$$

where $D_c' = \{\boldsymbol{\theta}\colon R(\boldsymbol{\theta}) \geq C\}$. At test step n set

$$\overline{R}_n = \min\{C\colon \nu_c' \leq n\}, \qquad n = 1, 2, \ldots. \tag{9.99}$$

The random sequence defined in such a way forms a sequential $(1 - \beta)$-UCL for $R(\boldsymbol{\theta})$. The proof of this is similar to the one given above for (9.95).

On the basis of the definition of the exclusion moment $\nu(\alpha, D)$ in (9.77), the sequential γ-LCL (9.95) and γ-UCL (9.99) can be written in more detail as

$$\underline{R}_n = \max_{1\leq k\leq n} \underline{R}_{k'}', \qquad \overline{R}_n = \min_{1\leq k\leq n} \overline{R}_{k'}', \tag{9.100}$$

where $\underline{R}_n'$ and $\overline{R}_n'$ are determined at test step n from the equations

$$V_{0n}(C) = U_n - \ln\left(\frac{1}{\alpha}\right), \qquad V_{1n}(C) = U_n - \ln\left(\frac{1}{\beta}\right), \tag{9.101}$$

where U_n is determined above in (9.85),

$$V_{0n}(C) = \max_{R(\theta)\le C} \sum_{1\le r\le n} \ln f(x_r, \boldsymbol{\theta}),$$
$$V_{1n}(C) = \max_{R(\theta)\ge C} \sum_{1\le r\le n} \ln f(x_r, \boldsymbol{\theta}). \tag{9.102}$$

Sequential confidence limits constructed in (9.100)–(9.102) are asymptotically optimal for $\alpha \to 0$ and $\beta \to 0$ in the sense of the average time to reach the specified accuracy (see Section 9.6.3).

9.4.2 Test Reliability Hypotheses with Sequential Confidence Limits

Suppose we need to test, on the basis of sequential tests (9.8), two composite hypotheses of the type

$$H_0: R(\boldsymbol{\theta}) \le R_0, \qquad H_1: R(\boldsymbol{\theta}) \ge R_1,$$

where $R_0 < R_1$ are given critical levels of index R. Let α and β be given risks of types I and II. One can see directly from (9.81)–(9.86), (9.94), (9.95), and (9.98)–(9.102) that the sequential criterion, constructed in Section 9.3, with risks α and β for hypotheses H_0 and H_1 can be written via sequential confidence limits as follows. The area of test continuation at step n is given by the two inequalities

$$\underline{R}_n < R_0, \qquad R_1 < \overline{R}_n, \tag{9.103}$$

where $\underline{R}_n$ is the sequential $(1 - \alpha)$-LCL and $\overline{R}_n$ is the sequential γ-UCL for $R(\boldsymbol{\theta})$ constructed in (9.100). The test continues, that is, we go to the next step $n + 1$ if at step n both inequalities in (9.103) hold. The test is stopped when at least one of the inequalities in (9.103) is violated. If the first inequality is violated, the hypothesis H_1 is accepted. If the second inequality is violated, the hypothesis H_0 is accepted.

In other words, based on sequential confidence limits, this decision rule means the following. Let, for instance, $R = R(\boldsymbol{\theta})$ be an index of type of failure probability or failure rate. Then the critical levels $R_0 < R_1$ are, respectively, equivalent to the levels of acceptance R_0 and rejection R_1. The test continues if the confidence interval $(\underline{R}_n, \overline{R}_n)$ at step n contains both critical levels R_0 and R_1. If the $(1 - \alpha)$-LCL $\underline{R}_n$ crosses the level of acceptance R_0, then the test is stopped and hypothesis H_1 is assumed true (rejection). If the $(1 - \beta)$-UCL $\overline{R}_n$ crosses the level of rejection R_1, then the test is stopped and hypothesis H_0 is assumed true (acceptance).

Remark 9.2 This sequential decision rule preserves its meaning in the case where the intermediate area of indifference I is absent, that is, if critical levels coincide, $R_0 = R_1$. In this case risks of types I and II, α and β, are preserved but the average test volume increases if the value of index $R(\boldsymbol{\theta}) \to R_0$, that is, $N(\boldsymbol{\theta}) \to \infty$ if $R(\boldsymbol{\theta}) \to R_0$, as can be seen from (9.87)–(9.89). (For details, see Section 9.6.3.) □

9.4.3 Scheme of Dependent Tests

Consider a case where the test process $x_0, x_1, \ldots, x_n, \ldots$ represents a homogeneous Markov chain "starting" from the initial state x_0. Let $f(x/y, \boldsymbol{\theta})$ be the transitive density of distribution, in other words, the density of r.v. x_{n+1} observed at step $n+1$ for a given value of parameter $\boldsymbol{\theta} = (\theta_1, \ldots, \theta_m)$ and under the condition that at step n the value of r.v. $x_n = y$ was observed.

Consider again the composite hypothesis

$$H\colon \boldsymbol{\theta} \in D, \tag{9.104}$$

where D is some area in the parameter space $\boldsymbol{\theta}$. The moment of exclusion of hypothesis (9.104) is determined analogously to (9.77) in the scheme with independent tests, namely, as

$$\nu(\alpha, D) = \min\left\{n\colon \prod_{1\le r\le n} f(x_r|x_{r-1}, \hat{\theta}_{r-1}) \ge \frac{1}{\alpha} \max_{\theta\in D} \prod_{1\le r\le n} f(x_r|x_{r-1}, \theta_{r-1})\right\}, \tag{9.105}$$

where $\hat{\theta}_n = \hat{\theta}(x_1, \ldots, x_n)$ is again the point estimate of parameter $\boldsymbol{\theta}$ based on the test results $x_1, \ldots, x_n$ at the first n steps. It is possible to show (see Section 9.6.4) that this moment, as in the scheme of independent tests above, satisfies the inequality

$$P_\theta\{\nu(\alpha, D) < \infty\} \le \alpha \quad \text{for all } \boldsymbol{\theta} \in D. \tag{9.106}$$

Then, sequential confidence limits for $R = R(\boldsymbol{\theta})$ are constructed on the basis of the moment of exclusion of composite hypotheses (9.105) by previous formulas (9.94) and (9.95) for the lower limit and by formulas (9.98) and (9.99) for the upper limit. The lower and upper limits at test step n are determined analogously to (9.100)–(9.102) by the formulas

$$\underline{R}_n = \max_{1\le k\le n} \underline{R}'_k, \qquad \overline{R}_n = \max_{1\le k\le n} \overline{R}'_k, \tag{9.107}$$

where $\underline{R}'_n$ and $\overline{R}'_n$ are determined at test step n, respectively, for the following levels of C:

$$V_{0n}(C) = U_n - \ln\left(\frac{1}{\alpha}\right), \qquad V_{1n}(C) = U_n - \ln\left(\frac{1}{\beta}\right), \tag{9.108}$$

where

$$U_n = \sum_{1 \le r \le n} \ln f(x_r | x_{r-1}, \hat{\theta}_{r-1}) \tag{9.109}$$

$$V_{0n}(C) = \max_{R(\theta) \le C} \sum_{1 \le r \le n} \ln f(x_r | x_{r-1}, \boldsymbol{\theta}), \tag{9.110}$$

$$V_{1n}(C) = \max_{R(\theta) \ge C} \sum_{1 \le r \le n} \ln f(x_r | x_{r-1}, \boldsymbol{\theta}).$$

The random sequences $\underline{R}_n$ and $\overline{R}_n$, $n = 1, 2, \ldots,$ constructed in such a way form sequential $(1 - \alpha)$-LCL and sequential $(1 - \beta)$-UCL for $R(\boldsymbol{\theta})$. (The proof of this fact coincides with that given in Section 9.4 for the scheme of independent tests.)

9.5 SEQUENTIAL CONFIDENCE LIMITS AND TESTS OF HYPOTHESES FOR SYSTEMS ON THE BASE OF UNIT TESTS

Consider now some application of the results obtained above. Begin with a simplest "binomial model" of system unit tests (see Example 9.7).

9.5.1 Binomial Model

Let a system consist of m units of different types. At the current nth test step (cycle) N_i units of type i are simultaneously tested. Each unit has reliability parameter (failure probability) q_i, $1 \le i \le m$. Failures of different units are independent. As the results of the test, we observe a sequence of i.i.d. random vectors

$$\mathbf{x}_1, \mathbf{x}_2, \ldots, \mathbf{x}_n, \ldots, \tag{9.111}$$

where random vector $\mathbf{x}_n$ represents a set of all numbers of failures for units of different types that were observed at step n:

$$\mathbf{x}_n = (d_{1n}, d_{2n}, \ldots, d_{mn}), \qquad n = 1, 2, \ldots,$$

where d_{in} is the number of failures of N_i tested units of type i at step (cycle) n. Thus, r.v. d_{in} has binomial distribution with parameters (N_i, q_i) where the value of parameter q_i is unknown. This model, obviously, is a multidimensional analogue of the classical scheme of sequential independent Bernoulli trials. The Bernoulli scheme is a particular case of this general scheme for $N_i = 1$ and $m = 1$.

We need, on the basis of test results (9.111), to construct confidence limits or to test hypotheses about some system reliability index

$$R = R(\mathbf{q}) = R(q_1 1, \ldots, q_m),$$

where $\mathbf{q} = (q_1, \ldots, q_m)$ is a vector of unknown parameters of unit reliability.

Denote $M_{in} = d_{i1} + \cdots + d_{in}$ the total number of units i failed during n test steps, and

$$\hat{q}_i(n) = \frac{M_{in} + r_i}{n + k_i} \tag{9.112}$$

is the point estimate of parameter q_i at step n. Values k_i and r_i take into account some prior information. For instance, if there is information about previous tests, then k_i might be the number of tested units and r_i the number of observed failures, $1 \le i \le m$. Denote $\hat{p}_i(n) = 1 - \hat{q}_i(n)$ the point estimate for parameter $p_i = 1 - q_i$ at test step n.

Applying general formulas (9.100)–(9.102), we obtain that sequential $(1 - \alpha)$-LCL $\underline{R}_n$ and $(1 - \beta)$-UCL $\overline{R}_n$ for parameter $R = R(\mathbf{q})$ can be found at the nth test step as

$$\underline{R}_n = \max_{1 \le l \le n} \underline{R}'_l, \qquad \overline{R}_n = \min_{1 \le l \le n} \overline{R}'_l, \tag{9.113}$$

where values $\underline{R}'_n$ and $\overline{R}'_n$ are found at step n from solution of the following equations with respect to C:

$$V_{0n}(C) = U_n - \ln\left(\frac{1}{\alpha}\right), \qquad V_{1n}(C) = U_n - \ln\left(\frac{1}{\beta}\right), \tag{9.114}$$

where

$$U_n = \sum_{1 \le i \le m} \sum_{1 \le r \le n} \{d_{ir} \ln \hat{q}_i \cdot (r - 1) + (N_i - d_{ir}) \ln \hat{p}_i \cdot (r - 1)\},$$

$$V_{0n}(C) = \max_{R(\mathbf{q}) \le C} H_n(\mathbf{q}), \qquad V_{1n}(C) = \max_{R(\mathbf{q}) \ge C} H_n(\mathbf{q}),$$

$$H_n(\mathbf{q}) = \sum_{1 \le i \le m} \{M_{in} \ln q_i + (nN_i - M_{in}) \ln(1 - q_i)\}.$$

Suppose we need to test two composite hypotheses on the basis of test results (9.111):

$$H_0: R(\mathbf{q}) \leq R_0, \qquad H_1: R(\mathbf{q}) \geq R_1.$$

The sequential criterion with given risks of types I and II (α and β) for hypotheses H_0 and H_1 is constructed in accordance with (9.84)–(9.86) as follows. The area of test continuation on step n is given by the inequalities

$$U_n - V_{0n}(R_0) < \ln\left(\frac{1}{\alpha}\right), \qquad U_n - V_{1n}(R_1) < \ln\left(\frac{1}{\beta}\right). \tag{9.115}$$

The test continues if both inequalities (9.15) hold. The test is stopped if at least one of the inequalities is violated. The violation of the first one leads to acceptance of hypothesis H_1, and of the second one to acceptance of hypothesis H_0.

In accordance with (9.113), this sequential criterion can also be written in a more compact form via sequential confidence limits (9.113). Namely, the area of test continuation at step n is given by the inequalities

$$\underline{R}_n < R_0, \qquad R_1 < \overline{R}_n, \tag{9.116}$$

which are equivalent to previous inequalities (9.115). The decision rule is formulated in this case similarly. The test continues if both inequalities (9.16) hold and stops if at least one of these inequalities is violated. The violation of the first one leads to acceptance of hypothesis H_1, and of the second one to acceptance of hypothesis H_0.

This sequential criterion is asymptotically optimal for $\alpha \to 0$ and $\beta \to 0$ in the sense of the average test volume (see Section 9.3.6, Theorem 9.2). Let us write formulas for the average test volume, assuming for the sake of simplicity that $\alpha = \beta$. Let ν be the moment of the first violation of the inequalities above and $N(\mathbf{q}) = E_q\nu$ the average test volume for a given vector of parameters $\mathbf{q}$. In accordance with (9.87)–(9.89), the average test volume, $N(\mathbf{q})$, can be estimated for $\alpha = \beta \to 0$ from the formulas

$$N(\mathbf{q}) \approx \begin{cases} \dfrac{\ln(1/\alpha)}{\Lambda_1(\mathbf{q})} & \text{for } R(\mathbf{q} \leq R_0, \\ \dfrac{\ln(1/\alpha)}{\Lambda(\mathbf{q})} & \text{for } R_0 < R(\mathbf{q}) < R_1, \\ \dfrac{\ln(1/\alpha)}{\Lambda_0(\mathbf{q})} & \text{for } R(\mathbf{q}) \geq R_1, \end{cases}$$

where

$$\Lambda(\mathbf{q}) = \max[\Lambda_0(\mathbf{q}), \Lambda_1(\mathbf{q})],$$

$$\Lambda_0(\mathbf{q}) = \max_{R(\mathbf{q}')\le R_0} \rho(\mathbf{q}, \mathbf{q}'), \qquad \Lambda_1(\mathbf{q}) = \max_{R(\mathbf{q}')\ge R_1} \rho(\mathbf{q}, \mathbf{q}'),$$

$$\rho(\mathbf{q}, \mathbf{q}') = \sum_{1\le i\le m} N_i \left[q_i \ln\left(\frac{q_i}{q_i'}\right) + (1 - q_i) \ln\left(\frac{1 - q_i}{1 - q_i'}\right)\right],$$

$$\mathbf{q} = (q_i, \ldots, q_m), \qquad \mathbf{q}' = (q_1', \ldots, q_m').$$

In practice, for often used $0.01 \le \alpha \le 0.2$, these approximate formulas give satisfactory estimates for average test volume. A more precise value of the average test volume $N(\mathbf{q})$ can be estimated with the Monte Carlo method (see Examples 9.13 and 9.14).

Gain in Average Test Volume with Unit Test Information Consider a series–parallel system (see Example 9.7) characterized by

$$R = R(\mathbf{q}) = 1 - \prod_{1\le i\le m} (1 - q_i^{N_i}), \tag{9.117}$$

where N_i is the number of parallel units in the ith redundant group, $1 \le i \le m$. The test of hypotheses H_0 and H_1 for (9.117) can be performed on the basis of test results for the system, that is, on the basis of the sequence

$$\delta_1, \delta_2, \ldots, \delta_n, \ldots, \tag{9.118}$$

where δ_n is the system failure indicator at step n: $\delta_n = 1$ if at least one redundant group has failed at the step n, that is, at least for one i, $1 \le i \le m$, $d_{in} = N_i$. Otherwise, $\delta_n = 0$.

In this case the problem is reduced to the standard test of hypothesis of type

$$H_0\colon R \le R_0, \qquad H_1\colon R \ge R_1$$

for binomial parameter (failure probability) R on the basis of results of Bernoulli trials (9.118). This problem can be solved on the basis of the sequential Wald criterion (see Section 9.2 and Example 9.9). The average test volume in this case is estimated with the known Wald's formulas (9.63).

The serious deficiency of such a simple method, based directly on system failures (9.118), is that all information about the unit's failures is not taken into account. In is clear on an intuitive level that this fact should lead to increase in the average test volume, especially for highly reliable systems. For instance, in this case we consider as equivalent cases where at the test step n we observe $d_{in} = 0$ or $d_{in} = N_i - 1$ though these two cases are quite different.

Below we suggest numerical examples and compare average test volume (for the same values of R_0, R_1, α, and β) for the sequential Wald's criterion based on the system's failures (9.118) with the sequential method (9.115)–(9.116) based on complete information (9.111) taking into account complete data of the unit's failures.

From these examples one can see that sequential criterion (9.115)–(9.116) gives a substantial gain in average test volume.

Example 9.13 (Series–Parallel System) Consider a series–parallel system with reliability index of form (9.117). The system consists of $m = 2$ parallel subsystems connected in series. The number of units within the subsystems are $N_1 = N_2 = 2$. The critical levels are $R_0 = 0.05$ and $R_1 = 0.2$, and risks of types I and II are $\alpha = \beta = 0.1$. Using formula (9.63), we obtain that for $R = R_0$ the average test volume for Wald's criterion based on the system's failures is

$$N_W(R_0) \approx \frac{\omega(\alpha, \beta)}{\rho(R_0, R_1)} = \frac{(1 - \alpha)\ln[(1 - \alpha)/\beta] + \alpha \ln[\alpha/(1 - \beta)]}{R_0 \ln(R_0/R_1) + (1 - R_0)\ln[(1 - R_0)/(1 - R_1)]}$$

$$= \frac{0.9 \ln 9 - 0.1 \ln 9}{0.95 \ln(0.95/0.80) - 0.05 \ln 4} = 18.9. \ \square$$

For the sequential criterion (9.115)–(9.116), based on complete information, the average test volume $N(\mathbf{q})$ from Monte Carlo simulation is 8.3 for the value of reliability index $R(\mathbf{q}) = R(q_1, q_2) = R_0$. [The average test volume was estimated for a symmetrical point $q_1 = q_2$, $R(q_1, q_2) = R_0$, that is, for equally reliable units for both subsystems.] Thus, in this case, the average test volume can be decreased more than two times due to use of sequential criterion (9.115)–(9.116) taking into account complete information. □

Example 9.14 (Parallel System) Consider a parallel system consisting of $m = 2$ units of different types. The reliability index (in this case the failure probability) has the form

$$R = R(\mathbf{q}) = q_1 q_2, \tag{9.119}$$

where q_i is the failure probability of the ith unit, $\mathbf{q} = (q_1, q_2)$ is the vector of unit parameters. As in the previous example, at each step a single system is tested. Let the number of tested units of the first and second types be $N_1 = N_2 = 1$. The vector of test results, x_n, at step n has the form

$$x_n = (d_{1n}, d_{2n}), \qquad n = 1, 2, \ldots, \tag{9.120}$$

where $d_{in} = \{0, 1\}$ is the failure indicator for a unit of the ith type at step n. The system failure indicator at step n equals $\delta = 1$ if $d_{1n} = d_{2n} = 1$—that is, both units have been failed—and $\delta = 0$ otherwise. Suppose we need to test two composite hypotheses for reliability index (9.119):

$$H_0: R \leq R_0, \qquad H_1: R \geq R_1,$$

where critical levels $R_0 = 0.01$ and $R_1 = 0.05$. The given risks of types I and II are $\alpha = \beta = 0.05$.

Tests of hypotheses H_0 and H_1 can be done either by Wald's criterion applied to the sequence of system failure indicators

$$\delta_1, \delta_2, \ldots, \delta_n, \ldots \tag{9.121}$$

or on the basis of criterion (9.115)–(9.116) applied to sequence (9.120). Using (9.63), we obtain the average test volume for Wald's criterion for $R = R_0$ as

$$N_W \approx \frac{\omega(\alpha, \beta)}{\rho(R_0, R_1)} = \frac{0.95 \ln 19 - 0.05 \ln 19}{0.99 \ln(0.99/0.95) - 0.01 \ln(0.05/0.01)} = 110.$$

For sequential criterion (9.115)–(9.116) the average test volume for $R(\mathbf{q}) = q_1 q_2 = R_0$ (i.e., at symmetrical point $q_1 = q_2 = \sqrt{R_0}$) gives, from Monte Carlo simulation, the value of 44. Again we have a substantial gain in the average test volume due to the use of complete information about unit failures and applying sequential criterion (9.115)–(9.116). □

Example 9.15 (Sequential UCL for Failure Probability of Parallel System) Consider a parallel system of m different units. The failure probability of such a system is in the form

$$R = R(\mathbf{q}) = \prod_{1 \leq i \leq m} q_i. \tag{9.122}$$

Construct $(1 - \beta)$-UCL for a reliability index of type (9.122) using formulas (9.113)–(9.114). In this case function $V_{1n}(C)$ has the form

$$V_{1n}(C) = \max H_n(\mathbf{q}), \tag{9.123}$$

where the maximum is taken under the following restrictions for vector of parameters $\mathbf{q} = (q_1, \ldots, q_m)$:

$$\prod_{1 \leq i \leq m} q_i \geq C, \qquad 0 \leq q_i \leq 1, \qquad 1 \leq i \leq m. \tag{9.124}$$

It is easy to see that the maximum in (9.123) is attained in the inner point of the area specified by restriction (9.124), since $H(\mathbf{q}) \to \infty$ for $q_i \to 1$. The necessary condition for the maximum (the system of Lagrange equations) in this case has the form

$$\frac{\partial H_n(\mathrm{q})}{\partial q_i} = \frac{M_{in}}{q_i} - \frac{n - M_{in}}{1 - q_i} = \frac{b}{q_i}, \qquad 1 \le i \le m,$$
$$\prod_{1 \le i \le m} q_i = C,$$

where $b > 0$ is a Lagrange multiplier. It gives, after simple transformations, that the $(1 - \beta)$-UCL for the probability of system failure at step n is determined by the formula

$$\overline{R}_n = \min_{1 \le l \le n} \overline{R}'_l, \tag{9.125}$$

where

$$\overline{R}'_n = \prod_{1 \le l \le m} \left(\frac{M_{in} + b}{n + b} \right) \tag{9.126}$$

and $b > 0$ is found from the equation

$$\sum_{1 \le l \le m} \left\{ M_{in} \ln \left(\frac{M_{in} + b}{n + b} \right) + (n - M_{in}) \ln \left(n - \frac{M_{in}}{n + b} \right) \right\}$$
$$= U_n - \ln \left(\frac{1}{\beta} \right). \; \square \tag{9.127}$$

Example 9.16 (Test of Parallel System up to First Failure) In the conditions of Example 9.17 consider a case where system tests are continued until such random step n when the first failure of at least one unit has occurred, that is, up to the step

$$\sigma = \min\{n\colon d_{in} > 0 \quad \text{at least for one unit } i, \quad 1 \le i \le m\}.$$

Take the point estimate (9.112) for parameter q_i (the probability of failure of unit i) as the standard estimate of maximum likelihood

$$\hat{q}_i(n) = \frac{M_{in}}{n}, \qquad 1 \le i \le m.$$

In this case, by the definition of the test stopping moment, σ for all numbers of failures of units equals $d_{in} = 0$ and, consequently, $\hat{q}_i(n) = 0$, $\hat{p}_i(n) = 1$. From (9.125)–(9.127) we have

$$\overline{R}'_n = \left(\frac{b}{n+b}\right)^m, \quad 1 \le n \le \sigma - 1,$$

where $b > 0$ can be found from the equation

$$mn \ln\left(\frac{n}{n+b}\right) = -\ln\left(\frac{1}{\beta}\right).$$

It follows that the sequential $(1 - \beta)$-UCL for the probability of system failure at test step n is determined by the formula

$$\overline{R}_n = (1 - \beta^{1/mn})^m, \qquad 1 \le n \le \sigma - 1. \tag{9.128}$$

The UCL (9.128) decreases fast with increasing n. At the test step σ this limit has the form

$$\overline{R}_\sigma = (1 - \beta^{1/m(\sigma-1)})^m. \ \square$$

9.5.2 General Parametric Model

Consider a system of m units of different types. Each unit has d.f. $F_i(x, \theta_i)$ and distribution density $f_i(x, \theta_i)$, where θ_i is an unknown parameter (in the general case, a vector). There are N_i units of type i, $1 \le i \le m$. All unit failures are independent. A failed unit is instantaneously replaced by a new one (see Example 9.6). During the test we observe $N_1 + \cdots + N_m$ independent renewal processes among which there are N_i processes of type i.

Let $R = R(\boldsymbol{\theta}) = R(\theta_1, \ldots, \theta_m)$ be some system reliability index depending on the vector of unit parameters. Using results of Section 9.4, it is easy to construct sequential confidence limits for $R = R(\boldsymbol{\theta})$ for more general model with continuous time. The following formulas for the sequential confidence limits can be obtained by using the results of Section 9.4.3 by dividing the time axis into intervals h and by then setting $h \to 0$ [see Pavlov (1983b, 1988) for details].

Let us introduce the following notation for test results:

- N_i is the number of units of type i tested at moment t (in this model N_i is constant for any t);
- t_{ij} is the moment of the jth failure of a unit of type i;
- s_{ij} is the total testing time of the ith unit at moment t_{ij};
- $D_i(t)$ is the number of failures of units of type i on time interval $(0, t]$, that is, the number of failure moments t_{ij} such that $t_{ij} \le t$; and
- $S_{ie}(t)$ is the testing time of eth unit of type i at a current moment t, $1 \le e \le N_i$.

We also introduce the following notation for the standard unit characteristics:

$$\lambda_i(x, \theta_i) = \frac{f_i(x, \theta_i)}{1 - F_i(x, \theta_i)}$$

is the failure rate of the ith type unit for a given value of parameter θ_i (we will further assume that this function is continuous in x for each $1 \le i \le m$:

$$\Lambda_i(x, \theta_i) = \int_0^x \lambda(u, \theta_i)\, du = -\ln[1 - F_i(x, \theta_i)]$$

is the "resource function" of the ith unit (in other terminology, "leading function").

Sequential $(1 - \alpha)$-LCL and $(1 - \beta)$-UCL for $R = R(\boldsymbol{\theta})$ are found for time moment t by the formulas

$$\underline{R}_t = \max_{u \le t} \underline{R}'_u, \qquad \overline{R}_t = \max_{u \le t} \overline{R}'_u, \tag{9.129}$$

where values of $\underline{R}'_t$ and $\overline{R}'_t$ are found from the following equations with respect to C:

$$V_{0t}(c) = U_t - \ln\left(\frac{1}{\alpha}\right), \qquad V_{1t}(c) = U_t - \ln\left(\frac{1}{\beta}\right), \tag{9.130}$$

where

$$V_{0t}(c) = \max_{R(\boldsymbol{\theta}) \le C} H_t(\boldsymbol{\theta}), \qquad V_{1t}(c) = \max_{R(\boldsymbol{\theta}) \ge C} H_t(\boldsymbol{\theta}), \tag{9.131}$$

$$H_t(\boldsymbol{\theta}) = \sum_{1 \le i \le m} \left\{ \sum_{1 \le j \le D_i(t)} \ln \lambda_i(s_{ij}, \theta_i) - \sum_{1 \le j \le D_i(t)} \Lambda_i(s_{ij}, \theta_i) - \sum_{1 \le e \le N_i} \Lambda_i[S_{ie}(t), \theta_i] \right\}, \tag{9.132}$$

$$U_t = \sum_{1 \le i \le m} \left\{ \sum_{1 \le j \le D_i(t)} \ln \lambda_i(s_{ij}, \hat{\theta}_i) - \int_0^t \sum_{1 \le e \le N_i} \lambda_i[S_{ie}(u), \hat{\theta}_i(u)]\, du \right\}, \tag{9.133}$$

where $\hat{\theta}_i(t)$ is the point estimate of parameter θ_i obtained by test results on time interval $(0, t]$ and $\hat{\theta}_{ij} = \hat{\theta}_i(t_{ij} - 0)$ is the left limit of function $\hat{\theta}_i(t)$ at point t_{ij}.

Example 9.17 (Exponential Model) Consider a particular case where units have exponential d.f. of TTF: $F_i(x, \lambda_i) = 1 - \exp(-\lambda_i x)$, $i = 1, \ldots, m$. In this case the reliability parameter of a type i unit is $\theta_i = \lambda_i$, the vector of unit reliability parameters $\boldsymbol{\theta} = \boldsymbol{\lambda} = (\lambda_1, \ldots, \lambda_m)$, and the reliability index of the system, $R = R(\boldsymbol{\lambda}) = R(\lambda_1, \ldots, \lambda_m)$, is some function of parameter vector $\boldsymbol{\lambda}$. Denote

$$\hat{\lambda}_i(t) = \frac{D_i(t) + r_i}{N_i t + T_i}$$

the point estimate of parameter λ_i obtained on the basis of test results up to moment t, where values of T_i and r_i allow us to take into account information about parameter λ_i if it exists (e.g., test duration T_i and number of observed failures r_i). On the basis of previous formulas (9.129)–(9.133), we obtain that in this case $R = R(\boldsymbol{\theta})$ sequential $(1 - \alpha)$-LCL and $(1 - \beta)$-UCL are determined at moment t by formulas (9.129) and (9.130), where

$$V_{0t}(c) = \max_{R(\boldsymbol{\lambda}) \le C} H_t(\boldsymbol{\lambda}), \qquad V_{1t}(c) = \max_{R(\boldsymbol{\lambda}) \ge C} H_t(\boldsymbol{\lambda}),$$

$$H_t(\boldsymbol{\lambda}) = \sum_{1 \le i \le m} [D_i(t) \ln \lambda_i - N_i \lambda_i t\},$$

$$U_t = \sum_{1 \le i \le m} \left\{ \sum_{1 \le j \le D_i(t)} \ln \hat{\lambda}_{ij} - N_i \int_0^t \hat{\lambda}_i(u)\, du \right\},$$

where $\hat{\lambda}_{ij} = \hat{\lambda}_i(t_{ij} - 0)$ is the point estimate of the left limit of estimate $\hat{\lambda}_i(t)$ at point t_{ij}. □

9.5.3 Markov Model of Tests with Censorship and Unit Renewal

We have considered particular cases of tests: Failed units are always instantaneously replaced and no unit tests are stopped before total test completion. However, these results can be extended to more general models.

In Section 3.6 we introduced a general Markov model [MMR] of tests of identical units with TTF d.f. $F(x, \boldsymbol{\theta})$. This model allows us to consider test censorship as well as unit renewal. Let us now introduce a more general model for which we use notation $[\text{MMR}]_m$ which differs from model [MMR] by following: There are units of m different types. In other words, $[\text{MMR}]_m$ represents a multidimensional analogue of [MMR].

At moment $t = 0$ we begin to test $N_i = N_i(0)$ identical units of type i. The TTF of each of them is a nonnegative r.v. ξ_i with d.f. $F_i(x, \theta_i)$, where θ_i is an unknown vector parameter, $1 \le i \le m$. The model is given by the sequences

$$(\tau_{i1}, n_{i1}), \ldots, (\tau_{ik}, n_{ik}), \ldots,$$

$$(\sigma_{i1}, \tilde{n}_{i1}), \ldots, (\sigma_{il}, \tilde{n}_{il}), \ldots,$$

where $\tau_{i1} < \tau_{i2} < \cdots < \tau_{ik} < \cdots$ are Markov (independent of the future) moments of termination of test of units of ith type; $\sigma_{i1} < \sigma_{i2} < \cdots < \sigma_{i1} < \cdots$ are Markov moments of placing on test new units of ith type; n_{ik} is the number of units of ith type whose test is terminated at moment τ_{ik}; and $\tilde{n}_{il}$ is the number of new units of ith type whose test begins at moment σ_{i1}. The r.v. $n_{ik} = 0, 1, 2, \ldots$ might depend on the prehistory of the test process before moment τ_{ik} but does not depend on the process for $t > \tau_{ik}$ developing in the future. An analogous r.v. $\tilde{n}_{il} = 0, 1, 2, \ldots$ might depend on the behavior of the test process before moment σ_{il} but does not depend on the process for $t > \sigma_{il}$ developing in the future. (More detailed and accurate formal definitions are given in Section 3.6 and in Section 3.8.1, 3.8.2 and 3.8.7.)

Denote

$$0 < t_{i1} < t_{i2} < \cdots < t_{ij} < \cdots$$

the sequential failure moments of units of the ith type, where t_{ij} is the moment of failure of unit j of type i.

Introduce the following notation:

$D_i(t)$ is the number of ith unit failures on interval $(0, t]$, that is, the number of failure moments t_{ij} such that $t_{ij} \le t$;

$N_i(t)$ is the number of units of the ith type that began to be tested and whose test has not been terminated before t:

$$N_i(t) = N_i(0) + B_i(t) - D_i(t) - L_i(t),$$

where $B_i(t) = \Sigma_{l:\sigma_{il} \le t} \tilde{n}_{il}$ is the number of units of the ith type that began to be tested on interval $(0, t]$ and $L_i(t) = \Sigma_{k:\tau_{ik} \le t} n_{ik}$ is the number of units of the ith type whose test has been terminated (before failure) on interval $(0, t]$;

u_{ir} is the test time of unit r of type i whose test has been terminated on interval $(0, t]$ before failure, $1 \le r \le L_i(t)$;

s_{ij} is the test time of the unit of type i that has failed at moment t_{ij}; and

$S_{ie}(t)$ is the total test time of unit e of type i that is on the test at moment t, $1 \le e \le N_i(t)$.

Let $R = R(\boldsymbol{\theta}) = R(\theta_1, \ldots, \theta_m)$ be a function (reliability index) from unit parameter vector $\boldsymbol{\theta} = (\theta_i, \ldots, \theta_m)$. The sequential $(1 - \alpha)$-LCL and $(1 - \beta)$-UCL for $R = R(\boldsymbol{\theta})$ are calculated for this model by formulas analogous to those in Section 9.5.2 (see also Pavlov, 1985 and 1988):

$$\underline{R}_t = \max_{u \le t} \underline{R}'_u, \qquad \overline{R}_t = \max_{u \le t} \overline{R}'_u, \tag{9.134}$$

where $\underline{R}'_t$ and $\overline{R}'_t$ are determined during the test process at each current time moment t from the corresponding equations with respect to C:

$$V_{0t} = U_t - \ln\left(\frac{1}{\alpha}\right), \qquad V_{1t} = U_t - \ln\left(\frac{1}{\beta}\right), \tag{9.135}$$

where

$$V_{0t} = \max_{R(\boldsymbol{\theta}) \le C} H_t(\boldsymbol{\theta}), \qquad V_{1t} = \max_{R(\boldsymbol{\theta}) \ge C} H_t(\boldsymbol{\theta}), \tag{9.136}$$

where, in turn,

$$H_t(\boldsymbol{\theta}) = \sum_{1 \le i \le m} \left\{ \sum_{1 \le j \le D_i(t)} \ln \lambda_i(s_{ij}, \theta_i) - \sum_{1 \le j \le D_i(t)} \Lambda_i(s_{ij}, \theta_i) \right.$$
$$\left. - \sum_{1 \le e \le L_i(t)} \Lambda_i[u_{ir}, \theta_i] - \sum_{1 \le e \le N_i(u)} \Lambda_i[S_{ie}(t), \theta_i] \right\}, \tag{9.137}$$

$$U_t = \sum_{1 \le i \le m} \left\{ \sum_{1 \le j \le D_i(t)} \ln \lambda_i(s_{ij}, \hat{\theta}_{ij}) \right.$$
$$\left. - \int_0^t \sum_{1 \le e \le N_i(u)} \lambda_i[S_{ie}(u), \hat{\theta}_i(u)]\, du \right\}, \tag{9.138}$$

where $\hat{\theta}_i(t)$ is, as before, the point estimate of parameter θ_i obtained by test results on time interval $(0, t]$, $\hat{\theta}_{ij} = \hat{\theta}_i(t_{ij} - 0)$ is the left limit of function $\hat{\theta}_i(t)$ at point t_{ij}.

Formulas (9.134)–(9.138) are similar to (9.129)–(9.133) and are their generalization.

Consider now some particular cases.

Example 9.18 (Markov Model [MMR] for Identical Units) Consider a particular case of model $[\text{MMR}]_m$ for $m = 1$, that is, where all tested units are identical with the same TTF d.f. $F(x, \boldsymbol{\theta})$, where $\boldsymbol{\theta}$ is some vector parameter. Remember that in Chapter 3 we considered the nonparametric case related to model [MMR]. Formulas (9.134)–(9.138) allow us to obtain corresponding results for the parametric case. Let $R = R(\boldsymbol{\theta})$ be some reliability index, for instance, the unit PFFO

$$R(\boldsymbol{\theta}) = 1 - F(t_0, \boldsymbol{\theta}) \tag{9.139}$$

or the MTBF

$$R(\boldsymbol{\theta}) = \int_0^{\infty} [1 - F(x, \boldsymbol{\theta})]\, dx. \tag{9.140}$$

Using formulas (9.134)–(9.138) with $m = 1$, we find that the sequential $(1 - \alpha)$-LCl and $(1 - \beta)$-UCL for $R = R(\boldsymbol{\theta})$ are calculated for each current moment of time t as

$$\underline{R}_t = \max_{u \le t} \underline{R}'_u, \qquad \overline{R}_t = \min_{u \le t} \overline{R}'_u, \tag{9.141}$$

where $\underline{R}'_t$ and $\overline{R}'_t$ for moment t are defined from the following equations with respect to C:

$$V_{0t} = U_t - \ln\left(\frac{1}{\alpha}\right), \qquad V_{1t} = U_t - \ln\left(\frac{1}{\beta}\right), \tag{9.142}$$

where

$$V_{0t}(c) = \max_{R(\boldsymbol{\theta}) \le C} H_t(\boldsymbol{\theta}), \qquad V_{1t}(c) = \max_{R(\boldsymbol{\theta}) \ge C} H_t(\boldsymbol{\theta}), \tag{9.143}$$

and, in turn,

$$\begin{aligned} H_t(\theta) = {} & \sum_{1 \le j \le D(t)} \ln \lambda(s_j, \theta) - \sum_{1 \le j \le D(t)} \Lambda(s_j, \theta) \\ & - \sum_{1 \le e \le L(t)} \Lambda[u_r, \theta] - \sum_{1 \le e \le N(t)} \Lambda[S_e(t), \theta] \end{aligned} \tag{9.144}$$

$$U_t = \sum_{1 \le j \le D(t)} \ln \lambda(s_j, \hat{\theta}_j) - \int_0^t \sum_{1 \le e \le N(u)} \lambda[S_e(u), \hat{\theta}(u)]\, du \tag{9.145}$$

where $\lambda(x, \boldsymbol{\theta})$ is the unit failure rate; $\Lambda(x, \boldsymbol{\theta}) = -\ln[1 - F(x, \boldsymbol{\theta})]$ is the unit resource function; t_j is the moment of the jth failure during the test; s_j is the test time of the unit failed at moment t_j; $D(t)$ is the number of failures on interval $(0, t]$; $N(t)$ is the number of tested units at current moment t; $L(t)$ is the number of units whose tests have been terminated (without failure) on interval $(0, t]$; u_r is time that unit r was tested on interval $(0, t]$ until stopping the test (without failure), $1 \le r \le L(t)$; $S_e(t)$ is the test time for unit e that is under test at the current moment t, $1 \le e \le N(t)$; $\hat{\theta}(t)$ is the point estimate of parameter $\boldsymbol{\theta}$ by test results on interval $(0, t]$; and $\hat{\theta}_j = \hat{\theta}(t_j - 0)$ is the left limit of estimate $\hat{\theta}(t)$ at moment t_j. □

Example 9.19 (Weibull–Gnedenko Distribution) In the conditions of the previous example, consider a case where the d.f. of unit TTF has a Weibull–Gnedenko distribution, that is,

$$F(x, \boldsymbol{\theta}) = F(x, \lambda, \alpha) = 1 - \exp(-\lambda x^{\alpha}).$$

In this $\boldsymbol{\theta}$ is a two-dimensional parameter; $\boldsymbol{\theta} = (\lambda, \alpha)$. Any chosen reliability index is a function $R = R(\lambda, \alpha)$ of this two-dimensional parameter. For instance, a reliability index of type (9.139), that is, the unit PFFO during time t_0, in this case has the form

$$R(\lambda, \alpha) = \exp(-\lambda t_0^{\alpha}).$$

Another standard reliability index of type (9.140), that is, the unit MTTF, is written in this case as

$$R(\lambda, \alpha) = \int_0^{\infty} e^{-\lambda x^{\alpha}}\, dx = \frac{\Gamma(1 + 1/\alpha)}{\lambda^{1/\alpha}}$$

where

$$\Gamma(\alpha) = \int_0^{\infty} x^{\alpha-1} e^{-x}\, dx$$

is the gamma function.

For this d.f. the failure rate and "resource function" have the form

$$\lambda(x, \boldsymbol{\theta}) = \lambda\alpha x^{\alpha-1} \qquad \text{and} \qquad \Lambda(x, \boldsymbol{\theta}) = \lambda x^{\alpha}.$$

By substituting these functions into (9.144) and (9.145), we find that confidence limits $\underline{R}_t$ and $\overline{R}_t$ for reliability index $R = R(\lambda, \alpha)$ for the Weibull–Gnedenko distribution are determined for current moment t by (9.141) and (9.142), where

$$V_{0t}(c) = \max_{R(\lambda,\alpha)\le C} H_1(\lambda, \alpha), \qquad V_{1t} = \max_{R(\lambda,\alpha)\ge C} H_t(\lambda, \alpha),$$

where, in turn,

$$H_t(\lambda, \alpha) = D(t)\ln(\lambda\alpha) + (\alpha - 1)\sum_{1\le j\le D(t)} \ln s_i - \lambda \sum_{1\le j\le D(t)} s_j^{\alpha} - \lambda \sum_{1\le r\le L(t)} u_r^{\alpha} - \lambda \sum_{1\le e\le N(t)} S_e^{\alpha}(t),$$

$$U_t = \sum_{1\le j\le D(t)} \ln(\hat{\lambda}_j\hat{\alpha}_j) + \sum_{1\le j\le D(t)} (\hat{\alpha}_j - 1)\ln s_j - \int_0^t \sum_{1\le e\le N(u)} \hat{\lambda}(u)\hat{\alpha}(u)[S_e(u)]^{\hat{\alpha}(u)-1}\, du$$

where $\hat{\lambda}(t)$ and $\hat{\alpha}(t)$ are point estimates of parameters λ and α obtained by test results on time period $(0, t]$ and $\hat{\lambda}_j = \hat{\lambda}(t_j - 0)$ and $\hat{\alpha}_j = \hat{\alpha}(t_j - 0)$ are left limits of these estimates at moment t_j. □

Example 9.20 (Renewal Process) In the conditions of Example 9.18 consider a particular case of the renewal process with d.f. $F(x, \boldsymbol{\theta})$.

In this case replacement is instant, that is, $L(t) = 0$ always. The number of units $N(t)$ under testing at any current moment t is also constant, $N(t) = 1$. The sequential LCL and UCL for reliability index $R = R(\boldsymbol{\theta})$ at moment t are determined by (9.141)–(9.145). In this case (9.144)–(9.145) can be simplified as

$$H_t(\boldsymbol{\theta}) = \sum_{1 \le j \le D(t)} \ln \lambda(s_i, \boldsymbol{\theta}) - \sum_{1 \le j \le D(t)} \Lambda(s_j, \boldsymbol{\theta}) - \Lambda[S(t), \boldsymbol{\theta}],$$

$$U_t = \sum_{1 \le j \le D(t)} \ln \lambda(s_j, \hat{\theta}_j) - \int_0^t \lambda[S(u), \hat{\theta}(u)]\, du,$$

where $S(t)$ is the test time of the unit under test at current moment t, that is, $S(t) = t - t_j$, where t_j is the moment of the last failure in time period $(0, t]$.

On the basis of sequential confidence limits, we can construct corresponding criteria for tests of hypotheses of types $H_0\colon R \le R_0$ and $H_1\colon R \ge R_1$ with respect to reliability index $R = R(\boldsymbol{\theta})$. □

9.5.4 Sequential Confidence Limits for Availability Coefficient of Renewable Unit

Consider a unit with d.f. of TTF equal to $F_1(x, \theta_1)$ and d.f. of renewal time equal to $F_2(x, \theta_2)$, where θ_1 and θ_2 are unknown parameters, in the general case, vectors. Consider a simple test scheme with test results in the form of alternating independent random intervals

$$(s_{1n}, s_{2n}), \qquad n = 1, 2, \ldots, \tag{9.146}$$

where s_{1n} is the TTF of the unit and s_{2n} is its random renewal time at the nth step of test. All r.v.'s are independent. Thus (9.146) represents an alternating renewal process. All r.v.'s s_{1n} have d.f. $F_1(x, \theta_1)$, and all s_{2n} have d.f. $F_2(x, \theta_2)$. The test results of such a test are sequential moments of failure and renewal of the unit:

$$t_{1n} = \sum_{1 \le j \le n-1} (s_{1j} + s_{2j}) + s_{1n}, \qquad t_{2n} = \sum_{1 \le j \le n} (s_{1j} + s_{2j}),$$

where t_{1n} is the moment of the nth failure and t_{2n} is the moment of the nth renewal, $n = 1, 2, \ldots$. Graphical illustration of this process is given in

Figure 9.11, where $I(t)$ is an indicator function such that $I(t) = 0$ if a unit is in the renewal state and $I(t) = 1$ if a unit is in the operational state at moment t.

A standard reliability index for a renewable unit is its availability coefficient, that is, the stationary probability of the operational state of the unit,

$$K = \lim_{t \to \infty} P\{I(t) = 0\}.$$

This reliability index can be found by the well-known formula

$$K = K(\theta_1, \theta_2) = \frac{T_1(\theta_1)}{T_1(\theta_1) + T_2(\theta_2)}, \tag{9.147}$$

where

$$T_1(\theta_1) = \int_0^\infty [1 - F_1(x, \theta_1)]\, dx$$

is the unit TTF, and

$$T_2(\theta_2) = \int_0^\infty [1 - F_2(x, \theta_2)]\, dx$$

is the unit mean renewal time.

Applying general formulas (9.134)–(9.138), we obtain that the sequential $(1 - \alpha)$-LCL $\underline{K}_t$ and the $(1 - \beta)$-UCL $\overline{K}_t$ for reliability index (9.147) can be calculated for a current moment t as

$$\underline{K}_t = \max_{u \le t} \underline{K}'_u, \qquad \overline{K}_t = \min_{u \le t} \overline{K}'_u, \tag{9.148}$$

where $\underline{K}'_t$ and $\overline{K}'_t$ are found from equations

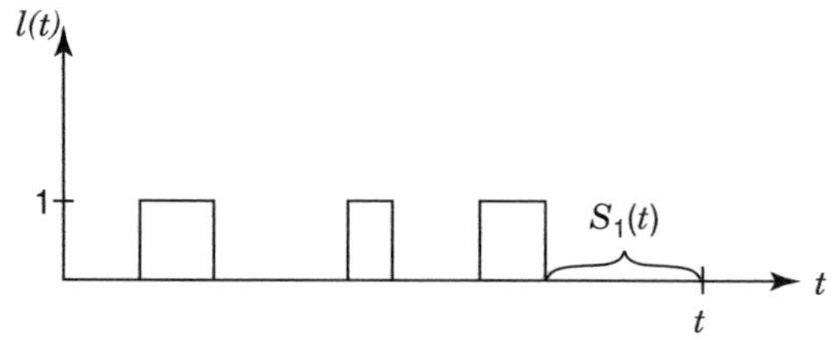

Figure 9.11 Sample of time diagram for renewal unit testing.

$$V_{0t}(C) = U_t - \ln\left(\frac{1}{\alpha}\right), \qquad V_{1t}(C) = U_t - \ln\left(\frac{1}{\beta}\right), \tag{9.149}$$

where

$$\begin{aligned} V_{0t}(C) &= \max_{K(\theta_1,\theta_2)\le C} H_t(\theta_1, \theta_2), \\ V_{1t}(C) &= \max_{K(\theta_1,\theta_2)\ge C} H_1(\theta_1, \theta_2) \end{aligned} \tag{9.150}$$

and, in turn,

$$\begin{aligned} H_t(\theta_1, \theta_2) = & \sum_{1\le j\le D_1(t)} \ln \lambda_1(s_{1j}, \theta_1) - \sum_{1\le j\le D_1(t)} \Lambda_1(s_{1j}, \theta_1) \\ & - [1 - I(t)]\Lambda_1[S_1(t), \theta_1] + \sum_{1\le j\le D_2(t)} \ln \lambda_2(s_{2j}, \theta_2) \\ & - \sum_{1\le j\le D_2(t)} \Lambda_2(s_{2j}, \theta_2) - I(t)\Lambda_2[s_2(t), \theta_2], \end{aligned} \tag{9.151}$$

$$\begin{aligned} U_t = & \sum_{1\le j\le D_1(t)} \ln \lambda_1(s_{1j}, \hat{\theta}_{1j}) + \sum_{1\le j\le D_2(t)} \ln \lambda_2(s_{2j}, \hat{\theta}_{2J}) \\ & - \int_0^t \lambda_1[S_1(u), \hat{\theta}_1(u)][1 - I(u)]\, du - \int_0^t \lambda_2[s_2(u), \hat{\theta}_2(u)]I(u)\, du \end{aligned} \tag{9.152}$$

where $\Lambda_i(x, \theta_i) = -\ln[1 - F_i(x, \theta_i)]$, $i = 1, 2$, are "leading functions" for failure and renewal,

$$\lambda_i(x, \theta_i) = \frac{d}{dx} \Lambda_i(x, \theta_i)$$

are corresponding failure and renewal rates, $D_i(t)$ are corresponding numbers of failures (renewals) on interval $(0, t]$, $S_i(t)$ are total times when a unit was in the operation (down) state if it is in the operational (down) state at moment t, $\hat{\theta}_i(t)$ is a point estimate of parameter θ_i on the basis of test results on the interval $(0, t]$, and $\hat{\theta}_{1j} = \hat{\theta}_1(t_{1j} - 0)$ is the left limit of estimate $\hat{\theta}_1(t)$ and $\hat{\theta}_{2j} = \hat{\theta}_2(t_{2j} - 0)$ is the left limit of estimate $\hat{\theta}_2(t)$ at the moment of the jth failure (renewal), that is, t_{1i} and t_{2j}, respectively.

Suppose we need to test two composite hypotheses

$$H_0\colon K \le K_0 \quad \text{and} \quad H_1\colon K \ge K_1$$

with respect to the availability coefficient $K = K(\theta_1, \theta_2)$. Then the sequential criterion for a test of hypotheses with risks of types I and II can be constructed on the basis of sequential confidence limits (9.148) in the same manner as it was done in Section 9.4.2. Namely, the area of test continuation at moment t is given by the inequalities

$$\underline{K}_t < K_0, \qquad K_1 > \overline{K}_t. \tag{9.153}$$

The test continues if both inequalities hold and are stopped at the moment when at least one of them is violated. If the first inequality is violated, hypothesis H_1 is accepted; if the second is violated, then hypothesis H_0 is accepted.

Notice that in this model, with continuous time, moments of renewals t_{2n} are moments of regeneration for random process $I(t)$, and pairs of r.v.'s (s_{1n}, s_{2n}), defined in (9.146), form i.i.d. "regeneration cycles." Thus, $I(t)$ is an alternating renewal process formed with two-dimensional r.v.'s (s_{1n}, s_{2n}). Therefore properties of asymptotic optimality (for $\alpha \to 0$ and $\beta \to 0$) of sequential confidence limits (9.148) and the sequential criterion of the test of hypotheses (9.153) follow from corresponding results for the scheme of independent and identical tests [see Section x.x.x and Pavlov (1988, 1990)]. In general case where one can not distinguish such regeneration moments, corresponding properties of sequential confidence limits above (9.134)–(9.138) need further investigations.

Exponential Model Take a particular case of the test scheme considered above, where the d.f. of the unit TTF is exponential, $F_1(x, \lambda_1) = 1 - \exp(-\lambda_1 x)$ with unknown parameter λ_1 (the failure rate), and the d.f. of unit's renewal time is also exponential, $F_2(x, \lambda_2) = 1 - \exp(-\lambda_2 x)$ with unknown parameter λ_2 (the renewal rate). In this case $\theta_1 = \lambda_1$ and $\theta_2 = \lambda_2$. The availability coefficient (9.147) has the form

$$K - K(\lambda_1, \lambda_2) = \frac{\lambda_2}{\lambda_1 + \lambda_2}.$$

Consider interval $(0, t]$. Denote the total time of unit operation by $W_1(t)$ and the total time of unit operation by $W_2(t)$. Obviously, $W_1(t) + W_2(t) = t$. These values are determined as

$$W_1(t) = \sum_{1 \le j \le D_1(t)} s_{1j} + s_1(t)[1 - I(t)],$$

$$W_2(t) = \sum_{1 \le j \le D_2(t)} s_{2j} + s_2(t)I(t).$$

Let us also introduce

$$\hat{\lambda}_1(t) = \frac{D_1(t) + D_1^0}{W_1(t) + W_1^0}, \qquad \hat{\Lambda}_2(t) = \frac{D_2(t) + D_2^0}{W_2(t) + W_2^0},$$

which are point estimates of parameters λ_1 and λ_2 based on the test results obtained on interval $(0, t]$. Values W_1^0 and D_1^0 allow us to take prior information into account (if available) about parameter λ_1. For instance, W_1^0 might be the total test time and D_1^0 the number of failures during some previous tests. Also, W_2^0 and D_2^0 might play the same role with respect to parameter λ_2.

Applying formulas (9.148)–(9.152), we obtain that in this case the $(1 - \alpha)$-LCL $\underline{K}_t$ and the $(1 - \beta)$-UCL $\overline{K}_t$ for the availability coefficient for moment t are calculated by formulas (9.148) and (9.149), where

$$V_{0t}(C) = \max_{K(\lambda_1, \lambda_2) \le C} H_t(\lambda_1, \lambda_2), \tag{9.154}$$

$$V_{1t}(C) = \max_{K(\lambda_1, \lambda_2) \ge C} H_t(\lambda_1, \lambda_2), \tag{9.155}$$

and, in turn,

$$H_t(\lambda_{12}, \lambda_2) = D_1(t) \ln \lambda_1 + D_2(t) \ln \lambda_2 - \lambda_1 W_1(t) - \lambda_2 W_2(t),$$

$$U_t = \sum_{1 \le j \le D_1(t)} \ln \hat{\lambda}_{1j} + \sum_{1 \le j \le D_2(t)} \ln \hat{\lambda}_{2j} - \int_0^t \hat{\lambda}_1(u)]1 - I(u)]\, du - \int_0^t \hat{\lambda}_2(u) I(u)\, du,$$

where $\hat{\lambda}_{1j} = \hat{\lambda}_1(t_{1j} - 0)$ is the left limit of estimate $\hat{\lambda}_1(t)$ and $\hat{\lambda}_{2j} = \hat{\lambda}_2(t_{2j} - 0)$ is the left limit of estimate $\hat{\lambda}_2(t)$ at the moment of the jth failure (renewal), that is, t_{1i} and t_{2j}, respectively.

In this case the maximum in (9.154) and (9.155) can be found analytically. First we will find the maximum in (9.154). The restriction $K(\lambda_1, \lambda_2) \le C$, for which maximum in (9.154) is being found, is equivalent to one of the inequalities

$$\frac{\lambda_1}{\lambda_2} \ge \frac{1 - C}{C}$$

or

$$\ln\left(\frac{\lambda_1}{\lambda_2}\right) \ge C',$$

where

$$C' = \ln \frac{1 - C}{C}.$$

Taking into account the concavity of the function $H_t(\lambda_1, \lambda_2)$ in (λ_1, λ_2) and compiling the Lagrange function for this problem,

$$L = H_t(\lambda_1, \lambda_2) + b \ln \frac{\lambda_1}{\lambda_2},$$

we obtain that the maximum in (9.154) is attained at the point that satisfies the following system of Lagrange equations:

$$\frac{D_1(t)}{\lambda_1} - W_1(t) = -\frac{b}{\lambda_1}, \qquad \frac{D_2(t)}{\lambda_2} - W_2(t) = \frac{b}{\lambda_2},$$

where b is the Lagrange multiplier. After simple transformations, this gives the $(1 - \alpha)$-LCL for the availability coefficient at moment t in the form

$$\underline{K}_t = \max_{u \le t} \underline{K}'_u,$$

where $\underline{K}'_t$ is found at moment t by the formula

$$\underline{K}'_t = \frac{W_1(D_2 - b)}{W_1(D_2 - b) + W_2(D_1 + b)},$$

where the value $b > 0$ is determined from the equation

$$D_1 \ln \left(\frac{D_1 + b}{W_1} \right) + D_2 \ln \left(\frac{D_2 - b}{W_2} \right) - D_1 - D_2 = U_t - \ln \left(\frac{1}{\alpha} \right).$$

In these formulas and below we use the shortened notation $W_1 = W_1(t)$, $W_2 = W_2(t)$, $D_1 = D_1(t)$, and $D_2 = D_2(t)$.

In an analogous way, by calculating the maximum in (9.155), we obtain the sequential $(1 - \beta)$-UCL for the availability coefficient at moment t,

$$\overline{K}_t = \min_{u \le t} \overline{K}'_u,$$

where $\overline{K}'_t$ is found at moment t by the formula

$$\overline{K}'_t = \frac{W_1(D_2 + b)}{W_1(D_2 + b) + W_2(D_1 - b)},$$

and the value $b > 0$ is determined from the equation

$$D_1 \ln\left(\frac{D_1 - b}{W_1}\right) + D_2 \ln\left(\frac{D_2 + b}{W_2}\right) - D_1 - D_2 = U_t - \ln\left(\frac{1}{\beta}\right).$$

In conclusion we emphasize that Wald's testing plans become an important part of modern test planning.

9.6 APPENDIX

9.6.1 Computation of Accurate Values of Characteristics of Sequential Test Plans

Assume that we observe a sequence of i.i.d. scalar r.v.'s $x_1, x_2, \ldots, x_n, \ldots$. Denote the sum of these r.v.'s observed at the first n steps by $S_n = x_1 + \cdots + x_n$. Consider some sequential criterion (control plan) for hypotheses of type (9.49) for which the stopping time has the form

$$\nu = \min\{n: S_n \in G_n\},$$

where G_n is the area of test stop at step n. Let $W_n = \overline{G}_n$ be the area of test continuation at step n, where $\overline{G}_n$ is the area complement to G_n. Thus at step n the test continues if

$$S_n \in W_n. \tag{9.156}$$

The test is stopped at the first violation of condition (9.156), that is, at such step $\nu = n$ when r.v. S_n first reaches stopping area G_n. The decision rule at time stop $\nu = n$ has the following form:

If $S_n \in G_{n0}$, then hypothesis H_0 is accepted.
If $S_n \in G_{n1}$, then hypothesis H_1 is accepted.

Here G_{n0} and G_{n1} are nonintersecting areas such that

$$G_n = G_{n0} + G_{n1}, \qquad n = 1, 2, \ldots.$$

In other words, the area of stopping G_n is divided into two subareas:

G_{n0} is the area of hypothesis H_0 acceptance and
G_{n1} is the area of hypothesis H_1 acceptance.

The main sequential control plans (Wald's, truncated Wald's, and others) have such a form.

Now consider the calculation of accurate characteristics such as risks of types I and II (α and β), operative characteristic $L(\boldsymbol{\theta})$, and average test volume

$N(\boldsymbol{\theta})$. At first consider a discrete case where r.v. x_n takes a finite set of values $0, 1, \ldots, l$. Let

$$\pi(j, \boldsymbol{\theta}) = P_\theta\{x_n = j\}$$

be the probability that r.v. $x_n = j$ for parameter value $\boldsymbol{\theta}$, where $0 \le j \le l$. Assume that the area of test continuation is restricted; that is, a finite n_m exists such that $W_n = \emptyset$ for $n \ge n_m$ and, besides, area W_n is restricted for each n, $1 \le n \le n_m - 1$. Value n_m represents the maximum possible value of stopping time ν.

Fix the value of parameter $\boldsymbol{\theta}$ and introduce a sequence of numbers

$$b_n(i, \boldsymbol{\theta}), \qquad 0 \le i \le n(l+1), \qquad n = 1, 2, \ldots, \tag{9.157}$$

which are recurrently calculated by

$$b_{n+1}(i, \boldsymbol{\theta}) = \sum_{j \in W_n} b_n(j, \boldsymbol{\theta})\pi(i - u, \boldsymbol{\theta}), \qquad n = 1, 2, \ldots, \tag{9.158}$$

where at the first step we set $b_1(j, \boldsymbol{\theta}) = \pi(j, \boldsymbol{\theta})$, $0 \le j \le 1$. Directly from construction of sequence (9.157) it follows that $b_n(i, \boldsymbol{\theta})$, determined as

$$b_n(i, \boldsymbol{\theta}) = P_\theta(\nu > n - 1, S_n = i),$$

is the probability that the test process will not be terminated before step n, and at step n, r.v. $S_n = i$. In other words,

$$b_n(i, \boldsymbol{\theta}) = P_\theta(S_1 \in W_1, S_2 \in W_2, \ldots, S_{n-1} \in W_{n-1}, S_n = i).$$

Notice that sum in (9.158) is taken over a finite number of subscripts j because W_n has finite elements. Notice also that, at step $n + 1$, the value of $b_{n+1}(i, \theta)$ differs from zero only for $i \in W_n$ and for i such that $i \le j + 1$ where $i \in W_n$. So, the number of subscripts i for the calculation of sum (9.158) at step $n + 1$ is equal to $k_n + 1$, where k_n is the number of integer points in the area of test continuation W_n at step n. Consequently, the volume of calculation at step $n + 1$ for this recurrent procedure, (9.158), is proportional to $(k_n + 1)(l + 1)$.

Let us introduce the events

$$A_n = \{\nu = n\},$$

$$A_{n0} = \{\nu = n \text{ and hypothesis } H_0 \text{ is accepted}\},$$

$$A_{n1} = \{\nu = n \text{ and hypothesis } H_1 \text{ is accepted}\},$$

where A_n is an event that the test is terminated at step n, A_{n0} is an event that the test is terminated at step n and hypothesis H_0 is accepted, and A_{n1} is an event that the test is terminated at step n and hypothesis H_1 is accepted. The probabilities of these events for fixed values of parameter $\boldsymbol{\theta}$ are expressed via values $b_n(i, \boldsymbol{\theta})$ as

$$P_\theta(A_n) = \sum_{i \in G_n} b_n(i, \boldsymbol{\theta}),$$

$$P_\theta(A_{n0}) = \sum_{i \in G_{n0}} b_n(i, \boldsymbol{\theta}),$$

$$P_\theta(A_{n1}) = \sum_{i \in G_{nl}} b_n(i, \boldsymbol{\theta}).$$

It gives us the following expression for operative characteristic $L(\boldsymbol{\theta})$ for a fixed value of parameter $\boldsymbol{\theta}$:

$$\begin{aligned} L(\boldsymbol{\theta}) &= P_\theta(\text{to accept hypothesis } H_0) \\ &= \sum_{1 \le n \le n_m} P_\theta(A_{n0}) = \sum_{1 \le n \le n_m} \sum_{i \in G_{n0}} b_n(i, \boldsymbol{\theta}). \end{aligned} \tag{9.159}$$

Thus, it is easy to calculate operative characteristic $L(\boldsymbol{\theta})$ with a computer using (9.158) and (9.159). The precise values of risks of types I and II can be found via operative characteristics from the formulas

$$\alpha = 1 - L(\theta_0) \quad \text{and} \quad \beta = L(\theta_1). \tag{9.160}$$

The average test volume $N(\theta_0)$ for a given value of parameter $\boldsymbol{\theta}$ can also be expressed via values $b_n(i, \boldsymbol{\theta})$ and found with the same recurrent procedure (9.158) by the formula

$$N(\boldsymbol{\theta}) = E_\theta \nu = \sum_{1 \le n \le n_m} n\, P_\theta(A_n) = \sum_{1 \le n \le n_m} n \sum_{i \in G_n} b_n(i, \boldsymbol{\theta}). \tag{9.161}$$

If the area of test continuation is not restricted (e.g., in Wald's plan without truncation), the characteristics mentioned above can be calculated on the basis of the same formulas (9.159)–(9.161) if n_m is sufficiently large.

Notice that the area of test continuation W_n, the area of test termination G_n, and areas G_{n0} and G_{n1} in most of cases have the form of intervals, or in other words, they are given as inequalities of the type

$$W_n\colon g_n < S_n < h_n,$$

$$G_n\colon g_n \ge S_n, < h_n \le S_n,$$

$$G_{n0}\colon g_n \ge S_n,$$

$$G_{n1}\colon h_n \le S_n,$$

where g_n and h_n are boundaries of the area of test continuation at step n, $n = 1, 2, \ldots$, and these values do not depend on the concrete test plan.

Example 9.21 (Binomial Plan) Consider a binomial test where the r.v. x_n is a failure indicator at step n, parameter $\boldsymbol{\theta} = q$ is the failure probability, and r.v. $S_n = x_1 + \cdots + x_n$ is a random number of failures during n steps. (See Example 9.9.) In this case $l = 1$, r.v. x_n only takes values 0 or 1, and values $\pi(j, \theta)$ have the form

$$\pi(j, \theta) = \pi(j, q) = \begin{cases} q & \text{if } j = 1, \\ 1 - q & \text{if } j = 0. \end{cases}$$

In this case the sum in (9.158) is taken over two subscripts and recurrent procedure (9.158) becomes extremely simple.

If r.v. x_n has continuous d.f. $F(x, \boldsymbol{\theta})$, then recurrent procedure (9.158) allows one also to calculate the main characteristics $L(\boldsymbol{\theta})$, $N(\boldsymbol{\theta})$, α, β with needed accuracy if we use a discrete approximation of continuous d.f. $F(x, \boldsymbol{\theta})$. Naturally, the average test volume increases with a more accurate approximation. For instance, if r.v. x_n is approximated by a discrete r.v. with $l + 1$ different states, then the computational burden on the basis of (9.158) roughly increases as l^2.

Now consider the case of continuous time for a Poisson process with parameter λ (see Example 9.11). In this case formulas (9.158)–(9.161) can be applied if we use a standard approximation. Divide the time axes into equal small intervals h. In this case we denote the number of failures on interval $[(n - 1)h, nh]$ by x_n, $n = 1, 2, \ldots$. In formulas (9.158)–(9.161) take $l = 2$ and $\boldsymbol{\theta} = \boldsymbol{\lambda}$, and define $\pi(j, \boldsymbol{\theta}) = \pi(j, \boldsymbol{\lambda})$, $j = 0, 1, 2$ as

$$\pi(0, \boldsymbol{\lambda}) = e^{-\lambda h}, \qquad \pi(1, \boldsymbol{\lambda}) = \lambda h e^{-\lambda h}, \qquad \pi(2, \boldsymbol{\lambda}) = 1 - e^{-\lambda h}(1 + \Lambda h).$$

Obviously, we neglect the probability of occurrence of two or more failures on the interval h. The error of such approximation has the order $o(h^2)$. The main characteristics of different sequential test plans can be calculated with recurrent procedure (9.158) and formulas (9.159)–(9.161). The accuracy of the calculation is defined by the size of value h. The corresponding average test volume and number of steps of the test increases proportionally to $(1/h)$. Of course, some more modifications for specific cases can allow us to gain even more. □

Test Plan Allowing Specified Risks α and β The sequential Wald's test plan is characterized by two parameters $a = \ln A$ and $b = \ln(1/B)$, which determine the boundaries of the area of test continuation. If these parameters are given and fixed, approximate values of risks α and β can be found by approximate Wald's formulas (9.33) and precise values can be found with a computer using the recurrent procedure described above. The inverse problem is more complicated: to find plan parameters a and b such that precise values of risks α and β coincide with needed values. Notice that, in the general case, the proof of the fact that a plan with precise risks α and β for any $\alpha + \beta < 1$ exists is not trivial. Nevertheless, the approximate solution is always available using a computer on the basis of the following simple heuristic arguments. Let α and β be given levels of risks. Consider the values

$$a_1 = \ln\left(\frac{1-\beta}{\alpha}\right), \qquad b_1 = \ln\left(\frac{1-\alpha}{\beta}\right)$$

found by Wald's approximate formulas (9.33) as the first iteration. Using a computer, calculate precise values of risks α_1 and β_1 for the plan with parameters a_1 and b_1. Then we change parameters a and b in such a way that the precise values of risks become close to the given values of α and β. Here we can use the property of the operative characteristic $L(\boldsymbol{\theta})$: For any fixed $\boldsymbol{\theta}$ this function is monotone increasing in a for fixed b, and, on the contrary, this function is monotone decreasing in b for fixed a. Notice that the type I risk, $1 - L(\theta_0)$, is sensitive to the variation of parameter a and substantially less sensitive to the variation of parameter b. In contrast, the type II risk, $L(\theta_1)$, is sensitive to the variation of parameter b and substantially less sensitive to the variation of parameter a. It means that if we increase parameter a for fixed b, then $1 - L(\theta_0)$ decreases and, simultaneously, $L(\theta_1)$ increases, though at a slower rate. However, if we increase parameter b for fixed a, then $L(\theta_1)$ decreases and $1 - L(\theta_0)$ increases slowly. Knowing about these monotone dependencies, we can alternately change parameters a and b and calculate corresponding precise values of risks on a computer and iteratively adjust plan parameters a and b in such a way that these risks coincide at last with given values α and β.

Analogously, the same procedure with a computer helps for other test plans (e.g., truncated Wald's plan).

9.6.2 Wald's Equivalency

Assume that we observe a sequence of i.i.d. r.v.'s $z_1, z_2, \ldots, z_n, \ldots$ with mathematical expectation $E|z_n| < \infty$. Let ν be some Markov moment (with respect to a system of σ-algebra $\mathcal{F}_n = \sigma(z_1, \ldots, z_n)$, $n = 1, 2, \ldots$, related to the sequence z_n) such that

$$P(\nu < \infty) = 1.$$

Consider the following mathematical expectation:

$$E \sum_{1 \le n \le \nu} z_n = E(z_1 + \cdots + z_\nu).$$

This value can be represented in the form

$$E \sum_{1 \le n \le \nu} z_n = E \sum_{1 \le n < \infty} z_n I(\nu \ge n),$$

where $I(\nu \ge n)$ is an indicator of the event $\{\nu \ge n\}$. Event $\{\nu \ge n\} \in \mathfrak{F}_{n-1}$; in other words, occurrence of this event depends only on values of $z_1, \ldots, z_{n-1}$ observed at the first $n - 1$ steps. It follows that r.v.'s z_n and $I(\nu \ge n)$ are independent. So,

$$E \sum_{1 \le n < \infty} z_n = \sum_{1 \le n < \infty} E\{z_n I(\nu \ge n)\} = \sum_{1 \le n < \infty} E z_n \cdot E I(\nu \ge n).$$

Since the mathematical expectation $Ez_n = Ez$ does not depend on n, it follows that

$$E \sum_{1 \le n \le \nu} z_n = Ez \sum_{1 \le n < \infty} EI(\nu \ge n0 = Ez \sum_{1 \le n < \infty} P(\nu \ge n) = EzE\nu,$$

which gives us Wald's equivalency.

9.6.3 Sequential Optimal Test Criterion for Composite Hypotheses

Observe a sequence of i.i.d. r.v.'s

$$x_1, x_2, \ldots, x_n, \ldots. \tag{9.162}$$

The r.v. x_n takes its values from a measurable space $(X, \mathfrak{B})$, where X is a completely separable metric space and $\mathfrak{B}$ is its Borel σ-algebra. Distribution of x_n is given by density $f(x, \boldsymbol{\theta})$ with respect to some σ-finite measure μ on $(X, \mathfrak{B})$. Here $\boldsymbol{\theta} = (\theta_1, \ldots, \theta_m)$ is a vector parameter taking its values from some subset Θ of m-dimensional Euclid space R_m. We assume that set $\{x: f(x, \boldsymbol{\theta}) > 0\}$ does not depend on $\boldsymbol{\theta} \in \Theta$.

Let $(\Omega, \mathfrak{F}, P_\theta)$ be the probabilistic space on which the process (9.162) is defined; $\mathfrak{F}_n = \sigma(x_1, \ldots, x_n)\ \mathfrak{F}_n \subset \mathfrak{F}_{n+1} \subset \mathfrak{F}$, $n = 1, 2, \ldots$, be a system of σ-algebras related to process (9.162); and P_θ and E_θ be the probabilistic measure and mathematical expectation for a given value of parameter $\boldsymbol{\theta} \in \Theta$. Denote the standard Kullback–Leiber "information distance" between points $\boldsymbol{\theta}$ and φ from Θ by

$$\rho(\boldsymbol{\theta}, \varphi) = E_\theta \ln \left[\frac{f(x_n, \boldsymbol{\theta})}{f(x_n, \varphi)} \right].$$

The value

$$\rho(\boldsymbol{\theta}, D) = \inf_{\varphi \in D} \rho(\boldsymbol{\theta}, \varphi)$$

is called the "distance" from point $\boldsymbol{\theta}$ to set $D \subset \Theta$.

Let the set Θ of all possible values of parameter $\boldsymbol{\theta}$ be divided into $l + 1$ nonintersected subsets:

$$\Theta = D_1 \cup D_2 \cup \cdots \cup D_l \cup I,$$

where I is the area of "indifference." Let there be a correspondence between each point $\boldsymbol{\theta}$ and index $K(\boldsymbol{\theta})$ of "closest" set D_i; that is, $K(\boldsymbol{\theta})$ is defined by the condition

$$\rho(\boldsymbol{\theta}, D_{K(\boldsymbol{\theta})}) = \min_{1 \le i \le l} \rho(\boldsymbol{\theta}, D_i).$$

Introduce the value

$$\Lambda(\boldsymbol{\theta}) = \min_{i \ne K(\theta)} \rho(\boldsymbol{\theta}, D_i) \tag{9.163}$$

assuming that $\Lambda(\boldsymbol{\theta}) > 0$ for all points $\boldsymbol{\theta} \in D_1 \cup D_2 \cup \cdots \cup D_l$. The sequential criterion for a test of l composite hypotheses

$$H_i\colon \boldsymbol{\theta} \in D_i, \qquad 1 \le i \le l \tag{9.164}$$

by pair (τ, d), where τ is Markovian (with respect to the system of σ-algebras $\mathfrak{F}_n$, $n = 1, 2, \ldots$) stopping time and d is a decision made at the stopping time τ, or in other words, d is a $\mathfrak{F}_\tau$-measurable function taking l values. Hypothesis H_i is accepted if $d = i$, $1 \le i \le l$.

Introduce $\boldsymbol{\alpha} = (\alpha_1, \ldots, \alpha_1)$. Denote by K_α a class of all (λ, d) such that

$$P_\theta(d \ne i) \le \alpha_i \quad \text{for all } \boldsymbol{\theta} \in D_1,\ 1 \le i \le l \tag{9.165}$$

In other words, K_α is a class of all criteria (τ, d) with the probabilities of errors not larger than the given values $\alpha_1, \ldots, \alpha_1$. Further we construct criterion $(\tau^*, d^*) \in K_\alpha$, which is asymptotically optimal in class K_α in the sense of the average test volume for $\alpha_i \to 0$, $1 \le i \le l$.

Further discussion is based on the following idea. At first, for any hypothesis H_i, we construct a Markovian (with respect to system $\mathfrak{F}_n$) moment of exclusion of this hypothesis ν_i such that

$$P_\theta(\nu_i < \infty) \le \alpha_i \quad \text{for all } \boldsymbol{\theta} \in D_i.$$

This inequality means that the probability of exclusion of hypothesis H_i (i.e., to decide that it is untrue) at some test step does not exceed α_i if the hypothesis is true.

Let $\nu(\boldsymbol{\alpha}, D)$ be the moment of exclusion of a composite hypothesis of a general type H: $\boldsymbol{\theta} \in D$. [This moment is defined in (9.77).] The moment of exclusion of hypothesis H_i: $\boldsymbol{\theta} \in D_i$ is defined as

$$\nu_i = \nu(\alpha_i, D_i), \qquad 1 \leq i \leq l. \tag{9.166}$$

Sequential criterion (τ^*, d^*) for the test of composite hypotheses (9.164) can be constructed on the basis of the moments of exclusion (9.166) as follows. Denote ordered-in-time moments of exclusion (9.166) by

$$\nu_{(1)} \leq \nu_{(2)} \leq \cdots \leq \nu_{(l-1)} \leq \nu_{(l)}.$$

Now define the time stop τ^* and the decision made at this moment, d^*, as follows:

$$\tau^* = \nu_{(l-1)}, \qquad d^* = \arg\max_{1 \leq i \leq l} \nu_i; \tag{9.167}$$

that is, the test continues until moment τ^*, when all hypotheses except one are excluded and a single one, which is not excluded, is accepted. For $l = 2$, the sequential criterion of composite hypotheses (9.167) takes the form

$$\tau^* = \min(\nu_1, \nu_2),$$
$$d^* = \arg\max_{1 \leq i \leq 2} \nu_i = \begin{cases} 1 & \text{if } \nu_2 < \nu_1, \\ 2 & \text{if } \nu_1 < \nu_2; \end{cases}$$

that is, the test continues until the moment of exclusion of one hypothesis and at the moment the remaining hypothesis is accepted. If $\nu_1 = \nu_2$ at some moment, then any of these two hypotheses is accepted.

Using results found elsewhere (Pavlov, 1985, 1990), we can show that for general enough conditions, the moment of exclusion $\nu(\alpha, D)$ is asymptotically optimal for $\alpha \to 0$. It follows that the sequential criterion (τ^*, d^*) is also optimal (see Theorems 9.1–9.3).

Lemma 9.1 Moment $\nu(\alpha, D)$ satisfies the inequality

$$P_\theta\{\nu(\alpha, D) < \infty\} \leq \alpha \quad \text{for all } \boldsymbol{\theta} \in D \; \square \tag{9.168}$$

Proof Let each $\boldsymbol{\theta}$ correspond to the moment

$$\nu_\theta = \min\left(n\colon \xi_n(\boldsymbol{\theta}) \geq \frac{1}{\alpha}\right),$$

where

$$\xi_n(\boldsymbol{\theta}) = \frac{\Pi_{1\leq r\leq n} f(x_r, \hat{\theta}_{r-1})}{\Pi_{1\leq r\leq n} f(x_r, \theta)}.$$

For each fixed $\boldsymbol{\theta}$, sequence $\xi_n(\boldsymbol{\theta})$, $n = 1, 2, \ldots,$ represents a nonnegative martingale with respect to system $\mathfrak{F}_n$, $n = 1, 2, \ldots,$ with mathematical expectation $E_\theta \xi_n(\boldsymbol{\theta}) = 1$. Applying a Doob–Kolmogorov inequality, for each $n = 1, 2, \ldots,$ we have

$$P_\theta(\nu_\theta \leq n) = P_\theta\left(\max_{l\leq r\leq n} \xi_r(\boldsymbol{\theta}) \geq \frac{1}{\alpha}\right) \leq \alpha,$$

and then

$$P_\theta(\nu_\theta < \infty) \leq \alpha$$

for all $\boldsymbol{\theta} \in \Theta$. Moreover,

$$\nu(\alpha, D) \leq \sup_{\boldsymbol{\theta}\in\Theta} \nu_\theta.$$

It follows that for each $\boldsymbol{\theta} \in D$ the inequality

$$P_\theta\{\nu(\alpha, D) < \infty\} \leq P_\theta\left\{\sup_{\boldsymbol{\theta}\in D} \nu_\theta < \infty\right\} \leq P_\theta\{\nu_\theta < \infty\} \leq \alpha)$$

holds. This proves (9.168). □

The following lemma gives the LCL for the mathematical expectation of the moment of exclusion of a composite hypothesis H: $\boldsymbol{\theta} \in D$ in the case where it is untrue, that is, $\boldsymbol{\theta} \notin D$. □

Lemma 9.2 Let D be some subset Θ and ν be a Markovian moment (with respect to system $\mathfrak{F}_n$, $n = 1, 2, \ldots$) such that inequality $P_\theta(\nu < \infty) \leq \alpha$ for any $\boldsymbol{\theta} \in D$, where $0 < \alpha < 1$. Then the mathematical expectation of this moment satisfies the inequality

$$E_\theta \nu = \geq \frac{\ln(1/\alpha)}{\rho(\boldsymbol{\theta}, D)} \tag{9.169}$$

for any $\boldsymbol{\theta} \in \overline{D}$ where $\overline{D} = \Theta / D$. □

Proof Let $\boldsymbol{\theta} \in \overline{D}$; that is, the hypothesis H: $\boldsymbol{\theta} \in D$ is untrue. If $P_\theta(\nu < \infty) < 1$, then $E_\theta \nu = \infty$ and inequality (9.169) is trivial. Let $P_\theta(\nu < \infty) = 1$. Let us choose some point $\theta_0 \in D$. Consider two simple hypotheses h_0 and h_1 that correspond to points θ_0 and $\boldsymbol{\theta}$. Let us fix some moment $n < \infty$ and consider the following sequential criterion for the test of these simple criteria. The test continues until the truncated Markovian moment

$$\nu_n = \min(\nu, n).$$

If $\nu_n < n$, then hypothesis h is accepted. If $\nu_n = n$, then hypothesis h_0 is accepted. For this criterion the probability of error of type I (i.e., to accept hypothesis h when hypothesis h_0 is true) satisfies the relations

$$P_{\theta_0}(\nu_n < n) = P_{\theta_0}(\nu < n) \leq P_{\theta_0}(\nu < \infty) \leq \alpha.$$

The probability of error of type II (i.e., to accept hypothesis h_0 when hypothesis h is true) is denoted by

$$\beta_n = P_\theta(\nu_n = n) = P_\theta(\nu \geq n).$$

Since $P_\theta(\nu < \infty) = 1$, sequence $\beta_n \to 0$ for $n \to \infty$. Using the lower limit (9.48) of the average test volume for any criterion for two simple hypotheses h_0 and h, we have

$$E_\theta \nu_n \geq \frac{\omega(\beta_n, \alpha)}{\rho(\boldsymbol{\theta}, \theta_0)}, \tag{9.170}$$

where

$$\omega(\beta, \alpha) = (1 - \beta) \ln\left(\frac{1 - \beta}{\alpha}\right) + \beta \cdot \ln\left(\frac{\beta}{1 - \alpha}\right).$$

For $n \to \infty$ we have $E_\theta \nu_n \to E_\theta \nu$ and $\omega(\beta_n, \alpha) \to \ln(1/\alpha)$. From here, we obtain, taking the limit in (9.170) for $n \to \infty$, that

$$E_\theta \nu \geq \frac{\ln(1/\alpha)}{\rho(\boldsymbol{\theta}, \theta_0)}.$$

Since θ_0 is an arbitrary point in D, (9.169) follows. This proves the lemma. □

Notice that from a formal viewpoint, Lemma 9.2 and inequality (9.169) are true if $\boldsymbol{\theta} \in D$. In this case inequality (9.169) is trivial, because $\rho(\boldsymbol{\theta}, D) = 0$ and $E_\theta \nu = \infty$.

Denote the estimate of maximum likelihood of parameter $\boldsymbol{\theta}$ on the basis of test results $x_1, \ldots, x_n$ at the first n steps by $\hat{\theta}'_n = \hat{\theta}'_n(x_1, \ldots, x_n)$. Let us introduce the following two groups of conditions. The first group is given as follows:

1. Θ is compact;
2. X is compact;
3. $f(x, \boldsymbol{\theta}) > 0$ for all $(x, \boldsymbol{\theta})$ and continuous in $(x, \boldsymbol{\theta})$, function $\rho(\boldsymbol{\theta}, \varphi) > 0$ for all $\boldsymbol{\theta} \neq \varphi$; and
4. for any estimate $\hat{\theta}_n$ equivalent to the maximum-likelihood estimate $\hat{\theta}'_n$, the following equality holds:

$$P\left\{\lim_{n\to\infty} |\hat{\theta}_n - \hat{\theta}'_n| = 0\right\} = 1.$$

The second group of conditions does not use the condition of X compactness. This group is as follows:

1′. $f(x, \boldsymbol{\theta})$ belongs to the exponential family of distribution densities of the type

$$f(x, \boldsymbol{\theta}) = \exp\left\{\sum_{1\le i\le m} \theta_i T_i(x) - b(\boldsymbol{\theta})\right\},$$

and set

$$\Phi = \left\{\boldsymbol{\theta}: \int \exp\left[\sum_{1\le i\le m} \theta_i T_i(x)\right] d\mu(x) < \infty\right\}$$

is an open subset of m-dimensional Euclidean space R_m;

2′. Θ is a compact subset of Φ; and

3′. this condition coincides with condition 4.

Conditions 1–4 or 1′–3′ listed above lead to the results presented below. These results state the asymptotic optimality of the moment of exclusion, $\nu(\alpha, D)$, and the corresponding sequential criteria and confidence limits (Theorems 9.1–9.4). For a detailed proof of these theorems see Pavlov (1985, 1990). One can also find results for weaker conditions there. The following theorem with inequality (9.169) gives the asymptotic optimality for $(\alpha \to 0)$

of the moment of exclusion $\nu(\alpha, D)$ constructed in (9.77) for a composite hypothesis H: $\boldsymbol{\theta} \in D$.

Theorem 9.1 If $\alpha \to 0$,

$$E_\theta \nu(\alpha, D) \leq \frac{\ln(1/\alpha)}{\rho(\boldsymbol{\theta}, D)} [1 + o(1)] \tag{9.171}$$

for any $\boldsymbol{\theta} \in \overline{D}$. □

The following inequality gives the system lower limit for the average test volume for any sequential criterion $(\tau, d) \in K_a$.

Lemma 9.3 Let $\mathbf{a} = (\alpha t_1, \ldots, \alpha t_l)$, where $t_1, \ldots, t_l$ are arbitrary positive constants. Then for any $(\tau, d) \in K_a$ inequality

$$E_\theta \tau \geq \frac{\ln(1/\alpha)}{\Lambda(\boldsymbol{\theta})} [1 + o(1)] \tag{9.172}$$

is valid for all $\boldsymbol{\theta} \in \Theta$ if $\alpha \to 0$, where $\Lambda(\boldsymbol{\theta})$ is a function defined in (9.163). □

It is easy to check that the sequential criterion (τ^*, d^*) belongs to class K_a; in other words, it satisfies inequalities (9.165). Indeed, taking into account that $\nu(\alpha, D)$ is monotone decreasing in α, it follows from (9.171) that, for any $0 < \alpha < 1$,

$$E_\theta \nu(\alpha, D) < \infty$$

for arbitrary $\boldsymbol{\theta}$ such that $\rho(\boldsymbol{\theta}, D) > 0$. It follows that for arbitrary $0 < \alpha_i < 1$, $1 \leq i \leq l$, the inequality

$$P_\theta(\tau^* < \infty) = 1 \tag{9.173}$$

holds for any $\boldsymbol{\theta}$ such that $\Lambda(\boldsymbol{\theta}) > 0$. Let $\boldsymbol{\theta} \in D_i$; then, due to Theorem 9.1, inequality

$$P_\theta(\nu_i < \infty) = P_\theta\{\nu(\alpha_i, D_i) < \infty\} \leq \alpha_i$$

follows. From here, taking into account (9.173), we obtain

$$P_\theta(d^* = i) \geq P_\theta(\tau^* < \nu_i) \geq P_\theta(\tau^* < \infty, \nu_i = \infty) = P_\theta(\nu_i = \infty) \geq 1 - \alpha_i$$

for all $1 \leq i \leq l$. Thus, the sequential criterion $(\tau^*, d^*) \in K_a$, or in other words, it has probability of errors not larger than specified values $\alpha_1, \ldots, \alpha_l$.

The following theorem accompanied by inequality (9.172) gives asymptotic optimality of sequential criterion (τ^*, d^*) in the sense of the average test volume within class K_a for all criteria with probability of errors not larger than specified values $\alpha_1, \ldots, \alpha_l$.

Theorem 9.2 Let $\mathbf{a} = (\alpha t_1, \ldots, \alpha t_l)$, where $t_1, \ldots, t_l$ are arbitrary positive constants. Then the average test volume for sequential criterion (τ^*, d^*) satisfies the inequality

$$E_\theta \tau^* \leq \frac{\ln(1/\alpha)}{\Lambda(\boldsymbol{\theta})} [1 + o(1)] \tag{9.174}$$

for all $\boldsymbol{\theta} \in \Theta$ if $\alpha \to 0$. □

Notice that Lemma 9.3 and Theorem 9.2 are also valid if the area of indifference I is absent, for instance, for the case $l = 2$ of composite hypotheses of the form

$$H_1: R(\boldsymbol{\theta}) \leq R_1, \qquad H_2: R(\boldsymbol{\theta}) \geq R_2,$$

where critical levels R_1 and R_2 coincide: $R_1 = R_2$. In this case $\Lambda(\boldsymbol{\theta}) \to 0$ and average test volume $E_\theta \tau \to \infty$ for $R(\boldsymbol{\theta}) \to R_1 = R_2$. □

Sequential Confidence Limits Let $\underline{R}_n = \underline{R}_n(x_1, \ldots, x_n)$, $n = 1, 2, \ldots,$ be the sequential γ-LCL for $R = R(\boldsymbol{\theta})$. (Definition of sequential confidence limits was given in Section 9.4.) Assume that $R(\boldsymbol{\theta})$ is continuous in $\boldsymbol{\theta} \in \Theta$. Introduce a random moment (step) T_c where the limit $\underline{R}_n$ crosses a fixed level c:

$$T_c = \min\{n: \underline{R}_n \geq c\}. \tag{9.175}$$

We assume that $T_c = \infty$ if $\underline{R}_n < c$ for any finite $n = 1, 2, \ldots$.

If level $c < R(\boldsymbol{\theta})$, then T_c characterizes the test step where $\underline{R}_n$ reaches the specified accuracy for the first time, in other words, the specified deviation $\Delta = R(\boldsymbol{\theta}) - c$ from the true value of $R(\boldsymbol{\theta})$. The mathematical expectation of this moment, $E_\theta T_c$, characterizes the speed of attaining the true $R(\boldsymbol{\theta})$ by the confidence limit $\underline{R}_n$ or $c < R(\boldsymbol{\theta})$. The smaller the value of $E_\theta T_c$, the more is the effective sequential γ-LCL $\underline{R}_n$. The following lemma gives the LCL for $E_\theta T_c$.

Lemma 9.4 Let $\underline{R}_n$, $n = 1, 2, \ldots,$ be the sequential γ-LCL for $R(\boldsymbol{\theta})$. Then the mathematical expectation of moment (9.175) satisfies the inequality

$$E_\theta T_c \geq \frac{\ln(1-\gamma)^{-1}}{\rho(\boldsymbol{\theta}, D_c)} \tag{9.176}$$

for all $\boldsymbol{\theta} \in \Theta$, where $D_c = \{\boldsymbol{\theta}: R(\boldsymbol{\theta}) < c\}$. □

Proof First consider the case where $R(\boldsymbol{\theta}) \leq c$. In this case (9.167) is trivial since $\rho(\boldsymbol{\theta}, D_c) = 0$, and from the definition of the sequential LCL it follows directly that $E_\theta T_c = \infty$. Let now $R(\boldsymbol{\theta}) > c$. From the definition of the LCL follows the relation

$$\{T_c < \infty\} \subset \{\underline{R}_{T_c} \geq c\}.$$

It follows that at any point $\boldsymbol{\theta} \in D_c$ the following chain of inequalities is valid:

$$P_\theta(T_c < \infty) \leq P_\theta(\underline{R}_{T_c} \geq c) \leq P_\theta\{\underline{R}_{T_c} \geq R(\boldsymbol{\theta})\}$$
$$= 1 - P_\theta\{\underline{R}_{T_c} \leq R(\boldsymbol{\theta})\} \leq 1 - \gamma.$$

The proof of Lemma 9.2 follows. □

Let the sequential $(1-\alpha)$-LCL for $R(\boldsymbol{\theta})$ be defined by (9.95). The following theorem accompanied by Lemma 9.4 shows that this limit is asymptotically optimal (for $\alpha \to 0$) in the sense of the average time, $E_\theta T_c$, of attaining the specified accuracy. Introduce the notation

$$T_c^* = \min\{n: \underline{R}_n \geq c\},$$

where $\underline{R}_n$ is the confidence limit defined in (9.95).

Theorem 9.3 Inequality

$$E_\theta T_c^* \leq \frac{\ln(1/\alpha)}{\rho(\boldsymbol{\theta}, D_c)}[1 + o(1)] \tag{9.177}$$

holds for all $\boldsymbol{\theta} \in \Theta$ if $\alpha \to 0$. □

Proof For $c \geq R(\boldsymbol{\theta})$ inequality (9.177) is trivial, since $\rho(\boldsymbol{\theta}, D_c) = 0$. Let $c < R(\boldsymbol{\theta})$. By construction of sequence $\underline{R}_n$, $n = 1, 2, \ldots,$ in (9.95), the following relations are valid:

$$\{\nu_c = n\} \subset \{\underline{R}_n \geq c\} \subset (T_c^* \leq n\}.$$

It follows that $T_c^* \leq \nu_c$. Since $c < R(\boldsymbol{\theta})$, $\boldsymbol{\theta} \in \overline{D}_c$. From Theorem 9.1 it follows that

$$E_\theta T_c^* \le E_\theta \nu_c \le \frac{\ln(1/\alpha)}{\rho(\boldsymbol{\theta}, D_c)} [1 + o(1)],$$

which proves (9.177). □

Analogous results for sequential UCLs are formulated and obtained in the same manner.

9.6.4 Scheme of Dependent Tests

Assume that we observe a random sequence

$$x_1, x_2, \ldots, x_n, \ldots, \tag{9.178}$$

where, in contrast with (9.162), r.v.'s at different steps can be dependent. Let $x^{(n)} = (x_1, \ldots, x_n)$ be the set of observations at first n steps and

$$p_n(x^{(n)}, \boldsymbol{\theta}) = p_n(x_1, \ldots, x_{n-1}, x_n, \boldsymbol{\theta}), \qquad n = 1, 2, \ldots, \tag{9.179}$$

is the density of the distribution of $x^{(n)}$ with respect to measure μ^n on $(X^n, \mathcal{B}^n)$, $n = 1, 2, \ldots$. System of finite-measurable densities (9.179) must satisfy standard conditions of "accordance":

$$\int p_n(x_1, \ldots, x_{n-1}, x_n, \boldsymbol{\theta})\, d\mu(x_n) = p_{n-1}(x_1, \ldots, x_{n-1}, \boldsymbol{\theta}) \tag{9.180}$$

for all $\boldsymbol{\theta} \in \Theta$, $n = 1, 2, \ldots$. Assume that set $\{x^{(n)}: p_n(x^n, \boldsymbol{\theta}) > 0\}$, on which the density is defined, coincides with X^n for all $\boldsymbol{\theta} \in \Theta$, $n = 1, 2, \ldots$. Let us introduce the following notation:

$$p_n(x_n | x_1, \ldots, x_{n-1}, \boldsymbol{\theta}) = \frac{p_n(x_1, \ldots, x_{n-1}, x_n, \boldsymbol{\theta})}{p_{n-1}(x_1, \ldots, x_{n-1}, \boldsymbol{\theta})}$$

is the conditional density of distribution of x_n under the condition that test results $x_1, \ldots, x_{n-1}$ at the first $n - 1$ steps are known, and

$$B_n = \prod_{1 \le r \le n} p_r(x_r | x_1, \ldots, x_{r-1}, \hat{\theta}_{r-1}),$$

where $\hat{\theta}_n = \hat{\theta}_n(x_1, \ldots, x_n)$ is the point estimate of parameter $\boldsymbol{\theta}$ based on observations on the first $n - 1$ steps.

Consider composite hypothesis H: $\boldsymbol{\theta} \in D$ with respect to parameter $\boldsymbol{\theta}$, where D is the subset of the parameter space Θ. Introduce the moment of exclusion of this hypothesis, $\nu(\alpha, D)$, which is analogous to the moment (9.77) in the scheme of independent tests:

$$\nu(\alpha, D) = \min \left\{ n: B_n \geq \frac{1}{\alpha} \sup_{\theta \in D} p_n(x^{(n)}, \boldsymbol{\theta}) \right\}.$$

Let us show that this moment satisfies the inequality

$$P_\theta\{\nu(\alpha, D) < \infty\} \leq \alpha \quad \text{for any } \boldsymbol{\theta} \in D. \tag{9.181}$$

Indeed, from the definition of this moment, it follows that

$$\nu(\alpha, D) \geq \sup_{\theta \in D} \nu_\theta, \tag{9.182}$$

where

$$\nu_\theta = \left\{ \min n: B_n \geq \frac{1}{\alpha} p_n(x^{(n)}, \boldsymbol{\theta}) \right\}.$$

Consider the random sequence

$$\xi_n(\boldsymbol{\theta}) = \frac{B_n}{p_n(x^{(n)}, \boldsymbol{\theta})}, \qquad n = 1, 2, \ldots .$$

For each fixed $\boldsymbol{\theta}$, sequence $\xi_n(\boldsymbol{\theta})$, $n = 1, 2, \ldots$, is a nonnegative martingale [with respect to system $\mathfrak{F}_n = \sigma(x_1, \ldots, x_n)$, $n = 1, 2, \ldots$] with mathematical expectation $E_\theta \xi_n(\boldsymbol{\theta}) = 1$ for all n. Applying the Doob–Kolmogorov inequality for nonnegative martingales, we obtain

$$P_\theta(\nu_\theta \leq n) = P_\theta \left\{ \max_{1 \leq r \leq n} \xi_r(\boldsymbol{\theta}) \geq \frac{1}{\alpha} \right\} \leq \alpha$$

for each $n = 1, 2, \ldots$, from where it follows that

$$P_\theta(\nu_\theta < \infty) = \lim_{n \to \infty} P_\theta(\nu_\theta \leq n) \leq \alpha. \tag{9.183}$$

From (9.182) and (9.183), it follows that, for any $\boldsymbol{\theta} \in D$, the inequality

$$P_\theta\{\nu(\alpha, D) < \infty\} \leq P_\theta \left\{ \sup_{\theta \in D} \nu_\theta < \infty \right\} \leq P_\theta\{\nu_\theta < \infty\} \leq \alpha$$

holds, which proves (9.181).

In the particular case, if sequence (9.178) is Markovian with "transitive density" $f_n(x_n | x_{n-1}, \boldsymbol{\theta})$, then the formulas above take the form

$$p_n(x^{(n)}, \boldsymbol{\theta}) = p_n(x_1, \ldots, x_n, \boldsymbol{\theta}) = \prod_{1 \le r \le n} f_r(x_r | x_{r-1}, \boldsymbol{\theta}),$$

$$p_n(x_n | x_1, \ldots, x_{n-1}, \boldsymbol{\theta}) = f_n(x_r | x_{n-1}, \boldsymbol{\theta}),$$

$$B_n = \prod_{1 \le r \le n} f_r(x_r | x_{r-1}, \hat{\theta}_{r-1}).$$

Moment $\nu(\alpha, D)$ in this case coincides with (9.105), and inequality (9.181) is equivalent to (9.106).

CHAPTER 10

MONTE CARLO SIMULATION

10.1 NATURE AND PURPOSE

Monte Carlo simulation is used to imitate the behavior of complex systems whose operational processes are difficult or impossible to describe using analytical models. Computer simulations may also be used if an analytical model is available but numerical solution for this model requires more time than a simulation.

After building a model formally (in terms of formal description of interrelations between system states and processes), we should develop appropriate software or adapt one that may be available. Sometimes the model itself needs to be modified and made compatible with available software. We could do this by introducing some reasonable assumptions that simplify or modify the initial model. Sometimes the formal model needs corrections due to the lack of appropriate input data.

After these essential steps, we are ready to perform actual simulations. Monte Carlo simulation (runs) is a statistical imitation of possible behaviors of the investigated system in the frame of accepted assumptions and constraints. Data obtained as the result of simulation are processed in the same way as real data obtained during the system operation or field tests.

A reliability simulation model is commonly a discrete model, describing sequences of discrete events and their interactions. In reliability analysis, these events are failures and repairs of a system's units, switching from main units to redundant units, and interaction with external events (traffic in telecommunication networks, lightning in electric power systems, floods and hurricanes for various terrestrial systems, etc.). Development of a formal model requires creation of an algorithm that transforms a set of initial data into a sequence of discrete output events that are subject to further analysis. Statistical inferences are made about the behavior of the system after the runs of the model are completed.

A Monte Carlo simulation has the following important components:

- strong algorithmic description of the behavior of the investigated system and the interrelation of a system's units;
- software allowing one to perform the process of imitation of the behavior of an investigated system (probably, including some special analytical blocks);
- input data characterizing time-probabilistic properties of the system's units;
- generators of different necessary random variables with needed properties; and
- computer tools for obtaining statistical inferences about output data.

Usually, system imitating, a generator of random variables, and tools for processing output data are combined into a united software tool.

10.2 GENERAL DESCRIPTION OF A DISCRETE SIMULATION MODEL

A discrete model of the investigated system replicates the structure and units' interaction. Any model reflects only some essential sides of a real system, which are interesting in the undertaken research. The model cannot be complete. As physicists joke, "The best model of a cat is a cat, but the very best model of a cat is the same cat".

We will consider a discrete model of a system consisting of some units operating in continuous time. In fact, in a computer model time is also discrete, but its discreetness can be neglected.

Thus, the modeled process is represented by the sequence of discrete events that are caused by the unit's transition from state to state. The moments of those transitions are determined by random values that are generated by computer.

We will denote the system's units as $A_1, A_2, \ldots$. Each unit A_j is characterized with a set of attributes a_j. Attributes include the state s_j that describes the dynamics of the system's transition in time and some special auxiliary variables π_j. Thus, $a_j = (s_j, \pi_j)$.

Specified events e_j may occur with unit A_j. For example, the unit might fail or repair/replacement might renew failed units. The state s_j defines the moment of the event, and its content is defined by attributes a_j. The event e_j may follow a change of attributes of units. A set of units determining the event depends on the nature of the event.

One of the attributes, a_j, is the real variable τ_j, which is the time until the occurrence of a new event for this unit if there are no intervening events in the system that could change the behavior of this particular unit. At the initial time $t = 0$, $s_j = s_j^0$ and $\tau = \tau_j^0$ for each unit A_j. The first event occurs after a period of time $\theta_1 = \min \tau_j$ over all j. If $\tau_{j_1} = \theta_1$, the template event occurs in unit A_{j1}. Event e_{ji} occurring at time $t_1 = \theta_1$ changes attributes of one or more units. A new state and new residual time τ_j^1 must be found for each affected unit. For each unaffected unit the residual time is given by $\tau_j^1 = \tau_j^0 - \theta_1$. The procedure then continues, and we find a new interoccurrence time $\theta_2 = \min_j \tau_j^1 = \tau_{j_2}^1$, that is, the second event is e_{j_2} and it occurs at time $t_1 + \theta_2$. Again, the attributes and the residual times of one or more affected units are changed depending on the nature of the event e_{j_2}. For unaffected units, the new residual time $\tau_j^2 = \tau_j^1 - \theta_2$. This procedure continues on.

As the result of this procedure, we have the so-called governing sequence (t_1, e_{j_1}), (t_2, e_{j_2}), . . . , with corresponding attributes that allow one to reconstruct the entire trajectory of the simulation process.

10.3 DETAILED EXAMPLE OF ALGORITHMIC DESCRIPTION

In this section we consider a simple example that illustrates some details of Monte Carlo simulation. Consider a system consisting of m main and n standby redundant units. All units are identical and independent. There are l servers for repair, $1 \le l \le n + m$. The FIFO (first in, first out) discipline is supposed. The time to failure of each unit is a r.v. with d.f. $F(x)$ and the repair time is a r.v. with d.f. $G(x)$.

Thus the system has $n + m$ units $A_1, A_2, \ldots, A_{n+m}$ with respective attributes $a_i = (s_i, \pi_i)$, $i = 1, 2, \ldots, n + m$. The unit state is defined as follows:

$$s_i = \begin{cases} \vdots \\ 3 \text{ if a unit is second in line for repair,} \\ 2 \text{ if a unit is first in line for repair,} \\ 1 \text{ if unit } i \text{ is in repair,} \\ 0 \text{ if unit } i \text{ is operating,} \\ -1 \text{ if unit } i \text{ is first in line for replacement,} \\ -2 \text{ if unit } i \text{ is second in line for replacement,} \\ \vdots \end{cases}$$

(Here assume that the line of spare units is arranged in accordance with order of completion of their repair.)

The residual time τ_i is defined only for states 0 or 1, and in the first case it is the residual time to failure and in the second case it is the residual repair time. In other cases assume that $\tau_i = \infty$.

Event e_i can take only values 0 (failure) and 1 (repair). Assume that for unit A_i the attribute π_i is defined by the number of failures that have occurred at the current moment of time.

Let the model be described by the vector $[(s_1^k, \pi_1^k, \tau_1^k), (s_2^k, \pi_2^k, \tau_2^k), \ldots, (s_{n+m}^k, \pi_{n+m}^k, \tau_{n+m}^k)]$ at the time t_k. The nearest occurrence time then is $t_{k+1} = t_k + \theta$, where $\theta = \min_{1\le i\le n+m} \tau_i^k$ is the shortest residual time for all four considered units at time t_{k+1}. Let $\theta = \tau_r^k$, which means that event e_r occurs at time t_{k+1}.

Consider the following two cases:

(a) For the first case, let $e_r = 0$ and, consequently, $s_r^k = 0$. Then for all standby units A_α (defined by the condition $s_\alpha < 0$), we can write

$$(s_\alpha^{k+1}, \pi_\alpha^{k+l}, \tau_\alpha^{k+1}) = \begin{cases} (s_\alpha^k + 1, \pi_\alpha^k, \infty) & \text{if } s_\alpha^k < -1, \\ (0, \pi_\alpha^k, \xi_k) & \text{if } s_\alpha^k = -1. \end{cases}$$

Thus one of the standby units (if any) becomes operating and others advance in the queue. Here $\{\xi_k\}$ are times to failure of the unit A_α (All these r.v.'s are i.i.d. with d.f. F.).

Let γ^k be the number of repair servers busy with the repair at moment t_k (and hence at time t_{k+1} because no repair is finished until this moment). Thus γ^k is the number of A_j for which $s_j^k = 1$. If there is at least one empty repair server, that is, $\gamma^k < 1$, then

$$(s_r^{k+1}, \pi_r^{k+1}, \tau_r^{k+1}) = (1, \pi_r^k + 1, \eta_k),$$

where $\{\eta_k\}$ are i.i.d. r.v.'s with d.f. $G(t)$ and η_k is the repair time for unit A_r. When a repair is completed, the auxiliary variable π_r increases by 1.

If $\gamma^k = 1$,

$$(s_r^{k+1}, \pi_r^{k+1}, \tau_r^{k+1}) = (\max_{1\le i\le n+m} s_i^k + 1, \pi_r^k + 1, \infty).$$

The unit joins the repair queue in position with the number $\max_{1\le i\le n+m} s_i^k + 1$. For units A_β ($\beta \ne r$) that are failed and a main unit (for which $s_\beta^k \ge 1$), we make the following changes:

$$(s_\beta^{k+1}, \pi_\beta^{k+1}, \tau_\beta^{k+1}) = (s_\beta^k, \pi_\beta^k, \tau_\beta^k - \theta).$$

Here we take into account that for units A_β in the queue for which $s_\beta^k > 1$, $\tau_\beta^k = \infty$.

(b) For the second case, let $e_r = 1$. It means that the repair of unit A_r has just been completed, that is, $s_r^k = 1$. In this case the repair server who just completed the repair takes the first unit in the queue (if any), and other units advance one position. The units under repair stay at their repair places, but all residual repair times decrease by θ. Thus

$$(s_\alpha^{k+1}, \pi_\alpha^{k+1}, \tau_\alpha^{k+1}) = \begin{cases} (s_\alpha^k - 1, \pi_\alpha^k, \infty) & \text{if } s_\alpha^k > 2, \\ (1, \pi_\alpha^k, \xi_k) & \text{if } s_\alpha^k = 2, \\ (s_\alpha^k, \pi_\alpha^k, \tau_\alpha - \theta) & \text{if } s_\alpha^k = 1. \end{cases}$$

Let ψ^k be the number of operable units at time t_k, that is, the number of units A_j for which $z_j^k \leq 0$. If $\psi^k < m$, then the repaired unit is directed to occupy the position of a main unit:

$$(s_r^{k+1}, \pi_r^{k+1}, \tau_r^{k+1}) = (0, \pi_r^{k+1}, \xi_k).$$

If $\psi^k = m$, unit A_i becomes the last unit in the queue of spare units:

$$(s_r^{k+1}, \pi_r^{k+1}, \tau_r^{k+1}) = \max_{1 \leq i \leq m+n} (s_i^k - 1, \pi_r^k, \infty).$$

All main and standby units stay in their places. The residual time of all main units decrease by θ:

$$(s_\beta^{k+1}, \pi_\beta^{k+1}, \tau_\beta^{k+1}) = (s_\beta^k, \pi_\beta^k, \tau_\beta^k - \theta) \quad \text{for } s_\beta^k \leq 0.$$

These relationships entirely describe the chosen model.

Notice that this description of the model does not give us an opportunity to analyze the behavior of repair servers because they were considered only as a group of objects without their individual attributes. This simplification of the model was done for a concise explanation of the essential features.

In Figure 10.1 we present a sample of trajectories of the system's units. The system under consideration consists of one main and two standby units with a single repair server. Let us comment on changing the system states represented in this figure.

The system has three units A_1, A_2, and A_3. The trajectories of changing the states for each unit are represented as three staircase functions.

Consider an initial state of units. The first unit A_1 is on working position (state $s_1 = 0$); its time to failure is generated by the counter of random values. This time is $\tau_1^1 = t_1$. The interval on which a unit is working is denoted by a double line on the upper trajectory. During period (t_0, t_1), the second unit is the first in the line for replacement (state $s_2 = -1$) and the third one is on the second place in this line ($s_2 = -2$). Intervals where a unit is waiting for installation into the working position ($s_i = -1$ or $s_1 = -2$) are denoted by thin lines.

A random value of TTF of unit A_1 is generated. Moment t_1 is defined as the moment of failure of unit A_1. At this moment unit A_1 is directed to a

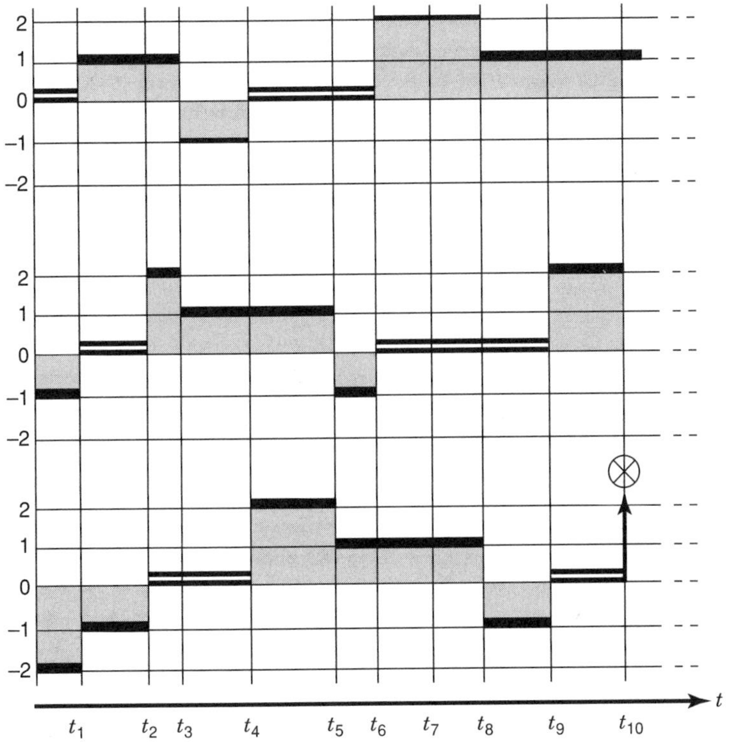

Figure 10.1 Sample trajectories of changing system states.

repair shop, and its state becomes $s_1 = 1$. The duration of the repair is generated equal to η_1^1. We denote the interval of repair by a bold line. Unit A_2 becomes a working one (state $s_2 = 0$) and its generated TTF equals τ_2^1. At the same moment, A_3 changes its state to $s_3 = -1$.

Moment t_2 is defined as $\min(\eta_1', \tau_2')$. (Here we use a prime to denote a residual value.) In our case, it happens that $\eta_1^1 > \tau_2^1$. At moment t_2, and A_2 has failed and moves to the queue for the repair shop (the failed unit A_1 is still repaired), that is, its state becomes $s_2 = 2$. The interval of waiting in a queue is denoted by a thin line. Unit A_3 becomes working (state $s_3 = 0$). The residual time of repair of unit A_1 equals $\eta_1^1 - (t_2 - t_1) = t_3 - t_2$.

Moment t_3 is defined as $\min(\eta'^1_1, \tau_3^1)$. (Here we use a prime to denote a residual value.) In our case, it happens that $\eta_1'^1 < \tau_3^1$. At the moment t_3, unit A_1 has been repaired and takes the state $s_1 = -1$ because unit A_3 keeps working. Unit A_2, which was waiting for repair, enters the repair shop (changes state $s_2 = 2$ for $s_2 = 1$). And so on.

We will not continue the description of the trajectories any further. Note only that at the moment t_{10} unit A_3 has failed and at the same time unit A_1 is under repair and unit A_2 is failed and waiting for repair. The failure of A_3

means that the system as a whole has failed (here we use conditional notation: a cross on the level s_3).

Thus realization of Monte Carlo simulation is represented in a form close to that obtained in a real testing: One records system failures, times to and between failures, and duration of repair. All these data can be used in an ordinary way to obtain different reliability indices.

Of course the description of the above model could be different. It might contain more or less details. For instance, we could only be interested in numbers of repaired and standby units (without description of their individual behavior). It makes sense if all units are identical and independent. If we have several repair servers, it might be interesting to obtain information about their loading. The reader can find detailed description of some related models in *Handbook of Reliability Engineering* (Ushakov, 1994, pp. 459–461).

10.4 RANDOM NUMBERS

Monte Carlo simulation uses random values with required distributions. Many modern software libraries have random number generators that produce uniformly distributed random numbers. In fact, a computer generates the so-called pseudo-random numbers. The procedure of such a generation is a recurrrent computation with withdrawal of some intermediate numbers, for instance, digits on the positions from k to $k + n$ in an N-digit number, $N >> k + n$. It is clear that a recurrent procedure generates cycles when the initial state repeats during a calculation. The main problem is to make such a cycle long enough to cover a Monte Carlo experiment with "independent" pseudo-random numbers. At the same time it is very important to avoid regular dependence on neighboring numbers.

The idea of generating pseudo-random numbers with the help of recurrent calculations belongs, most probably, to John von Neuman, who used it more than 50 years ago. Now this type of procedure of random number generation is standard.

However, Monte Carlo models need not just uniformly distributed random numbers. The problem of generating random numbers with an arbitrary given distribution can be done using the so-called probability integral transform. Let us explain this considering a simple example. Suppose we have a huge sample S_N from a distribution $F(x)$: $X_1, X_2, \ldots, X_N$. Order these realizations of an r.v. and draw a histogram (see Figure 10.2).

It is clear that, if $N \to \infty$ the ladder-like function $\hat{F}(x)$ will converge to $F(x)$ (this follows from Glivenko's theorem; see Gnedenko and Ushakov, 1995). Now let us take a sample of size n, S_n, from the original sample S_N, assuming that $N >> n$. To take this sample, one should use uniformly distributed random numbers in the following way.

Let N be a k-digit number (for the sake of simplicity) with a value from 0 to 10^k–1. Consider a set of n k-digit uniformly distributed random numbers. To form sample S_n, let us pick up values $X_{j_1}, X_{j_2}, \ldots, X_{j_n}$ whose subscripts

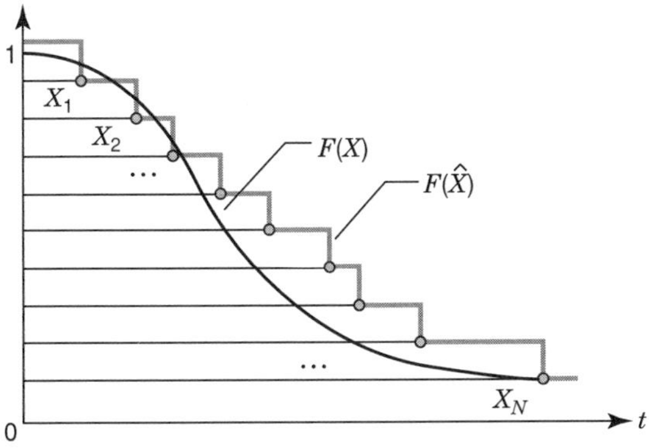

Figure 10.2 Distribution function $F(X)$ and empirical function $F(\hat{X})$.

coincide with random numbers chosen above (see Figure 10.1). In other words, we take a uniformly distributed ordinate of the function $\hat{F}(x)$ (which represents an empirical distribution) and obtain in response a realization of an r.v. having this distribution.

After these obvious explanations, we can formulate the following rule. If there is a given arbitrary d.f. $F(x)$ and a set of uniformly distributed r.v.'s X_1, $X_2, \ldots, X_n$, in [0, 1], then one can obtain n r.v.'s $Y_1, Y_2, \ldots, Y_n$ with distribution $F(x)$ by solving the equation

$$F(Y_j) = X_j,$$

or in an equivalent form,

$$Y_j = F^{-1}(X_j).$$

Let us now describe this problem in a more strict way. Suppose that an r.v. X has a d.f. F and let $Y = F(X)$. Let us show that the distribution of Y is uniform on the interval (0, 1). By the definition of the distribution function, $0 < F(x) < 1$ for $-\infty < x < \infty$. Let x_0 be a number such that $F(x_0) = y$, where $0 < y < 1$. Since, by definition, F is strictly increasing, there exists a unique number x_0 such that $F(x_0) = y$. However, if $F(x) = y$ over an entire interval of values of x, then we can choose x_0 arbitrarily from its interval. If G denotes the d.f. of Y, then

$$G(y) = Pr(Y \leq y) = Pr(X \leq x_0) = F(x_0) = y.$$

It follows that $G(y) = y$ for $0 < y < 1$. Since this function corresponds to the uniform distribution on the interval (0, 1), Y has this uniform distribution.

Now suppose that X is an r.v. with d.f. F and that G is some other d.f. It is required to construct a random variable $Z = r(X)$ for which the d.f. will be G.

For any value of y, $0 < y < 1$, let $z = G^{-1}(y)$ be any number such that $G(z) = y$. We can now define the random variable Z in the following way:

$$Z = G^{-1}[F(X)].$$

To verify that the d.f. of Z is actually G, we note that, for any number z such that $0 < G(z) < 1$,

$$Pr(Z \leq z) = Pr\{G^{-1}[F(X)] \leq z\} = Pr[F(X) \leq G(z)].$$

It follows from the probability integral transformation that $F(X)$ has a uniform distribution and, consequently, that

$$Pr[F(X) \leq G(z)] = G(z).$$

Hence, $P(Z \leq z) = G(z)$, that is, G is the distribution of Z. Thus, the theorem is proven.

Generation of Bernoulli's r.v.'s, which are frequently used in reliability modeling, in principle, is similar. However, it has a simpler explanation. Let us construct a sample of a Bernoulli's sequence with the probability of success (denoted as 0) equal to p and the probability of failures (denoted as 1) equal to $q = 1 - p$. Suppose that a random number generator produces uniformly distributed values ξ_j, $0 \leq \xi_j \leq C$. Introduce a threshold A such that $A = qC$. Bernoulli's random variable B_j is generated by the rule

$$B_j = \begin{cases} 0 & \text{if } \xi_j \leq A, \\ 1 & \text{if } \xi_j > A. \end{cases}$$

Obviously, this type of probability transform is not the only way of generating r.v.'s with a given distribution on the basis of uniform distributed r.v.'s. For instance, in many software implementations, a normally distributed r.v. is obtained as a sum of a very restricted number of uniformly distributed r.v.'s. This method is based on the central limit theorem (Gnedenko and Ushakov, 1995). If the number of uniformly distributed r.v.'s in the sum equals k, the generated quasi-normal distribution will have mean equal to $\frac{1}{2}$ and the variance equal to $\frac{1}{12}$.

10.5 SAMPLE COMPUTER SIMULATION

Monte Carlo simulation is especially important if we intend to analyze a nonstationary process. Analytical results for this case are available only for simple Markov models. A Markov model implies that all distributions within the model are assumed exponential.

Consider a simple model: a single renewable unit with the exponential distribution of time between failures and constant repair time. For an illustration let us choose MTBF and mean repair time equal. Consider the mechanism of Monte Carlo simulation on an example with two realizations.

Each realization of the unit operation process might be presented as an alternating process where 1, as above, corresponds to the failure state and 0 corresponds to the operational state. In Figure 10.3 we present two separate realizations for two units. There is also a superposition of these two processes in a form of a staircase-like function:

- 2 corresponds to moments of time where both units have been failed,
- 1 corresponds to moments where one unit is up and another is down, and
- 0 means that both units are operating.

It is clear that increasing the number of superimposed realizations makes the resulting function more "continuous;" that is, discrete increments become smaller and smaller.

In Figure 10.4 we present results of computer simulation for different numbers of realizations: 165 [see (*a*)], 1293 [see (*b*)], and 11,013 [see (*c*)]. (The duration of simulation period is 10 MTBF.) In Figure 10.5 the same curves are presented in a smoothed form: Each realization is presented in the form of a running average with time window about 0.25 of the unit of time.

This illustrative example is simple by its nature. However, it demonstrates dependence of result accuracy on the number of realizations.

10.6 MODELING NETWORK RELIABILITY

The advantage of Monte Carlo simulation for continuous processes of a complex nature is almost obvious: Sometimes we have no way of solving a huge

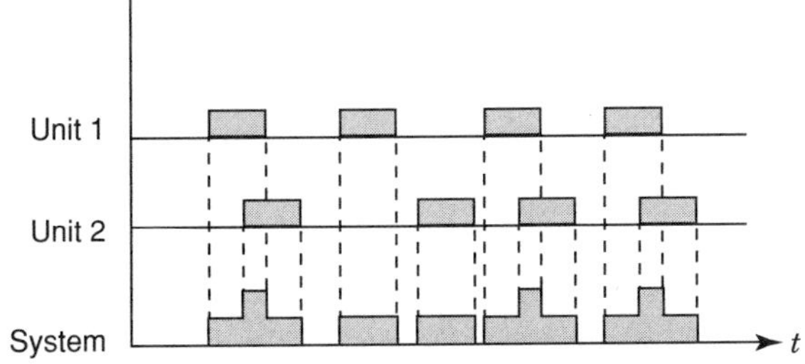

Figure 10.3 Example Monte Carlo simulation with two realizations.

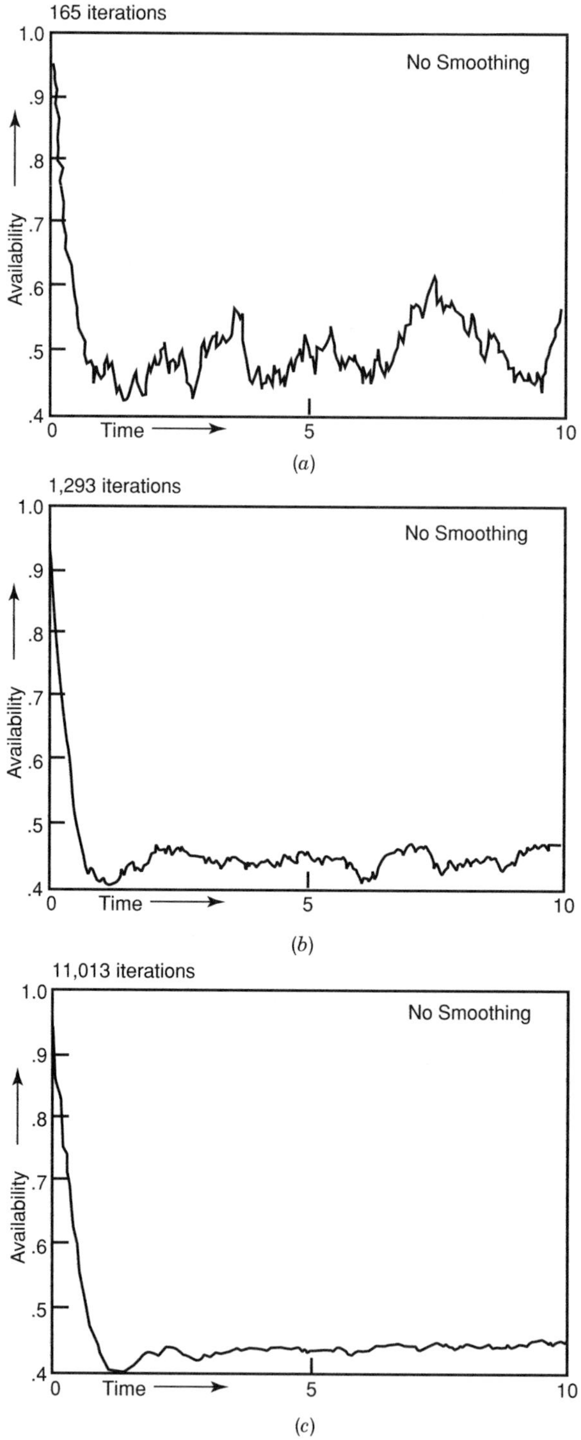

Figure 10.4 Results of Monte Carlo simulations for different sizes of samples: (*a*) small sample; (*b*) average sample; (*c*) large sample. (*Source:* From I. A. Ushakov (Ed.), *Handbook of Reliability Engineering,* Wiley, New York, 1994, p. 469–471. Copyright © 1994 John Wiley & Sons, Inc. Reprinted by permission.)

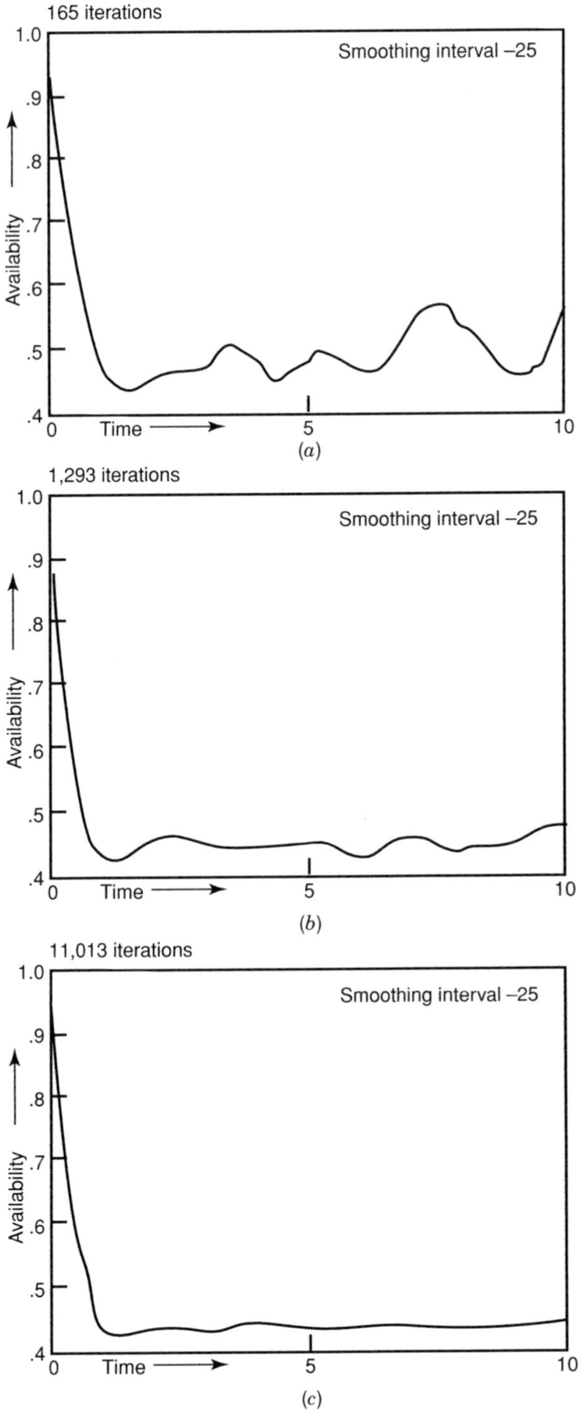

Figure 10.5 Smoothed results of Monte Carlo simulations for different sizes of samples: (*a*) small sample; (*b*) average sample; (*c*) large sample. (*Source:* From I. A. Ushakov (Ed.), *Handbook of Reliability Engineering,* Wiley, New York, 1994, p. 469–471. Copyright © 1994 John Wiley & Sons, Inc. Reprinted by permission.)

system of integro-differential equations even with the use of powerful interactive procedures.

Another case requiring the use of the Monte Carlo method is analysis of networks of complex structure. We will not consider a queuing network, that is, a network whose nodes can generate an input traffic of messages and process transit traffic messages (as servers of a queuing system). In general, such a model will be very close to that described in Section 10.3. For the sake of simplicity, we will consider a two-pole network with links and nodes subjected to failures. Let us focus on the probability of connectivity of the network.

In general, the procedure of simulation of a network in this case consists of the following steps:

- description of the two-pole network (a matrix of connectivity: what nodes are connected by direct links),
- definition of criteria of connectivity of the two distinguished poles,
- the procedure of generating random states of the network,
- the procedure of checking the connectivity of the network for each realization of state,
- collection of output data, and
- statistical inferences.

If the system's units are highly reliable, the direct "event-by-event" simulation takes too much computer time because a great number of realizations may be required before obtaining the necessary number of events of interest for getting the required level of confidence. (The planning of test volume is outside the scope of this book.)

Consider a two-pole network. Such a network is supposed to operate successfully if its input and output nodes are connected by some chain of links. The system is assumed to have n nodes and k links, each characterized by the probability of successful operation p_i. The system can be in 2^{n+k} different state. Of course, enumerating all states, evaluation of their probability, and checking for each stae if the network connectivity has taken place or not present enormous calculating problems. For example, even for a moderate-size network of 20 nodes and 50 links, the total number of states is about 10^{21}. In this case the Monte Carlo method is obviously more effective than a direct calculation. Usually several thousands of realizations are enough for statistical estimate. (In this particular case, each realization needs the generation of 70 Bernoulli r.v.'s and checking every time for network connectivity.)

For illustrative purpose, we consider a simple two-pole network, namely, the so-called bridge structure. All links are assumed independent and identical in their reliability parameters. Assume that the probabilities of the link's successful operation $p_i = 0.5$ for all i. (We take this condition only for simplicity of checking the result obtained.) Assume that the structure nodes are absolutely reliable. Table 10.1 lists three-digit random numbers ξ_i uniformly distributed.

Table 10.2 contains Bernoulli r.v.'s B_i for units of the bridge structure. These r.v.'s are formed in accordance with the following criteria:

$$B_i = \begin{cases} 1 & \text{if } \xi_i \le 1000p_i, \\ 0 & \text{if } \xi_i > 1000p_i; \end{cases}$$

that is, $B_i = 1$ corresponds to the operational state and $B_i = 0$ to the failure of the ith link.

Example 10.1 Each row of Table 10.2 contains five realizations of Bernoulli r.v.'s corresponding to links of the considered bridge structure. These realizations of link states are used to obtain the system state realizations B_{sys} (see column 6). These realizations are obtained in accordance with the rule

$$B_{\text{sys}} = (B_1 \wedge B_4) \vee (B_2 \wedge B_5) \vee (B_1 \wedge B_3 \wedge B_5) \vee (B_2 \wedge B_3 \wedge B_4), \tag{10.1}$$

Table 10.1 Uniformly Distributed Random Numbers by Unit of Bridge Structure

1	2	3	4	5
707	806	544	61	823
657	866	17	836	332
155	112	346	209	266
560	838	499	85	298
899	668	269	362	73
674	558	883	332	373
224	185	187	583	667
939	82	812	888	207
340	576	909	209	936
762	379	28	50	894
182	883	729	16	490
493	540	31	302	891
161	837	465	808	590
623	91	781	193	3
555	684	18	663	933
630	902	194	677	220
579	406	73	753	566
222	358	345	591	565
522	15	951	693	407
926	77	149	407	310

Table 10.2 Sample Monte Carlo Simulation for Bridge Structure

Bernoulli Random Variables Unit States					System State
0	0	0	1	0	0
0	0	1	0	1	0
1	1	1	1	1	1
0	0	1	1	1	0
0	0	1	1	1	0
0	0	0	1	1	0
1	1	1	0	0	0
0	1	0	0	1	1
1	0	0	1	0	1
0	1	1	1	0	1
1	0	0	1	1	1
1	0	1	1	0	1
1	0	1	0	0	0
0	1	0	1	1	1
0	0	1	0	0	0
0	0	1	0	1	0
0	1	1	0	0	0
1	1	1	0	0	0
0	1	0	0	1	1
0	1	1	1	1	1

that is, if at least one path between input and output exists, the system is supposed to be operational.

From the statistical experiment performed, we obtain the result $\hat{P}_{\text{sys}} = \frac{9}{20} = 0.45$. We intentionally choose $p_i = 0.5$ for all i because in this case the probability of bridge structure connectivity can be easily calculated. Indeed,

$$P_{\text{sys}} = \binom{5}{0}p^5 + \binom{5}{1}p^4q + \left[\binom{5}{2} - 2\right]p^3q^2 + 2p^2q^3$$
$$= (1 + 5 + 8 + 2) \times 0.5^5 = \tfrac{16}{32} = 0.5.$$

The reader can find more details about network structures and, in particular, about the bridge structure in Gnedenko and Ushakov (1995). □

Of course, where the units are equally or almost equally reliable, the procedure can be improved. Particularly, analytical and Monte Carlo methods can be fixed to reduce the total time of simulation. Such a combination of methods is especially useful in the analysis of highly reliable systems.

10.6.1 Simulating with Fixed Failed States

Let a network include many links that are highly reliable and almost identical (in a probabilistic sense). Besides, assume that the network is redundant, that is, system failure cannot occur if the number of link failures has not exceeded, say, M. In this case many realizations will have no useful information. For instance, if the number of failed links k in some realization is less than M, there is no real information in this particular test: It is clear that $k < M$ failures never leads to system failure.

Moreover, it may occur that "informative states" appear very seldom. In this case we can use the following procedure: First, compute analytically the probabilities of different system states with fixed numbers of failed units and, then, estimate the conditional probabilities of system failure under the condition that the number of links is fixed by simulation. The general procedure follows:

1. Analytically calculate probabilities $P_{(k)}$ of the states, which have exactly k failed units. Under condition of almost identical units, this probability is

$$P_{(k)} = \binom{n}{k} q^k p^{n-k}.$$

2. Use Monte Carlo simulation to estimate each conditional probability $\hat{\Phi}_{(k)}$ that the system in state $H_{(k)}$ is operational.
3. Compute the total probability of system successful operation:

$$P = \sum_{0 \le k \le M} P_{(k)} + \sum_{M+1 \le k \le n} \hat{\Phi}_{(k)} P_{(k)}.$$

Example 10.2 Consider the same bridge system and use the random numbers in Table 10.1. As we know, this system cannot fail if the number of failed units is less than 2. It means that $\hat{\Phi}_{(0)} = \hat{\Phi}_{(1)} = 1$. At the same time the system is failed if its four or five units are failed, that is, $\hat{\Phi}_{(4)} = \hat{\Phi}_{(5)} = 0.6$. Thus we need to estimate the conditional probabilities of the system failure for the cases where two or three units are failed.

Let us use the first 10 rows of Table 10.1 to simulate two failures in the bridge structure and the next 10 rows to simulate three failures. In the first case let us choose two smallest numbers within each row and consider them as failed units. As the result we construct Table 10.3, with exactly two failed links and using the same formula (10.1) to determine the sixth column (system state). Analyzing Table 10.3, we conclude that $\hat{\Phi}_{(2)} = 0.7$.

Table 10.4 contains analogous results of simulation with exactly three failed links. We choose three largest numbers within each row of the second part of Table 10.1 (rows from 11 to 20) and consider them as failed units. Analyzing Table 10.4, we conclude that $\hat{\Phi}_{(3)} = 0.4$. Thus the resulting mixed estimate, which was obtained using both analytical calculations and experimental estimation, is found as follows:

Table 10.3 Sample Monte Carlo Simulation for Bridge Structure with Two Failed Links

Bernoulli Random Variables for Unit States					System State
1	1	0	0	1	1
1	1	0	1	0	1
0	0	1	1	1	0
1	1	1	0	0	0
1	1	0	1	0	1
1	1	1	0	0	0
1	0	0	1	1	1
1	0	1	1	0	1
0	1	1	0	1	1
1	1	0	0	1	1

$$\tilde{P} = (0.5)^5 + 5 \times (0.5)^5 + (0.7) \times 10 \times (0.5)^5$$
$$+ (0.4) \times 10(0.5)^5 = 17 \times (0.5)^5 \approx \tfrac{0}{53}.$$

10.6.2 Modeling Link Failures Until System Failure

In this case links of the network are excluded one by one. The failure criterion is again the loss of network connectivity. When the system fails, the number of links k_s that were excluded in the sth realization is stored in computer memory. After a sufficient number of experiments N, we use the following estimate for the conditional probability of the loss of connectivity:

Table 10.4 Sample Monte Carlo Simulation for Bridge Structure with Three Failed Links

Bernoulli Random Variables for Unit States					System State
0	1	1	0	0	0
0	1	0	0	1	1
0	1	0	1	0	0
1	0	1	0	0	0
0	1	0	0	1	1
0	1	0	1	0	0
1	0	0	1	0	1
0	0	0	1	1	0
0	0	1	1	0	0
1	0	0	1	0	1

$$\Phi_{(k)} = \frac{1}{N} \sum_{1 \le s \le N} \delta(k_s),$$

where we use an indicator function

$$\delta(k_s) = \begin{cases} 1 & \text{if } k_s \le k, \\ 0 & \text{otherwise.} \end{cases}$$

Example 10.3 Let us illustrate the method on a bridge structure. We will use the same table of uniformly distributed numbers. Consider each row of the table and exclude links in order corresponding to increase in the numbers. The numbers in Table 10.5 show the order of the link's failures. From Table 10.5 we can see that in 10 cases of the total 20 the structure can stand three failure and fails only after the fourth failure has occurred. This gives us $\hat{\Phi}_{(3)} = \frac{10}{20}$. Only in 3 cases did system failure occur after failure of the second link. It means that $\hat{\Phi}_{(2)} = \frac{17}{20}$.

Table 10.5 Sample Monte Carlo Simulation for Order of Failed Links until System Failure

Order of Link Failure					Number of Failed Links Until System Failure
3	4	2	1	—	4
3	—	1	—	2	3
2	1	—	—	—	2
—	—	—	1	2	2
—	—	2	3	1	3
—	—	—	1	2	2
3	1	2	—	—	3
—	1	3	4	2	4
2	3	—	1	—	3
—	3	1	2	—	3
2	—	—	1	3	3
3	4	1	2	0	4
1	—	2	4	3	4
—	2	—	3	1	3
2	4	1	3	—	4
3	—	1	—	2	3
4	2	1	—	3	4
1	3	2	—	—	3
3	1	—	—	2	3
—	1	2	4	3	4

Note: All examples are purely illustrative. The reader should also keep in mind that the use of the same table of random numbers makes all results dependent on each other.

10.6.3 Accelerated Simulation

Direct Monte Carlo simulation is convenient due to simplicity of the model. But, as we mentioned above, if the system's units are highly reliable or/and the system's structure is highly redundant, direct simulation may be ineffective, since many realizations will be noninformative. The solution may be obtained by simulation of similar structures but with less reliable units and transformation of the results afterward. (Mizin, Bogatyrev and Kuleshov, 1986)

Direct Transform Consider a network state where there are z failed links belonging to some set u. For instance, it might be some network cut disjoining two specified nodes (say, input and output). The frequency of system failures due to this cut tends to its probability

$$P_u = \prod_{i \notin u} p_i \prod_{i \in u} (1 - p_i).$$

If n denotes the total number of network links, then

$$P_u = \prod_{1 \leq i \leq n} p_i \prod_{j \in u} \frac{1 - p_j}{p_j}.$$

Let is now introduce into the network units with lesser reliability, $p_{i'}$, for which the condition

$$\frac{1 - p_{i'}}{p_{i'}} = \gamma \frac{1 - p_i}{p_i}$$

holds. This condition can be rewritten in the form

$$p_{i'} = p_i[p_i + \gamma(1 - p_i)]^{-1}.$$

After substitution of new values for all network links, the probability of occurrence of network failure due to failure of cut u will change and be equal to

$$P_{u'} = \prod_{1 \leq i \leq n} p_i[p_i + \gamma(1 - p_i)]^{-1} \prod_{j \in u} \gamma \frac{1 - p_i}{p_j}.$$

Let us now introduce the coefficient of modeling acceleration, δ, which characterizes how the frequency of occurrence of a cut increases:

$$\delta = \frac{P_{u'}}{P_u} = \gamma^z \prod_{1 \le i \le n} [p_i + \gamma(1 - p_i)]^{-1}, \tag{10.2}$$

where z is the power of set u, that is, the number of links belonging to this cut. Notice that the second multiplier in (10.2) is a constant for a specified network:

$$K = \prod_{1 \le i \le n} [p_i + \gamma(1 - p_i)]^{-1}.$$

Thus, for cuts of the same power the relative frequency of occurrence is preserved. At the same time, the relative frequency of cuts of power $z + \omega$ increases in δ_ω times. Thus, during Monte Carlo simulation, we are able to collect satisfactory statistics for cuts of larger power that occur in regular simulation with a negligible probability.

Inverse Transform Inverse calculation of the probability of occurrence of failure of cut u can be performed by the formula

$$P_u = \frac{1}{\delta} P_{u'} = \frac{1}{\delta^z K} P_{u'}. \tag{10.3}$$

We emphasize that the power of a corresponding cut plays an essential role in the inverse transform. From (10.3), we see that the frequency of failure of cut u with power z in a transformed model is δ times higher than the same value in an initial model. If during the simulation process the transformed system has had m failures, then for the initial system the number of failures can be calculated by the formula

$$m' = \frac{1}{K} \sum_{1 \le i \le m} \frac{1}{\gamma^{z_i}},$$

where z_i is the number of units that cause the ith failure of the system. After a current failure of a transformed system, the probability P_m should be calculated as

$$P_m = 1 - \frac{1}{MK} \sum_{1 \le i \le m} \frac{1}{\gamma^{z_i}}$$

where M is the total number of realizations taken into account.

Naturally, parameter γ must be chosen in a way that allows one to accelerate the most critical events. Usually, such an event corresponds to a failure of a set of the system units whose power equals the power of the minimal cut, that is, the minimal set of the system's units that causes system failure.

CHAPTER 11

MONTE CARLO SIMULATION FOR OPTIMAL REDUNDANCY

11.1 INTRODUCTORY REMARKS

The problem of optimal redundancy plays a special role in reliability engineering and theory. On the one hand, this problem reflects a real practical need: optimal allocation of spare parts. On the other hand, this problem is extremely interesting from a mathematical viewpoint. This explains why several books are entirely or partially dedicated to this problem (Bellman, 1957; Bellman and Dreyfus, 1962; Ushakov, 1969; Barlow and Proschan, 1981; Tillman et al., 1980; Volkovich et al., 1992; Gnedenko and Ushakov, 1995). The readers can find most of these results in a compact form in Ushakov (1994).

A series system of independent unit is usually considered as a starting point for analysis. The reliability goal function is expressed as a product of probabilities (or a sum of their logarithms). More complex models (multipurpose systems) or more complex goal functions (mean time to failure) are very rare (Tillman et al., 1980; Volkovich et al., 1992; Ushakov, 1994; Gnedenko and Ushakov, (1995).

The goal of this section is to present a new numerical method of finding the optimal solution of the problem, including some nontraditional tasks.

The most important nontraditional case concerns the mutual dependence of the system units and their collective dependence on the prehistory of the system operating and/or a current environment.

Some general ideas of the proposed method were considered by Ushakov and Yasenovets (1977) and Ushakov and Gordienko (1978). The goal of this chapter is to present a new numerical method of solution of the optimal redundancy problem for obtaining results for nontraditional tasks. For instance, one can consider the problem of optimization of the system's mean time to failure (MTTF); there might be some kind of unit mutual dependence as well as dependence of unit reliability on different factors (current system's "age", number of replacements of this particular system's unit or the total number of replacements in the system, etc.).

Some general ideas of the proposed method were considered by Ushakov and Yasenovets (1977) and Ushakov and Gordienko (1978). Roughly speaking, the idea of the proposed approach is to observe the process of the spare unit expenditure (replacement of failed units) until specified restrictions allow one to do so. This may be a simulation process or an observation of the real exploitation of the system. After the stopping moment, we start another realization of the simulation process or observation of the real data. When the appropriate statistics are collected, we use an optimization technique for an empirical function in the obtained sample space.

Avoiding a formal description of the algorithm, let us demonstrate it on numerical examples that will clarify the idea of the method and its specific technique.

11.2 FORMULATION OF OPTIMAL REDUNDANCY PROBLEMS

Consider a series system of n units, that is, any failure of a system unit leads to system failure. To improve system reliability, each unit i can be supported by x_i redundant units.

We consider the following two goal functions: system cost

$$C(\mathbf{x}) = C(x_1, x_2, \ldots, x_n) \tag{11.1}$$

and system reliability index

$$R(\mathbf{x}) = R(x_1, x_2, \ldots, x_N), \tag{11.2}$$

where $\mathbf{x} = (x_1, x_2, \ldots, x_N)$ and x_i is the number of spare units of the ith type. Concrete forms of goal functions will be considered later. In this approach we have no special restrictions on them; they need only be known a priori.

The following two problems are under consideration:

1. *Direct Problem of Optimal Redundancy*

$$\max_{\mathbf{x}} \{R(\mathbf{x}) | C(\mathbf{x}) \leq C^*\}, \tag{11.3}$$

that is, to maximize the system reliability index under restriction on system cost.

2. *Inverse Problem of Optimal Redundancy*

$$\min_{\mathbf{x}} \{C(\mathbf{x})|R(\mathbf{x}) \geq R^*\}, \tag{11.4}$$

that is, to minimize system cost under restriction on the system reliability index.

The problem of optimal redundancy can easily be solved if the redundant groups are independent and the function itself can be presented in multiplicative (or additive) form, that is,

$$R(\mathbf{x}) = \prod_{1\leq i\leq n} R_i(x_i) \quad \text{or } L(\mathbf{x}) = \sum_{1\leq i\leq n} Li(x_i), \tag{11.5}$$

where $R_i(x_i)$ is the probability of failure-free operation of the ith redundant group of x_i units, $L_i(x_i) = \ln R_i(x_i)$, $i = 1, 2, \ldots,$ and

$$C(\mathbf{x}) = \sum_{1\leq i\leq n} c_i x_i, \tag{11.6}$$

where c_i is the cost of a unit of the ith type.

Standard methods do not give a solution if the goal function is the mean time to failure

$$T(\mathbf{x}) = \int_0^\infty \prod_{1\leq i\leq n} R_i(t|x_i)\, dt \tag{11.7}$$

or if units are dependent, for instance, via a vector of some external factors (temperature, humidity, vibration, etc.) **g**,

$$R(\mathbf{x}) = \int_{\mathbf{g}\in G} R(x_1, x_2, \ldots x_n)|\mathbf{g})\, dF(\mathbf{g}),$$

where $F(\mathbf{g})$ is the d.f. of **g** and G is its domain.

11.3 ALGORITHM FOR TRAJECTORY GENERATION

To solve the formulated problems, we need to have data obtained from real experiment (or system exploitation) or from Monte Carlo simulation of the system's model. Though this procedure is routine, we will briefly describe it here.

Consider a series system of n units. (For simplicity of explanation of the algorithm, we will assume that the units are independent. However, everything described below can be easily extended to the general case: It will only affect a mechanism of random sequence generation.)

Let us consider the process of spare unit expenditure as the process of changing the system states and the total cost of spare units at sequential replacement moments. After failure each unit is immediately replaced with a spare one. Let $t_k^{(j)}$ be the moment of the kth replacement during the jth Monte Carlo experiment. The number of spare units of type i spent at moment $t_k^{(j)}$ is denoted by $x_k^{(j)}$.

An initial state at $t_0^{(j)} = 0$ is given as

$$x_{i0}^{(j)} \quad \text{for all } i,\ i = \overline{1, n}, \text{ and } j,\ j = \overline{1, N}.$$

The total cost spent for spare units at the initial moment is $C_0 = 0$. (Sometimes it might be reasonable to consider the initial cost of the system with no spare units as C_0, that is, $C_0 = \Sigma\, c_i$.)

(For the sake of brevity, we will omit the superscript in a number of realizations in this section.)

Consider the step-by-step procedure of generating trajectories $\psi^{(s)}$, $s = 1, 2, \ldots$. We begin with $\psi^{(1)}$, but the corresponding subscript, (1), will be omitted for the sake of convenience.

Step 1. Generate random time to failure (TTF) for each unit, ξ_i, $i = \overline{1, n}$, and define $b_{i1} = \xi_i$; that is, b_{i1} is the moment of the earliest (to the current moment) failure and instantaneous replacement of the ith unit. The current moment (for every unit i) at the beginning of any trajectory $\psi^{(\mathbf{j})}$ is $b_{i0} = 0$.

Step 2. Determine the moment of the occurrence of the first event (first replacement) within the first realization $\psi^{(1)}$ as $t_1 = \min_{1 \le i \le n} b_{i1}$.

Step 3. Assign to the corresponding unit (for which the moment of failure is the earliest one) a specific number $i = i_1$.

Step 4. Put into the spare units counter a new value $x_{i_1 1} = x_{i_1 0} + 1$.

Step 5. Rename the remaining x_{i0} as $x_{i0} = x_{i1}$ for all $i \neq i_1$.

Step 6. Calculate a new value of system cost, $C_1 = C_0 + c_{i_1}$.

Step 7. Generate the next random TTF for unit i_1, ξ_{i_1}.

Step 8. Calculate the next event occurring due to unit i_1, $b_{i_1 2} = t_1 + \xi_{i_1}$.

Step 9. Rename the remaining values $b_{i1} = b_{i2}$ for all $i \neq i_1$.

Here the first cycle is finished; return to step 2; that is, find $t_2 = \min_{1 \le i \le n} b_{i2}$, and so on, until the first realization stops.

The type of problem to be solved determines the stopping rule of each realization. If we need to solve the direct problem of optimal allocation of spare units, then we have:

Stopping Rule 1: The process is stopped at moment t_N when the total cost of spare units exceeds the permitted C^*.

For the inverse problem, we have:

Stopping Rule 2: The simulation process for each realization stops at the moment $t_M \geq t^*$, where t^* equals the required operational time t_0 (if the reliability index is the probability of failure-free operation) or t^* is the required system's MTTF.

After generation of the first trajectory, $\psi^{(1)}$, is terminated, we start to generate $\psi^{(2)}$ by the same rules. The number of needed realizations N is determined in a usual way by the required accuracy of estimation.

Thus, each trajectory j represents a set of the following data:

$$\{t_1^{(j)};\ \mathbf{x}_1^{(j)};\ C(\mathbf{x}_1^{(j)})\},$$
$$\{t_2^{(j)};\ \mathbf{x}_2^{(j)};\ C(\mathbf{x}_2^{(j)})\},$$
$$\vdots$$
$$\{t_M^{(j)};\ \mathbf{x}_M^{(j)};\ C(\mathbf{x}_M^{(j)})\},$$

where $\mathbf{x}_s^{(j)}$ is the set of spare units at moment $t_s^{(j)}$, that is, $\mathbf{x}_s^{(j)} = \{x_{1s}^{(j)}, x_{2s}^{(j)}, \ldots, x_{ns}^{(j)}\}$.

After the description of the simulating process, we will consider the optimization problems themselves.

We make an important remark: All problems above were formulated in probabilistic terms, but we will deal rather with statistical (empirical) functions. Below these problems will be reformulated in an appropriate way.

11.4 DESCRIPTION OF THE IDEA OF THE SOLUTION

Assume that we need to supply some system with spare units for some specified period of time. We have no prior knowledge on unit reliability but have an opportunity to observe a real process (or simulation) of that type.

Consider the direct problem of optimal redundancy. What shall we do in this case? We observe the process of spare units expenditure N times and record the process trajectories $\psi^{(j)}, j = 1, 2, \ldots N$, in an n-dimensional space where n is the number of system units. Each realization is stopped when the total cost of spare units exceeds the permitted amount; that is, each trajectory reaches or even penetrates a hyperplane

$$\Gamma = \{c(\mathbf{x}) < C^*\} \tag{11.8}$$

determined by the restriction on total system cost (see Figure 11.1). Then, in the same n-dimensional space, we construct such hypercubes χ_r, $r = 1, 2, \ldots$, such that each of their vertices is lying under the hyperplane (11.8) (see Figure 11.1). Denote the maximum time the trajectory $\psi^{(j)}$ is spending within hypercube χ_r by $\tau_r^{(j)} = \tau(\psi(j), \tau_r)$. Introduce the notation

$$\delta_r^{(j)} = \begin{cases} 1 & \text{if } \tau_r^{(j)} \geq t_0, \\ 0 & \text{otherwise.} \end{cases} \tag{11.9}$$

Among all hypercubes above we choose such a hypercube χ_r that maximizes the frequency of failure free operation during interval t_0 on the generated sample of trajectories under the cost restrictions

$$\max_{\chi_r} \left\{ \frac{1}{N} \sum_{1 \leq j \leq N} \delta_r^{(j)} \,\middle|\, \sum_{\substack{1 \leq i \leq n \\ x_i \in \chi_r}} x_i c_i \leq C^* \right\}, \tag{11.10}$$

where C^* is the specified system cost.

Maximization of system average time to failure is reached by the hypercube χ'_r, which corresponds to solution of the following problem of conditional optimization:

$$\max_{\chi'_r} \left\{ \frac{1}{N} \sum_{1 \leq j \leq N} \tau_r^{(j)} \,\middle|\, \sum_{\substack{1 \leq i \leq n \\ x_i \in \chi'_r}} x_i c_i \leq C^* \right\}. \tag{11.10'}$$

Now consider the inverse problem of optimal redundancy. In this case, the

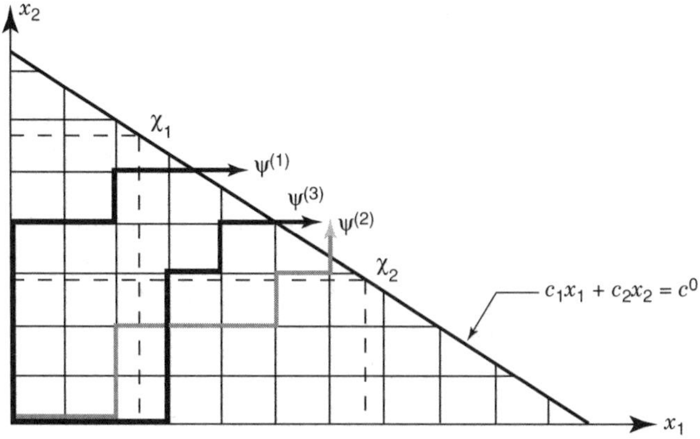

Figure 11.1 Sample of possible trajectories of unit replacement.

required time of system failure-free operation equals t_0. We observe the process of spare units expenditure N times until system time exceeds t_0 and record the process trajectories $\psi^{(j)}$, $j = 1, 2, \ldots, N$, in an n-dimensional space. Then we construct such hypercubes χ_r, $r = 1, 2, \ldots$, in the same n-dimensional space that include (cover) $R^* \times 100\%$ of all trajectories, where R^* is the specified level of the reliability index. Among all hypercubes described above, we choose the one that is characterized by minimum total cost. In other words, the hypercube $\tilde{\chi}_r$ must satisfy solution of the following problem:

$$\min_{\tilde{\chi}_r} \left\{ \sum_{\substack{1 \le i \le n \\ x_i \in \tilde{\chi}_r}} c_i x_i \,\middle|\, \frac{1}{N} \sum_{1 \le j \le N} \delta_r^{(j)} \ge R^* \right\}. \tag{11.11}$$

Now let the specified requirement be given for system average time to failure T^*. The hypercube $\tilde{\chi}'_r$ presenting the solution must be chosen corresponding to solution of the problem:

$$\min_{\tilde{\chi}_r} \left\{ \sum_{\substack{1 \le i \le n \\ x_i \in \tilde{\chi}_r}} c_i x_i \,\middle|\, \frac{1}{N} \sum_{1 \le j \le N} \tau_r^{(j)} \ge T^* \right\}. \tag{11.11$'$}$$

Of course, one should take into account that the operation with the frequency differs from the operation with the probability. The proposed solution is asymptotically accurate. Thus all these arguments are satisfactory only for a large enough sample size.

11.5 SOLUTION OF DIRECT OPTIMIZATION PROBLEM

Frequency of Successful Operation Versus System Cost We need to find χ_{opt} that satisfies the solution of (11.10). The solution algorithm is as follows:

Step 1. Choose a hypercube χ_1 whose diagonal vertex is lying on or under the hyperplane (11.8).

Step 2. Take the first realization, $\psi^{(1)}$ obtained with stopping rule 1. Find moment $\tau_1^{(1)}$ when this trajectory "punctures" the hypercube χ_1. This corresponds to the moment $t_k^{(1)}$, where

$$k = [k\colon (\forall x_{i,k-1} \le \chi_{i1}) \wedge (\exists x_{ik} > \chi_{i1}) \text{ is true}],$$

χ_{i1} a component of χ_1.

Step 3. Assign to $\tau_1^{(1)}$ a value 1 or 0 by using the indicator function (11.9).

Step 4. Add $\delta_1^{(1)}$ to the value in the counter (initial value equals 0) of successful trajectories.

Repeat the procedure from step 2 for the next realization, $\psi^{(2)}$.

After analyzing all trajectories, we calculate the frequency of successful trajectories

$$\hat{P}_1(t_0) = \frac{1}{N} \sum_{1 \le j \le N} \delta_1^{(j)}$$

for the chosen hypercube χ_1. Then we find such a hypercube χ_K that is characterized by the maximum value of $\hat{P}_k(t_0)$. For this purpose, we can use a random search, steepest descent, or another numerical method in the discrete space of the trajectories of the spare units expenditure.

Example 11.1 Consider a series system of $n = 3$ main units. For the sake of simplicity of illustrative calculations and to compare with analytical solution, assume that the system units are independent and $c_i = c$ for all i, $1 \le i \le 3$. Let the units be distributed exponentially with parameters $\lambda_1 = 1$, $\lambda_2 = 0.5$, and $\lambda_3 = 0.25$, respectively. The specified time of failure-free operation is $t_0 = 1$. The system is allowed to have at most four spare units in total.

In Table 11.1 (left column) there are random exponentially distributed TTFs ξ_i for each of the three system units in accordance with their λ_i. Moments of sequential replacements of failed units $\theta_{i(k)}$ are also in the table. In other words, θ_{ik} is a random survival time of the ith redundant group consisting of one main unit and $k - 1$ spare units. It is clear that $\theta_{i(k)} = \xi_{i1} + \xi_{i2} + \cdots + \xi_{ik}$, where ξ_{ij} is the random time to failure of the jth installed unit of the ith type. □

Remark: We will use the same random numbers for all examples below. This leads to dependence of the results obtained for different problems, but our main goal is to illustrate the algorithm of the solution with the use of a numerical example rather than to execute an accurate statistical experiment.

Numerical Solution

1. Consider realization 1 from Table 11.1. Take values ξ_i of the first row under TTF: $\xi_{11} = 0.07$, $\xi_{21} = 0.75$, and $\xi_{31} = 0.24$. Denote them $\theta_{1(1)}$, $\theta_{2(1)}$, and $\theta_{3(1)}$, respectively, and set them into the first row under Replacement Time. Find the minimum value: $\min \theta_{i(1)} = \theta_{1(1)} = 0.07$.
2. Next take the value $\xi_{12} = 0.53$ under TTF, Unit 1. Form a new value: $\theta_{1(2)} = \theta_{1(1)} + \xi_{12} = 0.07 + 0.53 = 0.6$. Rename $\theta_{2(1)} = \theta_{2(2)}$ and $\theta_{3(1)} = \theta_{3(2)}$. Set this value into the second place under Replacement Time, Unit 1.

Table 11.1 Random TTF and Replacement Time for 10 Monte Carlo Realizations

TTF			Replacement Time		
Unit 1	Unit 2	Unit 3	Unit 1	Unit 2	Unit 3
		Realization 1			
0.07	0.75	0.24	0.07	0.75	0.24
0.53	4.97	3.19	0.6	5.72	3.43
0.06	0.45	1.41	0.66	6.17	4.84
0.53	2.59	3.42	1.19	8.76	8.26
1.44	5	1.59	2.63	13.76	9.85
		Realization 2			
0.42	0.13	4.92	0.42	0.13	4.92
0.16	1.15	12.9	0.58	1.28	17.82
0.45	3.29	0.83	1.03	4.57	18.65
0.28	0.35	2.35	1.31	4.91	21
0.25	2.1	1.74	1.55	7.02	22.73
		Realization 3			
0.62	3.47	3.22	0.62	3.47	3.22
3.66	2.72	3.92	4.28	6.2	7.14
5.11	2.47	3.9	9.39	8.67	11.04
0.31	1.69	1.21	9.7	10.36	12.25
1.42	0.86	0.96	11.12	11.22	13.21
		Realization 4			
1.45	5.85	0.51	1.45	5.85	0.51
1.13	1.26	8.64	2.58	7.11	9.15
1.27	2.14	4.71	3.85	9.25	13.86
0.45	0.67	1.16	4.29	9.92	15.01
2.48	1.52	6.38	6.77	11.44	21.4
		Realization 5			
0.32	0.22	0.54	0.32	0.22	0.54
0.75	0.15	1.53	1.08	0.37	2.06
0.73	1.49	1.78	1.81	1.87	3.84
0	0.68	0.89	1.81	2.55	4.73
0.25	3.06	1.68	2.05	5.6	6.41
		Realization 6			
0.11	2.03	5.54	0.11	2.03	5.54
1.03	0.48	10.57	1.13	2.52	16.11
0.88	2.26	5.14	2.01	4.77	21.25
0.39	5.19	0.92	2.41	9.96	22.17
3.45	1.12	6.58	5.86	11.08	28.74

Table 11.1 ***(Continued)***

TTF			Replacement Time		
Unit 1	Unit 2	Unit 3	Unit 1	Unit 2	Unit 3
		Realization 7			
1.22	0.11	2.69	1.22	0.11	2.69
1.87	0.91	0.1	3.09	1.02	2.79
0.41	2.11	1.9	3.5	3.13	4.69
3.95	0.36	3.72	7.45	3.49	8.41
0.4	1.67	1.43	7.85	5.17	9.84
		Realization 8			
0.27	1.49	22.49	0.27	1.49	22.49
0.44	0.53	1.24	0.71	2.02	23.73
0.74	1.07	12.07	1.45	3.09	35.8
0.76	1.13	2.86	2.2	4.22	38.65
0.36	2.99	2.87	2.57	7.21	41.52
		Realization 9			
0.46	1.55	7.9	0.46	1.55	7.9
1.06	4.8	7.59	1.52	6.35	15.49
1.9	2.66	8.14	3.42	9.01	23.63
0.17	0.37	1.26	3.59	9.38	24.89
2.18	0.43	5.17	5.77	9.8	30.06
		Realization 10			
0.83	1.08	0.58	0.83	1.08	0.58
0.4	1.76	3.76	1.23	2.84	4.33
1	0.94	8.73	2.23	3.79	13.07
0.4	1.48	3.74	2.63	5.26	16.81
0.47	2.91	3.73	3.1	8.18	20.54

3. Find the minimum value: min $\theta_{i(2)} = \theta_{3(2)} = 0.24$. Repeat step 2 until the total cost of each system equals 7. (For the case $c_i = c$, it means that all seven units are spent.) As the result, we spent three units of type 1, no units of type 2, and two units of type 3 (see the left upper corner of Table 11.2). In this case, the system time to failure does not reach the specified time $t_0 = 1$.
4. Repeat steps 1–3 with the remaining realizations from Table 11.1 and fill out Table 11.2.
5. Notice that in Table 11.2 certain units are labeled auxiliary; that is, in each particular case they are not necessary because t_0 has been reached before all permitted resources were spent.

Table 11.2 Initial Experiment with Exclusion of "Extra Units"

Unit 1	Unit 2	Unit 3	Unit 1	Unit 2	Unit 3
	Realization 1			*Realization 2*	
0.07	0.75	0.24	0.42	0.13	4.92
0.6		3.43	0.58	1.28	
0.66			1.03		
1.19			1.31[a]		
	Realization 3			*Realization 4*	
0.62	3.47	3.22	1.45	5.85	0.51
4.28	6.2[a]	7.14[a]	2.58[a]	9.15	
9.39[a]			3.85[a]		
			4.29[a]		
	Realization 5			*Realization 6*	
0.32	0.22	0.54	0.11	2.03	5.54
1.08	0.37	2.06	1.13	2.52[a]	
	1.87		2.01[a]		
			2.41[a]		
	Realization 7			*Realization 8*	
1.22	0.11	2.69	0.27	1.49	22.49
3.09[a]	1.02	2.79[a]	0.71	2.02[a]	
	3.13[a]		1.45		
			2.2[a]		
	Realization 9			*Realization 10*	
0.46	1.55	7.9	0.83	1.08	0.58
1.52	6.35[a]		1.23	2.84[a]	4.33
3.42[a]			2.23[a]		
3.59[a]					

[a] = Unnecessary units.

6. List all vectors $\mathbf{x}^{(k)} = (\mathbf{x}_1^{(k)}, \mathbf{x}_2^{(k)}, \mathbf{x}_3^{(k)})$, $k = 1, 2, \ldots, 10$, which are obtained from Table 11.2 after exclusion of the marked units (see Table 11.3).
7. Order each component of these vectors separately (see Table 11.4). In other words, Table 11.4 shows the frequency with which a corresponding number of spare units of a type has been met during 10 realizations of Monte Carlo simulation. We see that the use of the vector (3, 3, 2) of spare units for this realization will lead to 1 failure in 10 experiments. However, total system cost equals eight units. So, the next step is to find the best way of reducing total system cost.

Table 11.3 Realization of Units Spent

Realization Number	x_1	x_2	x_3	System TTF
1	4	1	2	<1
2	3	2	1	1.03
3	2	1	1	3.22
4	1	1	2	1.45
5	2	3	2	1.08
6	2	1	1	1.13
7	1	2	1	1.02
8	3	1	1	1.45
9	2	1	1	1.55
10	2	1	2	1.08
Maximum	3	3	2	

Note: Corrected for $t_0 \geq 1$. Realization 1 is not taken into account because its TTF < 1.

8. Put the number of realizations for which TTF has not reached $t_0 = 1$ into the failure counter. In our case there is only one, that is, realization ≤ 1.
9. Excluded from Table 11.2 all vectors that correspond to the realizations mentioned in step 7.
10. Find which unit in Table 11.4 has the smallest number of the use of the largest values of $\max_{1 \leq k \leq 10} x_i^{(k)}$. In our example, three units of type 1 were used in three realizations, three units of type 1 were used once,

Table 11.4 Ordered Numbers of Use of Units of Different Types

x_1	x_2	x_3
1	1	1
1	1	1
2	1	1
2	1	1
2	1	1
2	1	1
2	(1)	2
3	2	2
3	2	2
3	2	2
(4)	3	(2)

Note: Numbers in parentheses correspond to the first realization, which was not considered.

and two units of type 3 were used four times. (We exclude from consideration realization 1 since it did not deliver TTF ≥ 1.)

In this case we exclude one unit of type 2 because in this case we gain one unit of cost and "loss" of only one realization.

Remark: If several types of units have the same number of the use of the maximum spare units, then we exclude a unit of a type whose cost is large than others. □

11. Add the number of units excluded at step 9 into the counter of system failures.
12. Check if system cost is equal to or less than $C^* = 7$. If the answer is no, correct Table 11.3 by excluding vector 1 and continue the procedure from step 6. If the answer is yes, stop the procedure and go to step 13.
13. Calculate the ratio of realization without failure (the total number of realizations minus the number of failures from the counter) to the total number of performed realizations. □

In the example considered the final solution is (3, 2, 2). As a direct calculation with the use of tables of Poisson distribution shows, this vector delivers the probability 0.804. Of course, such a coincidence with observed frequency equal to 0.8 in a particular statistical experiment is not proof of the method. However, multiple results obtained by the proposed method for other different examples show a proper closeness to the exact solution.

Notice that this coincidence has deep roots. Indeed, we used an analog of steepest descent to the empirical distribution function that converges with the theoretical function for a larger sample. We know that the steepest descent method delivers an appropriate solution for nondegenerate discrete optimization problems, in particular, for an optimal redundancy problem. As mentioned above, asymptotic convergence of the solution to the optimal one was proved by Ushakov and Gordienko (1978). □

Average Time to Failure Versus System Cost We need to find the χ'_{opt} that satisfies the solution of (11.10′). In this case the algorithm almost completely coincides with the one described above. The only difference is in the absence of step 3. At step 4 we put directly $\tau_1^{(i)}$ in a counter of the survival time. After analyzing all of trajectories, the estimate of the MTTF for the hypercube χ_1 is calculated as

$$T_1 = \frac{1}{N} \sum_{1 \leq i \leq N} \tau_1^{(i)}.$$

Then we perform analogous calculations for other hypercubes finding those that are characterized by the maximum estimate of MTTF. The search for the maximum can be performed in the same way as done previously.

Example 11.2 We will consider the same example as above. The system is again allowed to have at most seven units in total. For illustration, we will use the same numerical tables as above.

Numerical Solution

Repeat steps 1–4 in Section 11.3. In other words, assume that Table 11.2 is constructed. To solve this problem, we will use all data of Table 11.2. The continuation of the algorithm for this case is as follows.

5. Consider the vectors of Table 11.2. In this case, the auxiliary components are included. Those vectors are obtained in the imitation process until seven units of price are spent. Now extract the corresponding values from the right side of Table 11.1 (see Table 11.5). From Table 11.5 we can see how long each unit was operating.
6. On the basis of Table 11.5, we compose Table 11.6. In each position of Table 11.6 we have the total time spent during all realizations. First, for independent and identical units, these values depend on the number of realizations where this unit was observed. (In the general case where units are different and could be dependent, the number of such realizations might not be a dominant parameter.) From the bottom these values show how much we will lose by excluding a unit.

It is clear that the loss will be less if we leave $x_1 = 3$, $x_2 = 2$, $x_3 = 2$. By these eliminations we decrease the total system cost up to seven units of price.

The time to failure for the system in each realization is calculated as the minimal value among those that are restricted by vector ($x_1 = 3$, $x_2 = 2$, $x_3 = 2$), that is, for the kth realization

$$\xi^{(k)} = \min(\xi_{13}^{(k)}, \xi_{22}^{(k)}, \xi_{32}^{(k)}).$$

These values can be found on the right side of Table 11.6. The results are shown in Table 11.7. These values allow us to calculate the mean time to failure of the investigated system. □

11.6 INVERSE OPTIMIZATION PROBLEM

System's Cost Versus Successful Operation We need to find the $\tilde{\chi}_r$ that satisfies the solution of (11.11). The solution algorithm in this case is as follows.

Table 11.5 Random Time to Failure for Each Realization Until Expenditure of Seven Units

	Realization 1			*Realization 2*	
0.07	0.75	0.24	0.42	0.13	4.92
0.53		3.19	0.16	1.15	
0.06			0.45		
0.53			0.28		
	Realization 3			*Realization 4*	
0.62	3.47	3.22	1.45	5.85	0.51
3.66	2.72	3.92	1.13		8.64
5.11		3.9	1.27		
			0.45		
	Realization 5			*Realization 6*	
0.32	0.22	0.54	0.11	2.03	5.54
0.75	0.15	1.53	1.03	0.48	
	1.49		0.88		
			0.39		
	Realization 7			*Realization 8*	
1.22	0.11	2.69	0.27	1.49	22.49
1.87	0.91	0.1	0.44	0.53	
	2.11		0.74		
			0.76		
	Realization 9			*Realization 10*	
0.46	1.55	7.9	0.83	1.08	0.58
1.06	4.8		0.4	1.76	3.76
1.9			1		
0.17					

Table 11.6 Sum of Times Spent by Units on Specified Positions

Unit 1	Unit 2	Unit 3
5.77	16.68	48.63
11.03	12.5	21.14
11.41	3.6[a]	3.9[a]
2.58[a]		

[a]Units are eliminated ($x_1 = 4$, $x_2 = 3$, and $x_3 = 3$).

Table 11.7 Time to Failure for 10 Realizations Picked up for Vector (3, 2, 2)

Realization Number	TTF
1	0.66
2	1.03
3	6.2
4	3.85
5	0.37
6	2.01
7	1.02
8	1.45
9	3.42
10	2.23

Step 1. Construct a realization of the first trajectory of the spare units expenditure unit $t_1^{(1)}$ exceeds the specified operational time t_0. Memorize the number of spare units spent, $x_i^{(1)} = i = \overline{1, n}$. Continue this procedure until all N required trajectories are constructed.

Step 2. Construct a hypercube χ_1 whose edges χ_{i_1} are found as $\chi_{i_1} = \max_{1 \le j \le N} x_{ij}$; that is, χ_{i1} is the maximum number of spare units of type i observed during all N realizations. (It means that for this particular sample of realizations all of them will lie within such a hypercube; that is, with such a stock of spare units we would not observe any system failures.)

Step 3. Calculate system cost for the hypercube $\chi^{(1)}$ for which all trajectories have survival time not less than t_0,

$$C_{\max} = \sum_{1 \le i \le n} c_i \chi_{i1}.$$

Step 4. For each i calculate

$$\gamma_i^{(1)} = \frac{\nu_i^{(1)}}{\Delta c_i^{(1)}},$$

where $\nu_i^{(1)}$ shows how many numbers equal to χ_{i1} exist for a unit of type i and $\Delta c_i^{(1)}$ is the value of system cost decrease if we reject using $\max_{1 \le j \le N} \chi_{ij}$ and use the next value in descending order.

Step 5. Find unit type that corresponds to the maximum value of $\gamma_i^{(1)}$ and name it i_1; that is, this number corresponds to the following condition:

$$i_1 = \left\{ i\colon \gamma_i = \max_{1 \le j \le n} \gamma_j^{(1)} \right\}.$$

Step 6. Exclude $\nu_{i_1}^{(1)}$ units of type i_1 and form a new value

$$\chi_{i_1 2} = \chi_{i_1 1} - \nu_{i_1}^{(1)}.$$

Step 7. Rename the remaining numbers

$$\chi_{i_j 2} = \chi_{i_j 1} \quad \text{for all } i_j \neq i_1.$$

Step 8. Calculate the system successful operation index after exclusion of ν_{i_1} units of type i_1,

$$\hat{P}^{(2)} = 1 - \frac{\nu_{i_1}^{(1)}}{N}.$$

Step 9. Calculate system spare units cost after exclusion of ν_{i_1} units of type i_1,

$$C^{(2)} = C_{\max} - c_{i_1}\nu_{i_1}^{(1)}.$$

Then we have a new hypercube χ_2,

$$\chi_2 = \{\chi_{12}, \chi_{22}, \ldots, \chi_{n2}\}.$$

Repeat the procedure from step 5 until system spare units cost becomes equal to or smaller than the given restriction.

Example 11.3 Take $\hat{R}(9) \ge 0.9$. In the previous example we found that the vector of spare units (4, 3, 2) satisfies 100% of successful realizations. So, if we take a vector (3, 3, 2), it will satisfy the condition $\hat{R}^* = 0.9$. Now we need to find the lower 90% confidence limit for the frequency 0.9 obtained in 10 experiments. This limit can be found with the Clopper–Pearson method.

In this particular case it is easier to make direct calculations. If we choose the estimate of the searched probability equal to 0.9, then the probability that we will observe not less than eight successes will be

$$0.9^{10} + \binom{10}{1}(0.1)(0.9)^9 + \binom{10}{2}(0.1)^2(0.9)^8 = 0.9298.$$

So, in the process of decreasing the number of used spare units, we must stop after the first exclusion; that is, the solution in this case is (3, 3, 2).

Thus, the solutions of direct and inverse problems of optimal redundancy are different although they should coincide. The difference lies in the different approaches: having the restriction on system cost, we maximize the possible observed *frequency*; in the latter case we consider minimization of system cost under the condition that the level of *probability* is guaranteed. This difference will be smaller if the number of realizations will be larger.

Remark: We could solve the problem above with an interaction procedure using the solution of the direct problem of optimal redundancy. The "fork method" is convenient in this case. We find the solution χ_1^{opt} for some cost restriction, say C_1^*, and calculate the value $R_1 = R(\chi_1^{opt})$. If $R_1 < R^*$, we choose $C_2^* > C_1^*$ and continue the procedure; if $R_1 > R^*$, we choose $C_2^* < C_1^*$ and also continue the procedure. For the next steps, we can use a simple linear approximation.

$$C_{k+1}^* = C_k^* - (C_k^* - C_{k-1}^*)\frac{R_k - R^*}{R_k - R_{k-1}},$$

where k stands for the current step, $k - 1$ for the previous step, and $k + 1$ for the next step. □

System Cost Versus Average Time to Failure We need to find the $\tilde{\chi}_r$ that satisfies the solution of (11.10). We could not find a convenient procedure for solving this particular problem. One might consider using an iterative procedure using the sequential solution of the second direct problem considered above. For instance, we can fix some restriction on system cost, say, $C_{sys}^{(1)}$, and find the corresponding optimal solution for $\hat{T}_{sys}^{(1)}$. If this value is smaller than the required $\hat{T}_{sys}^*$, it means that the system cost must be increased, say, up to some $C_{sys}^{(2)} > C_{sys}^{(1)}$. If $\hat{T}_{sys}^{(1)} > \hat{T}_{sys}^*$, one must choose $C_{sys}^{(2)} < C_{sys}^{(1)}$. This procedure continues until a satisfactory solution is obtained. At an intermediate step L for choosing $\hat{T}_{isys}^{(L)}$, one can use some linear extrapolation method. For example, assume that in the first situation described above the value $\hat{T}_{sys}^{(2)}$ is still less than $\hat{T}_{sys}^*$. Then the value of $C_{sys}^{(3)}$ can be chosen from the equation

$$\frac{C_{sys}^{(3)} - C_{sys}^{(1)}}{C_{sys}^{(2)} - C_{sys}^{(1)}} = \frac{T_{sys}^* - T_{sys}^{(1)}}{T_{sys}^{(2)} - T_{sys}^{(1)}}.$$

Obviously, one can also use a procedure similar to the one used to solve the direct problem of type 1. However, if possible, one should find an initial hypercube and construct all trajectories within it. (There is no stopping rule in this case.) Then one should construct a system of embedded hypercubes and again use the steepest descent.

While solving this problem, remember that the condition $\hat{T}_{sys} \geq T^*$ can be considered only in a probabilistic sense.

Thus we have suggested a numerical method of optimal spare allocation. The proof of correctness of this method for a series structure is given in the following section.

11.7 APPENDIX[1]

11.7.1 Consistency and Asymptotic Unbiasedness of Solution to Direct Problem

Maximization of System PFFO Let $\mathbf{x}$ denote the set of all n-dimensional vectors $\mathbf{x} = (x_1, \ldots, x_n)$ with integer valued nonnegative components x_i, $i = 1, 2, \ldots, n$. For a given system cost $C^* > 0$, we define, as in (11.8),

$$\Gamma = \left\{ \mathbf{x} \in X: \sum c_i x_i \leq C^* \right\}.$$

A vector $\mathbf{x} \in \Gamma$ is maximal if there is no $\mathbf{y} \in \Gamma$ such that $\mathbf{x} \leq \mathbf{y}$ and $x_i < y_i$ at least for one i. Denote by $\Gamma_{\max}$ the set of maximal vectors from Γ.

Remark 11.1 We will suppose $c_i > 0$, $i = 1, \ldots, n_i$, so that both sets Γ and $\Gamma_{\max}$ are finite.

Let $\mathbf{x} \in X$ and

$$p(\mathbf{x}) = P(\tau_{\mathbf{x}} \geq t_0),$$

where $\tau_{\mathbf{x}}$ is the failure-free system operational time under the condition that there are $\mathbf{x}$ spares and t_0 is the required operational time.

Define

$$p_* = \max_{\mathbf{x} \in \Gamma_{\max}} p(\mathbf{x}), \tag{11.12}$$

$$\Gamma_* = \{\mathbf{x} \in \Gamma_{\max}: p(\mathbf{x}) = p_*\}, \tag{11.13}$$

$$p_N = \max_{\mathbf{x} \in \Gamma_{\max}} \frac{1}{N} \sum_{1 \leq j \leq N} \delta^{(j)}(\mathbf{x}), \tag{11.14}$$

$$\Gamma_N = \left\{ \mathbf{x} \in \Gamma_{\max}: \frac{1}{N} \sum_{1 \leq j \leq N} \delta^{(j)}(\mathbf{x}) = p_n \right\}, \tag{11.15}$$

[1]This material is prepared by Eugene Gordienko.

where $\delta^{(i)}(\mathbf{x})$ is used for the value $\delta_r^{(j)}$ defined in (11.9). In the same manner as in (11.9) and (11.10′), we will use $\tau^{(j)}(\mathbf{x})$ instead of $\tau_r^{(j)}$ for random variables (r.v.'s) distributed as $\tau_{\mathbf{x}}$.

Let

$$\varepsilon_* = p^* - \max_{\mathbf{x}\in\Gamma\setminus\Gamma^*} p(\mathbf{x}) > 0. \square \tag{11.16}$$

Consistency

Proposition 11.1 With probability 1

$$\lim_{N\to\infty} p_N = p_*. \square$$

Proof By the definition (11.9), $\delta_r^{(j)} = \delta^{(j)}(\mathbf{x})$ and the strong law of large numbers

$$\frac{1}{N}\sum_{1\le j\le N} \delta^{(j)}(\mathbf{x}) \to p(\mathbf{x}) \quad \text{as } N\to\infty \tag{11.17}$$

almost surely (a.s.). Since set $\Gamma_{\max}$ is finite, (11.17) yields

$$|p_N - p_*| \le \max_{\mathbf{x}\in\Gamma}\left|\frac{1}{N}\sum_{1\le j\le N}\delta^{(j)}(\mathbf{x}) - p(\mathbf{x})\right| \xrightarrow{\text{a.s.}} 0 \tag{11.18}$$

as $N\to\infty$. □

Corollary 11.1 There exists a.s. a finite r.v. ν such that, for $N \ge \nu(\omega)$,

$$\Gamma_{\nu(\omega)} \subset \Gamma_*.$$

In particular, if the set Γ_* consists of the unique vector $\mathbf{x}_*$, we have a.s. convergence of numbers of spare units to their optimal values. □

Proof For each $\mathbf{x} \in \Gamma_{\max}$ there is a.s. a finite r.v. $\nu_{\mathbf{x}}$ such that for $N \ge \nu_{\mathbf{x}}$ we have

$$\left|\frac{1}{N}\sum_{1\le j\le N}\delta^{(j)}(\mathbf{x}) - p(\mathbf{x})\right| < \frac{\varepsilon_*}{2}$$

with ε_* defined in (11.16). It suffices to set $\nu = \max_{\mathbf{x}\in\Gamma_{\max}} \nu_{\mathbf{x}}$. □

Estimate of Rate of Convergence

Proposition 11.2 Let $L(\Gamma_{\max})$ be some upper bound of the number of vectors in the set $\Gamma_{\max}$. Then for every $\varepsilon > 0$, $N = 1, 2, \ldots,$

$$\text{(a)}\quad P(\nu \leq N) \geq 1 - 2L(\Gamma_{\max})\exp(-\tfrac{1}{2}\varepsilon_*^2),$$

$$\text{(b)}\quad R(|p_N - p_*| > \varepsilon) \leq 2L(\Gamma_{\max})\exp(-2\varepsilon^2). \ \square$$

Proof (a) Using the notation of the above corollary, we obtain

$$\begin{aligned} P(\nu \leq N) &= P\left(\max_{\mathbf{x}\in\Gamma_{\max}} \nu_{\mathbf{x}} \leq N\right) = 1 - P\left(\bigcup_{\mathbf{x}\in\Gamma_{\max}} (\nu_{\mathbf{x}} > N)\right) \\ &\geq 1 - \sum_{\mathbf{x}\in\Gamma_{\max}} P(\nu_{\mathbf{x}} > N) \\ &\geq \sum_{\mathbf{x}\in\Gamma_{\max}} P\left(\left|\frac{1}{N}\sum_{1\leq j\leq N} \delta^{(j)}(\mathbf{x}) - p(\mathbf{x})\right| \geq \tfrac{1}{2}\varepsilon_*\right) \\ &\geq 1 - 2L'(\Gamma_{\max})\exp(-\tfrac{1}{2}\varepsilon_*^2), \end{aligned} \tag{11.19}$$

where $L'(\Gamma_{\max})$ denotes the number of vectors in $\Gamma_{\max}$. The latter inequality in (11.19) holds due to Hoeffling's inequality for the sum of i.i.d. bounded r.v.'s $\delta^{(j)}(\mathbf{x})$. (For instance, see Petrov, 1975, p. 58.)

(b) In view of inequality (11.18), this part is proved similarly. □

Asymptotic Unbiasedness

Proposition 11.3 $\lim_{N\to\infty} Ep_N = p_*$.

The proof follows from Proposition 11.2 and Lebesgue's convergence theorem for bounded r.v.'s p_N, $N = 1, 2, \ldots$. □

Remark 11.2 It is difficult to expect $Ep_N = p_*$ (unbiasedness) for finite N because of the nonlinear function "max" in the definition of p_N and p_*. □

Maximization of System MTBF Let us define the following values and sets (assuming finiteness of $E\tau_{\mathbf{x}}$, $\mathbf{x} \in \Gamma$):

$$T_0 = \max_{\mathbf{x}\in\Gamma_{\max}} E\tau_{\mathbf{x}}; \tag{11.20}$$

$$\Gamma_0 = \{\mathbf{x} \in \Gamma_{\max}: E\tau_{\mathbf{x}} = T_0\},$$

$$T_N = \max_{\mathbf{x}\in\Gamma_{\max}} \frac{1}{N}\sum_{1\leq j\leq N} \tau^{(j)}(\mathbf{x}), \tag{11.21}$$

$$\Gamma_N = \left\{ \mathbf{x} \in \Gamma_{\max}: \frac{1}{N} \sum_{1 \le j \le N} \delta^{(j)}(\mathbf{x}) = p_N \right\},$$

$$\varepsilon_0 = T_0 - \max_{\mathbf{x} \in \Gamma \backslash \Gamma^*} E\tau_{\mathbf{x}} > 0.$$

Remark 11.3 We assume that $\Gamma_{\max} \backslash \Gamma_0 \neq \emptyset$.

The next two assertions are proved in a similar manner as Proposition 11.1 and Corollary 11.1. □

Proposition 11.4 With probability 1

$$\lim_{N \to \infty} T_N = T_0. \ \square$$

Corollary 11.2 There exists a.s. a finite r.v. ν_T such that $N > \nu_T(\omega)$,

$$\Gamma_v^*(\omega) \subset \Gamma_0. \ \square$$

Rate of Convergence To estimate the rate of convergence of T_N to T_0, we need to assume boundedness for moments of order $\gamma > 1$ of the r.v.'s $\tau_{\mathbf{x}}$, $\mathbf{x} \in \Gamma_{\max}$. Let $L(\Gamma_{\max})$ be an upper bound of the number of vectors in $\Gamma_{\max}$ as in Proposition 11.2.

Proposition 11.5 Assume that, for some $\gamma > 1$,

$$\max_{\mathbf{x} \in \Gamma_{\max}} E\tau_{\mathbf{x}}^{\gamma} \le M \le \infty. \tag{11.22}$$

Then, for each $\varepsilon > 0$, $N = 1, 2, \ldots$:

(a) if $\gamma \le 2$, then

$$P(\nu_T \le N) \ge 1 - 2^{1+\gamma} L(\Gamma_{\max}) M \varepsilon_0^{-\gamma} N^{1-\gamma}, \tag{11.23}$$

(b) and if $\gamma g > 2$, then

$$p(\nu_T \le N) \ge 1 - L(\Gamma_{\max}) M K_{\gamma} \varepsilon_0^{-\gamma} N^{-\gamma/2}, \tag{11.24}$$

where

$$K_{\gamma} = 2^{\gamma-1} \gamma(\gamma - 1) \max(1, 2^{\gamma-3}) \left[1 + \frac{2}{\gamma} \left(\sum_{1 \le s \le m} \frac{s^{2m-1}}{(s-1)!} \right)^{(\gamma-2)/2m} \right],$$

and the integer m satisfies the condition $2m \le \gamma < 2m + 2$.

(c) If $\gamma \leq 2$, then

$$P(|T_N - T_0| > \varepsilon) \leq 2L(\Gamma_{\max})M\varepsilon^{-\gamma}N^{1-\gamma},$$

and if $\gamma > 2$, then

$$P(|T_N - T_0| > \varepsilon) \leq 2^{-\gamma}L(\Gamma_{\max})MK_\gamma\varepsilon^{-\gamma}N^{-\gamma/2}. \square$$

Proof As in the proof of Proposition 11.2, this proof is reduced to obtaining the upper bounds for

$$P\left(\left|\frac{1}{N}\sum_{1\leq j\leq N}\tau^{(j)}(\mathbf{x}) - E\tau_{\mathbf{x}}\right| > \varepsilon\right),$$

where $\mathbf{x} \in \Gamma_{\max}$. By Chebyshev's inequality, the latter probability does not exceed the quantity

$$(\varepsilon N)^{-\gamma}E\left|\frac{1}{N}\sum_{1\leq j\leq N}[\tau^{(j)}(\mathbf{x}) - E\tau_{\mathbf{x}}]\right|^{\gamma},$$

which is bounded by

$$(\varepsilon N)^{-\gamma}2\sum_{1\leq j\leq N}E[\tau^{(j)}(\mathbf{x})]^{\gamma}$$

for $\gamma \leq 2$ and by

$$(\varepsilon N)^{-\gamma}2^{-\gamma}K_\gamma N^{(\gamma/2)-1}\sum_{1\leq j\leq N}E[\tau^{(j)}(\mathbf{x})]^{\gamma}$$

for $\gamma > 2$ (see Petrov, 1975, p. 60). Taking into account $E[\tau^{(j)}(\mathbf{x})]^{\gamma} \equiv E\tau_x^{\gamma} \leq M$ by (11.22), we come to the desired inequalities. $\square$

Asymptotic Unbiasedness

Proposition 11.6 $\lim_{N\to\infty} E|T_N - T_0| = 0$ and $ET_N \to ET_0$ as $N \to \infty$.

Proof By (11.20) and (11.21),

$$|T_N - T_0| \leq \max_{\mathbf{x}\in\Gamma_{\max}} |\xi_{N,\mathbf{x}}| \leq \sum_{\mathbf{x}\in\Gamma_{\max}} |\xi_{N,\mathbf{x}}|,$$

where

$$\xi_{N,\mathbf{x}} = \frac{1}{N} \sum_{1 \le j \le N} [\tau^{(j)}(\mathbf{x}) - E\tau].$$

Thus,

$$E|T_N - T_0| \le \sum_{\mathbf{x} \in \Gamma_{\max}} E|\xi_{N,\mathbf{x}}|.$$

But, for each $\mathbf{x} \in \Gamma_{\max}$, $\Sigma_{N\to\infty} E|\xi_{N,\mathbf{x}}| = 0$ by virtue of the L_1 version of the strong law of large numbers (see, e.g., Williams, 1991, p. 136). □

11.7.2 Consistency of Inverse Problem of Optimal Redundancy

Restrictions on PFFO Let $R^* < 1$ be the specified level of the reliability index and let the set

$$Q = [\mathbf{x} \in X: p(\mathbf{x}) \ge R^*\}$$

be nonempty and also

$$\overline{Q} = \{\mathbf{x} \in X: p(\mathbf{x}) < R^*\}. \tag{11.25}$$

We say that $\mathbf{x} \in X$ is the minimal vector if there does not exist $\mathbf{y} \in Q$ such that $\mathbf{y} \le \mathbf{x}$ and $y_i < x_i$ for some $i = 1, \ldots, N$. Denote by $Q_{\min}$ the set of all minimal vectors from Q. It is easy to see that $Q_{\min}$ is finite. Let

$$C_0 = \min_{\mathbf{x} \in Q_{\min}} \sum_{1 \le i \le N} c_i x_i, \qquad Q_0 = \left\{ \mathbf{x} \in Q_{\min}: \sum_{1 \le i \le N} c_i x_i = C_0 \right\}. \tag{11.26}$$

Remark 11.4 In general, both statistical procedures defined in (11.11) and (11.11′) are not consistent. Indeed, assume that set Q_0 consists of unique vector $\mathbf{x}_*$ and $p(\mathbf{x}_*) = R^*$. Then it can be shown that

$$\frac{1}{N_k} \sum_{1 \le j \le N_k} \delta^{(j)}(\mathbf{x}_*) < R$$

for an unbounded sequence $\{N_k\}$ with positive probability (and even with probability 1). [Here $\delta^{(j)}(\mathbf{x})$ has the same meaning as $\delta_r^{(j)}$ in (11.11). Therefore the minimum in (11.11) does not converge with positive probability to the optimal value C_0 in (11.26).]

In view of this remark, we modify (11.11) to make it consistent, but to do this, we need some prior information about the system. □

Assumption 11.1

(a) The set $\overline{Q}$ in (11.25) is finite.
(b) Some lower bound $\Delta > 0$ of the value $\min_{\mathbf{x}\in Q} (R^* - p(\mathbf{x}))$ is known.
(c) We define the estimate of a minimal cost as the solution of the following problem:

$$\min_{\overline{\chi}_r} \left\{ \sum_{\substack{1\le i\le n \\ x_i\in\overline{\chi}_r}} c_i x_i \middle| \frac{1}{N} \sum_{1\le j\le N} \delta_r^{(j)} \ge R^* - \frac{\Delta}{2} \right\}. \tag{11.27}$$

In other words, we are looking for

$$C_N = \min_{\mathbf{x}\in Q_N} \sum_{1\le i\le n} c_i x_i,$$

where

$$Q_N = \left\{ \mathbf{x} \in X\colon \frac{1}{N} \sum \delta^{(j)} (\mathbf{x}) \ge R^* - \frac{\Delta}{2} \right\}, \tag{11.28}$$

where $\delta^{(j)} (\mathbf{x})$ is the new notation for $\delta_r^{(j)}$ in (11.27). Also we denote

$$Q_N^* = \left\{ \mathbf{x} \in Q_N\colon \sum_{1\le i\le n} c_i x_i = C_N \right\}. \ \square$$

Remark 11.5 If it is known that $p(\mathbf{x}) > R^*$ for every $\mathbf{x} \in Q$, then we do not need Assumption 11.2(b), and so the proof of the next proposition shows that in this case we are able to establish consistency of the procedure (11.11). □

Proposition 11.7 There exists a.s. a finite r.v. μ such that for $N \ge \mu(\omega)$ we have

$$\text{(i)} \quad Q_{\min} \subset Q_N(\omega) \subset Q, \tag{11.29}$$

$$\text{(ii)} \quad C_N(\omega) = C_0, \tag{11.30}$$

$$\text{(iii)} \quad p(\mathbf{x}) \ge R^* \text{ for every } \mathbf{x} \in Q_N^*(\omega). \tag{11.31}$$

Moreover, if $L(\overline{Q})$ and $L(Q_{\min})$ are some upper bounds of the number of vectors, correspondingly, in $\overline{Q}$ and $Q_{\min}$, then, for $N = 1, 2, \ldots,$

$$P(\mu \le N) \ge 1 - 2\,[L(\overline{Q}) + L(Q_{\min})] \exp(-\tfrac{1}{2}N\Delta^2). \ \square \tag{11.32}$$

Remark 11.6 The inequalities (11.30)–(11.32) mean that for N large enough we get the true value of the minimum cost and guarantee the given system reliability index R^* with probability almost equal to 1. □

Proof The relations (11.29) evidently yield the inequalities (11.30) and (11.31).

If $\mathbf{x} \in Q_{\min}$ and for ω fixed

$$\left| \frac{1}{N} \sum \delta^{(j)}(\mathbf{x}) - p(\mathbf{x}) \right| < \frac{\Delta}{2}, \tag{11.33}$$

then since $p(\mathbf{x}) \geq R^*$, it follows that $\mathbf{x} \in Q_N(\omega)$ [see (11.28)]. Denote by $\mu_{\mathbf{x}}^{(1)}$ an integer valued r.v. such that (11.33) holds for $N \geq \mu_{\mathbf{x}}^{(1)}$ and set $\mu^{(1)}$ $\max_{\mathbf{x} \in Q_{\min}} \mu_{\mathbf{x}}^{(1)}$. By the strong law of large numbers and finiteness of $Q_{\min}$, the r.v. $\mu^{(1)}$ is a.s. finite. Thus $Q_{\min} \subset Q_N(\omega)$ for $N \geq \mu^{(1)}(\omega)$.

On the other hand, for arbitrary $\mathbf{x} \in \overline{Q}$, $p(\mathbf{x}) \leq R^* - \Delta$ by Assumption 11.1(b). If we suppose (11.33) holds for this $\mathbf{x}$ and that $\mathbf{x} \in Q_N(\omega)$, then [see (11.28)] $p(\mathbf{x}) > R^* - \Delta$. But the strong law of large numbers, (11.33) is satisfied for $N \geq \mu_{\mathbf{x}}^{(2)}$ with some a.s. finite r.v. $\mu_{\mathbf{x}}^{(2)}$. It means $\mathbf{x} \notin Q_N(\varphi)$ for such N. Setting $\mu^{(2)} = \max_{\mathbf{x} \in \overline{Q}} \mu_{\mathbf{x}}^{(2)}$, we find that, for $N \geq \mu^{(2)}$ and almost all ω, $\overline{Q} \subset X \backslash Q_N(\omega)$ or $Q_N(\omega) \subset Q$. To complete the proof of (11.29), it is enough to set $\mu = \max\{\mu^{(1)}, \mu^{(2)}\}$.

We get inequality (11.32) by the above definition of the r.v. μ: $P(\mu > N)$ is equal to the probability $P^* \leq 2|L(\overline{Q}) + L(Q_{\min})|\exp(-N\Delta^{(2)}/2)$ that there is $\mathbf{x} \in Q_{\min} \cup \overline{Q}$ such that

$$P\left(\left| \frac{1}{N} \sum_{1 \leq j \leq N} \delta^{(j)}(\mathbf{x}) - p(\mathbf{x}) \right| \geq \frac{\Delta}{2} \right)$$
$$\leq \sum_{\mathbf{x} \in Q_{\min} \subset \overline{Q}} P\left(\frac{1}{N} \left| \sum_{1 \leq j \leq N} \delta^{(j)}(\mathbf{x}) - p(\mathbf{x}) \right| \geq \frac{\Delta}{2} \right)$$

due to Hoeffling's inequality as in the proof of Proposition 11.2. □

11.7.3 Restrictions on System MTBF

Let T^* be the specified average time to failure. Introduce the following notation:

$$S = \{\mathbf{x} \in X: E\tau_{\mathbf{x}} \geq T^*\}, \qquad \overline{S} = \{\mathbf{x} \in X: E\tau_{\mathbf{x}} < T^*\}; \tag{11.34}$$

$S_{\min}$ is the set of all minimal vectors from S;

$$\overline{C} = \min_{\mathbf{x} \in S_{\min}} \sum_{1 \le i \le n} c_i x_i.$$

We assume the set S to be nonempty.

If we want to obtain consistent statistical estimates of $\overline{C}$, we have to modify the problem (11.11′) using the following assumption (see Remark 11.4).

Assumption 11.2

(a) The set $\overline{S}$ in (11.34) is finite.
(b) The lower bound $\Delta_T > 0$ of $\min_{\mathbf{x} \in \overline{S}} (T^* - E\tau_{\mathbf{x}})$ exists.

Let us introduce the version of (11.11′) as follows. As an estimate of $\overline{C}$ we will use the statistic

$$\overline{C}_N = \min_{\mathbf{x} \in S_N} \sum_{1 \le i \le n} c_i x_i$$

with

$$S_N = \left\{ \mathbf{x} \in X: \frac{1}{N} \sum_{1 \le i \le n} \tau^{(j)}(\mathbf{x}) \ge T^* - \tfrac{1}{2}\Delta \right\}.$$

Let

$$S_N = \left\{ \mathbf{x} \in S_N: \frac{1}{N} \sum_{1 \le i \le n} c_i x_i = \overline{C}_N \right\}.$$

The next proposition is proved analogously to Proposition 11.7. □

Proposition 11.8 There exists a.s. finite r.v. ζ such that for $N \ge \zeta(\omega)$ we have

(i) $S_{\min} \subset S_N(\omega) \subset S$,

(ii) $\overline{C}_N(\omega) = \overline{C}$, and

(iii) $E\tau_{\mathbf{x}} \ge T^*$ for $\mathbf{x} \in S_N^*$. □

Remark 11.7

(a) One can show consistency of the procedure (11.11′) provided that $E\tau_{\mathbf{x}} > T^*$ for every $\mathbf{x} \in S$.
(b) Bounds similar to (11.23) and (11.24) hold for $P(\zeta \leq N)$ provided that

$$\max_{\mathbf{x} \in S_{\min} \cup \bar{S}} E\tau_{\mathbf{x}}^{\tau} < \infty. \ \Box$$

APPENDIX

STATISTICAL TABLES

Table 1 Quantiles of normal distribution

$$P = \frac{1}{\sqrt{2\pi}} \int_{-\infty}^{U_p} e^{-x^2/2}\, dx = \Phi(U_p)$$

P	U_p	P	U_p
0.50	0	0.98	2.054
0.55	0.126	0.99	2.326
0.60	0.253	$0.9^2 5$	2.58
0.65	0.385	0.9^3	3.09
0.70	0.524	$0.9^3 5$	3.29
0.75	0.674	0.9^4	3.72
0.80	0.842	$0.9^4 5$	3.89
0.85	1.036	0.9^5	4.27
0.90	1.282	$0.9^5 5$	4.42
0.91	1.341	0.9^6	4.75
0.92	1.405	$0.9^6 5$	4.89
0.93	1.476	0.9^7	5.20
0.94	1.555	$0.9^7 5$	5.33
0.95	1.645	0.9^8	5.61
0.96	1.751	$0.9^8 5$	5.73
0.97	1.881	0.9^9	6.00

Table 2 Critical values $\chi^2(r, q)$ for a chi-square distribution

$$q = \frac{1}{2^{r/2}\Gamma(r/2)} \int_{\chi^2(r,\, q)}^{\infty} x^{r/2-1} e^{-x/2}\, dx$$

	p													
r	0.99	0.98	0.95	0.90	0.80	0.70	0.50	0.30	0.20	0.10	0.05	0.02	0.01	0.001
1	0.000	0.001	0.004	0.016	0.064	0.148	0.455	1.074	1.642	2.71	3.84	5.41	6.63	10.83
2	0.020	0.040	0.103	0.211	0.446	0.713	1.386	2.41	3.22	4.60	5.99	7.82	9.21	13.82
3	0.115	0.185	0.352	0.584	1.005	1.424	2.37	3.66	4.64	6.25	7.81	9.84	11.34	16.27
4	0.297	0.429	0.711	1.064	1.649	2.20	3.36	4.88	5.99	7.78	9.49	11.67	13.28	18.46
5	0.554	0.752	1.145	1.610	2.34	3.00	4.35	6.06	7.29	9.24	11.07	13.39	15.09	20.5
6	0.872	1.134	1.635	2.20	3.07	3.83	5.35	7.23	8.56	10.64	12.59	15.03	16.81	22.5
7	1.239	1.564	2.17	2.83	3.82	4.67	6.35	8.38	9.80	12.02	14.07	16.62	18.48	24.3
8	1.646	2.03	2.73	3.49	4.59	5.53	7.34	9.52	11.03	13.36	15.51	18.17	20.1	26.1
9	2.09	2.53	3.33	4.17	5.38	6.39	8.34	10.66	12.24	14.68	16.92	19.68	21.7	27.9
10	2.56	3.06	3.94	4.86	6.18	7.27	9.34	11.78	13.44	15.99	18.31	21.2	23.2	29.6
11	3.05	3.61	4.58	5.58	6.99	8.15	10.34	12.90	14.63	17.28	19.68	22.6	24.7	31.3
12	3.57	4.18	5.23	6.30	7.81	9.03	11.34	14.01	15.81	18.55	21.0	24.1	26.2	32.9

13	4.11	4.76	5.89	7.04	8.63	9.93	12.34	15.12	16.98	19.81	22.4	25.5	27.7	34.6
14	4.66	5.37	6.57	7.79	9.47	10.82	13.34	16.22	18.15	21.1	23.7	26.9	29.1	36.1
15	5.23	5.98	7.26	8.55	10.31	11.72	14.34	17.32	19.31	22.3	25.0	28.3	30.6	37.7
16	5.81	6.61	7.96	9.31	11.15	12.62	15.34	18.42	20.5	23.5	26.3	29.6	32.0	39.3
17	6.41	7.26	8.67	10.08	12.00	13.53	16.34	19.51	21.6	24.8	27.6	31.0	33.4	40.8
18	7.01	7.91	9.39	10.86	12.86	14.44	17.34	20.6	22.8	26.0	28.9	32.3	34.8	42.3
19	7.63	8.57	10.11	11.65	13.72	15.35	18.34	21.7	23.9	27.2	30.1	33.7	36.2	43.8
20	8.26	9.24	10.85	12.44	14.58	16.27	19.34	22.8	25.0	28.4	31.4	35.0	37.6	45.3
21	8.90	9.92	11.59	13.24	15.44	17.18	20.3	23.9	26.2	29.6	32.7	36.3	38.9	46.8
22	9.54	10.60	12.34	14.04	16.31	18.10	21.3	24.9	27.3	30.8	33.9	37.7	40.3	48.3
23	10.20	11.29	13.09	14.85	17.19	19.02	22.3	26.0	28.4	32.0	35.2	39.0	41.6	49.7
24	10.86	11.99	13.85	15.66	18.06	19.94	23.3	27.1	29.6	33.2	36.4	40.3	43.0	51.2
25	11.52	12.70	14.61	16.47	18.94	20.9	24.3	28.2	30.7	34.4	37.7	41.7	44.3	52.6
26	12.20	13.41	15.38	17.29	19.82	21.8	25.3	29.2	31.8	35.6	38.9	42.9	45.6	54.1
27	12.88	14.12	16.15	18.11	20.7	22.7	26.3	30.3	32.9	36.7	40.1	44.1	47.0	55.5
28	13.56	14.85	16.93	18.94	21.6	23.6	27.3	31.4	34.0	37.9	41.3	45.4	48.3	56.9
29	14.26	15.57	17.71	19.77	22.5	24.6	28.3	32.5	35.1	39.1	42.6	46.6	49.6	58.3
30	14.95	16.31	18.49	20.6	23.4	25.5	29.3	33.5	36.2	40.3	43.8	48.0	50.9	59.7

Table 3 Values $\Delta'_{1-\alpha}(d)$ and $\Delta'_{1-\alpha}(d)$, $1-\alpha = 0.95$

c \ N	50	60	80	100	150	200
0	0.07378	0.06148	0.04611	0.03689	0.02459	0.01844
	0.00000	0.00000	0.00000	0.00000	0.00000	0.00000
1	0.11257	0.09365	0.07009	0.05600	0.03727	0.02793
	0.00051	0.00042	0.00032	0.00025	0.00017	0.00013
2	0.14750	0.12248	0.09146	0.07298	0.04849	0.03631
	0.00489	0.00407	0.00305	0.00243	0.00162	0.00121
3	0.18090	0.14994	0.11172	0.08902	0.05904	0.04417
	0.01263	0.01049	0.00783	0.00625	0.00415	0.00311
4	0.21362	0.17672	0.13136	0.10454	0.06921	0.05173
	0.02248	0.01863	0.01388	0.01106	0.00734	0.00549
5	0.24607	0.20317	0.15066	0.11972	0.07912	0.05909
	0.03384	0.02800	0.02082	0.01657	0.01097	0.00820
6	0.27853	0.22949	0.16975	0.13471	0.08886	0.06630
	0.04640	0.03832	0.02842	0.02259	0.01493	0.01115
7	0.31115	0.25582	0.18875	0.14957	0.09848	0.07341
	0.05995	0.04942	0.03657	0.02902	0.01915	0.01429
8	0.34408	0.28226	0.20772	0.16435	0.10801	0.08044
	0.07440	0.06120	0.04518	0.03580	0.02358	0.01758
9	0.37742	0.30889	0.22671	0.17910	0.11748	0.08741
	0.08966	0.07360	0.05419	0.04289	0.02819	0.02100
10	0.41126	0.33578	0.24576	0.19385	0.12691	0.09434
	0.10570	0.08657	0.06358	0.05025	0.03297	0.02453
11	0.44568	0.36297	0.26490	0.20863	0.13631	0.10123
	0.12247	0.10008	0.07331	0.05785	0.03788	0.02816
12	0.48077	0.39053	0.28417	0.22344	0.14569	0.10810
	0.13996	0.11411	0.08336	0.06568	0.04293	0.03189
13	0.51659	0.41818	0.30359	0.23831	0.15507	0.11495
	0.15818	0.12865	0.09372	0.07372	0.04809	0.03569
14	0.55322	0.44688	0.32318	0.25325	0.16445	0.12178
	0.17711	0.14368	0.10437	0.08198	0.05337	0.03957
15	0.59074	0.47577	0.34295	0.26828	0.17384	0.12861
	0.19677	0.15921	0.11531	0.09042	0.05875	0.04352
16	0.62922	0.50519	0.36294	0.28341	0.18325	0.13544
	0.21717	0.17524	0.12654	0.09906	0.06123	0.04753
17	0.66875	0.53518	0.38314	0.29864	0.19267	0.14226
	0.23833	0.19177	0.13804	0.10788	0.06980	0.05160
18	0.70942	0.56578	0.40860	0.31399	0.20213	0.14909
	0.26027	0.20881	0.14982	0.11688	0.07547	0.05574
19	0.75131	0.59704	0.42431	0.32947	0.21161	0.15592
	0.28303	0.22637	0.16188	0.12607	0.08123	0.05993

Table 4 Quantiles $\Delta_{1-\alpha}(d)$ of the Poisson distribution, $L_d[\Delta_{1-\alpha}(d)] = 1 - \alpha$

d \ $1-\alpha$	0.99993	0.9999	0.9993	0.999	0.993	0.99	0.95
0	0.000070	0.000100	0.000700	0.00100	0.00702	0.01005	0.05129
1	0.01188	0.01421	0.03789	0.04540	0.12326	0.14855	0.35536
2	0.07633	0.08618	0.16824	0.19053	0.38209	0.43604	0.81769
3	0.21115	0.23180	0.38894	0.42855	0.74108	0.82325	1.36632
4	0.41162	0.44446	0.68204	0.73937	1.17032	1.27911	1.97015
5	0.66825	0.71375	1.03236	1.10710	1.65152	1.78528	2.61301
6	0.97222	1.03040	1.42874	1.52034	2.17293	2.33021	3.28532
7	1.31628	1.38697	1.86297	1.97081	2.72659	2.90611	3.98082
8	1.69465	1.77758	2.32894	2.45242	3.30682	3.50746	4.69523
9	2.10271	2.19758	2.82197	2.96052	3.90942	4.13020	5.42541
10	2.53672	2.64323	3.33840	3.49148	4.53118	4.77125	6.16901
11	2.99367	3.11150	3.87531	4.04244	5.16960	5.42818	6.92421
12	3.47103	3.59988	4.43033	4.61106	5.82265	6.09907	7.68958
13	3.96672	4.10632	5.00152	5.19544	6.48871	6.78235	8.46394
14	4.47896	4.62904	5.58725	5.79398	7.16642	7.47673	9.24633
15	5.00626	5.16657	6.18615	6.40533	7.85464	8.18111	10.03596
16	5.54732	5.71762	6.79705	7.02835	8.55241	8.89457	10.83214

d \ $1-\alpha$	0.93	0.90	0.80	0.70	0.60	0.50	0.40
0	0.07257	0.10536	0.22314	0.35667	0.51082	0.69315	0.91629
1	0.43081	0.53181	0.82439	1.09735	1.37642	1.67835	2.02231
2	0.94230	1.10206	1.53504	1.91378	2.28508	2.67406	3.10538
3	1.53414	1.74477	2.29679	2.76371	3.21132	3.67206	4.17526
4	2.17670	2.43259	3.08954	3.63361	4.14774	4.67091	5.23662
5	2.85488	3.15190	3.90366	4.51714	5.09098	5.67016	6.29192
6	3.55984	3.89477	4.73366	5.41074	6.03924	6.66964	7.34265
7	4.28584	4.65612	5.57606	6.31217	6.99137	7.66925	8.38977
8	5.02895	5.43247	6.42848	7.21993	7.94661	8.66895	9.43395
9	5.78633	6.22130	7.28922	8.13293	8.90441	9.66871	10.47568
10	6.55583	7.02075	8.15702	9.05036	9.86440	10.66852	11.51533
11	7.33581	7.82934	9.03090	9.97161	10.82624	11.66836	12.55317
12	8.12496	8.64594	9.91010	10.89620	11.78972	12.66823	13.58944
13	8.92222	9.46962	10.79398	11.82373	12.75462	13.66811	14.62431
14	9.72672	10.29962	11.68206	12.75388	13.72081	14.66802	15.65793
15	10.53773	11.13530	12.57389	13.68639	14.68814	15.66793	16.69043
16	11.35465	11.97613	13.46913	14.62103	15.65651	16.66785	17.72191

d \ $1-\alpha$	0.30	0.20	0.10	0.05	0.025	0.01	0.005
0	1.20397	1.60944	2.30258	2.99573	3.68888	4.60517	5.29832
1	2.43922	2.99431	3.88972	4.74386	5.57164	6.63835	7.43013
2	3.61557	4.27903	5.32232	6.29579	7.22469	8.40595	9.27379
3	4.76223	5.51504	6.68078	7.75366	8.76727	10.04512	10.97748
4	5.89036	6.72098	7.99359	9.15352	10.24159	11.60462	12.59409
5	7.00555	7.90599	9.27467	10.51303	11.66833	13.10848	14.14976
6	8.11105	9.07538	10.53207	11.84240	13.05947	14.57062	15.65968
7	9.20895	10.23254	11.77091	13.14811	14.42268	15.99996	17.13359
8	10.30068	11.37977	12.99471	14.43465	15.76319	17.40265	18.57822
9	11.38727	12.51875	14.20599	15.70522	17.08480	18.78312	19.99842
10	12.46951	13.65073	15.40664	16.96222	18.39036	20.14468	21.39783
11	13.54798	14.77666	16.59812	18.20751	19.68204	21.48991	22.77926
12	14.62316	15.89731	17.78158	19.44257	20.96158	22.82084	24.14494
13	15.69544	17.01328	18.95796	20.66857	22.23040	24.13912	25.49669
14	16.78512	18.12509	20.12801	21.88648	23.48962	25.44609	26.83598
15	17.83246	19.23316	21.29237	23.09713	24.74022	26.74289	28.16406
16	18.89769	20.33782	22.45158	24.30118	25.98300	28.03045	29.48196

BIBLIOGRAPHY

Aalen, O. O. (1978). Non-Parametric Inference for a Family of Counting Processes. *Ann. Statist.,* Vol. 6.

Albert, A. E. (1961). Sequential Design of Experiments for Infinitely Many States of Natures. *Ann. Math. Statist.,* Vol. 32, No. 3.

Ayvazyan, S. A. (1965). Distinguishing Close Hypotheses of Density Function for the Generalized Sequential Criterion. *Theory Prob. Appl.,* Vol 10, No. 4.

Barlow, R. E. (1985). Combining Component and System Information in System Reliability Calculation. In *Probabilistic Methods in the Mechanics of Solids and Structures,* ed. By S. Eggwertz and N. L. Lind. Springer-Verlag, Berlin.

Barlow, R. E., and **Gupta, S.** (1966). Distribution-Free Life Testing Sampling Plans. *Technometrics,* Vol. 8, No. 4.

Barlow, R. E., and **Proschan, F.** (1965). *Mathematical Theory of Reliability.* Wiley, New York.

Barlow, R. E., and **Proschan, F.** (1966). Tolerance and Confidence Limits for Classes of Distributions Based on Failure Rate. *Ann. Math. Statist.,* Vol. 37, No. 6.

Barlow, R. E., and **Proschan, F.** (1967). Exponential Life Test Procedures When the Distribution Has Monotone Failure Rate. *J. Statist. Assoc.,* Vol. 62.

Barlow, R. E., and **Proschan, F.** (1975). *Statistical Theory of Reliability and Life Testing.* Holt, Rinehart and Winston, New York.

Barlow, R. E., and **Proschan, F.** (1981). *Statistical Theory of Reliability and Life Testing,* 2nd ed., Silver Spring, MD. To Begin With.

Bellman, R. E. (1957). *Dynamic Programming.* Princeton University Press. Princeton, NJ.

Bellman, R. E., and **Dreyfus, E.** (1962). *Applied Dynamic Programming.* Princeton University Press, Princeton, NJ.

Belyaev, Yu. K. (1966a). Construction of the Lower Confidence Limit for the System Reliability on the Basis of Testing Its Components (Russian). In *Reliability of Complex Systems.* Sovetskoe Radio, Moscow.

Belyaev, Yu. K. (1966b). Confidence Limits for Functions of Several Variables (Russian). *Proc. Sov. Acad. Sci.,* Vol. 169, No. 4.

Belyaev, Yu, K. (1968). On Simple Methods of Confidence Limit Construction. *Eng. Cybernet.* (*USA*), No. 5.

Belyaev, Yu. K. (1982). *Statistical Methods for Analysis of Reliability Test Results* (Russian). Znanie, Moscow.

Belyaev, Yu. K. (1984). *Nonparametrical Methods in Statistical Data Inferences* (Russian). Znanie, Moscow.

Belyaev, Yu. K. (1985). Multiplicative Estimates of the Probability of Failure Free Operation. *Sov. J. Comput. Syst. Sci.* (*USA*). No. 4.

Belyaev, Yu. K. (1987). *Statistical Methods of Inferences of Censored Reliability Data* (Russian). Znanie, Moscow.

Belyaev, Yu. K., Dugina, T. N., and **Chepurin, E. V.** (1967). Computation of the Lower Confidence Limit for the Complex System Reliability. *Eng. Cybernet.* (USA), Nos. 2 and 3.

Belyaev, Yu. K., Gnedenko, B. V., and **Ushakov, I. A.** (1983). Mathematical Problems in Queuing and Reliability Theory. *Eng. Cybernet.* (*USA*), Vol. 21, No. 6.

Birnbaum, Z. W., and **Saunders, S. C.** (1968). A Probabilistic Interpretation of Miner's Rule. *SIAM J. Appl. Math.,* Vol 16, No. 3.

Bogatyrev, V. A., and **Mizin, I. A.** (1985). Reliability of Information Systems. In *Handbook of Reliability of Technical Systems* (Russian), ed. By I. Ushakov. Sovietskoe Radio, Moscow.

Bogdanoff, J. L., and **Kozin, F.** (1985). *Probabilistic Models of Cumulative Damage.* Wiley, New York.

Bolotin, V. V. (1989). *Prediction of Service Life for Machines and Structures.* ASME Press, New York.

Bol'shev, L N. (1965). On Construction of Confidence Limits. *Theory Prob. Appl.,* Vol. 1, No. 1.

Bol'shev, L. N., and **Loginov, E. A.** (1966). Interval Estimates Under Noisy Parameters. *Theory Prob. Appl.,* Vol. 2, No. 1.

Bol'shev, L. N., and **Smirnov, N. V.** (1965). *Tables of Mathematical Statistics* (Russian). Nauka, Moscow.

Breslow, N., and **Crowley, J.** (1974). A Large Sample Study of the Life Table and Product Limit Estimates under Random Censorship. *Ann. Statist.,* Vol. 2.

Buehler, R. J. (1957). Confidence Intervals for the Product of Two Binomial parameters. *J. Am. Statist. Assoc.,* Vol. 52, No. 3.

Buslenko, N. P., Kalashnikov, V. V., and **Kovalenko, I. N.** (1973). *Lectures on Theory of Complex Systems* (Russian). Sovietskoe Radio, Moscow.

Chan, C. K. (1990). A Proportional Hazard Approach to SiO_2 Breakdown Voltage. *IEEE Trans. Reliabil.,* Vol. R-39, No. 2.

Chernoff, H. (1959). Sequential Design of Experiments. *Ann. Math. Statist.,* Vol. 30, No. 3.

Clopper, C. I., and **Pearson, E. S.** (1934). The Use of Confidence or Fiducial Limits Illustrated in the Case of the Binomial. *Biometrika,* Vol. 26.

Cohen, A. C. (1963). Progressively Censored Samples in Life Testing. *Technometrics,* Vol. 5, No. 3.

Cohen, A. C. (1965). Maximum Likelihood in the Weibull Distribution Based on Complete and on Censored Samples. *Technometrics,* Vol. 7, No. 4.

Cole, P. V. Z. (1975). A Bayesian Reliability Assessment of Complex System for Binomial Sampling. *IEEE Trans. Reliabil.,* Vol. 24.

Cox, D. R., and **Hinkley, D. V.** (1978). *Problems and Solutions in Theoretical Statistics.* Wiley, New York.

Cox, D. R., and **Oaks, D.** (1984). *The Analysis of Survival Data.* Chapman & Hall, London, New York.

Cramer, H. (1948). *Mathematical Methods in Statistics.* University Press, Princeton, NJ.

Crowder, M. J., Kimber, A. C., Smith, R. L., and **Sweeting, T. J.** (1991). *Statistical Analysis of Reliability Data.* Chapman & Hall, London, New York.

De Groot, M. (1970). *Optimal Statistical Decisions.* McGraw-Hill, New York.

Dempster, A. P. (1963). On the Difficulties Inherent in Fisher's Fiducial Argument. *Proc. Int. Statist. Conf.,* Ottawa.

Doob, J. L. (1953). *Stochastic Processes.* Wiley, New York.

Dostal, R. G., and **Iannuzzelli, L. M.** (1977). Confidence Limits for System Reliability When Testing Takes Place at the Component Level. In *The Theory and Applications of Reliability,* Vol. 2, ed. by C. P. Tsakos and I. N. Shimi. Academic, New York.

Dragalin, V. P., and **Novikov, A. A.** (1987). Asymptotical Solution of the Kiefer-Weiss Problem for the Process with Independent Increments. *Theory Prob. Appl.,* Vol. 32, No. 4.

Dragalin, V. P., and **Novikov, A. A.** (1996). Adaptive Sequential Tests for Complex Hypotheses, in *Statistics and Control of Stochastic Processes* (Russian). Theory Prob. And Its Appl., Moscow.

Draper, N. R., and **Smith, H.** (1981). *Applied Regression Analysis.* Wiley, New York.

Dvoretzky, A., Kiefer, J., and **Wolfowitz, J.** (1953). Sequential Decision Problems for Processes with Continuous Time Parameter. *Ann. Math. Statist.,* Vol. 24.

Easterling, R. G. (1972). Approximate Confidence Limits for the System Reliability. *J. Am. Statist. Assoc.,* Vol. 67.

Efron, B. (1979). Bootstrap Method: Another Look at the Jackknife. *Ann. Statist.,* Vol. 4, No. 1.

Efron, B. (1981). Nonparametric Standard Errors and Confidence Intervals. *Can. J. Statist.,* No. 9.

Efron, B. (1982). The Jackknife, the Bootstrap, and Other Resampling Plans, CBMS Regional Conference. *Ser. Appl. Math.,* Vol. 38.

Efron, B. (1985). Bootstrap Confidence Intervals for a Class of Parametric Problems. *Biometrics,* Vol. 72, No. 1.

Engelman, L., Roach, H., and **Schick, G.** (1967). Computer Program for Exact Confidence Bounds. *J. Ind. Eng.,* Vol. 18, No. 8.

Epstein, B. (1960a). Statistical Life Test Acceptance Procedures. *Technometrics,* Vol. 2, No. 4, pp. 435–446.

Epstein, B. (1960b). Estimation from Life Test Data. *Technometrics,* Vol. 2, No. 4.

Epstein, B. (1960c). Testing for the Validity of the Assumption That the Underlying Distribution of Life is Exponential. *Technometrics,* Vol 2, No. 1.

Epstein, B., and **Sobel, M.** (1953). Life Testing. *J. Am. Statist. Assoc.,* Vol. 48.

Epstein, B., and **Sobel, M.** (1955). Sequential Life Testing in Exponential Case. *Ann. Math. Statist.,* Vol. 26, No. 1.

Farkhad-Zadeh, E. M. (1979). About Difference of Limit Values of Confidence and Fiducial Intervals for Reliability Parameters of Systems. *Eng. Cybernet.,* (USA), No. 4.

Feller, W. (1957). *An Introduction to Probability Theory and Its Applications,* Vol. 1., Wiley, New York.

Ferguson, T. S. (1973). A Bayesian Analysis of Some Nonparametric Problems. *Ann. Statist.,* Vol. 1, No. 2.

Fisher, R. A. (1935). Theoretical Argument in Statistical Inference. *Ann. Eugen.* Vol. 5, No. 3.

Fraser, D. A. S. (1961). The Fiducial Method and Invariance. *Biometrika,* Vol. 48.

Gill, R. D. (1981). Testing with Replacement and the Product-Limit Estimator. *Ann. Statist.,* Vol. 9.

Gill, R. D. (1983). Large Sample Behavior of the Product-Limit Estimator on the Whole Line. *Ann. Statist.,* Vol. 1.

Gnedenko, B. V. (1941). Limit Theorems for Maximum Order Statistic (Russian). *Rep. Acad. Sci. USSR,* Vol. 32, No. 1.

Gnedenko, B. V. (1943). Sur la Distribution Limite du Terme Maximum d'Une Serie Aleatorie. *Ann. Math.,* Vol. 44, No. 3.

Gnedenko, B. V. Ed. (1983). *Aspects of Mathematical Theory of Reliability* (Russian). Radio i Svyaz, Moscow.

Gnedenko, B. V. (1998). *Course of the Probability Theory.* Gordon & Breach, New York.

Gnedenko, B. V. and **Ushakov, I. A.** (1995). *Probabilistic Reliability Engineering.* Wiley, New York.

Gnedenko, B. V., Belyaev, Yu. K. and **Solovyev, A. D.** (1969). *Mathematical Methods of Reliability Theory.* Academic, San Diego.

Goldman, A. Ya. (1994). *Prediction of the Deformation Properties of Polymeric and Composite Materials.* American Chemical Society, Washington, DC.

Gordienko, E. I. (1979) Statistical Solution of the Inverse Problem of Optimal Redundancy (Russian), *Cybernetics,* Kiev, No. 6.

Groisberg, L. B. (1980a). On Fiducial Approach for Reliability Estimation. *Sov. J. Comput. Syst. Sci. (USA),* No. 4.

Groisberg, L. B. (1980b). On Fiducial Approach in Reliability Estimating (Russian). *Eng. Cybernet.,* No. 4.

Gugushvili, D. F., Zgenti, L. D., and **Namiecheyshvili, O. M.** (1975). Accelerated Reliability Tests. *Eng. Cybernet. (USA),* No. 3.

Hall, W. J., and **Wellner, J. A.** (1980). Confidence Bounds for Survival Curve from Censored Data. *Biometrika,* Vol. 67.

Harris, B. (1971). Hypothesis Testing and Confidence Intervals for Products and Quotients of Poisson Parameters with Applications to Reliability. *J. Am. Statist. Assoc.,* Vol. 66.

Hoeffding, W. (1960). Lower Bounds for the Expected Sample Size and the Average Risk of Sequential Procedure. *Ann. Math. Statist.,* Vol. 31, No. 2.

Huffman, M. D. (1983). An Efficient Approximation Solution of the Kiefer-Weiss Problem. *Ann. Statist.,* Vol. 11, No. 1.

Johns, M. V., and **Lieberman, G. J.** (1966). An Exact Asymptotycally Efficient Confidence Bound for Reliability in the Case of the Weibull Distribution. *Technometrics,* Vol. 8, No. 1.

Kalashnikov, V. V. (1995). Monte Carlo Simulation. In *Handbook of Reliability Engineering,* ed. By I. Ushakov. Wiley, New York.

Kaminskiy, M. (1987). Accelerated Life Test Planning and Data Analysis (Russian). In *Electronica,* Moscow.

Kaminskiy, M. (1994). Accelerated Life Tests. In *Handbook of Reliability Engineering,* ed. by I. A. Ushakov, Wiley, New York.

Kaminskiy, M., and **Kristalinsky, L.** (1992). Reliability Prediction of Capacitors on the Basis of Accelerated Life Test Data (Russian). In *Electronica.* Moscow.

Kaminskiy, M. P., Ushakov, I. A., and **Hu, J.** (1995). Statistical Inference Concepts. In *Product Reliability, Maintainability, and Supportability Handbook,* ed. by M. Pecht, CRC Press, Boca Raton, FL.

Kao, J. H. K. (1956). Computer Methods for Estimating Weibull Parameters in Reliability Studies. *IRE Trans. Reliabil. Qual. Control.*

Kaplan, E. L., and **Meier, P.** (1958). Nonparametric Estimation from Incomplete Observations. *J. Am. Statist. Assoc.,* Vol. 53.

Kiefer, J., and **Sacks, J.** (1963). Asymptotically Optimal Sequential Inference and Design. *Ann. Math. Statist.,* Vol. 34, No. 3.

Kiefer, J., and **Weiss, L.** (1957). Some Properties of Generalized Sequential Probability Ratio Tests. *Ann. Math. Statist.,* Vol. 28, No. 1.

Kiefer, J., and **Wolfowitz, J.** (1957). Consistency of the Maximum Likelihood Estimator in the Presence of Infinitely Many Incidental Parameters. *Ann. Math. Statist.,* Vol. 27.

Kredentser, B. P., et al. (1967). Solution of Reliability Tasks on Computer (Russian). Sov. Radio, Moscow.

Krol', I. A. (1974). Using the Confidence Set Method for Interval Estimation of Reliability Indices. *Eng. Cybernet.* (*USA*), No. 1.

Krol', I. A. 1975). Interval Reliability Estimation for "Continuous" and Instant Failures. *Eng. Cybernet.* (*USA*), No. 5.

Kudlaev, E. M. (1986). Estimation of Parameters of the Weibull-Gnedenko Distribution (Review). *Sov. J. Cybernet. Comput. Sci.* (*USA*). No. 6.

Kullback, S. (1959). *Information Theory and Statistics.* Wiley, New York.

Kullback, S., and **Leiber, R. A.** (1951). On Information and Sufficiency. *Ann. Math. Statist.* Vol. 22, No. 1.

Lai, T. L. (1973). Optimal Stopping and Sequential Tests Which Minimize the Maximum Expected Sample Size. *Ann. Statist.,* Vol. 1, No. 4.

Landau, L. D. (1976). *Mechanics.* Pergamon, New York.

Lawless, J. F. (1982). *Statistical Models and Methods for Lifetime Data.* Wiley, New York.

Leemis, L. M. (1995). *Reliability: Probabilistic Models and Statistical Methods.* Prentice-Hall, Englewood Cliffs, NJ.

Lehman, E. L. (1959). *Testing Statistical Hypotheses.* Wiley, New York.

Lieberman, G. J. (1962). Estimates of the Parameters of the Weibull Distribution (Abstract). *Ann. Math. Statist.,* Vol. 33.

Lipow, M. (1958). Measurement of Overall Reliability Utilizing Results of Independent Subsystem Tests. Report No. GM-TR-0165-00506, Space Technology Labs.

Lipow, M. (1959). Tables of Upper Confidence Limits on Failure Probability of 1, 2, and 3 Component Serial Systems. Report No. TR-50-0000-00756, Space Technology Labs.

Liptser, R. A., and **Shiryaev, A. N.** (1981). *Statistics of Stochastic Processes* (Russian). Nauka, Moscow.

Lloyd, D. K., and **Lipow, M.** (1962). *Reliability Management, Methods and Mathematics.* Prentice-Hall, Englewood Cliffs, NJ.

Lorden, G. (1976). 2-SPRT's and the Modified Kiefer-Weiss Problem of Minimizing an Expected Sample Size. *Ann. Statist.,* Vol. 4, No. 2.

Madansky, A. (1965). Approximate Confidence Limits for the Reliability of Series and Parallel Systems. *Technometrics,* Vol. 7.

Mann, N. R. (1968). Point and Interval Estimation Procedures for the Two-Parameter Weibull and Extreme-Value Distributions. *Technometrics,* Vol. 10, No. 2.

Mann, N. R. (1974). Approximate Optimum Confidence Bounds on Series and Series-Parallel System Reliability for Systems with Binomial Subsystem Data. *IEEE Trans. Reliabil.,* Vol. R-23.

Mann, N. R., and **Grubbs, F. E.** (1974). Approximately Optimum Confidence Bounds for System Reliability Based on Component Test Plan. *Technometrics,* Vol 16.

Mann, N. R., Schaefer, R. E., and **Singpurwalla, N. D.** (1974). *Methods of Statistical Analysis of Reliability and Life Data.* Wiley, New York.

Martz, H. F., and **Duran, I. S.** (1985). A Comparison of Three Methods for Calculating Lower Confidence Limits on System Reliability Using Binomial Component Data. *IEEE Trans. Reliabil.,* Vol. R-34, No. 2.

Martz, H. F., and **Waller, R. A.** (1982). *Bayesian Reliability Analysis.* Wiley, New York.

Martz, H. F., and **Waller, R. A.** (1990). Bayesian Reliability Analysis of Complex Series/Parallel Systems of Binomial Subsystems and Components. *Technometrics,* Vol. 32.

Martz, H. F., Waller, R. A., and **Fickas, E. T.** (1988). Bayesian Reliability Analysis of Series Systems of Binomial Subsystems and Components. *Technometrics,* Vol. 30.

Mastran, D. V. (1976). Incorporating Component and System Test Data into the Same Assessment: A Bayesian Approach. *Operat. Res.,* Vol. 24.

Mastran, D. V., and **Singpurwalla, N. D.** (1978). A Bayesian Estimation of the Reliability of Coherent Structures. *Operat. Res.,* Vol. 26.

Mennon, M. V. (1963). Estimation of the Shape and Scale Parameters of the Weibull Distribution. *Technometrics,* Vol. 5, No. 2.

Mirnyi, R. A., and **Solovyev, A. D.** (1964). Estimation of System Reliability on the Basis of Its Units Tests (Russian). In *Cybernetics in Service for Communism,* Vol. 2, Energiya, Moscow.

Mizin, I. A., Bogatyrev, V. A., and **Kuleshov, A. P.** (1986). Telecommunication Networks with Packet Commutation, ed. by V. S. Semenikhin (Russian), Radio I Svyaz, Moscow.

Myhre, J., and **Saunders, S. C.** (1968). Comparison of Two Methods of Obtaining Approximate Confidence Intervals for System Reliability. *Technometrics,* Vol. 10.

National Bureau of Standards. (1950). Tables of the Binomial Probability Distribution. National Bureau of Standards. Applied Mathematics Series 6, Washington, DC.

Natvig, B., and **Eide, H.** (1987). Bayesian Estimation of System Reliability. *Scand. J. Statist.* Vol. 14.

Nelson, W. (1969). Hazard Plotting for Incomplete Failure Data. *J. Qual. Technol.,* Vol. 1.

Nelson, W. (1972). Theory and Application of Hazard Plotting for Censored Failure Data. *Technometrics,* Vol. 14.

Nelson, W. (1982). *Applied Life Data Analysis.* Wiley, New York.

Nelson, W. (1990). *Accelerated Testing: Statistical Models, Test Plans and Data Analysis.* Wiley, New York.

Neyman, J. (1935). On the Problem of Confidence Intervals. *Ann. Math. Statist.,* Vol. 6, No. 1.

Neyman, J. (1937). Outline of a Theory of Statistical Estimation Based on the Classical Theory of Probability. *Trans. Roy. Philos. Soc.,* Vol. A236.

Owen, D. B. (1962). *Handbook of Statistical Tables.* Addison-Wesley, Palo Alto, CA.

Pavlov, I. V. (1972). Computation of Confidence Limits for Functions of Several Unknown Variables. *Eng. Cybernet. (USA)*, No. 2.

Pavlov, I. V. (1973). Confidence Limits for System Reliability on the Basis of Its Components Testing. *Eng. Cybernet. (USA)*, No. 1.

Pavlov, I. V. (1974). System Reliability Estimation on the Basis of Aging Units Tests. *Eng. Cybernet. (USA)*, No. 3.

Pavlov, I. V. (1977a). Interval Estimation for Functions of Several Unknown Variables. *Eng. Cybernet. (USA)*, Nos. 3 and 5.

Pavlov, I. V. (1977b). Monotone Confidence Limits for the Class of Distributions with Increasing Failure Rate. *Eng. Cybernet. (USA)*, No. 6.

Pavlov, I. V. (1980). Fiducial Approach for Estimation of Complex System Reliability by Results of Its Component Testing. *Eng. Cybernet. (USA)*, No. 4.

Pavlov, I. V. (1981a). On Fiducial Approach for Construction of Confidence Limits for Functions of Several Unknown Parameters. *Proc. Sov. Acad. Sci.,* Vol. 258, No. 6.

Pavlov, I. V. (1981b). On Correctness of the Fiducial Approach for Construction of Confidence Limits for Complex System Reliability Indices. *Sov. J. Comput. Syst. Sci. (USA)*, No. 5.

Pavlov, I. V. (1982a). On Construction of Sequential Confidence Intervals and Sets. *Sov. J. Comput. Syst. Sci. (USA)*, No. 3.

Pavlov, I. V. (1982b). Statistical Methods of Reliability Estimation by Tests Results. (Russian). Radio i Svyaz, Moscow.

Pavlov, I. V. (1983a). Sequential Confidence Sets. Rep. Acad. Sci. USSR, Vol. 270, No. 2.

Pavlov, I. V. (1983b). *Sequential Confidence Intervals and Sets* (Russian). Comput. Center of the Academy of Science of the USSR, Moscow.

Pavlov, I. V. (1984a). Sequential Statistical Inferences for Complex Hypotheses. *Sov. J. Comput. Syst. Sci. (USA)*, No. 1.

Pavlov, I. V. (1984b). Sequential Decision Rule for the Case of Multiple Complex Hypotheses. *Sov. J. Comput. Syst. Sci. (USA)*, No. 3.

Pavlov, I. V. (1985). Optimal Sequential Decision Rules, *Comput. Center of Ac. Sci.,* Moscow.

Pavlov, I. V. (1986). Sequential Control and Estimation of System Reliability on the Basis of Unit Testing, *Sov. J. Comput. Syst. Sci. (USA)*, No. 6.

Pavlov, I. V. (1987a). Sequential Procedure of Distinguishing Multiple Complex Hypotheses. *Theory Prob. Appl.,* Vol. 32, No. 1.

Pavlov, I. V. (1987b). Sequential Procedure of Decision Making for Testing of Complex Systems, in *Cybernetics and Computers,* Nauka, Moscow.

Pavlov, I. V. (1988). Approximation of Optimal Confidence Limits for Repairable System Reliability. *Sov. J. Comput. Syst. Sci. (USA)*, No. 3.

Pavlov, I. V. (1990). Sequential Procedure of Complex System Testing with Application to the Kiefer-Weiss Problem. *Theory Prob. Appl.,* Vol. 35, No. 2

Pavlov, I. V. (1992a). Computation of the Confidence Limits for Series Systems (Russian). *Reliabil. Quality Control,* No. 5.

Pavlov, I. V. (1992b). Estimation and Prognosis of Systems Parameters on the Basis of Testing and Statistical Simulation. (Russian). Comput. Center Russian Ac. Sci. Moscow.

Pavlov, I. V. (1993). The Lower Limit of the Size of the Sample for Sequential Procedures of Item Recognition (Russian). *Proc. Steklov Math. Inst.,* Vol. 202, Moscow.

Pavlov, I. V. (1995). Confidence Limits of Reliability Indexes for Censored Samples (Russian). *Reliabil. Quality Control,* No. 7.

Pavlov, I. V., and **Ushakov, I. A.** (1984). Unbiased Estimator of Distribution Function Based on Multiple Censored Sample. *Theor. Prob. Appl.,* Vol. 29, No. 3.

Pavlov, I. V., and **Ushakov, I. A.** (1989). Calculation of Reliability Indices for Complex Repairable Systems. *Sov. J. Comput. Syst. Sci. (USA)*, No. 6.

Petrov, V. V. (1975). *Sums of Independent Random Variables.* Springer-Verlag, New York.

Pollyak, Yu. G. (1971). *Stochastic Modelling with Help of Computer* (Russian). Sovietskoe Radio, Moscow.

Raiffa, H., and **Schlaifer, R.** (1961). *Applied Statistical Decision Theory.* Harvard University Press, Boston.

Rao, C. R. (1965). *Linear Statistical Inference and Its Applications.* Wiley, New York.

Regel, V. R., Slutsker, A. I., and **Tomashevski, E. Ye.** (1974). *Kinetic Nature of the Strength of Solids.* Nauka, Moscow.

Robbins, H., and **Siegmund, D.** (1973). A Class of Stopping Rules for Testing Parametric Hypotheses. *Proc. 6th Berkeley Symp. Math. Stat. Prob.,* Vol. 4.

Robbins, H., and **Siegmund, D.** (1974). The Expected Sample Size of Some Tests of Power One. *Ann. Statist.,* Vol. 2, No. 3.

Romig, H. G. (1953). *Binomial Tables.* Wiley, New York.

Rukhin, A. L., and **Hsieh, H. K.** (1987). Survey of Soviet Work in Reliability. *Statist. Sci.,* Vol. 2, No. 4, pp. 484–503.

Savchuk, V. P. (1989). Bayesian Methods in Statistics (Russian). Nauka, Moscow.

Schick, G. J. (1959). Reliabilities, Confidence Limits and Their Improvements as Applied to Missile Reliability. Technical Publication, Aerojet General Corp.

Schick, G. J. (1967). A Comparison of Some Old and New Methods in Establishing Confidence Intervals of Serially Connected Systems. *J. Indust. Eng.,* Vol. 18, No. 8.

Schwarz, G. (1962). Asymptotic Shapes of Bayes Sequential Testing Regions. *Ann. Math. Statist.,* Vol. 33, No. 2.

Senetsky, S. A., and **Shishonok, N. A.** (1966). Estimation of Complex System by Statistical Simulation (Russian). In *On Reliability of Complex Systems.* Sovietskoe Radio, Moscow.

Shiryaev, A. N. (1965). Sequential Analysis and Controlled Stochastic Processes (Russian). *Cybernetics,* Vol. 3, No. 1.

Shiryaev, A. N. (1965). Statistical Sequential Analysis (Russian). Nauka, Moscow.

Shiryaev, A. N. (1980). *Probability.* Nauka, Moscow.

Smith, D. R., and **Springer, M. D.** (1976). Bayesian Limits for the Reliability of Pass/Fail Parallel Units. *IEEE Trans. Reliabil.,* Vol. R-25.

Sobczyk, K., and **Spencer, B. F., Jr.** (1992). *Random Fatigue: From Data to Theory.* Academic, New York.

Springer, M. D., and **Thompson, W. E.** (1964). The Distribution of Products of Independent Random Variables. Technical Report TR 64-46, General Motors Defense Labs, Santa Barbara, CA.

Springer, M. D., and **Thompson, W. E.** (1966). Bayesian Confidence Limits for the Product of N Binomial Parameters. *Biometrika,* Vol. 53.

Springer, M. D., and **Thompson, W. E.** (1967a). Bayesian Confidence Limits for the Reliability of Cascade Exponential Subsystems. *IEEE Trans. Reliabil.,* Vol. R-16.

Springer, M. D., and **Thompson, W. E.** (1967b). Bayesian Confidence Limits for Reliability of Redundant Systems When Tests are Terminated at First Failure. *Technometrics,* Vol. 10.

Steck, G. P. (1957). Upper Confidence Limits for Failure Probability of Complex Networks. Technical Report SC-4133(TR), Sabdia Corp.

Stein, C. (1959). An Example of Wide Discrepancy Between Fiducial and Confidence Intervals. *Ann. Math. Statist.,* Vol. 30.

Sudakov, R. S. (1974). About Interval Estimation of Reliability of Series System. *Eng. Cybernet* (*USA*), No. 3.

Sudakov, R. S. (1975). *Statistical Problems of System Testing and Tables for Numerical Calculation of Reliability Indexes* (Russian). Vys'shaya Shkola, Moscow.

Sudakov, R. S. (1986). *Non-Parametrical Methods in Testing Systems and Their Units* (Russian). Znanie, Moscow.

Tillman, F. A., Hwang, C. L., and **Kuo, W.** (1980). *Optimization of System Reliability.* Marcel Dekker, New York.

Tyoskin, O. I., and **Kursky, I. Yu.** (1986). Interval Estimation of System Reliability on the Basis of Its Units Testing (Russian). In *Problems of Transmitting and Processing Data for Testing of Aircraft* (Russian). Moscow Aviation Institute, Moscow.

U.S. Army (1952). Tables of Cumulative Binomial-Probabilities. U.S. Army Ordinance Corps. Pamphlet ORD P20-1, Washington, DC.

Ushakov, I. A. (1969). *Methods of Solving Simple Problems of Optimal Redundancy* (Russian). Sovietskoe Radio, Moscow.

Ushakov, I. A. (1980). Reliability Estimation Based on Truncated Tests. *Sov. J. Comput. Syst. Sci.* (*USA*), No. 5.

Ushakov, I. A. Ed. (1994). *Handbook of Reliability Engineering.* Wiley, New York.

Ushakov, I. A., and **Aliguliev, E. A.** (1989a). Optimization of Data-Transmitting Network Parameters Using the Method of Statistical Modeling. *Sov. J. Comput. Syst. Sci.* (*USA*), No. 1.

Ushakov, I. A., and **Aliguliev, E. A.** (1989b). Use of Method of Statistical Modeling for Optimizing the Number of Channels in a Communication Network. *Sov. J. Comp. Syst. Sci.* (*USA*), No. 3.

Ushakov, I. A., and **Gordienko, E. I.** (1978). Solution of Some Optimization Problems by Means of Statistical Simulation (Russian). *Electronosche Infdormationsverarbeitung und Kybernetik,* Vol. 14, No. 11.

Ushakov, I. A., and **Yasenovets, A. V.** (1977). Statistical Methods of Solving Problems of Optimal Standby. *Eng. Cybernet.,* No. 6.

Ushakov, I. A., Kozlov, M. V., and **Topolsky, M. V.** (1982). Point Estimates of Reliability Indexes on the Basis of Truncated Samples (Russian). *Reliabil. Quality Control,* No. 10.

Volkovich, V. L., Voloshin, A. F., Zaslavsky, V. A., and **Ushakov, I. A.** (1992). *Models and Methods of Optimization of Complex System Reliability* (Russian). Naukova Dumka, Kiev.

Wald, A. (1944). On Cumulative Sums of Random Variables. *Ann. Math. Statist.,* Vol. 15, No. 2.

Wald, A. (1947). *Sequential Analysis.* Wiley, New York.

Wald, A., and **Wolfowitz, J.** (1948). Optimum Character of the Sequential Probability Ratio Test. *Ann. Math. Statist.,* Vol. 19, No. 2.

Walsh, J. E. (1962). Nonparametric Confidence Intervals and Tolerance Regions, Chapter 8. In *Order Statistics,* ed. by A. E. Sarahan and B. G. Greenberg. Wiley, New York.

Weibull, W. (1939). A Statistical Theory of the Strength of Materials. *Ing. Vetenskaps Akad. Handl.,* No. 151.

Weibull, W. (1951). A Statistical Distribution Function of Wide Applicability. *J. Appl. Mech.,* Vol. 18, No. 3, pp. 293–297.

Weiss, L. (1962). On Sequential Tests Which Minimize the Maximum Expected Sample Size. *J. Am. Statist. Assoc.,* Vol. 57, No. 3.

Wilks, S. S. (1938). The Large-Sample Distribution of the Likelihood Ratio for Testing Composite Hypotheses. *Ann. Math. Statist.,* Vol. 9.

Williams, D. (1991). *Probability with Martingales.* University Press, Cambridge.

Winterbottom, A. (1974). Lower Confidence Limits for Series System Reliability from Binomial Subsystem Data. *J. Am. Statist. Assoc.,* Vol. 69.

Zacks, S. (1971). *The Theory of Statistical Inference.* Wiley, New York.

Zamyatin, A. A. (1986). Statistical Inferences for Renewable System with Multiple Failures. *Sov. J. Comput. Syst. Sci. (USA),* No. 3.

Zelen, M., and **Dannemiller, M. C.** (1961). The Robustness of Life Testing Procedures Derived from the Exponential Distribution. *Technometrics,* Vol. 3.

Zhuravlev, Yu. I. (1978). On Algebraic Approach for Recognition and Classification Problems (Russian). *Cybernet. Prob.,* No. 33.

SOLUTIONS TO PROBLEMS

CHAPTER 1

1.1. Express $P\{|\hat{\Theta}_n - \Theta| > \varepsilon\}$ in the form $P\{|\xi_n - \delta| > \varepsilon\}$, where $\xi_n = \Theta_n - b_n(\Theta)$, $b_n(\Theta) = E\{\hat{\Theta}_n\})$, and $\delta_n = b_n(\Theta) - \Theta$. Then the proof follows from the Chebyshev inequality (1.5):

$$P(|\xi_n - \delta_n > \varepsilon) < P(|\xi_n| > \tfrac{1}{2}\varepsilon) \quad \text{for } |\delta_n| < \tfrac{1}{2}\varepsilon.$$

1.2. This follows from the fact that $E\{x_i^r\} = E\{\xi^r\} = \mu_r$.

1.3. The likelihood equations have the form

$$\frac{\partial \ln L}{\partial \lambda} = \frac{n}{\lambda} - \sum_{1 \le i \le n} x_i^{\alpha} = 0,$$

$$\frac{\partial \ln L}{\partial \alpha} = \frac{n}{\alpha} + \sum_{1 \le i \le n} \ln x_i - \lambda \sum_{1 \le i \le n} X_i^{\alpha} \ln X_i = 0.$$

1.4. $\hat{a} = x_{(1)}$, $\hat{b} = x_{(n)}$.

1.5. Let x_i be an indicator of success in the ith trial, i.e., $x_i = 1$ if a success has occurred and $x_i = 0$ if it was a failure. Then the total number of successes is a likelihood function. In other words, the probability that the series $X_1, \ldots, X_n$ in n trials will be observed is

$$L(x_1, \ldots, x_n, p) = \prod_{1 \le i \le n} p^{X_i}(1 - p)^{1 - X_i} = p^m(1 - p)^{n-m},$$

where $m = x + \cdots + x_n$ is the total number of successes. Then the proof follows from the factorization criterion (Theorem 1.2).

1.6. Take as a central statistic $T = x_{(n)}/\Theta$. Show that the d.f. of this statistic has the form $F(t) = t^n$, $0 \le t \le 1$, an then apply the general procedure considered in Section 1.4.2 using Equation (1.17).

1.7. Take as a central statistic $T = x_{(1)} - \Theta$. Show that the d.f. of this statistic has the form $F(t) = n\lambda e^{-n\lambda t}$. Then apply Equation (1.17).

1.8. Take as a central statistic

$$T = \frac{(\mu_1 - \mu_2) - (\overline{X} - \overline{Y})}{\sqrt{\sigma_1^2/n - \sigma_2^2/m}}.$$

Using the procedure of Section 1.2, show that this statistic has a standard normal distribution with mean 0 and variance 1. Then, apply the general procedure considered in Section 1.4.2 using Equation (1.17).

1.9. Take as a central statistic

$$T = \frac{T_1}{\sqrt{T_2}} \sqrt{n + m - 2},$$

where

$$T_1 = \frac{(\mu_1 - \mu_2) - (\overline{X} - \overline{Y})}{\sigma\sqrt{1/n + 1/m}}, \qquad T_2 = \frac{nS_1^2}{\sigma^2} + \frac{mS_2^2}{\sigma^2},$$

and $\overline{X}$, $\overline{Y}$ are sample means and S_1^2, S_2^2 are sample variances for the first and second samples, respectively. Using the results from Section 1.2, check that statistic T_1 has the standard normal distribution and that statistic T_2 has the χ^2 distribution with $n + m - 2$ degrees of freedom. So, statistic T has the Student distribution with $n + m - 2$ degrees of freedom. After this, apply the general procedure.

1.10. Take as the central statistic

$$T = \frac{T_1(m - 1)}{T_2(n - 1)},$$

where statistics $T_1 = nS_1^2/\sigma_1^2$ and $T_2 = nS_2^2/\sigma_2^2$ are independent and have a χ^2 distribution with $n - 1$ and $m - 1$ degrees of freedom. Thus, statistic T has Fisher distribution with $n - 1$ and $m - 1$ degrees of freedom (see Section 1.2). Then, applying the general procedure, it is easy to obtain the expressions for the lower and upper limits of the confidence interval for parameter $K = \sigma_1^2/\sigma_2^2$:

$$\underline{K} = f_\alpha(n-1, m-1) = \frac{S_2^2 m(n-1)}{S_1^2 n(m-1)},$$

$$\overline{K} = f_{1-\beta}(n-1, m-1) = \frac{S_2^2 m(n-1)}{S_1^2 n(m-1)},$$

where $f_q(n-1, m-1)$ is the quantile of level q of the Fisher distribution with $n-1$ and $m-1$ degrees of freedom.

1.11. Take as the central statistic $T = T_1/T_2$, where statistics $T_1 = 2\lambda(x_1 + \cdots + x_n)$ and $T_2 = 2\mu(y_1 + \cdots + y_m)$ have χ^2 distributions with $2n$ an $2m$ degrees of freedom, respectively (see Section 1.2). Thus, statistic T has Fisher distribution with $2n$ and $2m$ degrees of freedom. Then, applying the general procedure, we obtain the lower and upper limits of the confidence interval for parameter $r = 1/m$:

$$\underline{\rho} = f_\alpha(2n, 2m)\,\frac{y_1 + \cdots + y_m}{x_1 + \cdots x_n},$$

$$\overline{\rho} = f_{1-\beta}(2n, 2m)\,\frac{y_1 + \cdots + y_m}{x_1 + \cdots x_n}.$$

1.12. Let ν be the number of cycles before the first failure has occurred. This r.v. has negative binomial distribution

$$P(\nu = n) = (1-p)p^{n-1}, \qquad n = 1, 2, \ldots.$$

Thus, the problem is reduced to constructing the confidence interval for parameter p of the negative binomial distribution on the basis of the observed value ν. The distribution function of r.v. ν is defined by the expression

$$F(n, p) = (1-p)(1 + p + \cdots + p^{n-1}) = 1 - p^n.$$

Applying further the general equations (1.23), we obtain the lower and upper limits of the confidence interval with confidence coefficient $\gamma = 1 - \alpha - \beta$ for parameter p:

$$\underline{p} = \sqrt[n-1]{\beta}, \qquad \overline{p} = \sqrt[n]{1-\alpha}.$$

Thus the lower limit coincides with the standard Clopper–Pearson limit for a binomial scheme of trial for the case $d = 0$ in a series of $n - 1$ trials. The upper limit also coincides with the standard Clopper–Pearson limit but for the case $d = 1$ in a series of n trials. This result responds to the nature of the test.

1.13. Using the results of Section 1.5.3 (also see Example 1.26), the critical region is given by

$$\sum_{1 \le i \le n} X_i \ge C, \tag{s1.1}$$

where constant C is chosen so that it satisfies the significance level $\alpha = 0.1$:

$$C = n\mu_0 + U_{1-\alpha}\sigma\sqrt{n} = 9 \times 53 + 1.28 \times 4 \times 3 = 492.4.$$

For the constructed criterion with critical region (s1.1), the error β is determined as

$$\beta = \Phi\left(\frac{C - n\mu_1}{\sigma\sqrt{n}}\right) = \Phi\left(\frac{492.4 - 9 \times 54}{4 \cdot 3}\right) = 0.76.$$

The power of the criterion is $1 - \beta = 0.24$. This value is small because of the small size of the sample.

1.14. Using the results of Section 1.5.4 (also see Example 1.27), we obtain

$$n^* = \frac{\sigma^2(U_{1-\alpha} + U_{1-\beta})^2}{(\mu_1 - \mu_0)^2} = \frac{16 \times (1.28 + 1.28)^2}{1^2} = 105.$$

The corresponding optimal Neyman–Pearson criterion in this case is given by the critical region of the form

$$\sum_{1 \le i \le n^*} X_i \ge C,$$

where constant C is determined from (1.34):

$$C = n^*\mu_0 + U_{1-\alpha}\sigma\sqrt{n^*} = 570.4.$$

1.15. Using the results of Section 1.5.5, we obtain the following:

(a) The necessary sample size is

$$n^* = \frac{\sigma^2(U_{1-\alpha} + U_{1-\beta})^2}{(\mu_1 - \mu_0)^2} = 5.$$

(b) The uniformly most powerful criterion coincides with the Neyman–Pearson criterion for the two simple hypotheses H_0: $\mu = 14$ and H_1: $\mu = 10$. The corresponding critical region is given by the inequality

$$\sum_{1 \le i \le n^*} X_i \le C,$$

where $C = n^*\mu_0 - U_{1-\alpha}\sigma\sqrt{n^*}$, $\mu_0 = 14$. Thus, $C = 61.4$.

(c) The power function (the probability to reject the null hypothesis) in this case has the form

$$E(\mu) = P\left\{\sum_{1 \le i \le n^*} X_i \le C|\mu\right\} = \Phi\left(\frac{C - n^*\mu}{\sigma\sqrt{n^*}}\right).$$

The operative characteristic of the criterion $S(\mu) = 1 - E(\mu)$.

CHAPTER 2

2.1. The total testing time of all units is given as

$$S = t_1 + \cdots + t_{r-1} + (N - r + 1)t_r = 60 + 7 \times 110 = 830 \text{ hr}.$$

Setting $\alpha = \beta = 0.05$ and applying formulas (2.8) with Table 12.4 in the Appendix, we obtain for l the following lower and upper γ-confidence limits with confidence coefficient 0.9:

$$\underline{\lambda} = \frac{\Delta_{0.95}(1)}{830} = \frac{0.355}{830} = 4.27 \times 10^{-4} \text{ hr}^{-1},$$

$$\overline{\lambda} = \frac{\Delta_{0.05}(1)}{830} = \frac{4.74}{830} = 57.1 \times 10^{-4} \text{ hr}.$$

This allows us to find the following confidence limits with confidence coefficient $\gamma = 0.9$ for reliability indices $P(t_0) = e^{-\lambda t_0}$ and $\tau = 1/\lambda$:

$$\underline{P}(t_0) = e^{-\overline{\lambda} t_0} = e^{-(57.1)\times 10^{-4}\times 5} = 0.972,$$

$$\overline{P}(t_0) = e^{-\underline{\lambda} t_0} = e^{-(4.27)\times 10^{-4}\times 5} = 0.998,$$

$$\underline{\tau} = \frac{1}{\overline{\lambda}} = 175 \text{ hr}, \qquad \overline{\tau} = \frac{1}{\underline{\lambda}} = 2340 \text{ hr}.$$

2.2. In this case the total testing time of units is

$$S = t_1 + \cdots + t_d + (N - d)T = 15 + 72 + 8 \times 100 = 887 \text{ hr}.$$

The point estimate of parameter λ is

$$\hat{\lambda} = \frac{d}{S} = \frac{2}{887} = 22.5 \times 10^{-4} \text{ hr}^{-1}.$$

Applying formulas (2.11) and (2.12), we obtain the following upper γ-confidence limit for parameter λ:

$$\overline{\lambda} = \frac{\overline{\Delta}'_{\gamma}(d)}{T} = \frac{-\ln(1 - \overline{p})}{T},$$

where $\overline{p}$ is the upper γ-confidence for parameter p (the failure probability) of the binomial distribution based on the results of the test: $d = 2$ and $N = 10$. Applying Table E.15 from the Appendix, we obtain the upper confidence limit with confidence coefficient 0.975 for parameter λ:

$$\overline{\lambda} = \frac{-\ln(1 - 0.556)}{100} = 81.1 \times 10^{-4} \text{ hr}^{-1}.$$

This gives us the corresponding point estimate and the lower confidence limit with confidence coefficient 0.975 for the reliability indices $P(t_0) = e^{-\lambda t_0}$ and $\tau = 1/\lambda$:

$$\hat{P}(t_0) = e^{-\hat{\lambda} t_0} = e^{-22.5 \times 10^{-4} \times 10} = 0.978,$$

$$\underline{P}(t_0) = e^{-\overline{\lambda} t_0} = e^{-81.1 \times 10^{-4} \times 10} = 0.920,$$

$$\hat{\tau} = \frac{1}{\hat{\lambda}} = 443 \text{ hr}, \qquad \underline{\tau} = \frac{1}{\overline{\lambda}} = 124 \text{ hr}.$$

2.3. In this case the total testing time of the units is

$$S = t_1 + \cdots + t_r + (N - r)t_r = 150 + 250 + 5 \times 400 = 2400 \text{ hr}.$$

The required testing time $t_0 = 20$ hr, for which we estimate that $P(t_0)$ satisfies the inequality $t_0 < S/N = 2400/7$. Applying formula (2.22) and using Table 12.4 from the Appendix, we obtain the following γ-confidence with $\gamma = 0.975$ for $P(t_0)$:

$$\underline{P}(t_0) = e^{-[\Delta_{1-\gamma}(r-1)/S]} = e^{[D_{0.025}(2)/2400]\times 20} = e^{-(7.22/2400)\times 20} = 0.942.$$

This lower confidence limit coincides with the analogous confidence limit for the exponential d.f.

2.4. Let $d = d(t_0)$ be the number of failures that have occurred up to moment t_0. We need to construct the lower confidence limit of Clopper–Pearson type for parameter $p = P(t_0)$ of the binomial distribution by the known value of d. In this case for $t_0 = 20$ we have $d = d(t_0) = 0$. By Table E.15 from the Appendix we find the following lower confidence limit with confidence coefficient 0.975 for $P(t_0)$:

$$\underline{P}(t_0) = 1 - 0.410 = 0.590.$$

Comparison of this result with that of problem 2.3 shows us that the lower confidence limit for an IFR distribution is significantly higher. Thus prior information that the distribution function of the unit's TTF is IFR permits to improve the lower confidence limit essentially [at least at the time interval for which $P(t)$ is close to 1].

2.5. Applying formula (2.30), we obtain for the IFR distribution the following lower γ-confidence limit with $\gamma = 0.975$ for the MTTF, t: $\underline{t}^* = C_r\underline{t}$, where

$$\underline{T} = \frac{S}{\Delta_{1-\gamma}(r-1)} = \frac{2400}{D_{0.025}(2)} = \frac{2400}{7.22} = 332 \text{ hr}$$

is the lower γ-confidence limit with confidence coefficient 0.975 for t found under the assumption that $F(t)$ is the exponential distribution, and

$$C_r = 1 - \exp\left(-\frac{\Delta_{1-\gamma}(r-1)}{N}\right) = 1 - \exp\left(\frac{\Delta_{0.025}(2)}{7}\right)$$
$$= 1 - \exp\left(-\frac{7.22}{7}\right) = 0.643$$

is the coefficient that indicates how much the lower confidence limit for t, found with the assumption of exponentiality, must be decreased for an IFR distribution. So,

$$\underline{t}^* = 0.643\underline{t} = 0.643 \times 332 = 214 \text{ hr}.$$

Thus, for plan $[N\ U\ r]$, using the exponential methods for the IFR class of distributions decreases the lower confidence limit for the MTTF. This is especially significant for a high level of censorship, i.e., where $r << N$. At the same time this influence on the lower confidence limits is sometimes negligible for reliability indices of type $P(t_0)$ or t_q (see problem 2.4 and Section 2.3).

2.6. The operational interval $t_0 = 1$ hr for which the unit reliability $P(t_0)$ is estimated satisfies the inequality

$$1 = t_0 \leq \frac{NT}{N + d} = \frac{4 \times 100}{4 + 6} = 40.$$

Therefore applying formula (2.34) and Table 12.4 from the Appendix, we find the following γ-confience limit with $\gamma = 0.95$ for $P(t_0)$:

$$\underline{P}(t_0) = \exp\left(-\frac{\Delta_{1-\gamma}(d)}{NT}\, t_0\right) = \exp\left(-\frac{\Delta_{0.05}(6)}{4 \cdot 100}\right)$$
$$= \exp\left(-\frac{11.84}{400}\right) = 0.971.$$

This lower confidence limit coincides with the analogous confidence limit for the case where $F(t)$ is exponential.

2.7. Applying formula (2.36) obtained for an IFR distribution and plan $[N\ R\ T]$, we obtain the following lower confidence limit with $\gamma = 0.95$ for reliability index t_q:

$$\underline{t}_q^* = \min\left\{\frac{(-\ln q)NT}{\Delta_{1-\gamma}(d)}, \frac{NT}{N + d}\right\}$$
$$= \min\left\{\frac{(-\ln 0.9) \times 4 \times 100}{\Delta_{0.05}(6)}, \frac{4 \times 100}{4 + 6}\right\}$$
$$= \frac{(0.11) \times 400}{11.84} = 3.75 \text{ hr.} \qquad \text{(s2.1)}$$

The lower confidence limit (s2.1) coincides with the corresponding confidence limit for exponential distribution of the unit's TTF.

2.8. Applying formula (2.37), we find the following lower γ-confience limit with $\gamma = 0.95$ for the unit's MTTF t:

$$\underline{T}^* = C_d \frac{NT}{\Delta_{1-\gamma}(d)},$$

where

$$\frac{NT}{\Delta_{1-\gamma}(d)} = \frac{4 \times 100}{D_{0.06}(6)} = \frac{400}{11.84} = 33.8 \text{ hr}$$

is the lower confidence limit for the MTTF found for the exponential distribution, and C_d is the correction coefficient for exponential methods for plan [N R T]. This coefficient equals

$$C_d = 1 - \exp\left(-\frac{\Delta_{1-\gamma}(d)}{N+d}\right) = 1 - \exp\left(-\frac{\Delta_{0.05}(6)}{4+6}\right)$$
$$= 1 - \exp\left(\frac{11.84}{10}\right) = 0.695.$$

(Notice that this coefficient is not too small for sufficiently small N). So, finally, we have

$$\underline{t}^* = (0.695) \times (33.8) = 23.5 \text{ hr}.$$

Thus, for plan [N R T], using exponential methods for IFR distributions leads to the decrease of the lower confidence limit for the MTTF. At the same time for reliability indices of type $P(t_0)$ and t_q, this method in many cases does not influence the lower confidence limit (see problem 2.8).

2.9. Denote the number of failures during a period of time t_0 by $d = d(t_0)$. The problem is equivalent to finding the standard lower confidence limit of the Clopper–Pearson type for the parameter of a binomial distribution $p = P(t_0)$ on the basis of the given result d (see Section 1.4.6). In this case no failures have been observed up to $t_0 = 40$ hours, i.e., $d = d(t_0) = 0$. From the Clopper–Pearson equation, we obtain the lower confidence limit with $\gamma = 0.9$ for $P(t_0)$:

$$\underline{P}(t_0) = \sqrt[N]{1-\gamma} = \sqrt[10]{0.1} = 0.794.$$

In Examples 2.11 and 2.12 we considered the analogous lower confience limits for the Weibull–Gnedenko and IFR distributions (for the same case), which were 0.796 and 0.869, respectively. Thus, in this case the assumption that the distribution $F(t)$ belongs to the Weibull–Gnedenko distribution does not practically improve the confidence limit. At the same time, the prior knowledge that $F(t)$ is IFR essentially improves the lower confidence limit.

CHAPTER 3

3.1. Let $A = \{x > t\}$ be the event that a unit has not failed up to the moment t and $B_j = \{jh < x \leq jh + h\}$ be the event that the unit has failed on the interval $(jh, jh + h]$. These events are related via the expression

$$A = \overline{B}_0 \times \overline{B}_1 \times \cdots \times \overline{B}_{M-1}.$$

Using the standard formula for the probability of the product of events, we have

$$P(A) = P(\overline{B}_0) \times P(\overline{B}_1|\overline{B}_0) \times P(\overline{B}_2||\overline{B}_0 \cdot \overline{B}_1) \times \cdots$$
$$\times P(\overline{B}_{M-1}|\overline{B}_0 \times \overline{B}_1 \times \cdots \times \overline{B}_{M-2}),$$

which gives us (3.74) if we take into account the conditional probability

$$P(\overline{B}_j|\overline{B}_0 \times \overline{B}_1 \times \cdots \times \overline{B}_{j-1}) = P(\xi > jh + h|\xi > jh) = 1 - q_j.$$

3.2. The moment of time for which we estimate the reliability function satisfies the inequalities

$$t_1 < t_2 < t < t_3 < t_4,$$

in this case $m = 2$. All values $d_m(t)$, d_k, and N_k have been found in Example 3.8. Using the recurrent procedure (3.11), we have

$$r_2 = d_2(t)\left(1 + \frac{n_2}{N_2}\right) = 1 \times (1 + \tfrac{2}{10}) = \tfrac{12}{10},$$

$$r_1 = (r_2 + d_1)\left(1 + \frac{n_1}{N_1}\right) = (\tfrac{12}{10} + 1)(1 + \tfrac{3}{13}) = \tfrac{22}{10} \times \tfrac{16}{13},$$

$$r_0 = (r_1 + d_0)\left(1 + \frac{n_0}{N_0}\right) = (\tfrac{22}{10} \times \tfrac{16}{13} + 3) \times 1.$$

From here we have

$$\hat{P}(t = 300) = 1 - \frac{r_0}{N_0} = 0.698.$$

This estimate coincides with the Kaplan–Meier estimate obtained in Example 3.5.

3.3. In this case the moment $t = 375$ satisfies the inequalities

$$t_1 < t_2 < t_3 < t < t_4,$$

that is, $m = 3$. Using values $d_m(t)$, d_k, and N_k found in Example 3.6 and using the recurrent procedure (3.11), we have

$$r_3 = d_3(t)\left(1 + \frac{n_3}{N_3}\right) = 0,$$

$$r_2 = (r_3 + d_2)\left(1 + \frac{n_2}{N_2}\right) = 2 \times (1 + \tfrac{2}{10}) = \tfrac{24}{10},$$

$$r_1 = (r_2 + d_1)\left(1 + \frac{n_1}{N_1}\right) = (\tfrac{24}{10} + 1)(1 + \tfrac{3}{13}) = \tfrac{34}{10} \times \tfrac{16}{13},$$

$$r_0 = (r_1 + d_0)\left(1 + \frac{n_0}{N_0}\right) = (\tfrac{34}{10} \times \tfrac{16}{13} + 3) \times 1.$$

From here we have

$$\hat{P}(t = 375) = 1 - \frac{r_0}{N} = 0.621.$$

This estimate coincides with the Kaplan–Meier estimate obtained in Example 3.6.

3.4. Use the Kaplan–Meier formula (3.3). In this case the moment of time $t = 200$ satisfies the inequalities

$$t_1 < t_2 < t_3 < t < t_4,$$

that is, $m = 3$. The number of failures on the intervals $(t_k, t_{k+1}]$ between sequential moments of intermediate terminations are given as

$$d_0 = D(t_1) - D(0) = D(t_1) = 2,$$

$$d_1 = D(t_2) - D(t_1) = 4 - 2 = 2,$$

$$d_2 = D(t_3) - D(t_2) = 4 - 4 = 0,$$

where $D(u)$ is the number of failures on the interval $(0, u]$, that is, including the moment u. The number of failures $d_3(t)$ on the interval $(t_3, t] = (180, 200]$ equals

$$d_3 = D(t) - D(t_3) = 5 - 4 = 1.$$

The number of units tested at the beginning of these intervals are

$$N_0 = N(0) = 15,$$

$$N_1 = N(t_1) = N_0 - d_0 - n_1 = 15 - 2 - 2 = 11,$$

$$N_2 = N(t_2) = N_1 - d_1 - n_2 = 15 - 2 - 1 = 8,$$

$$N_3 = N(t_3) = N_2 - d_2 - n_3 = 8 - 4 = 4.$$

Thus the estimate (3.3) equals

$$\hat{P}(t = 200) = \left(1 - \frac{d_0}{N_0}\right)\left(1 - \frac{d_1}{N_1}\right)\left(1 - \frac{d_2}{N_2}\right)\left(1 - \frac{d_3(t)}{N_3}\right)$$

$$= (1 - \tfrac{2}{15})(1 - \tfrac{2}{11})(1 - \tfrac{1}{4}) = \tfrac{13}{15} \times \tfrac{9}{11} \times \tfrac{3}{4} = 0.53.$$

Calculation using (3.5) gives the same result:

$$\hat{P}(t = 200) = \prod_{1 \leq j \leq D(t)} \left(1 - \frac{1}{N(t_j^-)}\right)$$

$$= (1 - \tfrac{1}{15})(1 - \tfrac{1}{14})(1 - \tfrac{1}{11})(1 - \tfrac{1}{10})(1 - \tfrac{1}{4}) = 0.53.$$

3.5. The moment of time $t = 120$ for which we estimate the reliability function satisfies the inequalities

$$t_1 < t < t_2 < t_3 < t_4,$$

that is, $m = 1$. The values $d_m(t)$, d_k, and N_k are given as

$$d_0 = D(t_1) - D(0) = D(t_1) = 2,$$

$$d_1(t) = D(t) - D(t_1) = 3 - 2 = 1,$$

$$N_0 = N(0) = 15, \qquad N_1 = N(t_1) = 11.$$

Formula (3.3) gives the estimate

$$\hat{P}(t = 200) = \left(1 - \frac{d_0}{N_0}\right)\left(1 - \frac{d_1(t)}{N_1}\right) = (1 - \tfrac{2}{15})(1 - \tfrac{1}{11}) = 0.788.$$

Calculation by (3.5) gives the same result:

$$\hat{P}(t = 200) = \prod_{1 \le j \le D(t)} \left(1 - \frac{1}{N(t_j^-)}\right) = (1 - \tfrac{1}{15})(1 - \tfrac{1}{14})(1 - \tfrac{1}{11})$$
$$= 0.788.$$

3.6. Using formula (3.20), we have the following expression for the resource function

$$\hat{\Lambda}(t = 150) = \sum_{1 \le j \le 4} \frac{1}{N(t_j^-)} = \frac{1}{N(41^-)} + \frac{1}{N(87^-)} + \frac{1}{N(104^-)} + \frac{1}{N(146^-)}$$
$$= \tfrac{1}{15} + \tfrac{1}{14} + \tfrac{1}{11} + \tfrac{1}{10} = 0.329.$$

The corresponding estimate of the reliability function is

$$\hat{P}(t = 150) = e^{-\hat{\Lambda}(150)} = e^{-0.329} = 0.718.$$

3.7. Using formulas (3.20) and (3.22), we find estimates of the reliability function and variance:

$$\hat{\Lambda}(t = 160) = \sum_{1 \le j \le 3} \frac{1}{N(t_j^-)} = \frac{1}{N(34^-)} + \frac{1}{N(79^-)} + \frac{1}{N(107^-)}$$
$$= \tfrac{1}{19} + \tfrac{1}{18} + \tfrac{1}{17} = 0.167,$$
$$\hat{V}(t = 160) = \sum_{1 \le j \le 3} \frac{1}{[N(t_j^-)]^2} = \frac{1}{19^2} + \frac{1}{18^2} + \frac{1}{17^2} = 0.0093.$$

Using (3.21), we obtain the approximate upper 0.95 confidence limit for the resource function at moment $t = 160$:

$$\overline{\Lambda}(t = 169) = \hat{\Lambda}(160) + u_{0.95}\sqrt{\hat{V}(160)}$$
$$= 0.167 + 1.64\sqrt{0.0093} = 0.325.$$

The corresponding approximate lower 0.95 confidence limit for the reliability function is

$$\underline{P}(t = 160) = e^{-\overline{\Lambda}(160)} = e^{-0.325} = 0.722.$$

3.8. Using (3.29) with $b = 5$, we find the following upper 0.9 confidence limit for the resource function:

$$\overline{\Lambda}(t = 150) = \frac{|\ln 0.1|}{5} - \frac{1}{5}\sum_{1 \leq j \leq 4} \ln\left[1 - \frac{5}{N(t_j^-)}\right]$$

$$= \frac{2.3}{5} - \frac{1}{5}\left[\ln\left(1 - \frac{5}{15}\right) + \ln\left(1 - \frac{5}{14}\right) + \ln\left(1 - \frac{5}{11}\right)\right.$$

$$\left. + \ln\left(1 - \frac{5}{10}\right)\right] = 0.881.$$

The corresponding lower 0.95 confidence limit for the reliability function $P(t = 150)$ is

$$\underline{P}(t = 150) = e^{-\overline{\Lambda}(150)} = e^{-0.881} = 0.414.$$

CHAPTER 4

4.1. The measure of quality of estimate $\hat{R} = \hat{R}(\mathbf{z})$ in this case is the posterior mathematical expectation (4.27):

$$E|\hat{R} - R| = \int_{\Theta} h(\boldsymbol{\theta})\, d\boldsymbol{\theta} \int_{Z} |\hat{R}(\mathbf{z}) - r(\boldsymbol{\theta})| L(\mathbf{z}|\boldsymbol{\theta})\, d\mathbf{z}$$

$$= \int_{Z} \varphi(\mathbf{z})\, d\mathbf{z} \int_{\Theta} |\hat{R}(\mathbf{z}) - R(\boldsymbol{\theta})| h(\boldsymbol{\theta}|\mathbf{z})\, d\boldsymbol{\theta}.$$

The problem is reduced to finding (for each fixed test results z) such a value $\hat{R} = \hat{R}(\mathbf{z})$ for which the inner integral in (4.27), that is, the function

$$H(\hat{R}) = \int_{\Theta} |\hat{R}(\mathbf{z}) - R(\boldsymbol{\theta})| h(\boldsymbol{\theta}|\mathbf{z})\, d\boldsymbol{\theta}, \qquad \text{(s4.1)}$$

turns minimum. The value of $H(\hat{R})$ is the posterior mathematical expectation of module $|\hat{R} - R(\boldsymbol{\theta})|$ of the deviation of estimate $\hat{R}$ from $R(\boldsymbol{\theta})$. Find a minimum of the function (s4.1) in $\hat{R}$. This function can be represented in the form

$$H(\hat{R}) = \int_{-\infty}^{\infty} |\hat{R} - R| g(R|\mathbf{z})\, dR = \int_{-\infty}^{\hat{R}} (\hat{R} - R) g(R|\mathbf{z})\, dR$$

$$+ \int_{\hat{R}}^{\infty} (r - \hat{R}) g(R|\mathbf{z})\, dR,$$

where $g(R|\mathbf{z})$ is the posterior distribution density of $R = R(\boldsymbol{\theta})$. Calculating the derivative of $H(\hat{R})$, after simple transformations, we have

$$H'(\hat{R}) = \int_{-\infty}^{\hat{R}} g(R|\mathbf{z})\, dR - \int_{R}^{\infty} g(R|\mathbf{z})\, dR.$$

It follows that a minimum of the function $H(\hat{R})$ is attained at point $\hat{R}$ satisfying the equation

$$\int_{-\infty}^{R} g(R|\mathbf{z})\, dR = \int_{R}^{\infty} g(R|\mathbf{z})\, dR.$$

This means that the best estimate $\hat{R} = \hat{R}(\mathbf{z})$ in the sense of the chosen criterion is the median of the posterior distribution of index $R(\boldsymbol{\theta})$. In other words, $\hat{R}$ coincides with the Bayes lower (or upper) confidence limit with confidence coefficient 0.5.

4.2. It is known that the uniform distribution is a particular case of the beta distribution with parameters $a = b = 1$. In (4.8)–(4.10) setting $a = b = 1$, we obtain that the posterior distribution density of parameter q is given by

$$h(q|d) = \frac{q^d(1 - q)^{N-d}}{B(d + 1, N - d + 1)}. \tag{s4.2}$$

The Bayes point estimate (the posterior mean) for parameter q has the form

$$\hat{q} = \frac{d + 1}{N + 2}. \tag{s4.3}$$

In accordance with (4.11) and (4.12), the Bayes $(1 - \alpha)$-LCL $\underline{q}$ and the $(1 - \beta)$-UCL $\overline{q}$ for parameter q can be found from the equations

$$\begin{aligned} I_{\underline{q}}(d + 1, N - d + 1) &= \alpha, \\ I_{\overline{q}}(d + 1, N - d + 1) &= 1 - \beta. \end{aligned} \tag{s4.4}$$

In accordance with (4.16) and (4.17), the Bayes confidence limits $\underline{q}$ and $\overline{q}$ can also be expressed via Clopper–Pearson confidence limits as

$$\underline{q} = \underline{q}_{1-\alpha}(N + 1, d + 1), \qquad \overline{q} = \overline{q}_{1-\beta}(N + 1, d), \tag{s4.5}$$

where $\underline{q}_\gamma(N, d)$ and $\overline{q}_\gamma(N, d)$ are the Clopper–Pearson γ-LCL and γ-UCL calculated on the basis of d failure in N trials. Thus, for the uniform distribution of parameter q, Bayes confidence limits for q are close to the standard Clopper–Pearson confidence limits, but, for instance, the Bayes UCL $\overline{q}$ is more "optimistic," in comparison with the Clopper–Pearson UCL, since

$$\overline{q} = \overline{q}_{1-\beta}(N + 1, d) < \overline{q}_{1-\beta}(N, d).$$

4.3. Using formula (4.1), we obtain the following posterior distribution density of parameter q:

$$h(q|d) = \frac{q^d(1 - q)^{N-d}}{\int_0^{q^*} q^d(1 - q)^{N-d}\, dq}, \tag{s4.6}$$

where $0 \le q \le q^*$. Thus, the Bayes point estimate (the posterior mean) of parameter q is

$$\hat{q} = \int_0^{q^*} qh(q|d)\, dq = \frac{\int_0^{q^*} q^{d+1}(1 - q)^{N-d}\, dq}{\int_0^{q^*} q^d(1 - q)^{N-d}\, dq}.$$

Using the incomplete beta function (4.13), after some simple transformations, we obtain

$$\begin{aligned} \hat{q} &= \frac{B(d + 2, N - d + 1)I_{q^*}(d + 2, N - d + 1)}{B(d + 1, N - d + 1)I_{q^*}(d + 1, N - d + 1)} \\ &= \frac{d + 1}{N + 2} \frac{I_q(d + 2, N - d + 1)}{I_q(d + 1, N - d + 1)}. \end{aligned} \tag{s4.7}$$

For $q^* = 1$ the formula (s4.7) gives (s4.3).

The Bayes $(1 - \alpha)$-LCL $\underline{q}$ and $(1 - \beta)$-UCL $\overline{q}$ for parameter q can be found from the equations

$$\int_0^{\underline{q}} h(q|d)\, dq = \alpha, \qquad \int_0^{\overline{q}} h(q|d)\, dq = 1 - \beta.$$

Taking into account (s4.6), we obtain that $\underline{q}$ and $\overline{q}$ follow from the equations

$$\int_0^{\underline{q}} q^d(1-q)^{N-d}\,dq = \alpha \int_0^{q^*} q^d(1-q)^{N-d}\,dq,$$

$$\int_0^{\overline{q}} q^d(1-q)^{N-d}\,dq = (1-\beta) \int_0^{q^*} q^d(1-q)^{N-d}\,dq,$$

or, using the incomplete beta function (4.13), these equations can be rewritten as

$$\begin{aligned} I_{\underline{q}}(d+1, N-d+1) &= \alpha I_{q^*}(d+1, N-d+1), \\ I_{\overline{q}}(d+1, N-d+1) &= (1-\beta) I_{q^*}(d+1, N-d+1). \end{aligned} \tag{s4.8}$$

Comparison of equations (s4.4), (s4.5), and (s4.8) shows that confidence limits $\underline{q}$ and $\overline{q}$ can be directly expressed via Clopper–Pearson confidence limits as

$$\underline{q} = \underline{q}_{\gamma_1}(N+1, d+1), \qquad \overline{q} = \overline{q}_{\gamma_2}(N+1, d+1), \tag{s4.9}$$

where

$$\begin{aligned} \gamma_1 &= 1 - \alpha I_{q^*}(d+1, N-d+1), \\ \gamma_2 &= (1-\beta) I_{q^*}(d+1, N-d+1). \end{aligned} \tag{s4.10}$$

These formulas can be interpreted in the following way. Prior information about parameter q in the form $q \le q^* < 1$ leads to the fact that Bayes confidence limits (s4.9) are calculated with "deformed" confidence coefficients γ_1 and γ_2, determined by (s4.10). For $q^* = 1$, formulas (s4.8) and (s4.9) produce (s4.4) and (s4.5), respectively (see problem 4.3).

4.4. Applying formulas (s4.3), we obtain the following Bayes point estimate (the posterior mean of parameter q):

$$\hat{q} = \frac{d+1}{N+2} = \frac{2}{12} = 0.166. \tag{s4.11}$$

If the measure of the quality of estimate $\hat{q}$ is the posterior mathematical expectation of $|\hat{q} - q|$, then $\hat{q}$ is the median of the posterior distribution of parameter q (see problem 4.1). In other words, the Bayes lower (upper) 0.5 confidence limits for q, on the basis of (s4.5), are

$$\hat{q} = \underline{q}_{0.5}(11, 2) = \overline{q}_{0.5}(11, 1) = 0.148. \tag{s4.12}$$

Thus, the Bayes estimate (s4.12), calculated as a median of the posterior distribution of q in this example, is more "optimistic" for failure probability q in comparison with (s4.11).

Furthermore, applying (s4.5) and Table E.15 from the Appendix, we find the following Bayes 0.95 confidence limits $(\underline{q}, \overline{q})$ for parameter q:

$$\underline{q} = \underline{q}_{0.975}(11, 2) = 0.023, \qquad \overline{q} = \overline{q}_{0.975}(11, 1) = 0.413 \quad \text{(s4.13)}$$

4.5. Applying (s4.7), we find the following Bayes point estimate (the posterior mean) of parameter q:

$$\hat{q} = \frac{d+1}{N+2}\frac{I_{q^*}(d+2, N-d+1)}{I_{q^*}(d+1, N-d+1)} = \frac{2}{12}\frac{I_{0.5}(3, 10)}{I_{0.5}(2, 10)}$$

$$= \frac{2}{12}\frac{0.981}{0.994} = 0.164.$$

Now find the Bayes 0.95 confidence interval for parameter q. Using (s4.8)–(s4.10), the Bayes $(1 - \alpha)$-LCL and $(1 - \beta)$-UCL for q are defined from the equations

$$I_{\underline{q}}(d+1, N-d+1) = 1 - \gamma_1, \qquad I_{\overline{q}}(d+1, N-d+1) = \gamma_2,$$

where

$$\gamma_1 = 1 - \alpha I_{q^*}(d+1, N-d+1) = 0.025 I_{0.5}(2, 10)$$

$$= 1 - 0.025 \times 0.994 = 0.9752,$$

$$\gamma_2 = (1 - \beta) I_{q^*}(d+1, N-d+1) = 0.975 I_{0.5}(2, 10)$$

$$= 0.975 \times 0.994 = 0.969.$$

Then we have

$$\underline{q} = \underline{q}_{\gamma_1}(N+1, d+1) = \underline{q}_{0.9752}(11, 2) = 0.022,$$

$$\overline{q} = \overline{q}_{\gamma_2}(11, 1) = \overline{q}_{0.969}(11, 1) = 0.402.$$

Comparison of values of $\hat{q}$, $\underline{q}$, and $\overline{q}$ with analogous values found in problem 4.4 for the same test results shows that additional prior information about parameter q (failure probability) leads to some decreasing Bayes UCL $\overline{q}$ and almost does not influence the point estimate $\hat{q}$ and the lower limit $\underline{q}$.

4.6. Applying (4.18) and (4.19) from Section 4.4, we obtain that the posterior distribution density of parameter λ is the density of the gamma distribution with parameters $(u + S, a + d) = (600, 3)$. Further, using (4.20) an (4.21) for $d = r$ and Table E.16 from the Appendix, we obtain the following Bayes point estimate for parameter λ:

$$\hat{\lambda} = \frac{a + d}{u + S} = \frac{2 + 1}{200 + 400} = 0.005 \text{ hr}^{-1}.$$

The Bayes 0.95 confidence lower and upper limits $\underline{\lambda}$ and $\overline{\lambda}$ for parameter λ are given as

$$\underline{\lambda} = \frac{\chi^2_{0.05}(6)}{1200} = \frac{1.635}{1200} = 0.0014 \text{ hr}^{-1},$$

$$\overline{\lambda} = \frac{\chi^2_{0.95}(6)}{1200} = \frac{12.59}{1200} = 0.0105 \text{ hr}^{-1}.$$

CHAPTER 6

6.1. Since the system has a single unit of each type (i, j), the number of tests of each type of units is $N_{ij} = N = 8, j = 1, 2, 3, i = 1, 2$. Using formulas (6.22)–(6.24), we have the following lower 0.9 confidence limit for the system's reliability function:

$$\underline{R} = \min(\underline{r}_1, \underline{r}_2),$$

where $\underline{r}_i$ is the lower 0.9 confidence limit for the ith redundant group. These values can be found from the formulas

$$\underline{r}_1 = 1 - \varphi_1, \qquad \underline{r}_2 = 1 - \varphi_2,$$

where

$$\varphi_1 = \prod_{1 \le j \le 2} \frac{t_1}{t_1 + N_{1j}} = \left(\frac{t_1}{t_1 + N}\right)^2,$$

$$\varphi_2 = \prod_{1 \le j \le 3} \frac{t_2}{t_2 + N_{2j}} = \left(\frac{t_2}{t_2 + N}\right)^3,$$

and, in turn, t_1 an t_2 are found from the equations

$$\sum_{1\le j\le 2} N_{1j} \ln\left(1 + \frac{t_1}{N_{1j}}\right) = 2N \ln\left(1 + \frac{t_1}{N}\right) = -\ln(1-\gamma),$$

$$\sum_{1\le j\le 3} N_{2j} \ln\left(1 + \frac{t_2}{N_{2j}}\right) = 3N \ln\left(1 + \frac{t_2}{N}\right) = -\ln(1-\gamma).$$

This gives us

$$\underline{r}_1 = 1 - [1 - (1-\gamma)^{1/2N}]^2 = 1 - [1 - (0.1)^{1/16}]^2 = 0.982,$$

$$\underline{r}_2 = 1 - [1 - (1-\gamma)^{1/3N}]^3 = 1 - [1 - (0.1)^{1/24}]^3 = 0.999.$$

Thus, the lower 0.9 confidence limit of the system's reliability function is

$$\underline{R} = \min(0.982, 0.999) = 0.982.$$

6.2. We need to use formula (6.14) for a series–parallel system with identical units,

$$\underline{R} = \min_{1\le i\le 2} \{1 - [1 - (1-\gamma)^{1/N_i}]^{n_i}\}, \tag{s6.1}$$

where N_i is the test volume for the ith unit type (the total number of units tested). In this particular task these values are $N_1 = Nn_1 = 8 \times 2 = 16$ and $N_2 = Nn_2 = 8 \times 3 = 24$. From (s6.1) we obtain the lower 0.9 confidence limit for the system's reliability function

$$\underline{R} = \min(0.982, 0.999) = 0.982,$$

which coincides with the answer to problem 6.1. Thus, the information about identity of the system units does not help to improve the confidence estimate. [From the viewpoint of formal calculations, it can be explained by the fact that the minimum of the reliability function (6.49) in the corresponding confidence set is attained at the point $\mathbf{p} = (p_{11}, p_{12}, p_{21}, p_{22}, p_{23})$ satisfying conditions (6.50).]

6.3. Using formulas (6.4)–(6.6), we can find the lower γ-confidence limit for the system's reliability function as

$$\underline{R} = \min\,(R(p) = \min[R_1(p_1) \times R_2(p_2)], \tag{s6.2}$$

where $R_i(p_i) = 1 - (1 - p_i)^{n_i}$ is the reliability function of the ith redundant group, $n_1 = 2$ and $n_2 = 3$. The minimum of function (s6.2) is found for the conditions

$$p_1^{N_1} p_2^{N_2} \geq 1 - \gamma,$$

$$0 \leq p_i \leq 1, \qquad i = 1, 2, \tag{s6.3}$$

where N_i is the number of units tested, i.e., $N_1 = Nn_1 = 8 \times 2 = 16$ and $N_2 = Nn_2 = 8 \times 3 = 24$. It is more convenient to rewrite this problem using the notation $z_i = -\ln p_i$:

$$\underline{R} = e^{-\overline{f}},$$

where

$$\overline{f} = \max f(z) \tag{s6.4}$$

and in turn

$$f(z) = f_1(z_1) + f_2(z_2),$$

$$f_i(z_i) = -\ln\lfloor 1 - (1 - e^{z_i})^{n_i} \rfloor, \qquad 1 \leq i \leq 2.$$

The maximum in (s6.4) is taken under the conditions

$$N_1 z_1 + N_2 z_2 \leq A,$$

$$z_1 \geq 0, \qquad z_2 \geq 0, \qquad z_1 \leq z_2 \tag{s6.5}$$

where $A = -\ln(1 - \gamma)$ and the latter restriction in (s6.5) corresponds to the mentioned prior information. The region G corresponding to (s6.5) is shadowed in Figure s6.1.

Function $f(z)$ is monotone increasing in each variable and convex

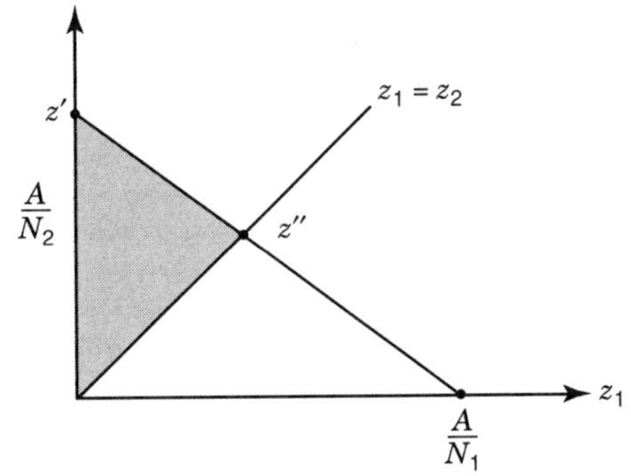

Figure s6.1 Region G described by (s6.5).

(downward) in $\mathbf{z} = (z_1, z_2)$. It follows (see Section 6.7.3) that the maximum of (s6.4) is attained at one of the border points z', z'' of the region G that are marked in Figure s6.1. These points have coordinates $z' = (0, A/N_2)$, $z'' = (z_1'', z_2'')$, where $z_1'' = z_2'' = A/(N_1 + N_2)$. It is easy to show that the maximum of (s6.4) is attained at the point z'' that corresponds to the case of identically reliable units of both types. After simple transformations it gives the lower γ-confidence limit of the system's reliability function,

$$\underline{R} = e^{-\bar{f}} = e^{-f(z'')} = \exp[-f_1(z_1'') - f_2(z_2'')] = R_1(\underline{p}_1)R_2(\underline{p}_2), \quad \text{(s6.6)}$$

where $\underline{p}_1 = \underline{p}_2 = (1 - \gamma)^{1/(N_1+N_2)}$. Formula (s6.6) corresponds to the following procedure. Calculating the lower confidence limit of the system's reliability function, we assume that all units of both types are identical. It allows us to sum the number of tested units N_1 and N_2. Then using standard Clopper–Pearson formulas, we find the lower γ-confidence limits for $\underline{p}_1$ and $\underline{p}_2$ and the corresponding confidence limit for the system. From (s6.6) for $N_1 + N_2 = Nn_1 + Nn_2 = 16 + 24 = 40$, we obtain the following lower 0.9 confidence limit for the system reliability function:

$$\underline{R} = \prod_{1 \le i \le 2} \{1 - [1 - (1 - \gamma)^{1/(N_1+N_2)}]^{n_i}\},$$

$$\{1 - [1 - (0.1)^{1/40}]^2\}\{1 - [1 - (0.1)^{1/40}]^3\} = 0.998.$$

Comparing this solution with the solution of problem 6.2 ($\underline{R} = 0.982$), we see that the prior information of type (6.52) leads to substantial improvement of the confidence limits.

6.4. The number of tested units of the first and second types are $N_1 = 2N = 16$ and $N_2 = 3N = 24$, respectively. Using (6.14), we obtain the following lower 0.9 confidence limit for the reliability function of the system of type (3, 2):

$$\underline{R} = \min(\underline{r}_1, \underline{r}_2),$$

where $\underline{r}_1$ and $\underline{r}_2$ are lower 0.9 confidence limits for the first and second redundant groups, respectively:

$$\underline{r}_1 = 1 - [1 - (1 - \gamma)^{1/N_1}]^{n_1} = 1 - [1 - (0.1)^{1/16}]^3 = 0.998,$$

$$\underline{r}_2 = 1 - [1 - (1 - \gamma)^{1/N_2}]^{n_2} = 1 - [1 - (0.1)^{1/24}]^2 = 0.992.$$

Thus, $\underline{R} = \min(0.998, 0.992) = 0.992$.

6.5. Systems with structures (2, 3) and (3, 2) were estimated in problem 6.2. Now find the lower confidence limits for systems with structures (1, 4) and (4, 1). [There is no variant (n_1, n_2) satisfying restrictions (6.53).] For variant (1, 4) we have the following lower 0.9 confidence limit for the system's reliability function:

$$\underline{R} = \min(\underline{r}_1, \underline{r}_2) = \underline{r}_1 = 1 - [1 - (1 - \gamma)^{1/N_1}]^{n_1}$$
$$= 1 - [1 - (0.1)^{1/16}]^1 = 0.866.$$

For the variant (4, 1), we have the analogous limit in the form

$$\underline{R} = \min(\underline{r}_1, \underline{r}_2) = \underline{r}_2 = 1 - [1 - (1 - \gamma)^{1/N_2}]^{n_2}$$
$$= 1 - [1 - (0.1)^{1/24}]^1 = 0.908.$$

The results are provided in Table s6.1. The largest value of the 0.9 confidence limit has variant (3, 2).

6.6. The number of units in the minimum simple cut b and the number of units in the minimum simple path a are obtained above and equal $a = b = 2$. Using (6.38) for the systems with complex structures, we have the following lower 0.95 confidence limit for the system's reliability function:

$$\underline{R} = 1 - [1 - (1 - \gamma)^{1/N_b}]^b = 1 - [1 - (0.05)^{1/4.2}]^2 = 0.902.$$

CHAPTER 7

7.1. The point estimate of the system reliability is

$$\hat{R} = \hat{p}_1\hat{p}_2 = \left(1 - \frac{d_1}{N_1}\right)\left(1 - \frac{d_2}{N_2}\right) = 1 - \frac{4}{200} = 0.980.$$

The minimum test number in this case is $N_1 = 100$. Utilizing formulas (7.56) and (7.57), we obtain the equivalent number of failures D_1:

$$D_1 = N_1(1 - \hat{R}) = 100(1 - 0.980) = 2.$$

Table s6.1 Lower 0.9 Confidence Limits for Different Variants

(n_1, n_2)	(1, 4)	(4, 1)	(2, 3)	(3, 2)
$\underline{R}$	0.866	0.908	0.982	0.992

The LCL with confidence coefficient 0.9 for the system PFFO is

$$\underline{R} = \underline{P}_g(N_1, D_1) = \underline{P}_{0.9}(100, 2) = 0.948,$$

where $\underline{P}_g(N, d)$ is a standard lower Clopper–Pearson γ-confidence limit (see Section 7.9.2).

7.2. The point estimate of the system PFFO is

$$\hat{R} = \hat{p}_1^2\hat{p}_2\hat{p}_3 = \left(1 - \frac{d_1}{N_1}\right)^2\left(1 - \frac{d_2}{N_2}\right)\left(1 - \frac{d_3}{N_3}\right)$$

$$= \left(1 - \frac{2}{200}\right)^2\left(1 - \frac{4}{250}\right) = 0.964.$$

The minimum relative test "volume" (N_i/l_i) corresponds to a unit of the first type:

$$\min_i \frac{N_i}{l_i} = \frac{N_1}{l_1} = \frac{200}{2} = 100.$$

Using formulas (7.58) and (7.59), we obtain the equivalent number of failures D_1 for units of the first type:

$$\left(1 - \frac{D_1}{N_1}\right)^2 = \hat{R},$$

from which

$$D_1 = N_1(1 - \sqrt{\hat{R}}) = 200(1 - \sqrt{0.964}) = 3.6.$$

The lower 0.95 confidence limit for the system PFFO is equal to

$$\underline{R} = [\underline{P}_\gamma(N_1, D_1)]^2 = [\underline{P}_{0.95}(200, 3.6)]^2 = (0.957)^2 = 0.916.$$

7.3. Consider an auxiliary (imaginary) series system consisting of the same two units as the initial parallel system. The PFFO of this system is denoted by

$$R' = p_1p_2.$$

Construct the confidence limit for the system PFFO R' of the series system. The point estimate for R' has the form

$$\hat{R}' = \left(1 - \frac{d_1}{N_1}\right)\left(1 - \frac{d_2}{N_2}\right) = 1 - \frac{1}{20} = 0.95.$$

The minimum test volume for this set of units is equal to $N_1 = 10$. Applying formulas (7.76) and (7.68) for a series system, we find the equivalent number of failures D_1:

$$D_1 = N_1(1 - \hat{R}') = 10(1 - 0.95) = 0.5.$$

The lower 0.9 confidence limit for the PFFO R of the auxiliary series system is equal to

$$\underline{R}' = \underline{P}_g(N_1, D_1) = \underline{P}^{0.9}(10, 0.5) = 0.725.$$

Further, applying formula (7.71), we obtain the lower 0.9 confidence limit for the reliability index R of the initial parallel system:

$$\underline{R} = 1 - (1 - \sqrt[m]{\underline{R}'})^m = 1 - (1 - \sqrt{0.725})^2 = 0.978.$$

7.4. Consider an auxiliary (imaginary) series system consisting of the same two units as the initial parallel system. The PFFO of this auxiliary system is denoted by

$$R' = p_1^2 p_2.$$

First construct the confidence limit for the system PFFO R' of the series system. The point estimate for R' has the form

$$\hat{R}' = \hat{p}_1^2 \hat{p}_2 = \left(1 - \frac{d_1}{N_1}\right)\left(1 - \frac{d_2}{N_2}\right) = 1 - \tfrac{1}{6} = 0.834.$$

The equivalent number of failures D_1 and D_2 determined from condition (7.75) are

$$D_1 = N_1(1 - \sqrt[n_1]{\hat{R}'}) = 12(1 - \sqrt{0.834}) = 1.06,$$

$$D_2 = N_2(1 - \sqrt[n_2]{\hat{R}'}) = 6(1 - 0.834) = 1.00.$$

Using formula (7.74), we obtain the lower 0.95 confidence limit for the PFFO of the auxiliary series system R':

$$\underline{R}' = \min\{\underline{P}_\gamma^2(N_1, D_1), \underline{P}_\gamma(N_2, D_2)\} = \min\{\underline{P}_{0.95}^2(12, 1.06), \underline{P}_{0.95}(6, 1)\}$$

$$= \min\{(0.655)^2, 0.418\} = 0.418.$$

The total number of units in the system is $n = n_1 + n_2 = 3$. By formula (7.82) the lower 0.95 confidence limit for the PFFO of the initial parallel system is

$$\underline{R} = 1 - (1 - \sqrt[n]{\underline{R}'})^n = 1 - (1 - \sqrt[3]{0.418})^3 = 0.984.$$

7.5. The point estimate for the system PFFO in this case equals

$$\hat{R} = \prod_{1 \le i \le m} [1 - (1 - \hat{p}_i)^{n_i}] = 0.999954.$$

Applying formulas (7.112) and (7.113), the equivalent number of failures D_i for the "weakest links," namely, for subsystems 4 and 7, is found from the condition

$$1 - \left(\frac{D_i}{N_i}\right)^{n_i} = \hat{R},$$

which gives

$$D_4 = N_4 \sqrt[n_4]{1 - \hat{R}} = 500\sqrt{46 \times 10^{-6}} = 3.38,$$
$$D_7 = N_7 \sqrt[n_7]{1 - \hat{R}} = 200\sqrt[3]{46 \times 10^{-6}} = 7.14.$$

Further, using formula (7.114), we obtain the lower 0.9 confidence limit for the system PFFO:

$$\underline{R} = \min_{1 \le i \le m} \underline{R}_i = \min(\underline{R}_4, \underline{R}_7),$$

where

$$\underline{R}_i = 1 - [1 - \underline{P}_\gamma(N_i, D_i)]^{n_i}.$$

Now we can find

$$\underline{R}_4 = 1 - [1 - \underline{P}_{0.9}(500, 3.38)]^2 = 1 - (1 - 0.9857)^2 = 0.9998,$$
$$\underline{R}_7 = 1 - [1 - \underline{P}_{0.9}(200, 7.14)]^3 = 1 - (1 - 0.9410)^3 = 0.9998.$$

Thus the method of equivalent tests produces the LCL for the system PFFO

$$\underline{R} = 0.9998,$$

which is higher in comparison with the accurate but "conservative" methods for which the value of the confidence coefficient g is guaranteed precisely (see Table 7.4).

7.6. The point estimate of the system PFFO is

$$\hat{R} = \prod_{1 \le i \le m} [1 - (1 - \hat{p}_i)^{n_i}] = 0.9792.$$

Applying formulas (7.112) and (7.113) gives the equivalent failures D_1 and D_6 for the first and sixth subsystems:

$$D_1 = N_1 \sqrt[n_1]{1 - \hat{R}} = 200\sqrt{0.0208} = 28.8,$$

$$D_6 = N_6 \sqrt[n_6]{1 - \hat{R}} = 300 \times (1 - \hat{R}) = 300 \times (1 - 0.9792) = 6.24.$$

It is easy to see that for the given system the LCL (7.114) has the form

$$\underline{R} = \min_{1 \le i \le m} \underline{R}_i = \min(\underline{R}_1, \underline{R}_6),$$

where

$$\underline{R}_1 = 1 - [1 - \underline{P}_\gamma(N_1, D_1)]^{n_1} = 1 - [1 - \underline{P}_{0.98}(200, 28.8)]^2$$
$$= 1 - (1 - 0.796)^2 = 0.958,$$
$$\underline{R}_6 = 1 - [1 - \underline{P}_\gamma(N_6, D_6)]^{n_6}$$
$$= \underline{P}_\gamma(N_6, D_6) = \underline{P}_{0.98}(300, 6.24) = 0.954.$$

Thus the lower 0.98 confidence limit for the system PFFO, computed by the method of equivalent tests, produces higher system LCL than those considered in Section 7.5 (see Table 7.4).

INDEX